Carpentry and Joinery

Carpentry and Joinery
Bench and Site Skills

Brian Porter LCG, FIOC

Reg Rose MCIOB, DMS, DASTE, FIOC
formerly Assistant Principal, Leeds College of Building, UK

OXFORD AMSTERDAM BOSTON LONDON NEW YORK PARIS
SAN DIEGO SAN FRANCISCO SINGAPORE SYDNEY TOKYO

Butterworth-Heinemann
An imprint of Elsevier Science
Linacre House, Jordan Hill, Oxford OX2 8DP
200 Wheeler Road, Burlington, MA 01803

First published 1997
Reprinted 2001 (twice), 2002

British Library Cataloguing in Publication Data
A catalogue record for this book is available from the British Library

ISBN 0 340 64528 8

For information on all Butterworth-Heinemann Publications visit our website at www.bh.com

Printed and bound in Great Britain by The Bath Press, Bath

Contents

Foreword

As a consultant timber technologist, I have known of the work of both Brian Porter and Reg Rose for over a dozen years: our paths first crossed when I was the TRADA Regional Officer based in Wetherby and they were both then working at Leeds College of Building. This textbook, which is intended to be a companion to the NVQ process, contains the wealth of knowledge and detail which Brian's readers have come to expect from his many years' experience as both a practitioner and teacher of carpentry and joinery skills, and the collaboration of Reg Rose in this present volume has enhanced that detail even further.

The 1990s have seen great changes in the specification and use of timber and wood-based materials, not least in the ousting of many old and cherished British Standards by the much-maligned 'EuroNorms' (or BSENs). This book will guide both the newcomer and the more experienced woodworker through the fascinating subject that is wood, and it will enable them to gain a much better appreciation of what to do and – most importantly – *how* to do it properly.

I am delighted to have been asked to provide this Foreword to what I know will be a valuable and much-used book. I await the next collaborative work of Brian and Reg with interest.

Read and learn!

James C. Coulson, AIWSc FFB
Senior Partner, Technology For Timber
Ripon, Yorkshire

Acknowledgements

The authors wish to thank Mr James C. Coulson for proof reading the text and writing the Foreword; Eric Cannell for editing and contributing material for Chapter 9; Alan Wilson for his help with previous works; Peter Kershaw (Managing Director North Yorkshire Timber Co Ltd) for compiling Table 1.11; Mr Stewart J. Kennmar – Glenhill (Imperial College of Science and Technology, London) and David Kerr and Barrie Juniper of the Plant Science Department, Oxford, for contributing Figures 1.99, 1.102 and 1.103; Mr John Common (Kiln Services Ltd), Timber Research and Development Association (TRADA), for information gleaned from 'TRADA Wood Information Sheets'.

The following companies and organisations are acknowledged for supplying artwork and/or technical information:

American Plywood Association (APA), Atlas Copco Tools Ltd, Black & Decker, Robert Bosch Ltd (Power Tools Division), British Gypsum Ltd, Cape Boards Ltd, Cembrit Building Products, Council of Forest Industries (COFI), CSC Forest Products (Sterling) Ltd, Denford Machine Tools Co Ltd, Dominion Machinery Tools Ltd, Elu Power Tools Ltd, English Abrasive & Chemicals Ltd, Eternit UK Ltd, Fibre Building Board Organisation, Forestor – Forest and Sawmill – Equipment (Engineers) Ltd, Finnish Plywood International, Forestor – Forest and Sawmill – Equipment (Engineers) Ltd, Formica Ltd, Fosroc Ltd, G.F. Wells Ltd (Timber Drying Engineers), Hicksons Timber Products Ltd, Hilti Ltd, ITW Paslode, Kiln Services Ltd, Lydney Products Ltd, Makita UK Ltd, Neill Tools Ltd, Norboard Industries (UK), Nordic Timber Council, Optical Measuring Instruments Cowley Ltd, Perstorp Warerite Ltd, Protim Ltd, Protimeter PLC, , Rabone Chesterman Ltd, Rapesco Ltd, Recod Tools Ltd, Record Marples (Woodworking Tools) Ltd, Rentokil Ltd, Rockwool Ltd, Stanley Tools, Stanley Works Ltd, Stenner of Tiverton Ltd, Thomas Robinson Group Plc, TRADA, Wadkin Group of Companies PLC.

Tables 1.4, 1.5 and 1.14 are extracted from British Standard EN336: 1995 – Structural Timber – Coniferous and Poplar Sizes – Permissions Deviations (Tables NA.1 Customary Lengths of Structural Timber, Table NA.2 Customary target sizes of sawn structural timber and Table NA.4 Customary target sizes of structural timber machined on all four sides.) Copies of this complete standard can be obtained from British Standards Institute, 389 Chiswick High Road, London W4 4AL. Copyright is held by the Crown and reproduced with kind permission of the British Standards Institute.

1

Timber

When handling timber do you ever stop to think which sort of tree it might have been cut from, or do you instinctively go ahead and use it for whatever purpose you think fit? Before a craftsperson makes the decision as to whether any piece of timber is suitable for a particular job, that person should be aware of its origin (i.e. the type of tree from which it was cut), and any subsequent treatment it may have been subjected to before it reaches his or her hand.

It is by studying the wood of the tree that the craftsperson becomes selective. After all, the piece of timber in question could have been processed from one of hundreds of different kinds of wood, most of which are as different as you are from the person next to you.

In many cases, however, others make decisions for you. For example, you could be asked to use a particular named wood; you would then have to identify a timber which had the same name as the tree from which it was cut. Look at Figure 1.1. You may be asked to advise on which 'wood' might be most suitable to use in a given situation, or how it would respond to being:

- **a** cut, by hand tools, or woodworking machines
- **b** subjected to different kinds of loads
- **c** bent
- **d** nailed and screwed into
- **e** glued
- **f** subjected to dampness
- **g** attacked by fungi
- **h** attacked by wood boring insects
- **i** subjected to fire
- **j** treated with wood preservative, paints, polishes, sealants, flame retardants, etc.
- **k** in contact with metals.

Knowing which wood to use, and where, is very important, and without the knowledge of how to identify different types, important decisions cannot be made with confidence. So I think we should start at the beginning – with the growing tree.

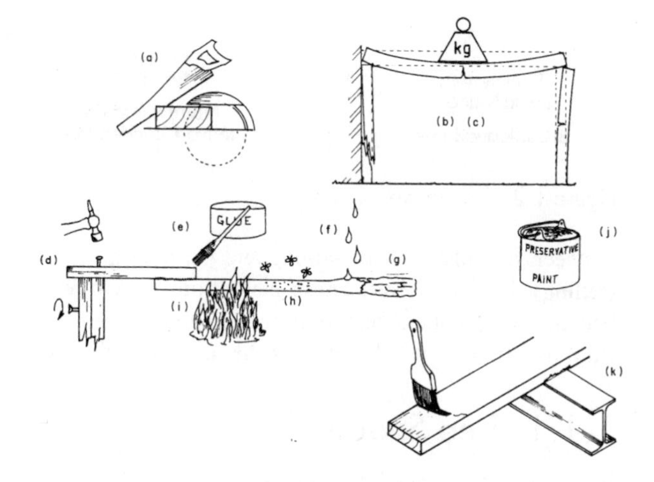

Figure 1.1 Timber may respond in different ways when being subjected to any of these treatments

1.1 THE TREE

Trees, like any other naturally regenerating resource, can, when demands outweigh available supply, be depleted by over felling and this can have devastating effects upon the environment. Trees in particular play a vital role in keeping our atmosphere in balance with nature. So, with the gradual depletion of many of the world's natural forests, the source of supply which the timber industry has relied upon for many years has now been drastically reduced. The timber industry is having to rely more heavily upon specially planted and managed forests, rather than on naturally occurring forests, as the industry's contribution towards world conservation.

A tree may take between 30 and 100 years to reach the stage when it is regarded as being suitable for felling for use as timber – this time will vary according to the type of tree and its growing conditions.

Sustainable (or renewable) supplies of timber have for many years been taken from managed forests. These forests are planted, cared for and harvested like any other crop; the main difference is the length of time

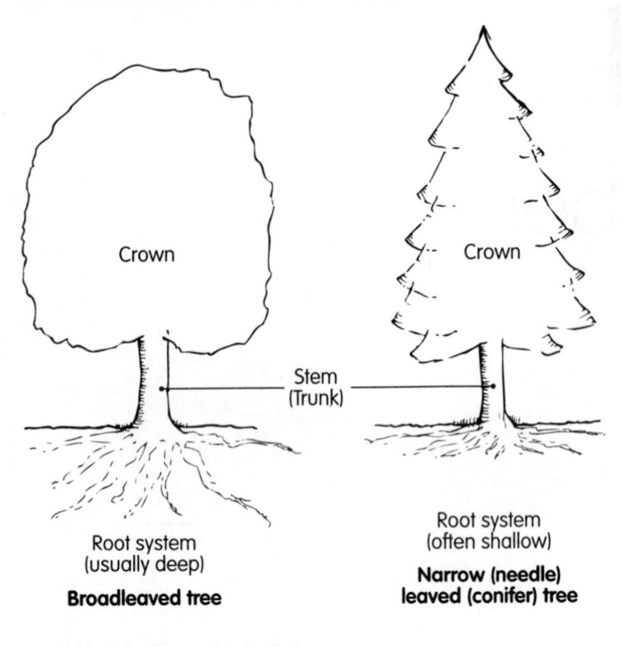

Figure 1.2 General tree shape

between first planting the sapling and harvesting (felling) the mature tree. These 'plantations' have now become an acceptable feature in the countryside, and they are often regarded as part of the natural landscape.

1.1.1 Tree Shape

By shape, trees generally belong to two basic types (Figure 1.2):

- trees with broad leaves (broadleaf trees)
- trees with narrow, or needle leaves (conifer trees).

If you look around open parkland or along any tree-lined roadways, or wherever trees are set within an open aspect, you should see that their overall shapes will take forms similar to those shown in Figure 1.3. These trees are known as 'open grown trees'. The main feature of these is their wide branching habit.

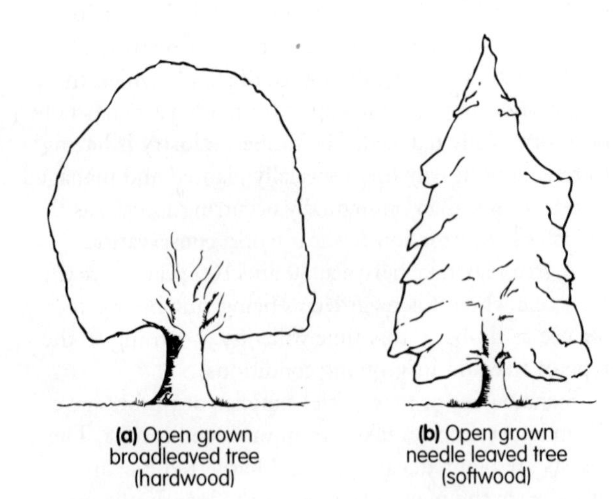

Figure 1.3 Open grown trees

Alternatively, if you were to visit a forest or dense woodland you would probably find that the trees would be shaped differently. Figure 1.4 shows 'forest grown trees', which have developed long trunks with the minimum of branches. The reason for this is that the trees are in competition with each other in trying to reach up above the forest canopy to reach the life-giving light energy – without which they would die. We could therefore say it is a fight for survival. In situations like this there would be no value in producing branches – so why bother?

However, within a man-made forest, tree survival is more a question of selection by the forester: it is not just left to nature. Selection starts from the day the young trees (saplings) are first planted out in the field in close proximity to each other for protection from competing ground vegetation (weeds etc.). Some years later they will be thinned out to allow more air and light to filter in, but not enough to encourage side branches, then thinned again after several years as the trees start to mature.

Forest thinnings are not wasted: they may go as fuel, or to the chipboard or paper industries. Eventually the remaining 'standing trees' may well end up as long lengths of timber with the minimum of knots – we shall be discussing knots in Section 1.6.10.

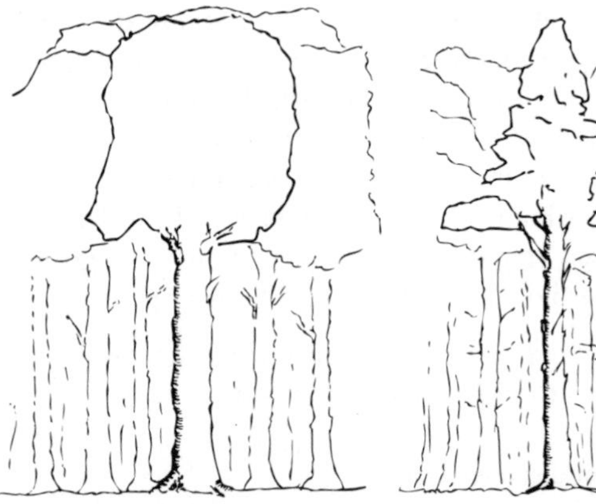

Figure 1.4 Forest grown trees

1.2 TREE COMPONENTS

All trees have three things in common; that is to say they all have, as shown in Figure 1.2:

- a root system
- a trunk (main stem)
- a crown.

The most important of these is the trunk – but we need to understand how the root system and crown operate to enable us to understand more about the quality of the timber we derive from it – so read on.

1.2.1 Root System

Roots have two main functions (Figure 1.5); they are:

1 to anchor the tree to the ground – size and spread will depend on type and size of tree; the radius of the roots can exceed the radius of the crown. Generally, conifers tend to have shallow root systems well adapted to cope with shallow soil layers over rocky sub-strata (layers below top soil), whereas broadleaf trees are inclined to set down deep tap roots
2 to extract water and the required dilute minerals from the ground via fine hairs which surround the root ends.

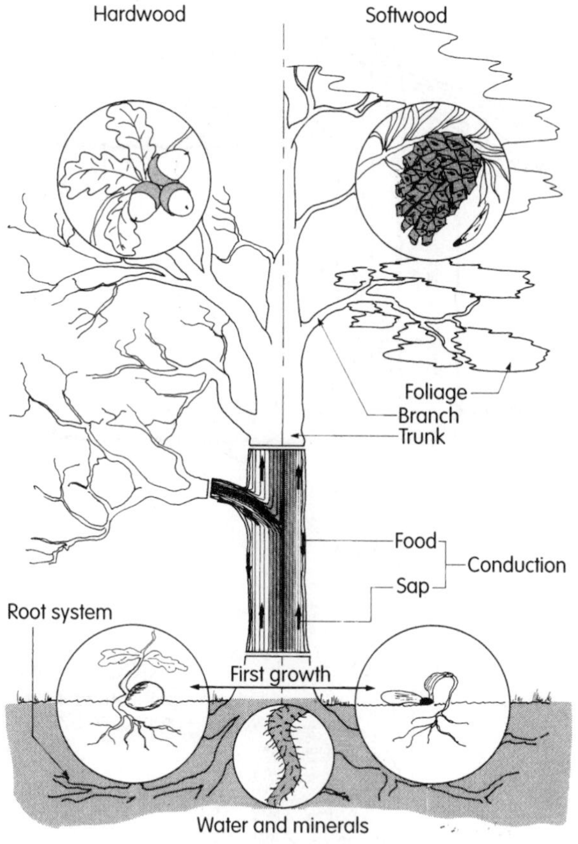

Hardwood
Softwood
Foliage
Branch
Trunk
Food
Conduction
Sap
Root system
First growth
Water and minerals

Figure 1.5 Tree growth

At certain times of the year, some trees may take up as much as 450 litres (100 gallons) per day from the ground – most of which, after travelling up through the trunk and branches to the leaves, will evaporate as water vapour into the atmosphere.

1.2.2 Trunk

The trunk is the main stem of the tree from which timber is cut (Figure 1.6). Once the trunk is cut into lengths these may be called **logs** or **boules**.

It is important that we know how this part of the tree is made up. Figure 1.6 identifies the main elements which affect the way we cut and select timber. So, starting from the centre of the tree, then moving outwards, we will take each in turn.

1.2.3 Pith (or Medulla)

The pith is shown in the illustration as the centre of the tree. When the pith is extremely off-centre it is usually an indication of the tree's unsuitability as timber – we will be dealing with this later under the heading of **Reaction Wood.**The pith is the result of the tree's earliest growth (a sapling) – wood immediately surrounding it is called 'juvenile wood,' which is not very desirable as timber.

1.2.4 Growth Ring

You might be more familiar with the term 'annual ring'. This implies that the rings of wood cells, which appear

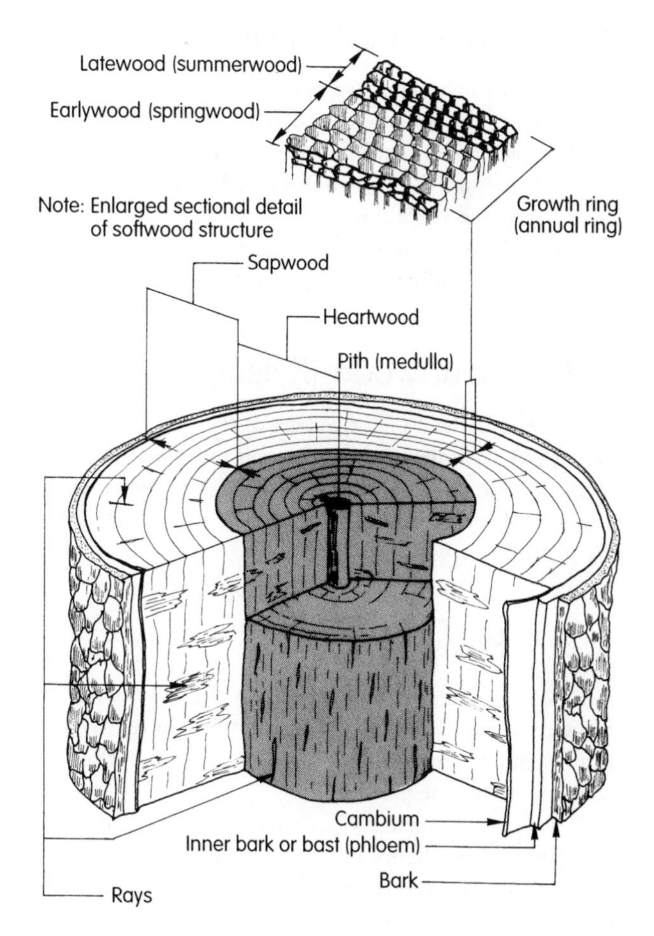

Figure 1.6 Section through the trunk (main stem)

as bands, are reproduced each year, and that by counting the number of rings we know how old the tree was when cut down. This is true of many trees which grow in regions of the world with a definite growth season each year, such as in the UK, but in other countries climate conditions may be different. For example, growth may be continuous, and therefore the growth pattern of the ring may not be annual, so we use the term 'growth ring.'

Not all trees have visible growth rings, but with those that do, it is usually possible to see that each consists of two bands – one lighter in colour than the other. The lighter one we call **earlywood** (originally called 'springwood'), and the darker one **latewood** (originally 'summerwood'). Latewood appears darker because its cell walls are thicker and more dense, as a result of being produced later in the growing season, whereas the earlywood would have been formed more rapidly when growing conditions were at their best.

Growth rings are important because they enable the woodworker to decide on the suitability of the wood as a whole for either its joinery qualities (appearance and workability) or its structural properties (strength and stability). As will be seen later in Section 1.8 one of the most important factors affecting the quality of timber is the rate at which the tree grows.

1.2.5 Sapwood (Xylem)

Sapwood is the outer active part of the tree responsible for receiving the water and minerals from the roots and conducting them around the tree to the leaves, as well as for food storage. The width of the sapwood band may be as narrow as 13–50 mm, or in some tropical countries as wide as 200 mm – depending on the type and species of tree and its growing conditions.

1.2.6 Heartwood (Xylem)

Heartwood is the non-active part of the tree – usually darker in colour than sapwood – and provides the tree with the rigidity often necessary to support the crown. The most durable wood (see Section 1.10.4e) for timber is cut from this part of the tree.

1.2.7 Rays

A ray consists of a strip of wood cells that allows sap to move transversely (across the trunk) through the wood, and provide for the storage of surplus food and its movement when required.

Rays may appear to radiate from the pith (medulla) to the inner bark; hence the name medullary rays is often used. However, the majority of rays start life further away from the pith, so it is best to generalise by calling them all rays.

Rays can be very important to the woodworker, who can use them to provide a decorative effect on the planed surface of several hardwoods – probably the most common of these is oak, which we will be studying in more detail later in Section 1.10.3j.

1.2.8 Cambium (Figure 1.7)

Cambium is the thin sleeve of cells located between the sapwood and the inner bark which covers the whole of the tree – that is its trunk, branches and twigs. These cells are responsible for the growth of the tree – both its girth (distance around the trunk) and height. As the cells are formed during the growing season they become subdivided in such a way that new cells are added to both the sapwood and the inner bark.

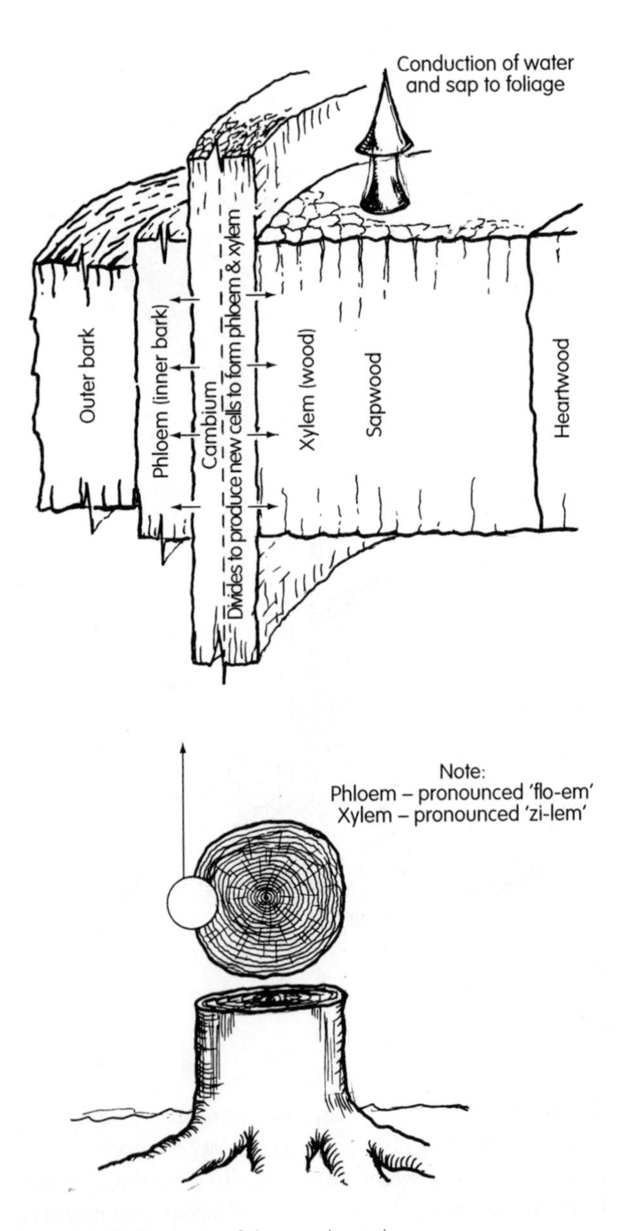

Figure 1.7 Function of the cambium layer

1.2.9 Inner Bark – Phloem

The inner bark distributes the food substance provided by the leaves throughout the whole of the tree. After one growing season it will become inactive as a means of conducting food; it takes on a new role as the outer bark – it will then be replaced by a new inner bark.

1.2.10 Bark (Figure 1.7)

The bark forms the outer sheath of the whole tree. Its function is to protect the inner parts of the tree from:

- extremes of temperature – both hot and cold
- fungal attack
- insect attack (although some species are capable of piercing the outer bark)
- mechanical abrasion by animals or other agents.

For the tree's general well-being the bark must also act as a moisture barrier and thermal insulator.

1.2.11 Crown

Branches, twigs, and foliage (leaves) all go to make up the crown of the tree – the top of which we call its canopy – under which both man and animals have often sought shelter from the rain, or shade from the sunlight.

This is the area where food for the tree is processed. Figure 1.8 shows that it is the leaves that make this possible. Within each leaf is a substance called chlorophyll which makes the leaves appear green in colour.

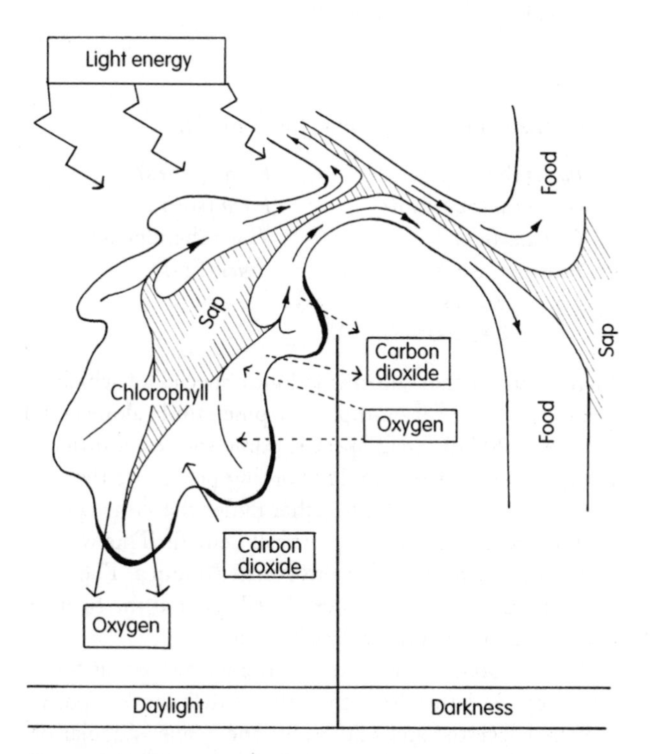

Figure 1.8 The process of photosynthesis

When the leaf absorbs daylight energy via this chlorophyll it converts a mixture of carbon dioxide from the air with sap from the roots into a food substance – mostly carbohydrates (varying amounts of sugars and starches), whilst releasing oxygen into the atmosphere as a waste product. This process is known as photosynthesis. During the hours of darkness it appears that to some extent the process is reversed by the tree taking in oxygen and releasing carbon dioxide – this is known as respiration (breathing).

Once the foods are made and processed they must be circulated back down and around the whole of the tree via its inner bark [also known as phloem (pronounced FLO–EM) or bast].

But how does the sap arrive at the leaves in the first place? It would seem that this is due to suction induced by the reaction of transpiration (when leaves lose moisture by evaporation – about 90% of the water taken from the ground via the roots is lost in this way), and/or to capillarity within the tubular cell structure of the wood.

1.3 HARDWOOD AND SOFTWOOD TREES

It has already been mentioned that trees generally belong to one of two types:

- broadleafs
- conifers.

You may have heard the term hardwood used to describe broadleaf trees, and conversely softwood used to describe conifer trees. These terms are botanically and commercially correct. In fact they refer to their structural differences, that is the make-up of the wood, not the trees' physical hardness or softness, which can be very confusing even for the most experienced woodworker. For example, many students will have heard of, if not built models made from, **balsa wood,** which is so soft and light one can easily indent the surface with a finger nail – believe it or not, that wood is classified commercially as a hardwood. Conversely the wood from a **yew** tree, which can be very hard to indent, is classed as a softwood. To add to this confusion, it is generally stated that hardwoods are deciduous (shed their leaves after their growing season – in temperate zones around the world that would mean those months preceding winter, known as autumn), but many thick leaved varieties of trees retain their leaves for two or more years. Hence such trees will always appear to take on an evergreen appearance – a condition usually reserved for conifers. Even then, the larch tree, which is a well known conifer and therefore a softwood, sheds its leaves in autumn. Confused? Don't be: Table 1.1 should help you sort out any confusion.

Table 1.1 Guide to recognising hardwood and softwood trees and their use

	Hardwoods	Softwoods (conifers)
Botanical grouping	**Angiosperms**	**Gymnosperms**
Leaf group	Deciduous* and evergreen	Evergreen†
Leaf shape	Broadleaf	Needle leaf or scale like
Seed	Encased	Naked via a cone
General usage	Paper and card Plywood (veneers and core) Particle board Timber – heavy structural, decorative joinery	Paper and card Plywood (veneer and core) Particle board Fibre board Timber – general structural joinery
Trade use	Purpose made joinery Shopfitting	Carpentry and joinery

Note: *Within temperate regions around the world; †Not always, for example: larch trees are deciduous

1.3.1 Tree and Timber Names

Most, if not all, of the trees we see and the timber we use will be referred to by name, so that we can identify one from another. The name generally used is its 'common English name'. Unfortunately, in the commercial world of the timber trade, this name may give insufficient information about the wood, or it may have another name common in other countries. So, its true, or scientific/botanical (Latin) name, may be written alongside its common name on an order form so that it can be identified no matter which country we are in or from whence the timber came.

Now look at Table 1.2 where you will see how the family groups are formed. The first division is, as we have already discussed, into hardwoods and softwoods, then come the 'family groupings', for example:

Hardwoods	**Softwoods**
Beech family	Pine family
Birch family	Cypress family
Elm family	Yew family

Then there are their 'genera', for example:

Fagus (beeches)	*Pinus* (pines)
Castanea (chestnuts)	*Picea* (spruces)
Quercus (oaks)	*Tsuga* (hemlocks)
Faxinus (ashes)	*Abies* (firs)
Larix (larches)	
Cedrus (cedars)	

Genera are further sub-divided into 'species' of which there are many. For example, the pine family alone could have over 200 different species. Some species of different genera have such similar working properties that they are sometimes sold together under the same common name, such as a group we call hem-fir. This wood is imported into the UK from North America. The hem is an abbreviation for Western hemlock, and the fir in this case could stand for Amabilis fir.

On the other hand if you see the written genus followed by the letters '*spp.*' this tells us that similar species may be harvested and sold under one genus (singular of genera) (for a preview of this, see Table 1.19.)

Table 1.2 The family tree

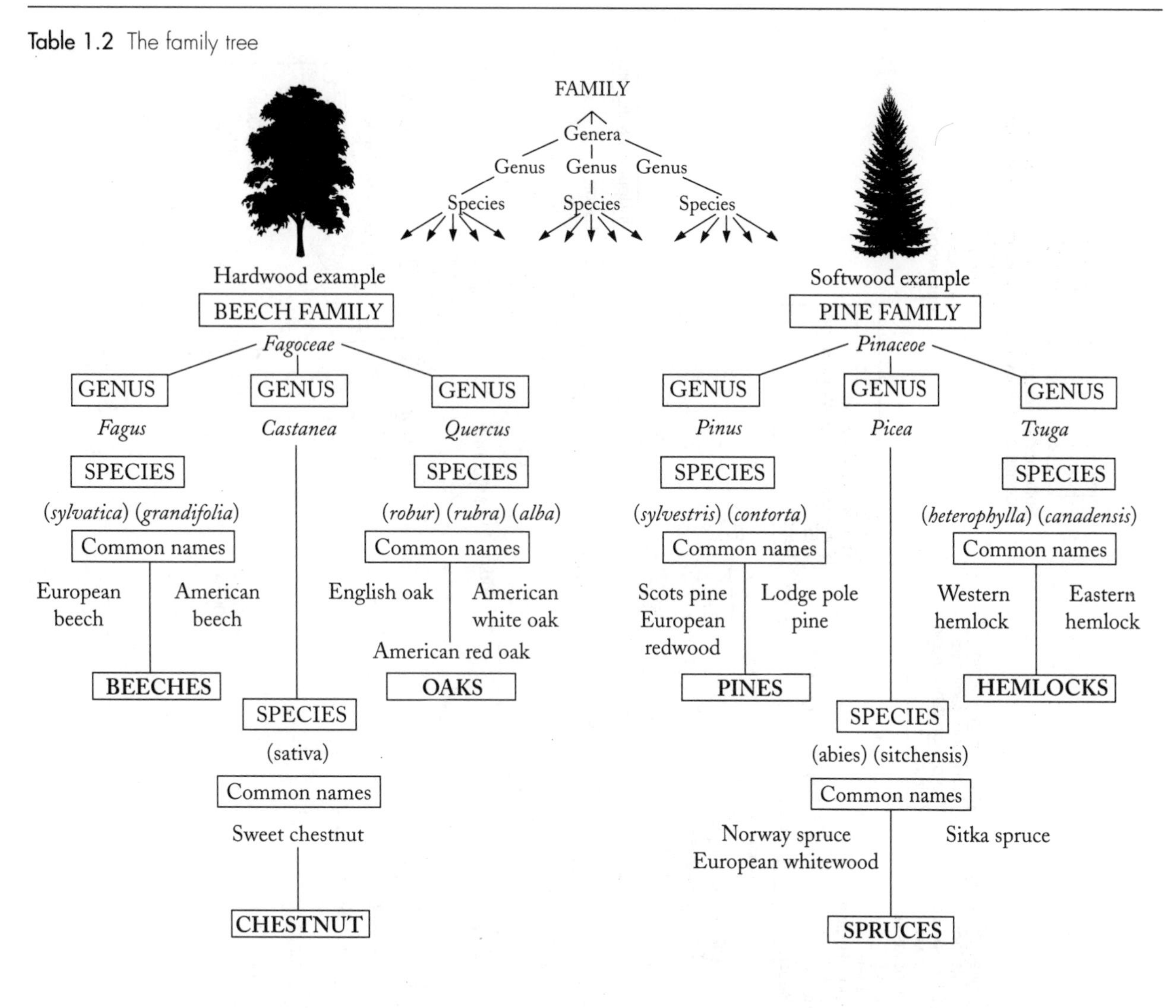

1.3.2 Forest Distribution

The map of the world shown in Figure 1.9 should give you some idea of where the main forests of the world are located. Notice how the conifer forests are mainly found within the northern hemisphere. This is because they are better able to cope with the cooler climates by having small leaves whereby transpiration is reduced, and their conical shape helps in shedding snow. However, even hot countries can support conifer growth within their colder mountainous regions (at high altitudes).

Broadleaved hardwoods prefer warmer regions where adequate supplies of water are available, such as the tropical rain forests, where they will remain evergreen like conifers. However, within the more temperate regions (neither extremes of heat nor cold), they will shed their leaves (deciduous) and become dormant during the colder months of the year.

You can also see areas where a mixture of hardwoods and softwoods can be found.

The main commercial regions have also been labelled so that you can cross reference the following species with those listed within Tables 1.11, 1.13, 1.20 and 1.21 to find from which country or region of the world they originated.

Hardwoods	**Softwoods**
European oak	European redwood
Red meranti	European whitewood
Lauan	European larch
Mahogany (Brazil)*	Douglas fir
Alder	Western hemlock
Chestnut	Western red cedar
Ash	Parana pine
Afrormosia*	
Beech	
European elm	
Teak*	
Iroko*	
Ramin*	

Note: * These species may have originated from tropical rain forest regions which the timber industry is trying to control as conservation areas.

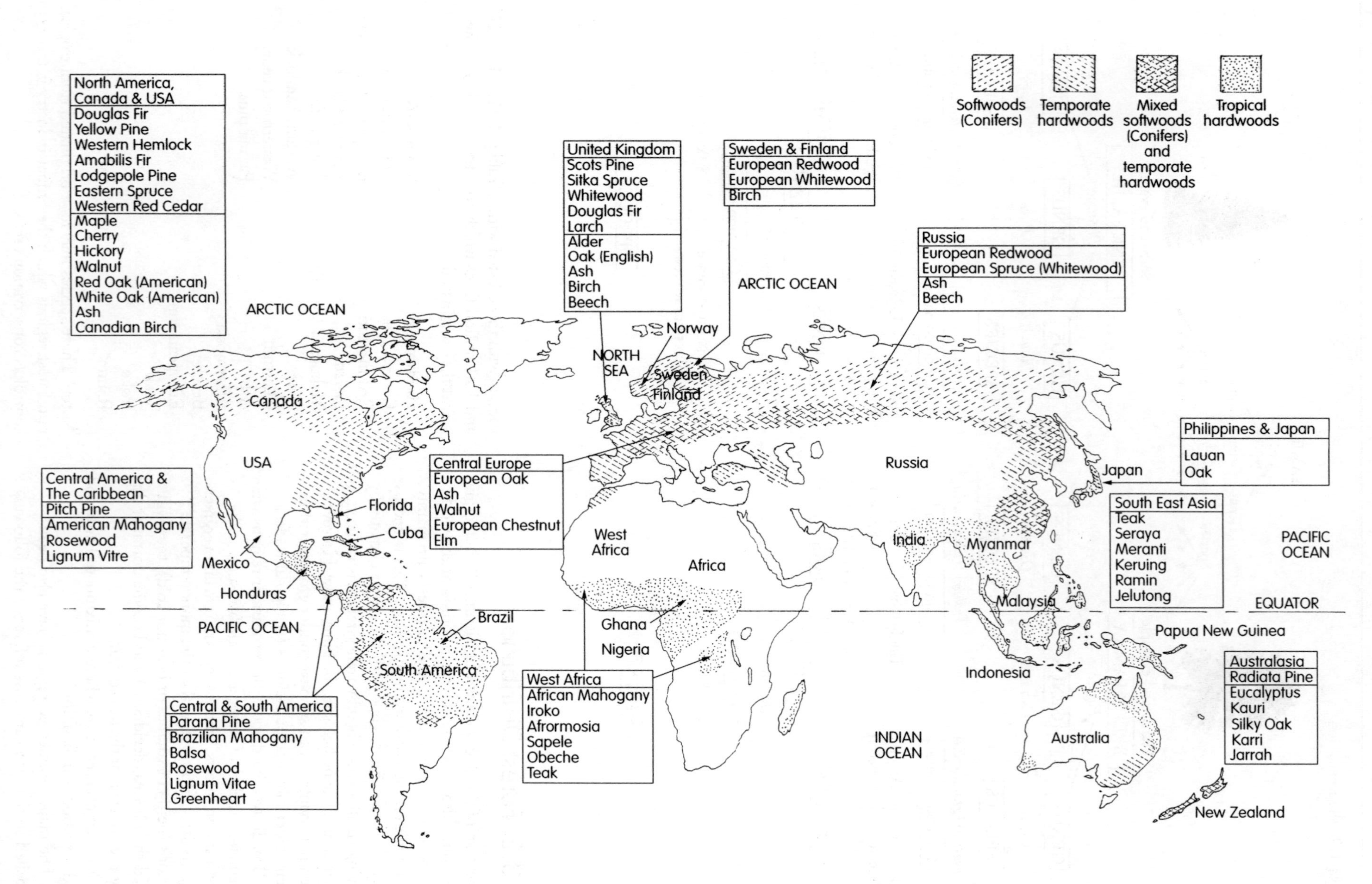

Figure 1.9 Forest regions of the world

1.4 CONVERSION INTO TIMBER

The division of the log into timber sections is called conversion. This operation involves the use of heavy machinery, much of which is very sophisticated, relying heavily on modern automated computerised systems to reduce manpower at the operation end to a minimum.

These modern systems can reduce human error and provide a safer working environment for the sawyer (machine operator) who still makes the final decision by controlling, and where necessary overriding, the computer. But before looking at this operation in more detail, let us consider the whole process, starting with the standing tree.

Figure 1.10 should give you some idea of the following stages in the conversion process.

1.4.1 Conversion Process

a) Tree Selection or Forest Clearance

With natural forests of hardwood and some softwoods, selective felling may be employed by taking out specific trees. This selection may be by size or species. In both cases the forester would use a hand operated chain saw.

Softwood plantations may have whole areas felled at once, again by using the chain saw. Or, where conditions permit (that is on land with a suitable terrain), a forest harvester may be employed. This is a motorised vehicle which, under the control of the operative, will fell the tree, remove side branches, debark the trunk, and then cut the log into manageable lengths ready for transportation to the saw mill.

b) Transportation to Sawmill

The simplest and most effective route will be chosen. This may be via road, rail, or water; it will depend on

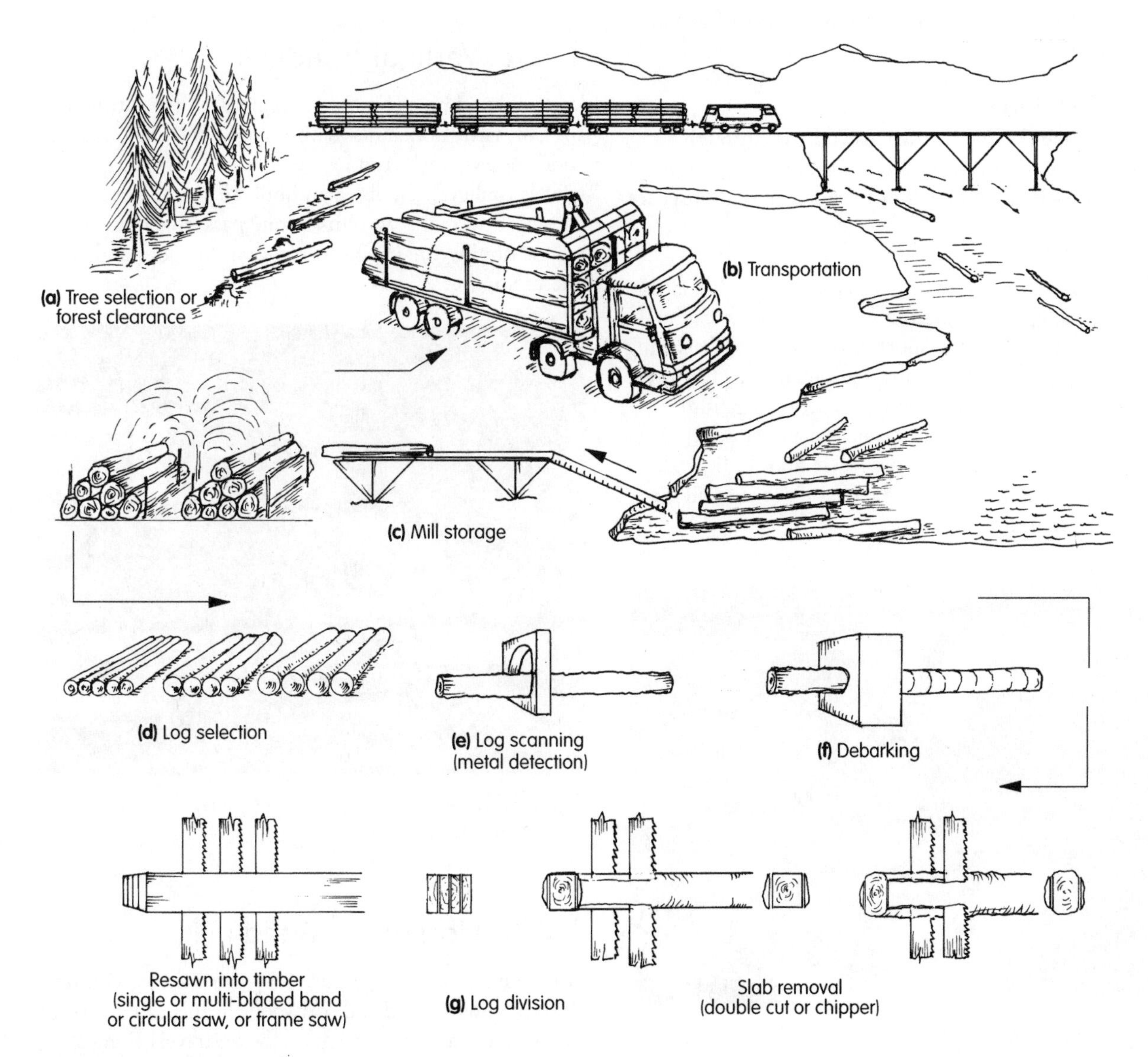

Figure 1.10 Conversion process

the forest location and accessibility to the various means of transportation.

c) Mill Storage

Where possible, mill ponds are used to keep the logs wet. In this way, drying out can be avoided and the logs prevented from shrinking prematurely, which could otherwise degrade the resulting timber. Failing this, during summer months, stockpiles of logs on land would be kept wet by the use of automatic sprinklers.

d) Log Selection

Logs will be lifted into bays according to girth size (circumference) and usually species.

e) Log Scanning

Each log must be scanned with a metal detector before it enters the saw mill to protect saw teeth from any foreign elements which may have become encapsulated within the growing tree – possibly as a result of notices being nailed to the tree, wire fencing, or even fragments of shrapnel.

f) Debarking

Softwoods and some hardwoods will pass through a debarking process before entering the mill; the bark residue may be used as fuel or for horticultural purposes.

g) Log Division

Log division is the stage which transforms the log into timber. It is worth noting that in some cases it can make commercial sense to transport heavy yet portable sawing machinery and accompanying equipment into the forest to carry out initial conversion. By cutting logs into squared sections, transportation costs are reduced, and wood waste can remain on the forest floor to degrade naturally, thereby adding to the fertility of the soil for the next generation of trees.

Figure 1.11 Circular saw (with kind permission, Stenner of Tiverton Ltd)

1.4.2 Sawing Machinery

The size and type of equipment will depend on the size of the mill and the kind of logs it would be expected to handle.

Small mills could have one or more of the following machines, with simple mechanical means of conveying the logs through the mill, whereas larger mills are usually semi-automated and computerised to remove a lot of guesswork and to speed up production. All control would be done from within an enclosed room or cabin in full view of the whole sawing and sorting operation. Closed circuit T.V. monitors are used to cover areas outside the operator's range of vision.

1.4.3 Circular Saws

Circular saws (Figure 1.11) are generally capable of cutting hardwood and softwood logs of small to medium diameter – blades could be as large as 1.8 m in diameter.

1.4.4 Vertical Bandmill

Vertical bandmills (Figure 1.12) are used for cutting all sizes of hardwood and softwood logs. The blade is a wide endless steel band revolving around two large wheels (pulleys). Double bandmills (Fig 1.13) may be employed to make two cuts in one pass; this is achieved by positioning the machines either in line, or parallel one to another.

Figure 1.12 Vertical bandmill (with kind permission, Stenner of Tiverton Ltd)

1.4.5 Horizontal Bandmill

Horizontal bandmills (Figure 1.14) are used for cutting all sizes of hardwood and softwood. In this case horizontal cuts are made. The machine shown travels along a track, allowing the log to remain static as it is being cut.

Figure 1.13 Double bandmill (with kind permission, Stenner of Tiverton Ltd)

Figure 1.14 'Forestor –150' horizontal bandmill – through and through sawing (with kind permission, Forestor Equipment Engineers Ltd)

Many of these machines can be taken into the forest where a temporary track is laid down.

1.4.6 Vertical Frame Saw or Gang Saw

Vertical frame saws or gang saws (Figure 1.15) are used for cutting small to medium diameter softwood logs. The logs are cut by being pushed towards a series of reciprocating (upward and downward movement) saw blades. These blades can be arranged to suit the width (thickness) of the required timber. Figure 1.16 shows how two slabs (outer segments of the log, see Figure 1.22) and two boards may be cut on the first pass, and then, after the log is rotated, two further slabs are removed, together with any number of sized boards, depending on how the blades are arranged.

Figure 1.15 Vertical frame saw – gang saw (with kind permission, Nordic Timber Council)

1.4.7 Methods of Conversion

The way the cuts are made will depend on several factors, some of which are:

- the type of sawing machine
- the size of log (diameter or girth)
- the condition of the log
- the wood species
- economy
- the end use of the resulting timber – for its appearance, decorative or structural requirements.

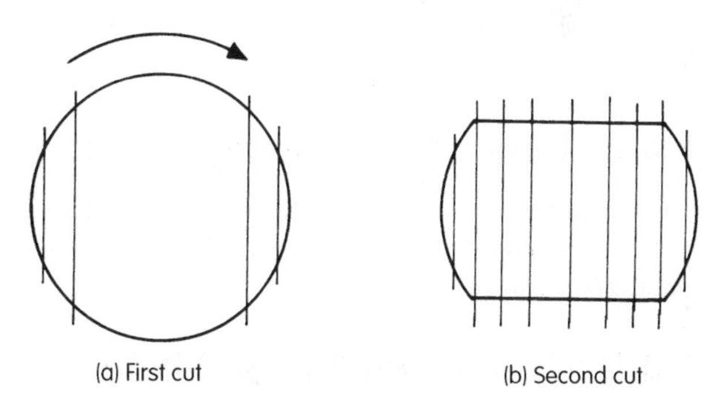

Figure 1.16 Possible cuts of a frame saw

The resulting timber section (with the exception of that which surrounds the pith – the 'heart') will be either:

- plain or tangential sawn
- quarter or rift sawn.

Figure 1.17 shows the resulting section through plain sawn timber – note that its growth rings meet the widest face at an angle of less than 45°.

Figure 1.18 shows a section through quarter sawn timber. Its growth rings meet its widest face at an angle of not less than 45°.

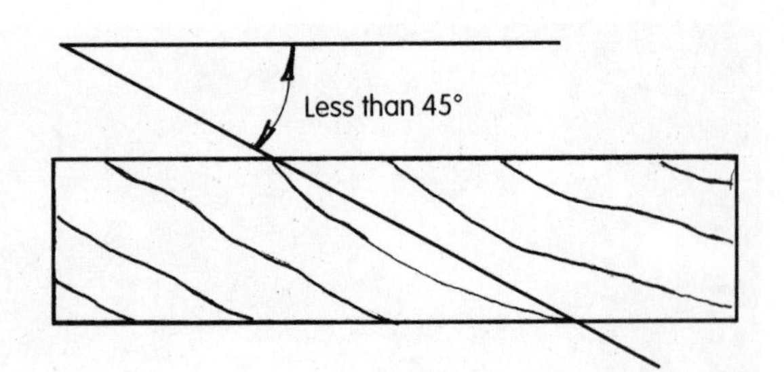

Figure 1.17 Plain sawn – growth rings meet the wide face of the board at an angle of less than 45°

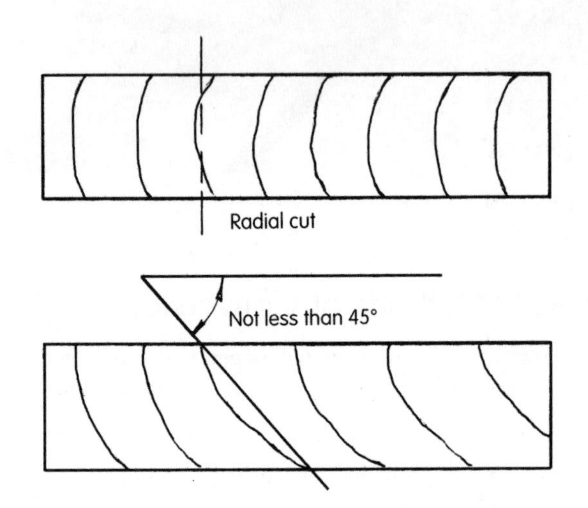

Figure 1.18 Quarter sawn – growth rings meet the wide face of the board at an angle of not less than 45°

Table 1.3 Comparison between 'plain' and 'quarter' sawn timber

Advantages	Disadvantages
Plain sawn 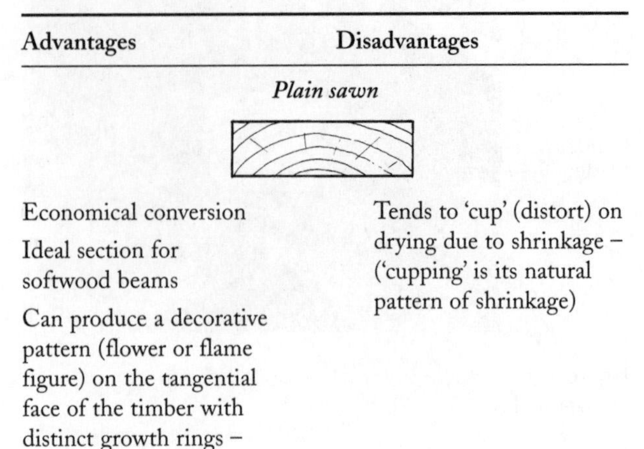	
Economical conversion Ideal section for softwood beams Can produce a decorative pattern (flower or flame figure) on the tangential face of the timber with distinct growth rings – see Figures 1.23, 1.106	Tends to 'cup' (distort) on drying due to shrinkage – ('cupping' is its natural pattern of shrinkage)
Quarter sawn 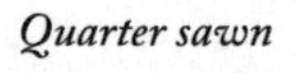 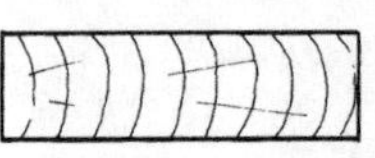	
Retains its shape better during drying Shrinkage across its width half of that of plain sawn timber Ideal selection for flooring with good surface wearing properties Produces a decorative radial face (e.g. Silver Figure) on hardwoods with broad ray tissue, see Figures 1.23, 1.106	Expensive form of conversion Conversion methods can be wasteful

What then is the importance of these two sections? Table 1.3 shows their possible advantages and disadvantages.

1.4.8 Through and Through Sawing

Large logs can be quickly converted in this way (Figure 1.19). The resulting timber will be part plain sawn and part quarter sawn.

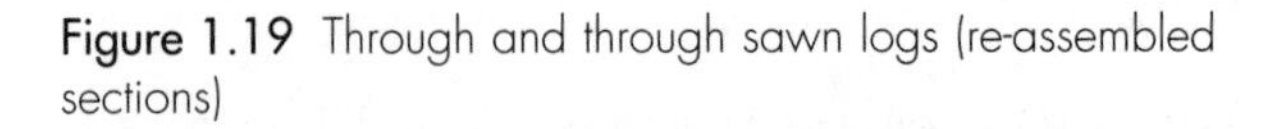

Figure 1.19 Through and through sawn logs (re-assembled sections)

1.4.9 Quarter (Radial) or Rift Sawn

Quarter or rift sawing can be wasteful and expensive, but it is necessary where a large number of radial, or near radial, sawn boards are required. Certain hardwoods cut in this way can produce beautiful figured boards (see Figure 1.23). An example is silver figured oak – in this case ray tissue has become exposed onto the surface of the board.

Quarter sawn boards retain their shape much better than plain sawn boards and will shrink less in service, making them well suited to good class joinery work and quality decking for floors. Figure 1.20 shows that, after the log is cut into quarters (quartered), a variety of conversion methods may be considered.

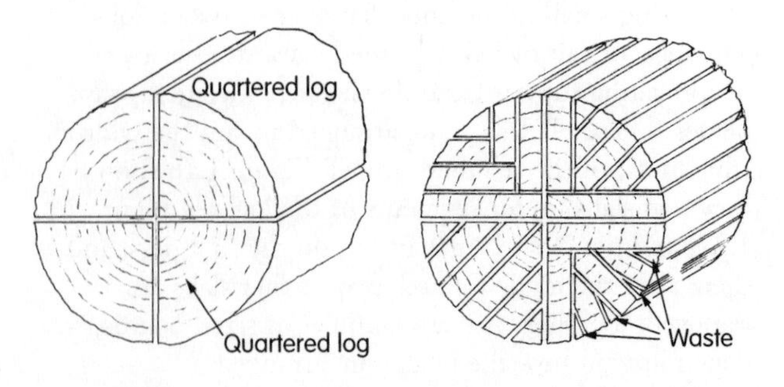

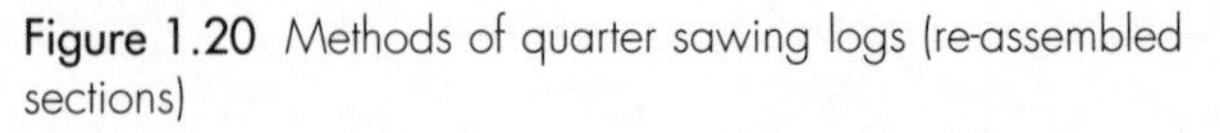

Figure 1.20 Methods of quarter sawing logs (re-assembled sections)

1.4.10 Plain Sawing (Plain or Tangential Sawn)

Logs cut in this way, will, with the exception of the heart (centre portion of the log) section, provide timber with a tangential face.

This method of conversion is used when cutting up large diameter logs of hardwood and softwood. Two of the many methods of producing plain sawn timber are shown in Figure 1.21.

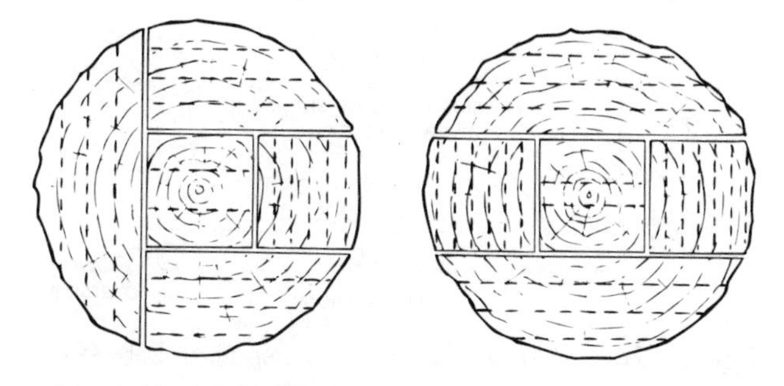

Figure 1.21 Plain sawn logs

1.4.11 Small Softwood Log Conversion

During this process the slabs (outer segments) are first removed by either a wood chipper or double bladed frame saw, double bandmill, or double bladed circular sawing machine.

Once the wood is flat faced or squared, further processing can be carried out as required. Figure 1.10 shows these final stages.

1.4.12 Conversion Geometry

Probably the simplest way to remember the various cuts is to imagine that the cross section of the log is circular, and that all its growth rings are concentric (having a common centre) to the pith.

By looking at Figure 1.22 you will see why timber sawn from a radius line is called radial or quarter sawn (look back to Figure 1.18) and that the quartered log is derived from four quadrants of a circle. Similarly, any cuts made tangential to a growth ring would be classified as being tangential or plain sawn.

The chords are straight lines that start and finish at the circumference. A parallel series of these would produce the pattern used for through and through sawing; the first of these chords when encompassed by an arc will form a segment, which we can now relate to the two outer slabs.

Chord lines are also used when making any cut tangential to a growth ring. The longest chord would normally be used to cut the log in half across its diameter.

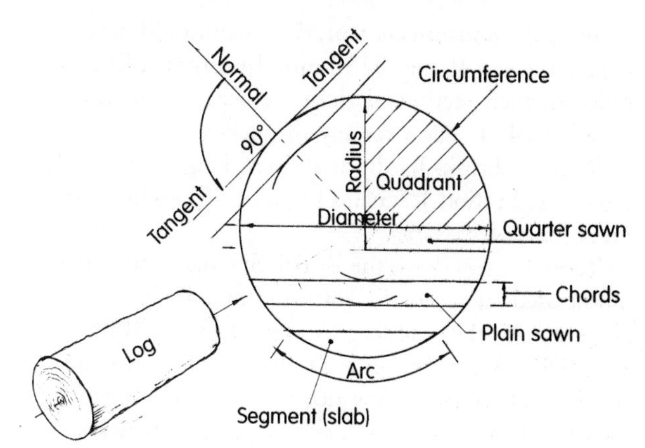

Figure 1.22 Conversion geometry

1.4.13 Decorative Boards

Figure 1.23 shows two examples of decorative boards, one hardwood and one softwood. As stated previously quarter sawn hardwoods with broad rays can produce nicely figured boards. European oak is well known for its silver figure. A similar pattern can be seen on beech but the patterning is generally much smaller.

When softwood with distinct growth rings, like Douglas fir and to a slightly lesser extent European redwood are tangentially sawn, a flame like pattern is produced on their surface. We call this patterning 'flame figure'.

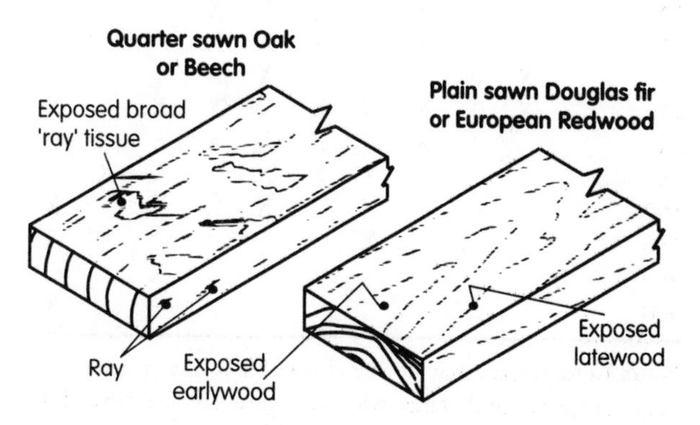

Figure 1.23 Decorative boards (other examples are shown in Figure 1.106)

1.5 SAWN (UNWROT) SIZES OF TIMBER

1.5.1 Softwoods

Depending on whether the supplies are from North America or Europe, the stated cross sectional size of timber can vary. Canadian mills may make no allowance in their sizes for any shrinkage which will take place when the wood is dried. For example, as shown in Figure 1.24a, if sections of 2 inches by 4 inches (50 mm by 100 mm) were to be produced then they would be sawn to that size knowing that, when they are dried to a marketable moisture content, their size would have reduced to say 46 mm by 96 mm. In contrast, European mills cut their sections slightly over size as shown in Figure 1.24b to allow for any shrinkage.

Timber shrinks hardly at all in its length when it has dried (see Section 1.7.5 and Figure 1.45), so little or no provision is necessary.

Table 1.4 sets down the British Standards basic recommended sawn sizes for softwood, and Table 1.5 (Figure 1.25) their cut lengths as produced by BSEN 336: 1995. Where timber has to be resawn to produce smaller section sizes, allowance will have to be made for the saw cut (known as its 'kerf'), which is usually about 2.5 mm wide. Otherwise, as shown in Figure 1.26, those sizes would be under size to those stated in Tables 1.4 and 1.6. In some cases this undersizing is accepted (see Table 1.12 within Section 1.8.4 dealing with stress-graded timber).

Figure 1.26 illustrates how undersizing can come about. For example if a 100 mm by 50 mm section is cut to produce four pieces at 25 mm by 50 mm, this would

Table 1.4 Customary target sizes of sawn structural timber (Extracted from BSEN 336: 1995)

Thickness (mm) to tolerance class 1	Width (mm) to tolerance class 1									
	75	100	125	150	175	200	225	250	275	300
22		X	X	X	X	X	X			
25	X	X	X	X	X	X	X			
38	X	X	X	X	X	X	X			
47	X	X	X	X	X	X	X	X		X
63		X	X	X	X	X	X			
75		X	X	X	X	X	X	X	X	X
100		X		X		X	X	X		X
150				X		X				X
250								X		
300										X

Note: Certain sizes may not be obtainable in the customary range of species and grades which are generally available. This standard has a lower limit of 24 mm. However, as thinner material is used in the UK, the customary sizes of each material are also listed here. Tolerance class 1
(a) for thicknesses and widths ≤100 mm $\left[{}^{+3}_{-1}\right]$ mm;
(b) for thicknesses and widths >100 mm $\left[{}^{+4}_{-2}\right]$ mm.
Target size of 20% moisture content.

Table 1.5 Customary lengths of structural timber. (Extracted from BSEN 336: 1995)

1.80	2.10	3.00	4.20	5.10	6.00	7.20
	2.40	3.30	4.50	5.40	6.30	
	2.70	3.60	4.80	5.70	6.60	
		3.90			6.90	

Note: Lengths of 5.70 m and over may not be readily available without finger jointing

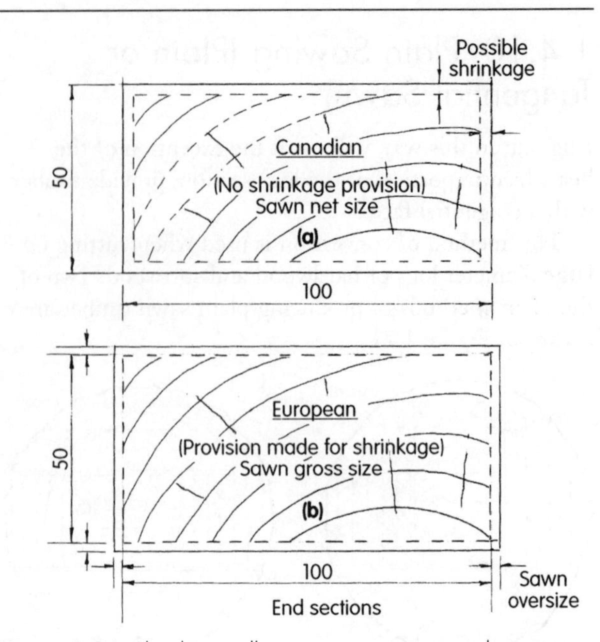

Figure 1.24 Shrinkage allowance – comparison between Canadian and European timber sections before any drying process

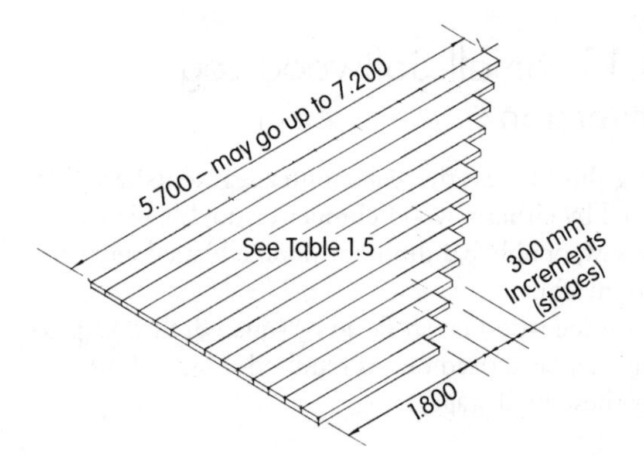

Figure 1.25 Available lengths of softwood – many mills limit supplies to maximum lengths of 5.7 m

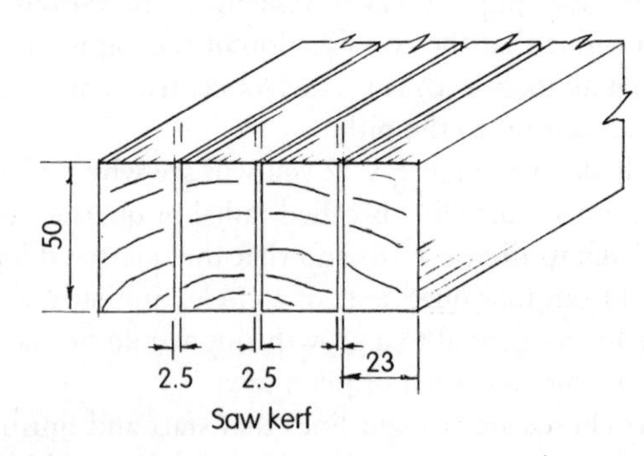

Figure 1.26 Stock (50 × 100) resawn to produce undersized (25 × 50) timber

entail making three saw cuts at 2.5 mm each, resulting in the section being undersized to ≃23 mm by 50 mm. Table 1.6 – produced by The Nordic Timber Council – shows the basic sizes as sawn from the log in Scandanavia from which sections may be resawn even smaller than shown.

Table 1.6 Guide to basic sawn sizes as sawn from the log (extracted from *How to Specify Redwood and Whitewood* (Nordic Timber Council))

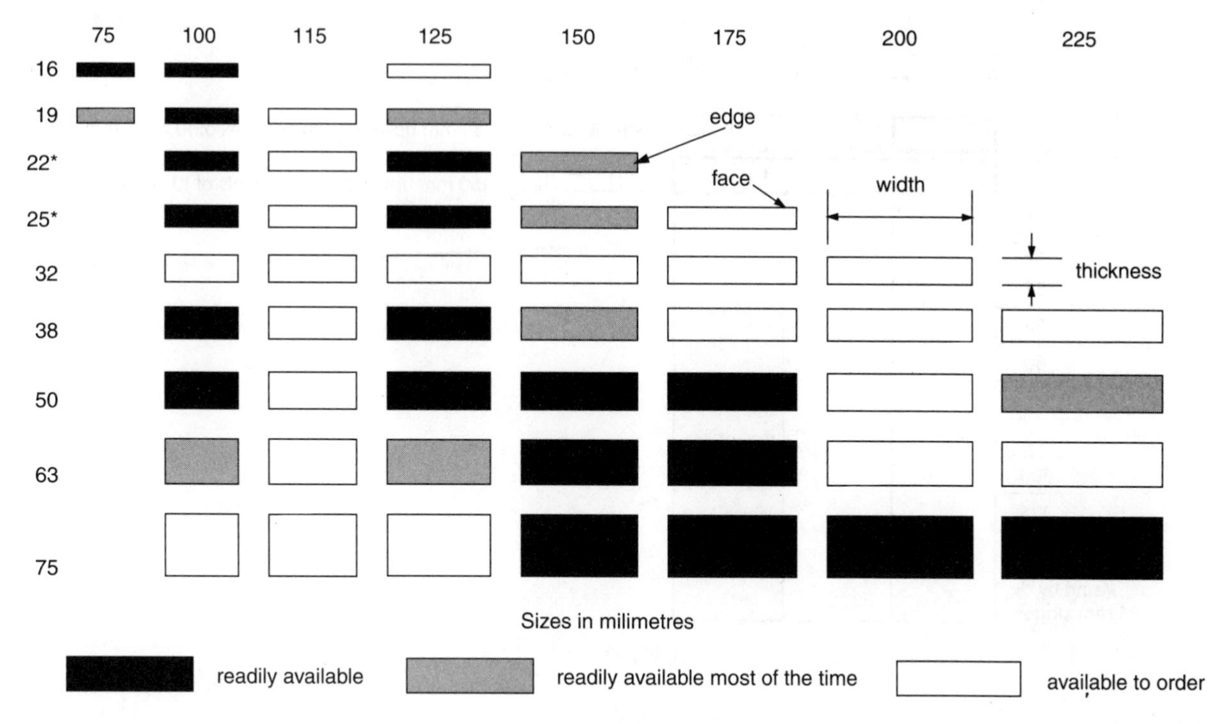

1.5.2 Hardwoods

Hardwoods are generally regarded as decorative material which is sawn into boards of varying thickness, thereby allowing the end user to recut the wood to suit individual needs.

The number of hardwood species imported into the UK is far greater than the number of softwoods, reflecting the many different types of sawmill situated around the world, ranging through the temperate, sub-tropical and tropical regions – timber sections and sizes will vary accordingly. Figure 1.27 gives some indication of how board profiles can vary. Table 1.7 shows the basic sizes of sawn hardwood according to BS 5450: 1977. Table 1.8, on the other hand, is based on sawn thicknesses where you can see that the emphasis on width is either **normal, strip** or **narrows.** Lengths are based on **normal** or **shorts.** (NB the term 'shorts' also applies to some softwood lengths).

It is important that we as end users are fully aware of all the basic sizes available to us in our locality, whether they are softwood or hardwood. By adopting standard sizes we can dramatically reduce:

- time spent on further conversion
- wastage
- build-up of short ends (incidentally, the term 'off-cuts' usually refers to waste pieces of sheet material)

making it possible to plan jobs more cost effectively.

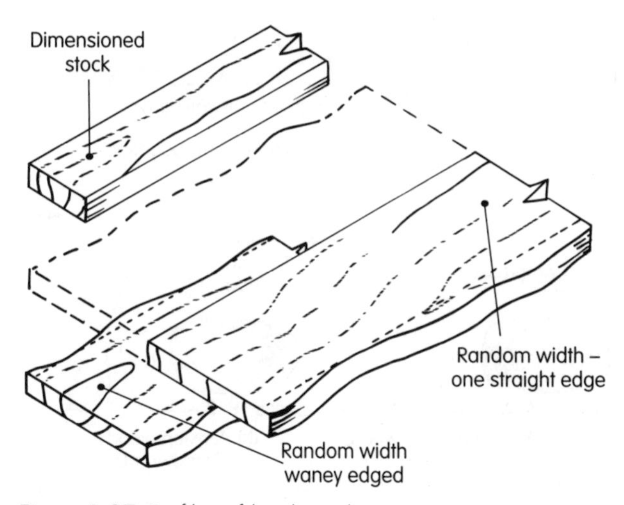

Figure 1.27 Profiles of hardwood sections

Table 1.7 Basic sizes of sawn hardwoods. (Extracted from BS 5450: 1977)

Thickness (mm)	Width (mm)										
	50	63	75	100	125	150	175	200	225	250	300
19			X	X	X	X	X				
25	X	X	X	X	X	X	X	X	X	X	X
32			X	X	X	X	X	X	X	X	X
38			X	X	X	X	X	X	X	X	X
50				X	X	X	X	X	X	X	X
63						X	X	X	X	X	X
75						X	X	X	X	X	X
100						X	X	X	X	X	X

Note: Designers and users should check the availability of specified sizes in any particular species

Table 1.8 Standard sawn sizes of hardwood

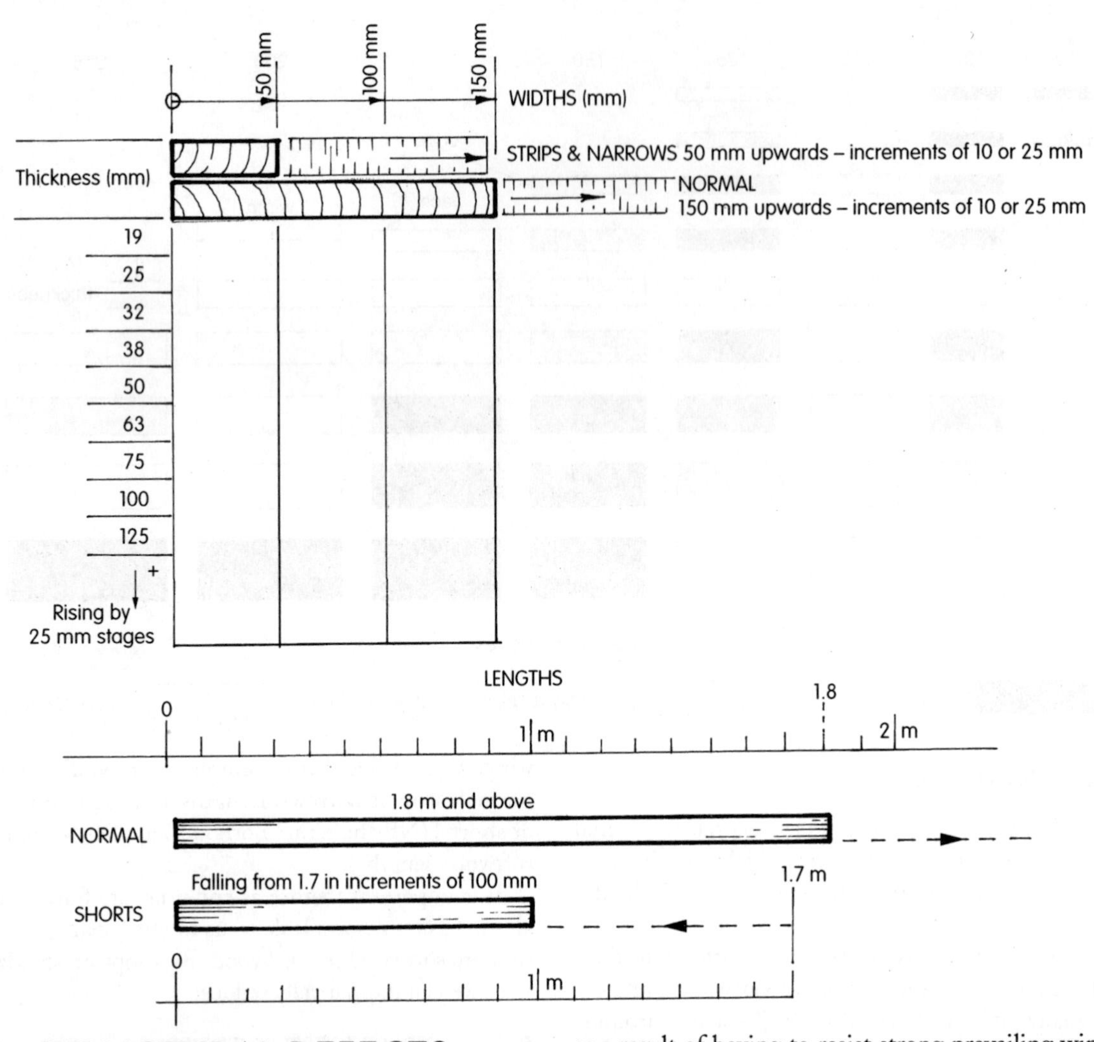

1.6 STRUCTURAL DEFECTS (NATURAL DEFECTS)

What then do we mean by structural defects? For the purpose of this section we shall regard them as those elements within the natural make-up of the growing tree which could influence the strength factors of the resulting timber, or possibly to a lesser extent the visual appearance of the surface of the timber.

Most of these defects have little, if any, detrimental effect on the growing tree, but they can degrade the timber cut from it, thereby lowering its market value. For the reasons described in Section 1.1.1, not all trees are suitable for use as timber. Remember the comparison we made between 'forest close grown' trees and 'open grown' trees, with their natural large branching habit.

1.6.1 Reaction Wood

This is another growth pattern which can seriously affect timber. The defect is the result of any tree which has had to grow with a natural leaning posture, usually as a result of having to resist strong prevailing winds, or having established itself on steep sloping ground. All such trees will resist any pressure exerted upon them by attempting to grow vertically, with added supportive

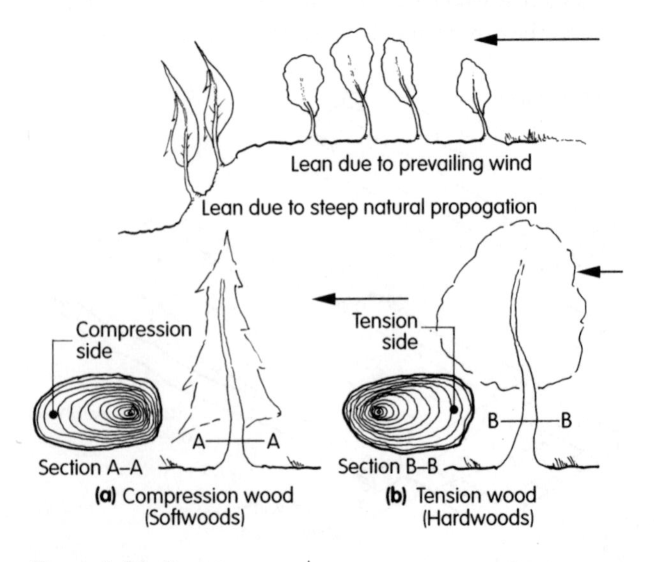

Figure 1.28 Reaction wood

wood growth to their stem. Any extra growth will be formed in such a way that the stem will take on an eccentric appearance around the pith, which, with softwood (Figure 1.28a), is pronounced on the side of the tree which is being subjected to compressive forces. This wood is known as **compression wood.** Hardwoods produce extra wood on the side likely to be stretched. Since this is the tensile side (Figure 1.28b), this wood is known as **tensile wood.** In both of these cases, such wood is unsuitable as timber, since it would be unstable when dried, and particularly hazardous when processed. Compression and tension wood are known collectively as **reaction wood.**

When we think about it, we realise that all tree branches to some extent will contain some reaction wood because of their drooping habit.

1.6.2 Heart Shake

The defects illustrated in Figure 1.29 are detailed below.

This defect shows as a shake (parting of wood tissue along the grain) within the centre region of the trunk, caused by uneven stresses being set up, which may increase as the wood dries. Where several of these shakes radiate from this region collectively they would be known as a star shake.

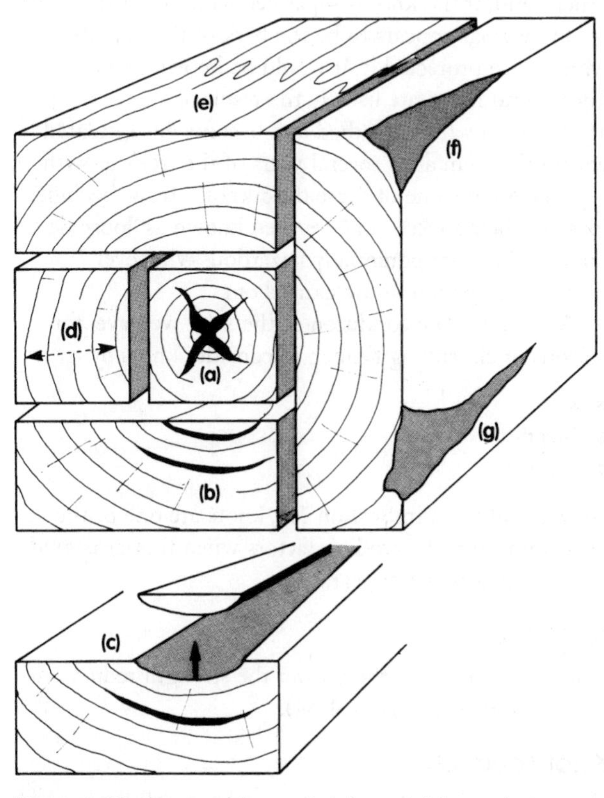

Figure 1.29 Structural (natural) defects: **(a)** heart shake (star shake); **(b)** cup shake; **(c)** possible result of cup shake; **(d)** rate of growth; **(e)** compression failure; **(f)** wane (Waney-edge); **(g)** enclosed bark

1.6.3 Ring Shake (Cup Shake)

This is a shake which follows the path of the growth ring. Using timber with this type of shake can result in a very dangerous situation if, as shown in the illustration, this portion of wood becomes dislodged during a machining operation.

1.6.4 Brittleheart

Wood is described as brittleheart when the juvenile wood (see Section 1.2.3) of largely tropical and sub-tropical species of low density (when dried) hardwoods becomes subjected to heavy stresses. Possibly because of bending etc., the immature fibres compress when trying to support the heavy crown. These fibres may end up becoming compounded (multiple compression fractures) into a compression failure.

Timber processed from this region will break quite easily, because the fibres are unable to respond to being bent. The result could well are an almost clean break across the grain of the wood; this would be known as a **brash failure.** Figure 1.30 compares a normal fracture with a brash failure.

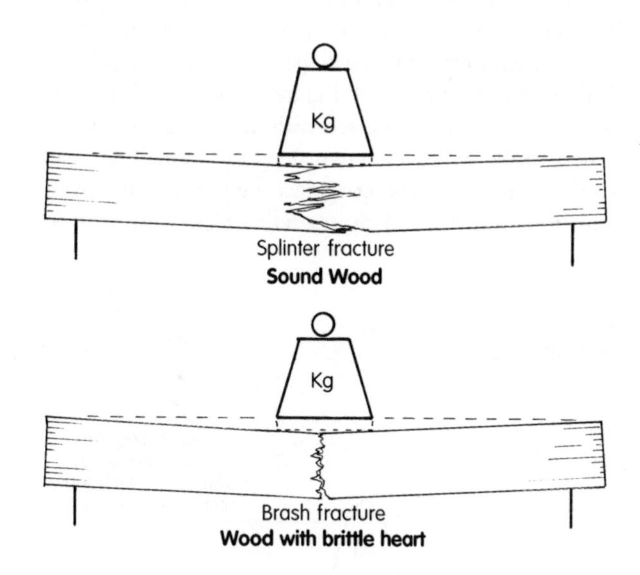

Figure 1.30 Possible effect of using wood with brittleheart

1.6.5 Natural Compression Failure (Upset)

Natural compression failure is the fracturing of the wood fibres across the grain; it is thought to be caused by sudden shock at the time of felling or by the tree becoming overstressed during growth – possibly by being bent by strong winds. Other terms used to describe this defect are 'thundershake' and 'lightning shake', which, in reality, have no bearing on the defect whatsoever.

1.6.6 Rate of Growth

The average spacing of growth rings (mm) can, in the main, determine timber's strength (see Sections 1.8.4 and 1.10.4g).

1.6.7 Wane (Waney-edge)

This refers to the edge of a piece of timber which has retained part of the tree's outer rounded surface (it may have also retained some of the bark).

1.6.8 Encased Bark

Encased bark is a condition which may appear on the face or edge of a piece of timber.

1.6.9 Sloping Grain

The grain (direction of the bulk of the wood tissue) slopes at an angle across the face or edge of a piece of timber. The steeper the angle the more severe the condition, which, as shown in Figure 1.31a, makes load bearing timbers unsafe. This condition is often brought about when timber is cut from a bent log (Figure 1.31b). Timber used for structural purposes has strict controls put on it with regard to the slope of its grain – this is dealt with in Section 1.8.4. Figure 1.31c shows how the slope of grain can be assessed with the use of a 'swivel handle scribe', which is drawn along the grain. The resulting angle of track set against the parallel sides of the timber is measured and given a ratio of 'gradient of slope'.

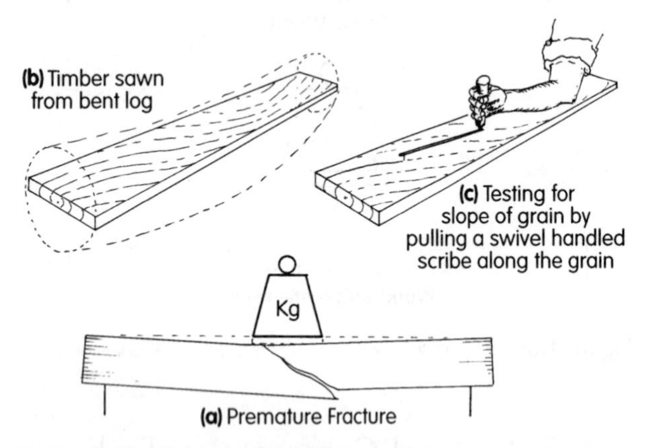

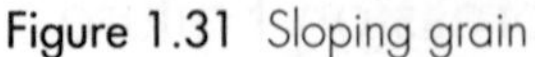
Figure 1.31 Sloping grain

1.6.10 Knots

Figure 1.32 shows how knots are formed as a result of trees' branching habit. All branches are an integral part of the stem, and therefore when the stem is converted into timber knots are exposed on its faces. Their condition (Figure 1.33) will depend on whether they are thoroughly bonded into the surrounding wood, in which case we call a knot a live, sound, or tight knot. On the other hand, if the knot is separated from its surrounding wood by what appears to be the bark of the branch which once protected it, it would be known as a dead knot. Dead knots are usually the result of a once damaged branch which has died and become encapsulated within the stem after several years of the tree's growth.

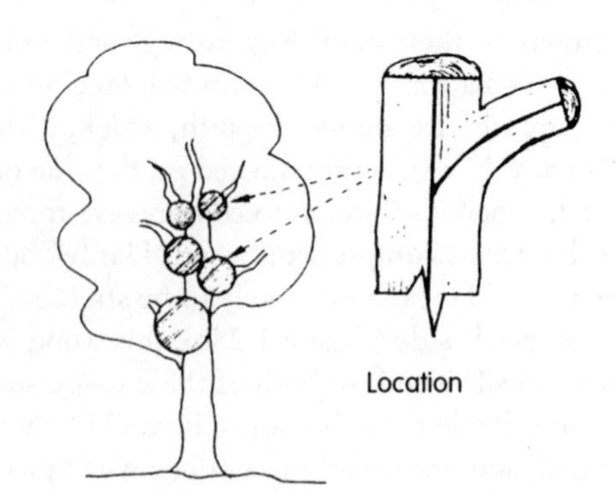

Figure 1.32 Knot location in relation to branches

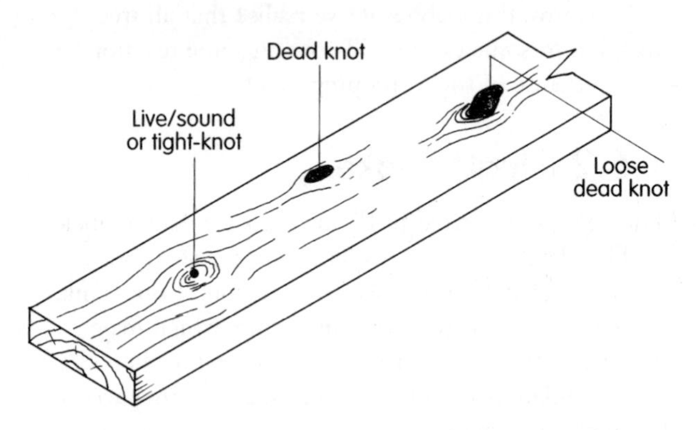

Figure 1.33 Knot condition

Very often some of these dead knots are, or become, loose in their sockets and are then known as 'loose dead knots', which are potentially hazardous whenever machining operations are carried out.

Apart from the condition of the knot we have, for reasons of classifying timber, to consider knots by their:

- size
- location
- number.

As you will learn in Section 1.8, knots are one of the most important controlling factors when timber is graded for strength and appearance.

Knot Size

The larger the knot, the greater the strength reduction of the timber (see Figure 1.34).

Knot Location

Knots nearer the edges of beams are generally going to reduce the strength properties of timber more than those nearer their centre. See Figure 1.35 and the section below on knot type.

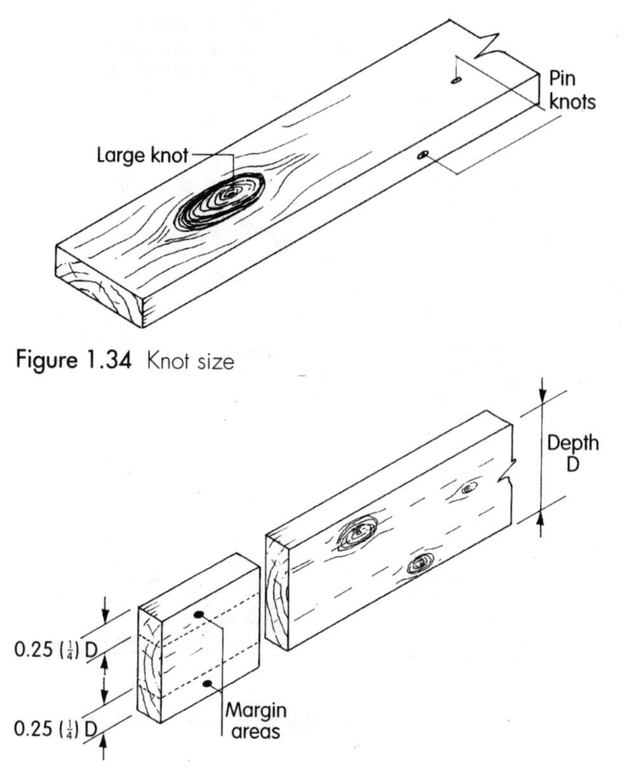

Figure 1.34 Knot size

Figure 1.35 Knot location

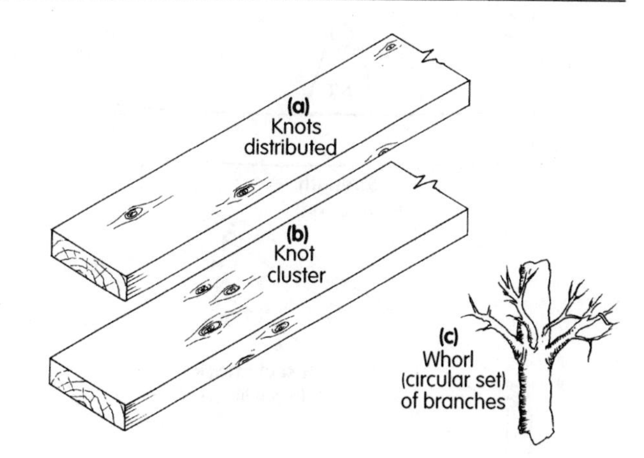

Figure 1.36 Number of knots

Number of Knots

Generally the greater the distance between knots the better. Depending on the age and species of tree, knots may appear well distributed (Figure 1.36a), or as clusters (Figure 1.36b), which is the result of a cut being made through a whorl (circular set) of branches (Figure 1.36c). With largish knots this can be a problem.

Knot Type

Knots appear in many forms on the surface of the timber (Figure 1.37); it is important to try to relate the position of the knot with its name. For example, face knots appear on the widest sides (faces) of the timber, but if they are situated within the upper or lower margin areas (Figure 1.35), they will also be termed **margin knots** or **margin face knots**. Knots on the narrow sides (edges) of the timber are called **edge knots**. A knot that falls across a corner between a face and an edge is called an **arris knot**, and knots which are cut through their length, either fully or partially at an angle are known as **splay knots**.

Other factors, including wood-boring insects and fungal attack, which can also be responsible for degrading timber, are dealt with in Section 3.19.

1.7 DRYING TIMBER

Green timber cut from a freshly felled log will, as far as the carpenter and joiner are concerned, be almost unusable. As we already know, wood in this condition houses considerable amounts of water. The water will affect timber in many ways, for example (see Figure 1.38):

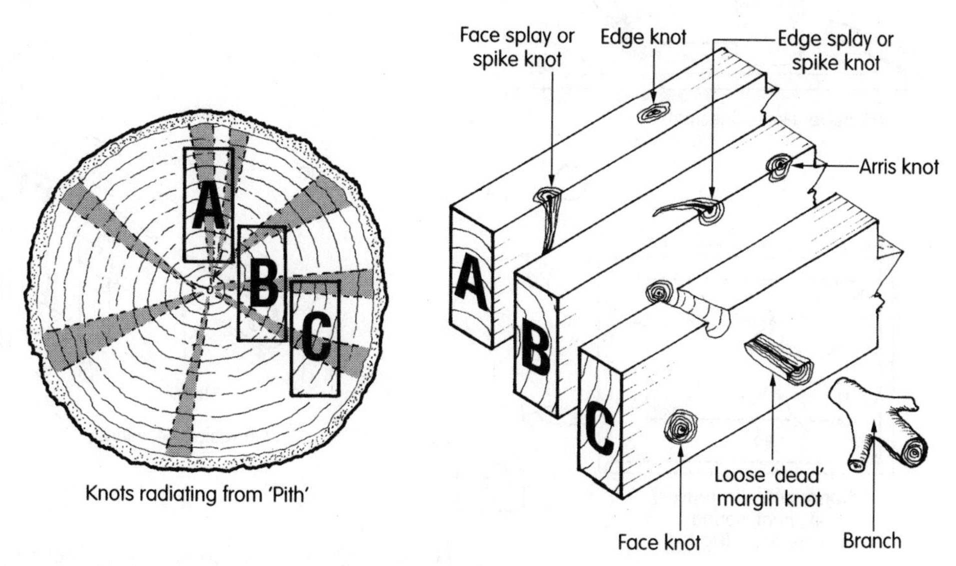

Figure 1.37 Knot type in relation to main stem of tree

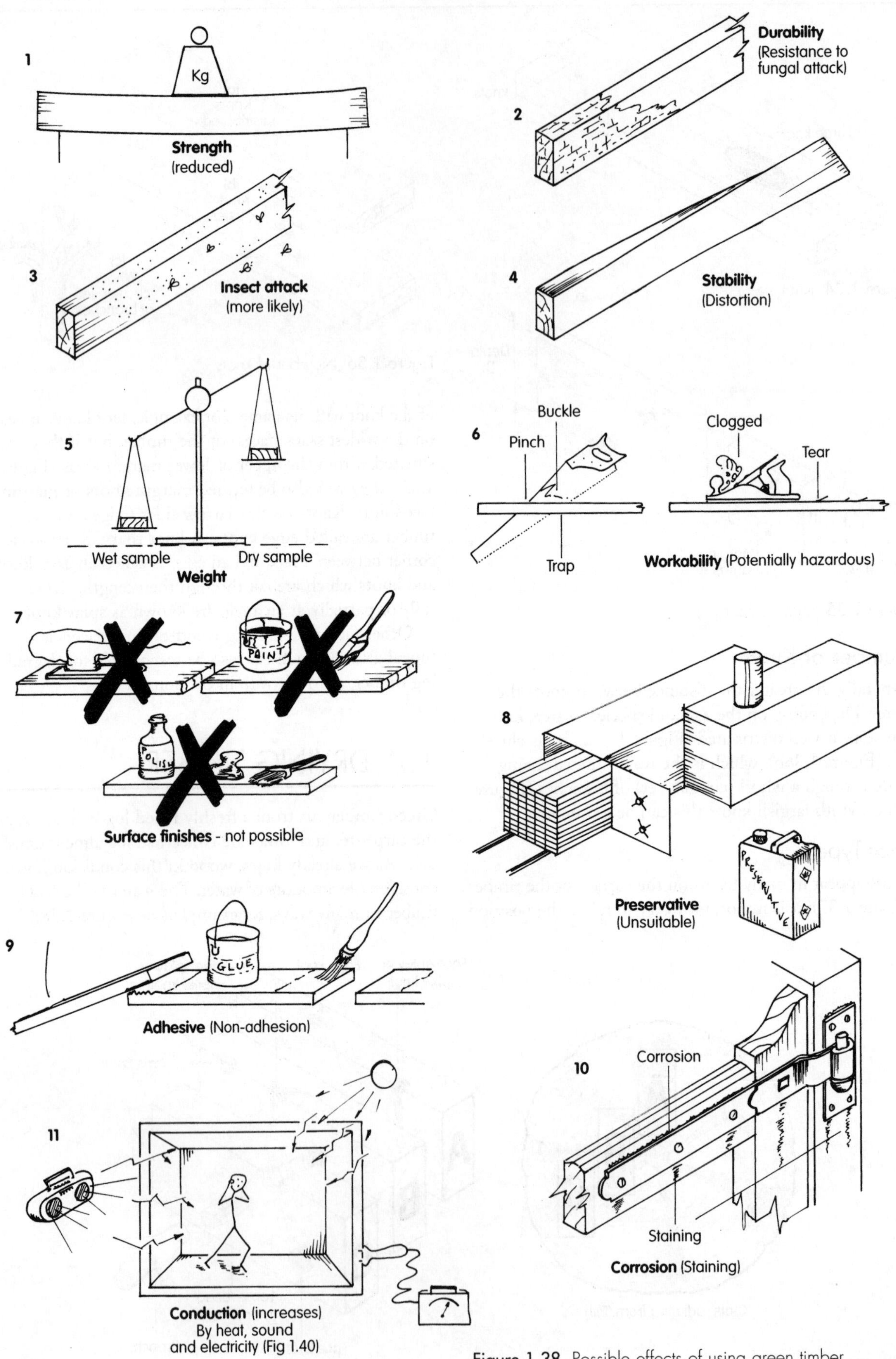

Figure 1.38 Possible effects of using green timber

1 strength properties will be reduced
2 fungal attack is more likely
3 the timber is vulnerable to some species of wood destroying insects
4 if moisture is allowed to leave the wood in an uncontrolled way it could, because of shrinkage and movement, result in the timber becoming distorted
5 there is excessive weight due to the amount of water present
6 the wood is unworkable by hand or machine tools (dangerous practices could result) – special saws are used for cutting green wood
7 the surface finish is unsuitable to receive either paint or polish
8 the wood is generally unsuitable for treatment with wood preservative, or fire retardant (except when the method used is by diffusion (see Section 4.4.1)
9 adhesives will have minimal holding properties
10 corrosive properties increase (see Section 1.10.3g), particularly upon ferrous metals (those metals containing iron)
11 conductivity increases, and thereby reduces thermal, sound and electrical insulation properties.

Therefore the only way that timber can be regarded as a stable material will be when the amount of water it contains is reduced to suit its particular end use as shown in Table 1.9. There are many methods used to reduce this moisture; the main techniques we are concerned with are:

- air drying methods (Section 1.7.6)
- kiln drying (Section 1.7.11).

However, a combination of the two can be used by starting with a short spell of air drying followed by kiln drying – the drying process as a whole is often called 'seasoning'. The degree of water contained within wood at any given time is measurable and termed 'moisture content'.

1.7.1 Moisture Content (MC)

Moisture content is measured as a percentage value of the dry weight of the wood. But don't be alarmed if you find quotations of over 100% MC. The reason for this is that the weight of the water within the wood can be greater than the weight of the wood substance; in other words moisture content should be seen as a percentage of the wood's weight when it is absolutely dry. For example, if the weight of the water was equal to that of the wood substance, then the wood's MC would be 100%, and likewise if the weight of the water was twice that of the wood substance this would produce a value of 200% MC.

There are several methods of obtaining moisture content values, but we will consider the two most popular:

- the oven dry method
- the moisture meter method.

1.7.2 Oven Dry Method

A small sample is cut, usually from a sample board, which is part of a batch of timber to be dried (Figure 1.56). Figure 1.39 illustrates the procedure, whereby the sample is first weighed to determine its *initial* or 'wet' weight (Figure 1.39a) – which must be accurately recorded. Then it is placed into a special drying oven (Figure 1.39b) set at 103°C ±2°C and left until no further weight loss can be recorded (Figure 1.39c). This will mean that over a period of possibly two days the sample will have to be systematically weighed. Once its 'final' or 'dry' weight is achieved its original moisture content can be calculated by using one of the following formulae (a simplified version of achieving this is shown in Figure 1.39):

Moisture content % =

$$\frac{\text{Initial (wet weight } (A)) - \text{Final (dry weight } (B))}{\text{Final or dry weight } (B)} \times 100$$

Or

$$\text{MC \%} \frac{\text{Inital (wet weight } (A))}{\text{Final or dry weight } (B)} - 1 \times 100$$

For example, if a sample has a wet weight of 25.24 g (A) and a dry weight of 19.12 g (B), then:

$$\text{MC \%} = \frac{A - B}{B} \times 100$$

$$\frac{25.24\text{ g} - 19.12\text{ g}}{19.12\text{ g}} \times 100 = 32\%$$

Or

$$\text{MC \%} = \left[\left(\frac{A}{B}\right) - 1\right] \times 100$$

$$\text{MC \%} = \left[\left(\frac{25.24\text{ g}}{19.12\text{ g}}\right) - 1\right] \times 100 = 32\%$$

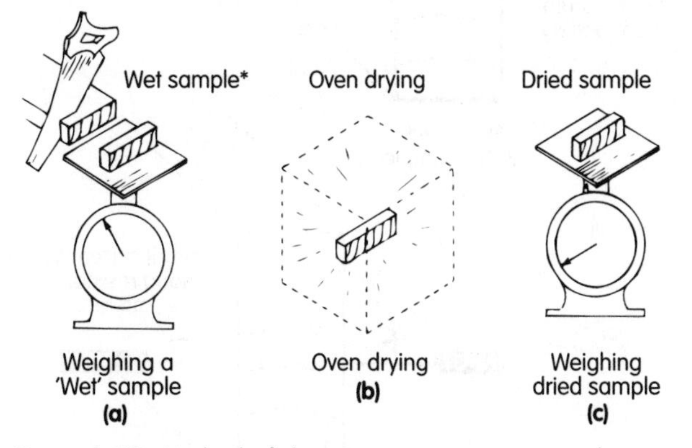

Figure 1.39 Method of determining moisture content by oven drying a small sample of timber.
Simple to remember formula for obtaining moisture content:

$$\text{MC \%} = \frac{A - B}{B} \times 100$$

1.7.3 Moisture Meters

Moisture meters are battery operated instruments which work by off-setting the electrical properties of dry wood relative to wet wood. In other words, because water conducts electricity, the resistance normally offered by dry wood becomes less as its moisture content increases. An example of this is shown in Figure 1.40.

With the exception of the small hand-held meters (Figure 1.41) moisture meters have two parts (Figure 1.42):

- the meter with either a pointer over a numerical scale, or digital read out – provision will be made for adjustment to suit different wood species; this part will also have a compartment to house the batteries
- spiked electrodes set into an insulated handpiece, with provision for connecting it to the meter via a detachable cable. The type of handpiece with its differing lengths of electrodes (needles) will, as Figure 1.43 shows, depend upon the sectional size of the timber being checked. For example, the short needle push-in types will have a needle length of about 10 mm, whereas the heavier hammer-in action types will have longer needles.

Moisture meters have a working range from about 6% MC to about 30% MC. They are, however, more than just useful aids for making on-site spot checks. One of their main functions is to be used in conjunction with

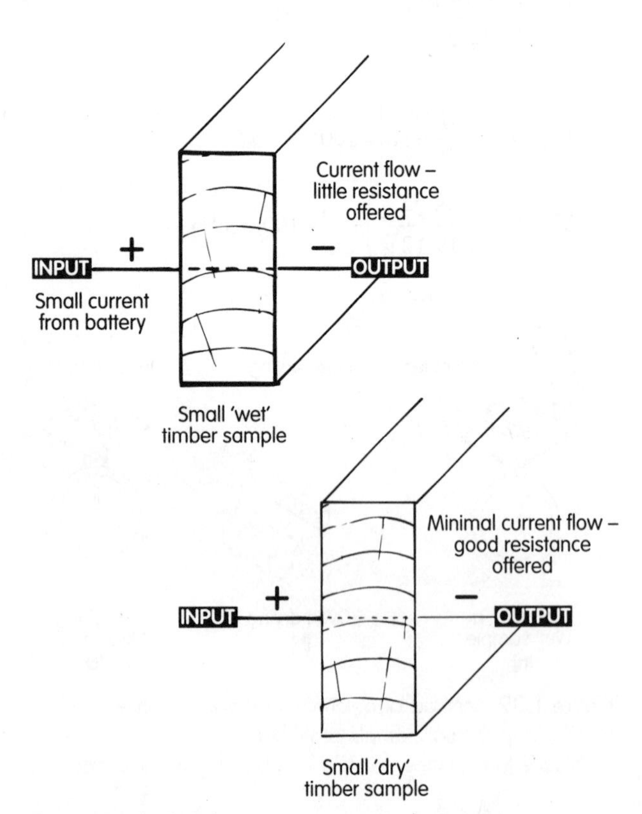

Figure 1.40 Conductivity increases with any increase in moisture content

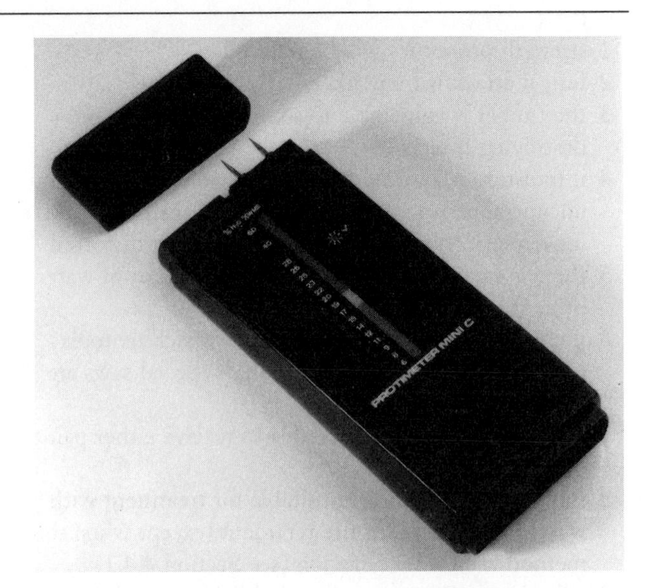

Figure 1.41 Hand held 'Mini' moisture meter by Protimeter (with kind permission from Protimeter Ltd)

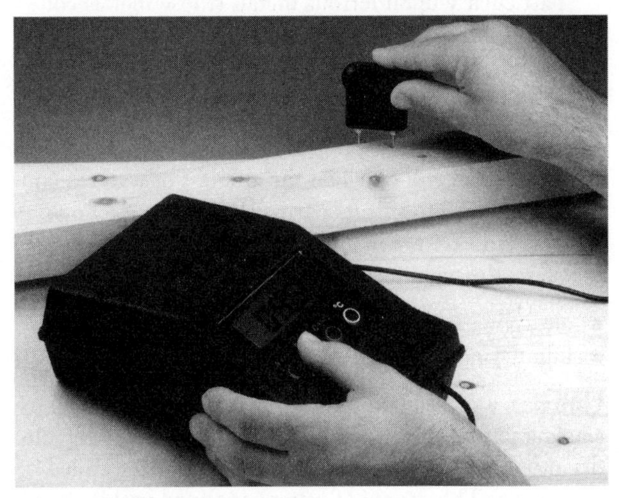

Figure 1.42 Protimeter's 'Diagnostic Timber Master' – a two part moisture meter (with kind permission from Protimeter Ltd)

timber drying procedures, be it by air, or kiln drying (Section 1.7.6 and 1.7.11).

Whenever a moisture meter is used it is important to consider the following points:

- the electrodes can reach the appropriate depth
- allowance can be made for different wood species, because inherent electrical properties of the wood can affect the meter reading
- the temperature of the timber is known, since meter readings can vary with temperature
- certain chemicals must not be present on the timber or within the wood, e.g. wood preservatives, or flame retardant solutions.

As far as the carpenter and joiner are concerned the main use of a moisture meter would probably be for sorting and checking large batches of delivered timber, or for checking the condition of assembled or fixed

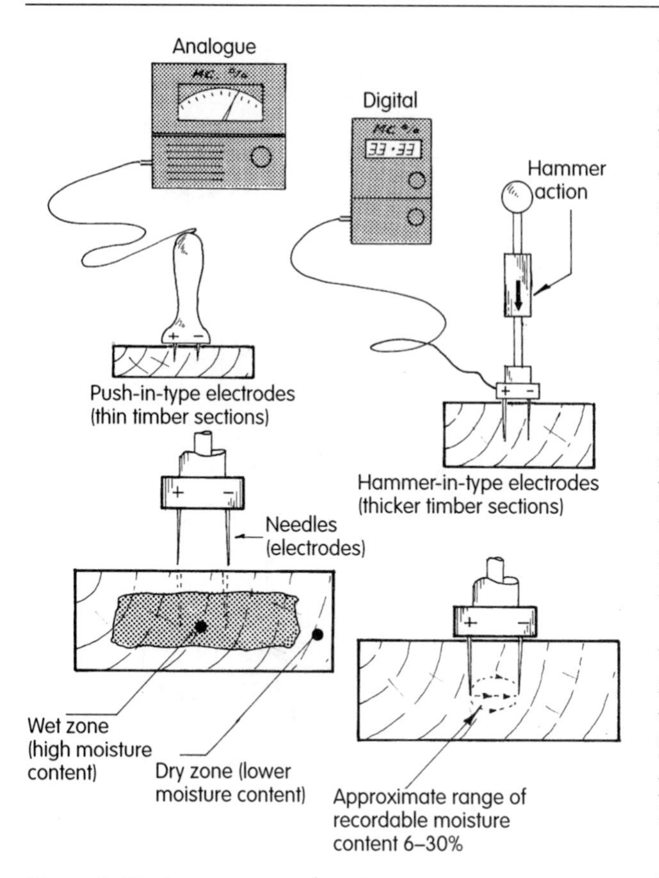

Figure 1.43 Battery operated moisture meter

items of carpentry and joinery. If a fungal attack is in evidence, or even suspected, a moisture meter can be invaluable in such situations (see Chapter 3).

1.7.4 Moisture Removal

In this section we will see how water can be effectively removed from the timber. The effect of this and the relevance to the timber's end use are outlined in Table 1.9.

Let us first try to understand where and how this water is held within the wood itself. In Figure 1.44, you will see an exaggerated view of the end grain of a piece of timber, as seen through a high powered lens. Notice that this end is made up of a series of minute cells (open ended tubes); we shall be studying these cells in greater detail in Section 1.10. We already know that green timber contains a great amount of water. This water is contained within these cell cavities and we call this *free water*, because it is free to move around from cell to cell, as originally required by the living tree. The water contained within the cell walls is fixed (chemically bound to them) and therefore known as *bound water*.

Depending on where the timber is sited, drying could, to a certain degree, commence almost immediately. As you will know the air we breathe contains varying amounts of moisture: the amount will depend on how much is suspended in the air as vapour at that point in time, which in turn will depend on the surrounding air temperature. As the air temperature increases, so does its capacity to absorb more moisture as vapour until the air becomes saturated, at which point we become very aware of how humid it has become. It is therefore this relationship between air temperature and the amount of moisture that the air can hold that we call relative humidity.

So, if the air surrounding the timber has a vacant capacity for moisture, it will take up any spare moisture from the wood until, eventually, the moisture capacity of the air is in balance, or equilibrium, with that of the timber, and when stable conditions are reached we can say an **equilibrium moisture content** (EMC) has been achieved. This process will, of course, act in reverse because wood is a **hygroscopic material**, which means that it has the means, provided the conditions (those mentioned above) are suitable, to pick up and shed moisture to its surrounding environment.

Free water will leave first via tiny perforations within the cell walls. As the outer cells start to dry, they will be replenished by the contents of the inner cells, and so on, until only the cell walls remain saturated. It is at this important stage of drying, known as the **fibre saturation point** (FSP), which will be reached at about 25% MC to 30% MC, that the timber will start to shrink.

Beyond this point (FSP), drying out bound water can be a very lengthy process if left to take place naturally. To speed up the process artificial drying techniques will need to be employed.

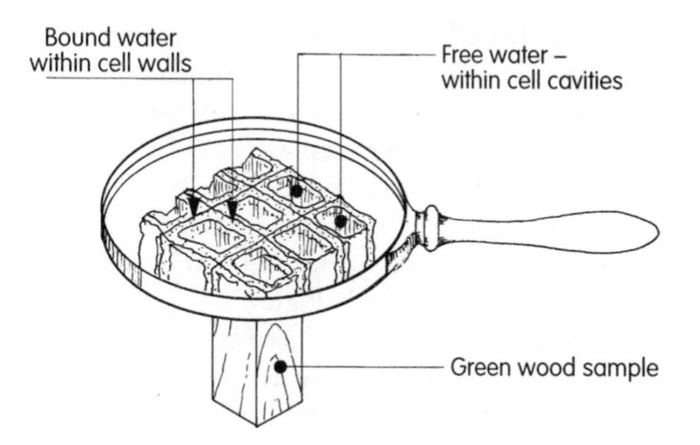

Figure 1.44 Constituents of moisture content (exaggerated view of end grain)

1.7.5 Shrinkage

Whether natural or artificial means of reducing the moisture content are used, timber sections will inevitably shrink. The amount by which they shrink will depend on the reduction below FSP, which we discussed in Section 1.7.4. But a more important factor is the relationship between the differing amounts of shrinkage compared with the timber's length (longitudinally), and its cross section (transverse section whether it is 'plain sawn' (tangentially) or 'quarter sawn' (radially). Figure 1.45 shows how we can view the differing proportions of shrinkage, for example:

a tangentially – responsible for the greatest amount of shrinkage
b radially – shrinkage of about half that of tangential shrinkage
c longitudinally – hardly any shrinkage.

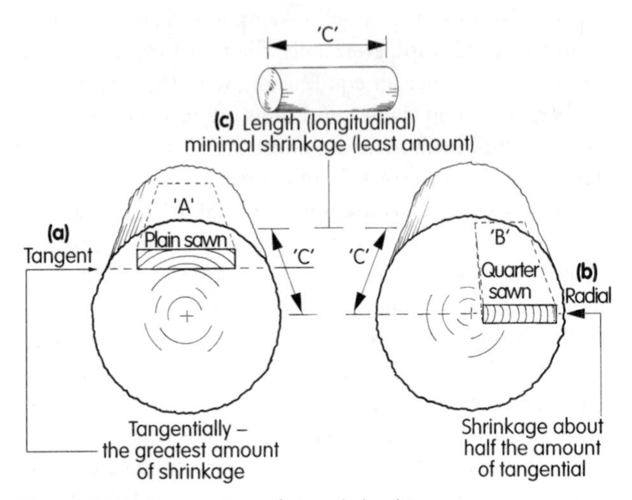

Figure 1.45 Proportions of wood shrinkage

We call these varying amounts of shrinkage **differential shrinkage**.

These variations come about because the tree is made up of different types of wood cells, which at this stage we can regard as being tubular in shape. These are distributed either lengthwise (longitudinally up the stem), or around the tree's axis (pith) – these we shall call **axil cells**. Those which go to make up the 'rays' which radiate outwards from the pith are known as **radial cells**. Figure 1.94 shows both axial and radial cells.

So it follows that the greatest amount of shrinkage we have mentioned will take place across the cells, not along their length. If you study Figure 1.46 you should notice that radial shrinkage is reduced because ray cells are positioned lengthwise, thereby helping to restrict the movement of those cells which run perpendicular to them in the radial direction.

As a result of this movement we can expect some sections of timber to distort in some way as extra moisture (below FSP) is lost. The shapes of distorted sections will depend on where the timber was cut out of the log during its conversion into timber. Figure 1.47 gives some idea of how certain sections of timber may end up after being dried – you will meet these again in Section 1.7.14.

Don't forget! When timber has reduced in section as a result of drying, this process can, if the timber is allowed to pick up moisture, be easily reversed – the timber will then swell. Remember that wood is a 'hygroscopic material' (see Section 1.7.4). The simplified diagram shown in Figure 1.48 should help you to remember the principle as a whole.

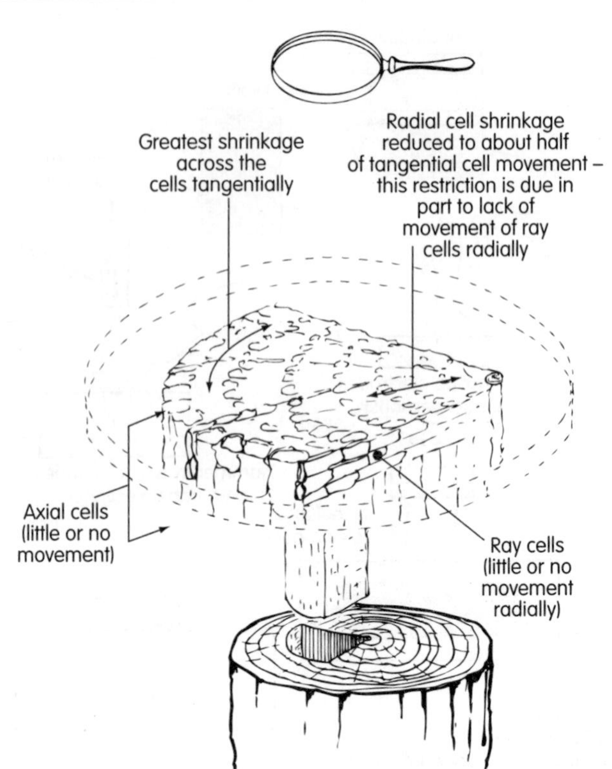

Figure 1.46 Shrinkage movement in relation to direction of wood cells (exaggerated view of end grain)

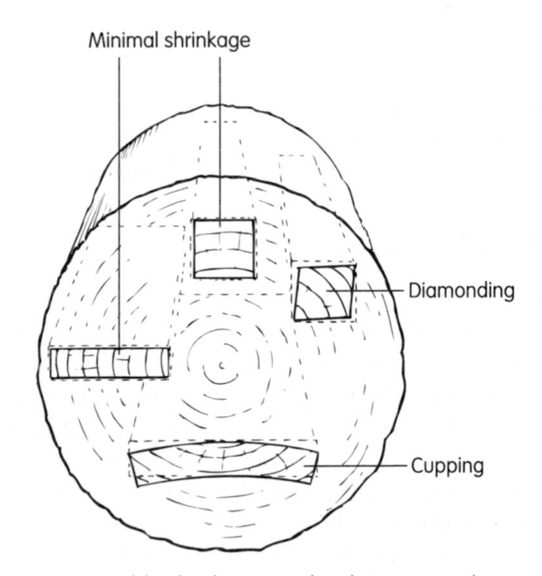

Figure 1.47 Possible shrinkage, and timber section distortion as a result of differential shrinkage after drying

1.7.6 Air Drying (Natural Drying)

Timber is stacked under controlled conditions in the open air in order to reduce its moisture content naturally. This system relies entirely on the sun and wind to provide drying. The sun provides the heat to raise the air temperature, and the air in turn picks up the moisture, whilst ideally the wind carries away the warm moist air, and prevents it from becoming saturated

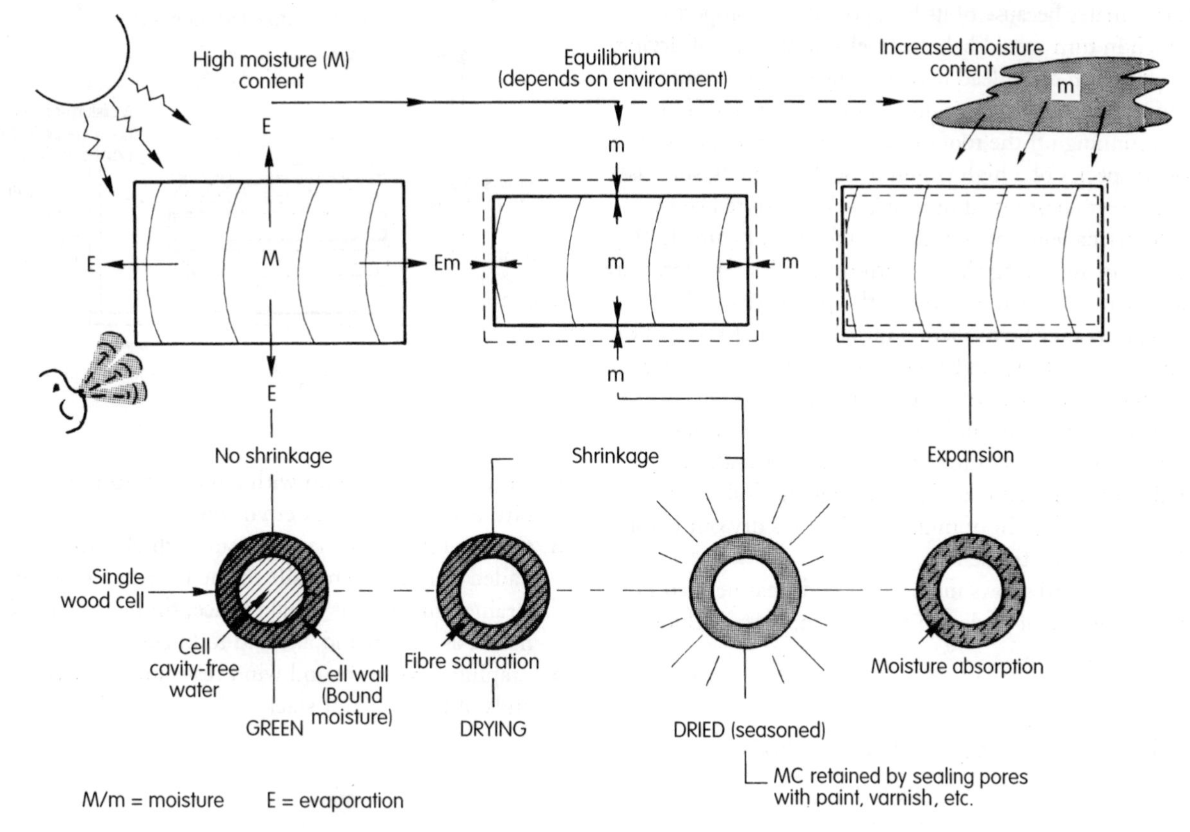

Figure 1.48 Basic principles of moisture movement

(similar to the process used when drying clothing outside on a clothes line). Because of climate fluctuations between seasons and different countries, the rate of drying is slow and dependent upon atmospheric conditions. Thus, this process can be most unreliable.

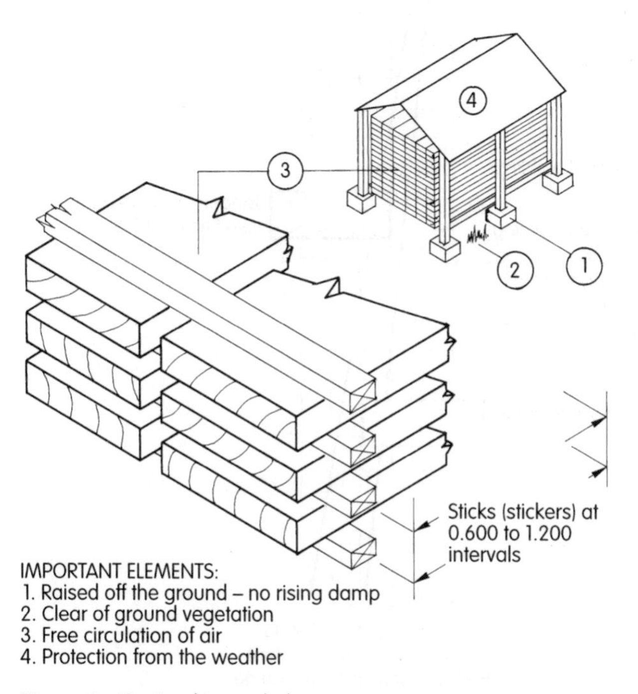

Figure 1.49 Air drying shelter

Where time is not important, the moisture content of timber in the UK can be as low as 15% during summer months, but on average we could be talking about 20%. The time span however, will depend on the amount of exposure, time of year when stacked, whether the wood is hardwood or softwood and the timber species, but most important of all on the thickness of the timber. For example, under good drying conditions, 50 mm hardwood could take in the region of one year to reach 20% moisture content, whereas softwood of the same thickness could take three to four months.

The success of air drying timber will mainly depend on its:

- weather protection, as and when necessary
- site conditions
- stacking
- climatic conditions for drying.

1.7.7 Weather Protection

Weather protection can be used except when drying certain hardwoods which can be dried as open-pile 'boule' (the log is through and through sawn then reassembled into its original form as shown later in Figure 1.52). A roof is used to protect the whole of the stack from direct rain, snow and extremes in temperature. The roof shape is unimportant provided it functions well, and is securely anchored down. Corrugated steel should be avoided in hot climates because of its heat conducting properties, which in turn would help to accelerate the rate of drying. This in turn could lead to one of the drying degrades. In addition coverings containing iron can rust, and if any rain running off the roof came in contact with any of the wood species of a high tannin or acidic content, such as oak, sweet chestnut, afrormosia, or Western red cedar, then permanent iron-staining could result. Figure 1.49 shows an air drying shelter with open sides. Shelters may also be in the form of sheds with open slatted sides; these slats may be fixed, or adjustable so that the air flow through the stack can be regulated, thereby achieving greater control over the rate of drying.

End protection can be very important, especially with some species of hardwood such as oak and beech, which are very liable to end splitting (see Table 1.9), since the end grain of timber will always dry out before the bulk of the timber.

Figure 1.50 shows methods of end treatment to slow down the rate of drying in this region; these include:

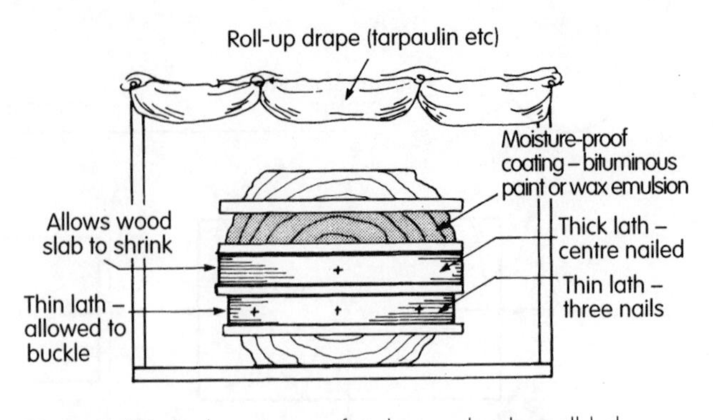

Figure 1.50 End treatment of timber, or boule, will help prevent end splitting

- treating the end grain with a moisture-proof sealer, bitumen paint, or wax emulsion
- tacking laths over the end grain – thick laths are nailed only in the middle of the board to allow unrestrained movement to take place; thin laths can be nailed as shown and allowed to buckle
- draping a moisture and wind resistant cover over the ends of the boule or stack.

Table 1.9 Moisture content of timber in relation to its end use

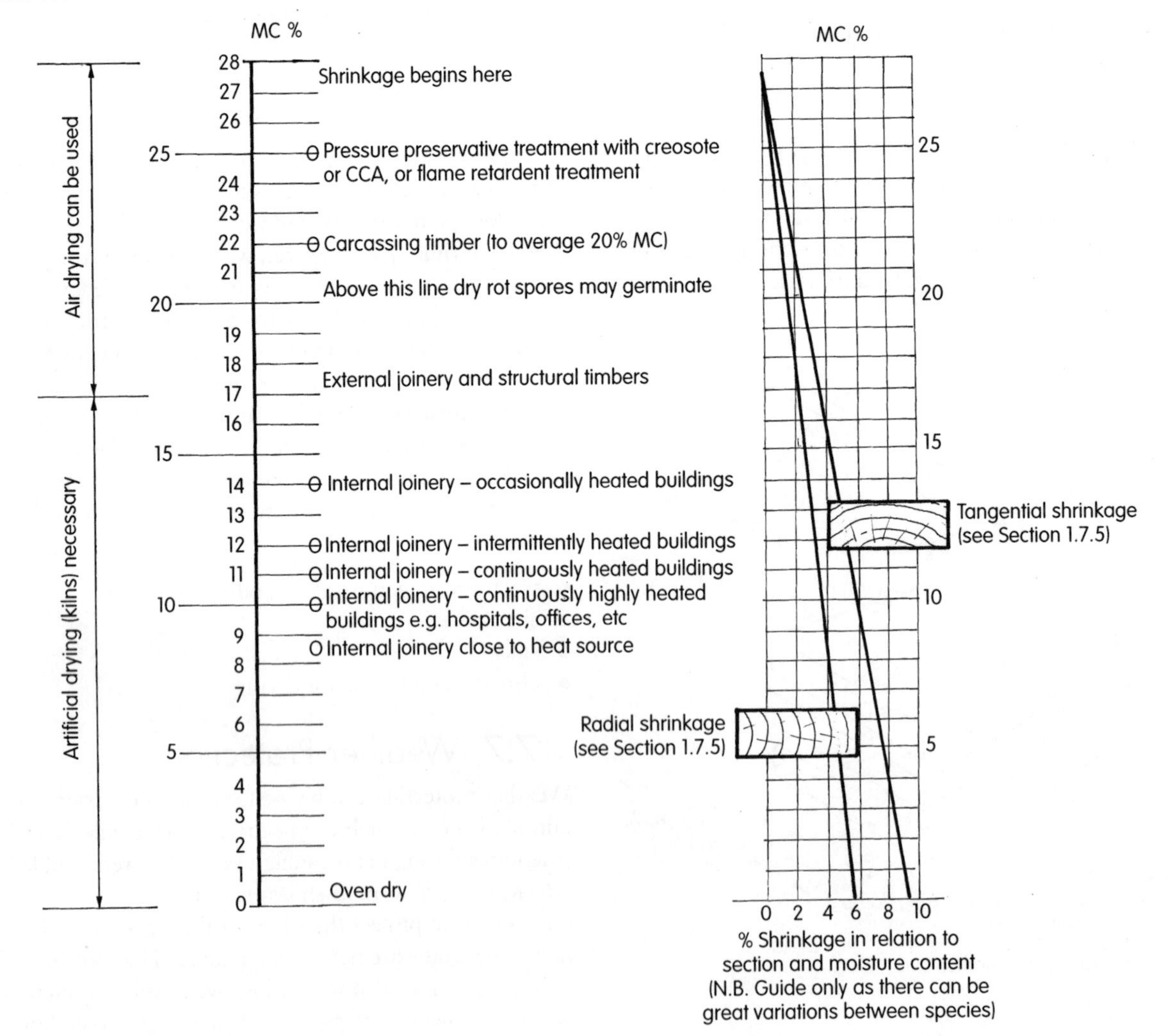

1.7.8 Site Conditions

The whole of the site surrounding the shelters (sheds) or stacks should be well drained, free from ground vegetation by blinding it with a covering of concrete etc., and levelled to allow for vehicle access. The surrounding floor area should at all times be kept tidy and clear of spent 'piling sticks' and 'short ends' of timber so that fungal or insect attack are discouraged.

Enough room should be left between shelters and stacks to allow for the loading and unloading of the timber and also to allow for carrying out routine moisture content checks, etc.

1.7.9 Stacking the Timber

The length of stack will be unlimited (depending on the timber lengths), but the height must be predetermined to ensure its stability and the stack (Figure 1.51) must be built to withstand wind. The width should not exceed two metres, otherwise cross air flow may well be restricted to only part of the stack. However, adjacent stacks can be as close as 300 mm to each other.

1.7.10 Piling Sticks (Stickers)

Hardwood piling sticks should be avoided as they can leave dark marks (stains) across the surfaces of the timber (Figure 1.65). The section sizes of the stickers can be 25 mm by 13 mm, 25 mm by 20 mm, or 25 mm by 25 mm (can be placed on either face) and their distance apart will vary according to board thickness, species of timber, and of course the rate at which it will dry. The greater the spacing between the boards the greater the air flow, but, as you will see in Table 1.10 this is not always desirable, because it could be a contributing factor towards drying degradation. However, stickers must always be positioned vertically in line one above another, otherwise boards may sag (bow), as shown in Figure 1.51a. Figure 1.51b shows how, by using short stickers, the stack can accommodate boards of random length. This will mean that overhanging boards will be restrained from drooping (bending).

Certain hardwoods, such as some of the temperate species grown in the U.K. can be dried in the 'boule' (log form). Figure 1.52 shows how these logs have been sawn through and through and reassembled with stickers between each board. This method of drying also ensures that dried boards are not separated from their original position in the log, and therefore will match one another in colour and figure. This is an important factor for many end users of decorative hardwood.

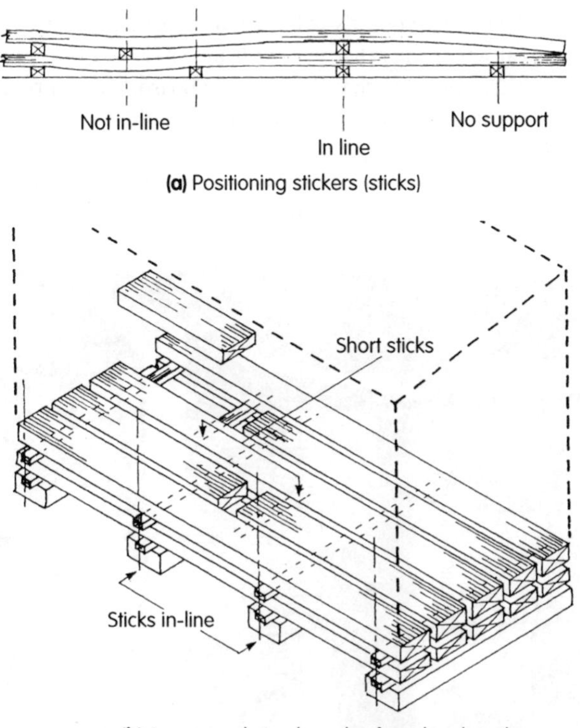

Figure 1.51 Build-up of stack

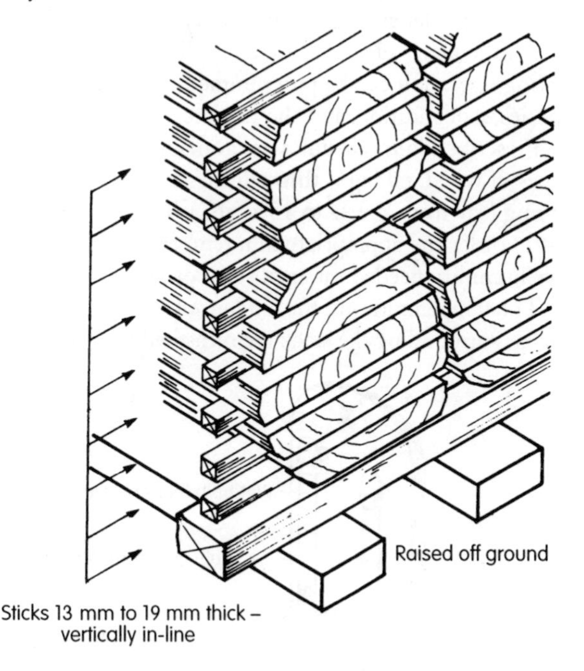

Figure 1.52 Hardwood boules – piled in log form

1.7.11 Kiln Drying (Artificial Drying)

The drying kilns used in our industry are generally large closeable chambers into which stacks of green timber are manoeuvred via a system of trolleys to undergo a controlled method of drying. These kilns dramatically reduce the drying time compared to air drying methods, by taking a matter of days instead of months. The kilns can vary in their construction, size and function. There are those where the stacks of timber remain static (stationary) until the required moisture content level is reached; these are known as *compartment kilns*. Then there is a method where the timber is moved in stages through a tunnel dryer known as a *progressive kiln*. Both types of kiln will require means of providing controlled:

- heat
- ventilation
- humidification
- air circulation.

Heat is often provided via steam or hot water pipes. The fuel used to fire the boiler may be wood waste, coal, oil or gas. Ventilation is achieved by adjustable openings strategically positioned in the kiln wall or roof. Alternatively, a dehumidifier can be used to extract unwanted moisture and channel it outside the kiln in the form of water, thus conserving heat and reducing fuel costs.

When the amount of moisture leaving the wood is insufficient to keep the humidity to the required level, jets of steam or water droplets may be introduced into the kiln chamber.

Air circulation is provided by a large single fan, or a series of smaller fans, located either above or to the side of the stack, depending on the kiln type.

All the above elements are controlled such that the drying operation can be programmed to suit the wood species, its condition, and the thickness of the timber.

Prescribed kiln schedules are available to suit most types of wood, taking it through the various stages of drying (say from 'green' to 15% MC). These schedules list in order the appropriate kiln temperature and relative humidity needed for each stage of drying.

The temperature and the amount of water vapour in the air surrounding the stack are measured with a kiln hygrometer, which will assess the relative humidity of the air. This will determine the rate at which the wood dries.

The principle by which a simple hygrometer works is through two thermometers (Figure 1.53). One measures the air temperature (dry-bulb thermometer). The other is attached to a fabric wick suspended in a container of distilled water (wet-bulb thermometer). Any evaporation of the water from the wick lowers the temperature of the wet-bulb by an amount governed by the rate of evaporation, which in turn depends on the relative humidity of the air. The lower the humidity of the air, the greater the rate of evaporation and therefore the lower the wet-bulb temperature. The greater the difference between the wet-bulb and dry-bulb temperatures, the lower the humidity of the air.

Relative humidity (RH) at a particular temperature is expressed as a percentage (fully saturated air has a value of 100% RH). Less water vapour at the same temperature means a lower relative humidity; therefore, by lowering the relative humidity, the drying potential is increased. It must also be remembered that the higher the air temperature, the greater its vapour holding capacity.

However, the kiln adjustments required to satisfy the schedules cannot be made until the current moisture content of the whole stack is known. This may mean 'sample testing', i.e. the removal and testing of several boards from different locations within the stack.

Figure 1.54 shows this operation being carried out (notice also the control panel with its relative humidity recorder). These boards will have been arranged for easy removal when the stack was built up, as shown in Figure 1.55; these boards are called 'sample boards'. This provision enables the kiln operator to remove the boards at will for testing.

These moisture content tests may be carried out by using either a sample weighing method, or by using a

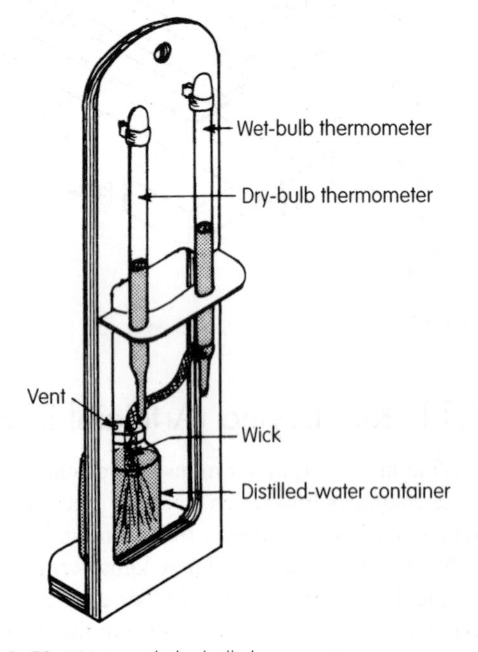

Figure 1.53 Wet- and dry-bulb hygrometer

Figure 1.54 Inspection and removal of kiln samples (with kind permission from G.F. Wells Ltd)

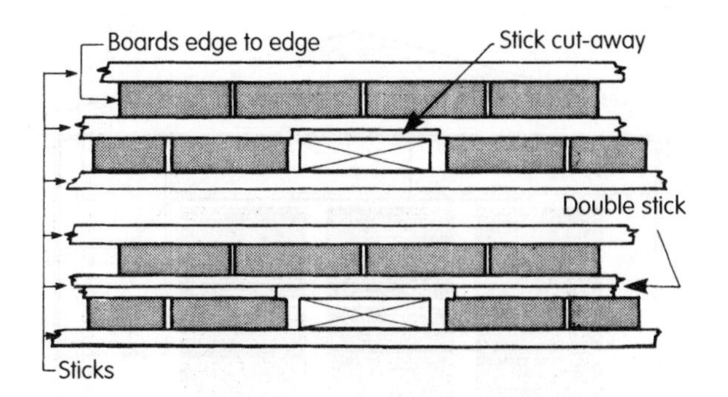

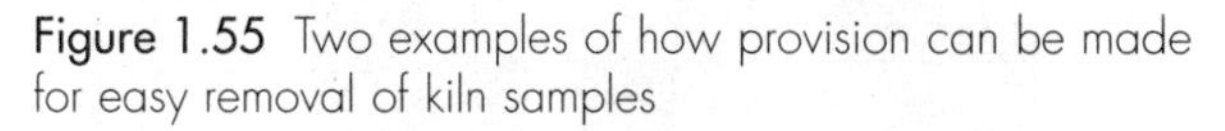

Figure 1.55 Two examples of how provision can be made for easy removal of kiln samples

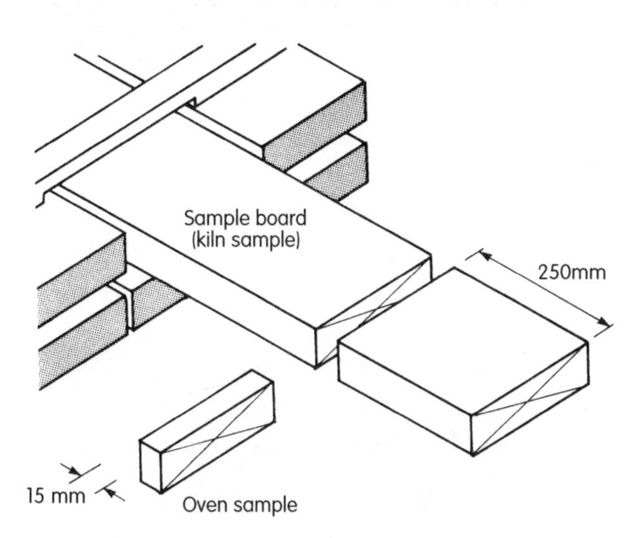

Figure 1.56 Oven sample cut from sample board so that moisture content of sample board can be determined

modern type of battery operated moisture meter as previously described (Section 1.7.3).

The weighing method would involve cutting out oven samples from each sample board (Figure 1.56), then finding the board's moisture content by using the oven dry method previously described (Section 1.7.2). The kiln sample is then weighed to obtain its 'wet weight'; its 'dry weight' can be estimated by using a simple calculation. Any future moisture content changes which occur during the drying cycle can be determined by re-weighing the kiln sample and carrying out another simple calculation or by referring to a drying table chart. However, by using control equipment that is fully automatic, the MC in the timber can be continually monitored. This will enable the Equilibrium Moisture Content (EMC) to be controlled according to the wood species and sectional size of the timber.

1.7.12 Compartment Kilns

Compartment kilns comprise sealable drying chambers (compartments) which receive batches of timber loaded

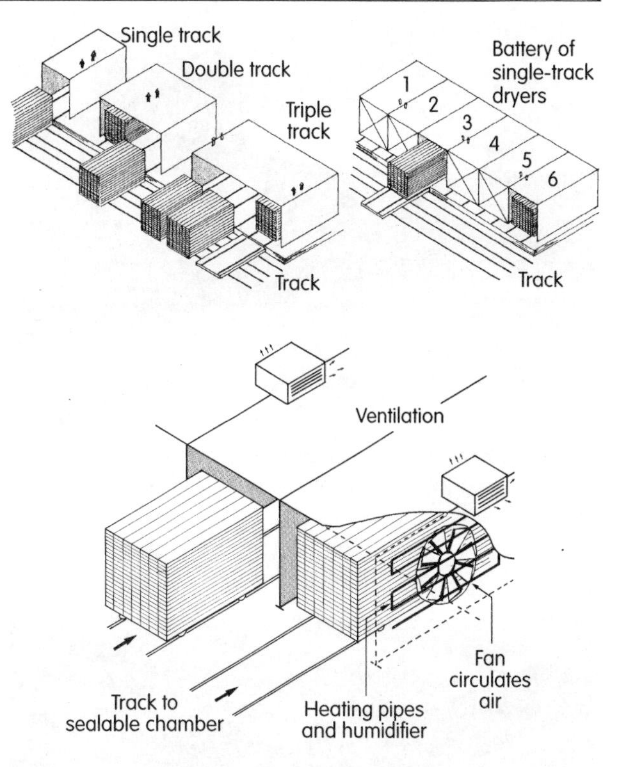

Figure 1.57 Different types of compartment kilns

on trolleys, which remain there until their drying schedule is complete.

Figure 1.57 shows how these kilns can be arranged, separately with single, double or triple tracks, or joined together in rows (battery). They may be sited outside like the one shown in Figure 1.58, or loaded into one of a battery (Figure 1.59). Alternatively they can be sited undercover like the battery of driers shown in Figure 1.60. Figure 1.61 shows how stacks of timber are arranged in a double track dryer with a central drying unit consisting of a large circulating fan, heating coils

Figure 1.58 Externally sited compartment dryer with 80 cubic metre capacity (with kind permission from Kiln Services Ltd)

Figure 1.59 Externally sited battery of compartment dryers – one kiln being loaded with timber (with kind permission from G.F. Wells Ltd)

Figure 1.60 Undercover battery of compartment dryers – one kiln being loaded with timber (with kind permission from G.F. Wells Ltd)

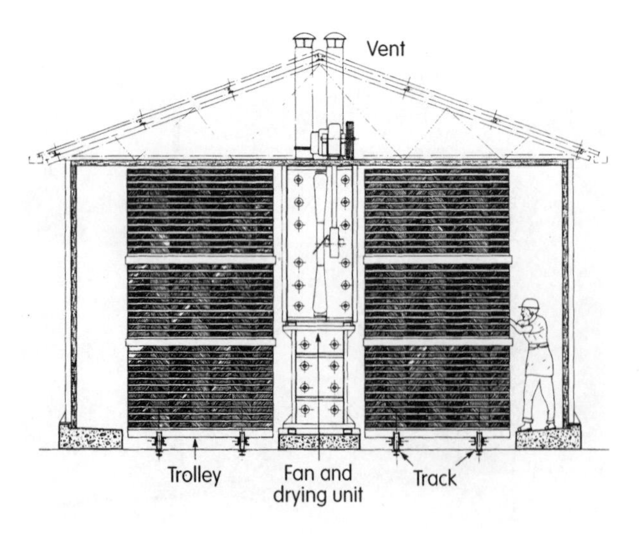

Figure 1.61 Cross-section through a 'Wells' double track high stacking prefabricated timber dryer

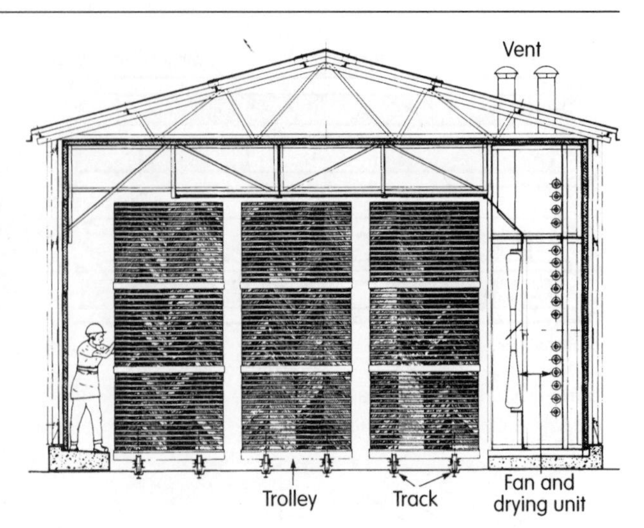

Figure 1.62 Cross-section through 'Wells' triple track high stacking prefabricated timber dryer

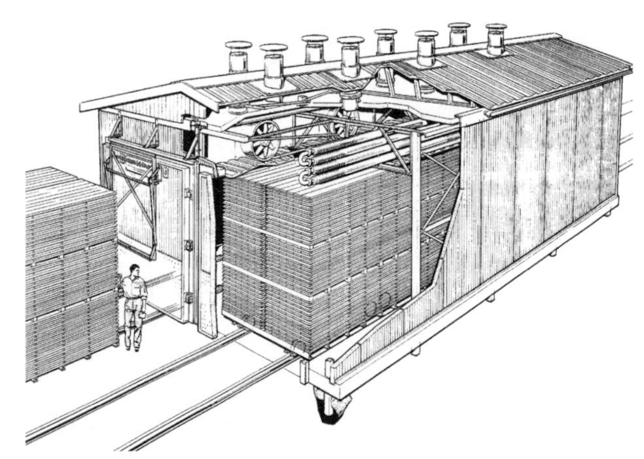

Figure 1.63 Cutaway view of a double stack rail entry kiln (kindly supplied by Kiln Services Ltd)

and a humidifier; provision for ventilation is also shown. Figure 1.62 shows a drying unit positioned to one side to accommodate three tracks. Figure 1.63 shows a cutaway view of a compartment kiln with a system of overhead fans.

1.7.13 Progressive Kilns (Continuous Driers)

Green timber enters the kiln tunnel at one end, then after a period of time, say 3–4 days (depending on the species of the timber, its cross-sectional area, and condition), it will emerge from the exit end much dryer.

Figure 1.64 shows how a batch of timber is being lined up on a trolleyed track outside the kiln ready to

Table 1.10 Analysis of drying defects

Defect term	Definition	Cause	Possible preventative measures
Stains			
Stick marks (Fig. 1.65)	Evidence of where sticks have been laid across a board	Using dirty sticks or sticks with an acid content, e.g. certain hardwoods	Use only softwood sticks – see 'Piling sticks' (1.7.10)
Sap stain (blue stain)	Bluish discoloration of the timber (fungal growth)	Close piling sapwood with over 25% MC – poor air circulation	Ventilate around each board
Distortion (warping)			
Cupping (Fig. 1.66)	Curvature in the cross-section	Differential shrinkage due to the position of growth rings	Use a low-temperature schedule when kiln seasoning
Diamonding (Fig. 1.67)	Square-sectioned timber becomes diamond-shaped	Diagonally positioned growth rings induce the greatest amount of shrinkage	As above
Spring (Fig. 1.68)	Curvature along a boards edge	Differential shrinkage longitudinally along irregular grain	As above
Bow (Fig. 1.69)	Curvature along a boards width	Sagging in the pile and/or differential shrinkage as above	As above
Twist (Fig. 1.70)	Spiral deformity – propeller-shaped	Irregular grain – spiral and/or interlocking	As above
Collapse (washboarding) (Fig. 1.71)	Buckling of the timber's surface – corrugated effect	Uneven shrinkage – drying too rapidly and collapsing the spring wood	As above
Checking and splitting			
Checks (surface and end checks) (Fig. 1.72)	Parting of the grain, producing cracks (fissures on the surface and/or end of timber)	Surfaces drying much quicker than the core	Use high humidity in the early drying stages
Splits (end splitting) (Fig. 1.73)	As above, but cracks extend through the timber from face to face	End grain drying quicker than the bulk of the wood	See 'Stacking', sealing end grain, etc.
Honeycombing (Fig. 1.74)	Parting of the grain internally	Shrinkage of the inner zone after outer zone has become case-hardened	Use low-temperature schedules; use high humidity early on
Case-hardening			
Case-hardening (Fig. 1.75)	Outer zone of the timber dries and 'sets' before inner zone, setting up internal stresses between the two	Rapid surface drying due to low humidity early on, or high temperature in the later stages of drying	High humidity early on, keeping a check on temperature throughout

Figure 1.75a shows how the different zones could appear. Figure 1.75b shows how a test-piece can be cut to test the extent of case-hardening – releasing tension will cause it to distort as shown. The danger of cutting case-hardened material should be apparent, e.g. trapping saw blades etc. (Figure 1.75c)

follow those already inside. On entry, each batch goes through a series of stationary drying stages; the entry end will be cool and humid, whereas the final stage will be warm and dry. As each batch leaves the tunnel, a new batch will enter from the other end to take its place.

To be cost effective this type of kiln requires a continuous run of timber of similar species with common drying characteristics and sectional size, like those you would find in large saw mills specialising in softwoods.

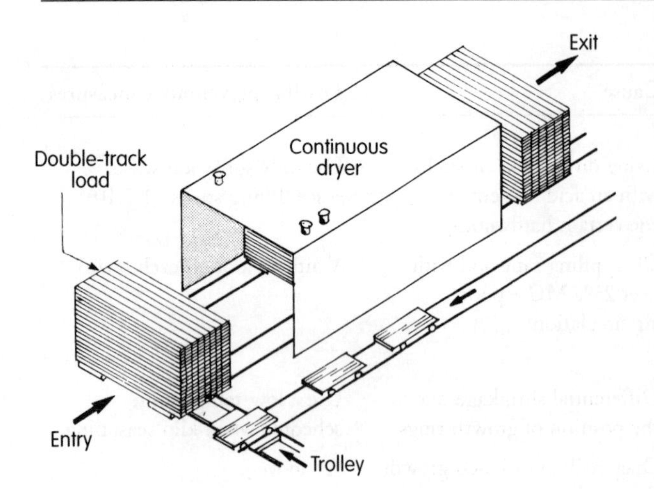

Figure 1.64 Progressive kilns (continuous dryer) – doors not shown

1.7.14 Drying Defects

We have already considered the effects of natural defects (Section 1.6), in other words those defects that can occur in the living tree, and as a result may materialise within the timber cut from it.

However, defects which occur as a result of drying the green timber can, in the majority of cases, be controlled. Successful drying will depend on how drying preparations are made and how the drying operation as a whole is conducted. This is also true of natural drying.

Green timber is generally pliable, but, after drying, it stiffens, hardens and sets. For example, if you were to bend a green twig and retain it in that position until dry it would set and remain partially bent. Therefore, if we were to allow green or partly dried timber to become distorted by incorrect piling, or to be subjected to unbalanced shrinkage during drying, permanent degradation could result. By using Table 1.10 and the accompanying illustrations (Figures 1.65–1.75) you

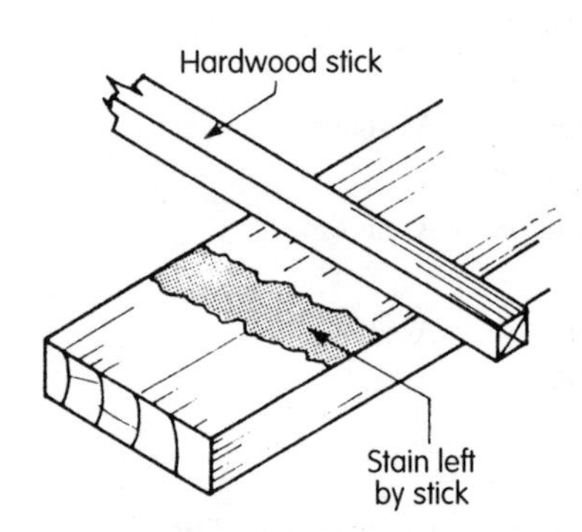

Figure 1.65 Stick mark (staining)

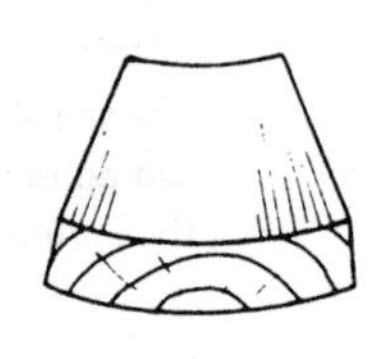

Figure 1.66 Cupping (distortion)

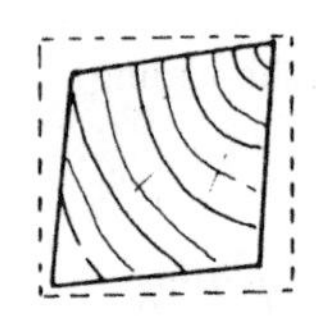

Figure 1.67 Diamonding (distortion)

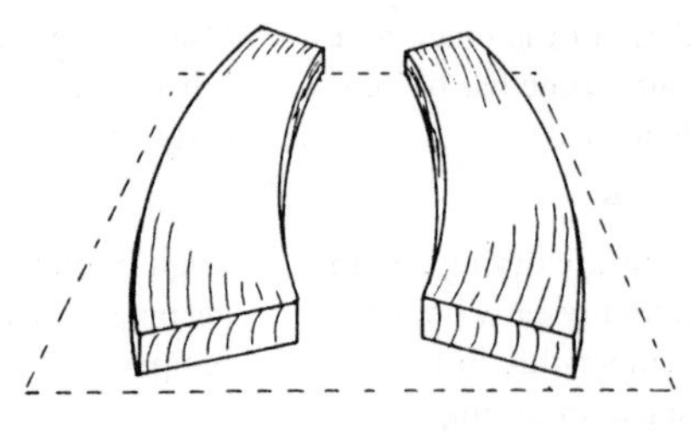

Figure 1.68 Spring (distortion)

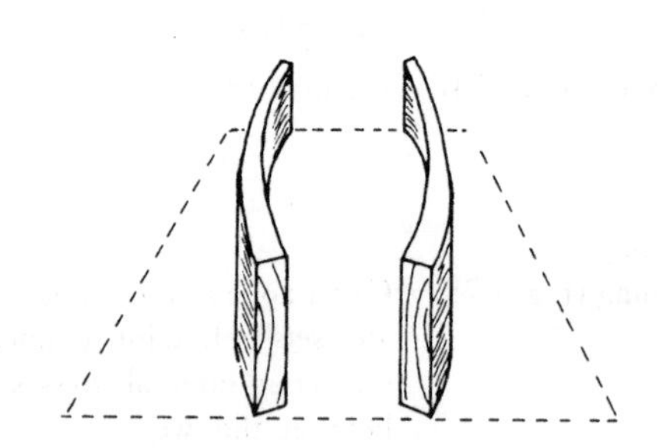

Figure 1.69 Bow (distortion)

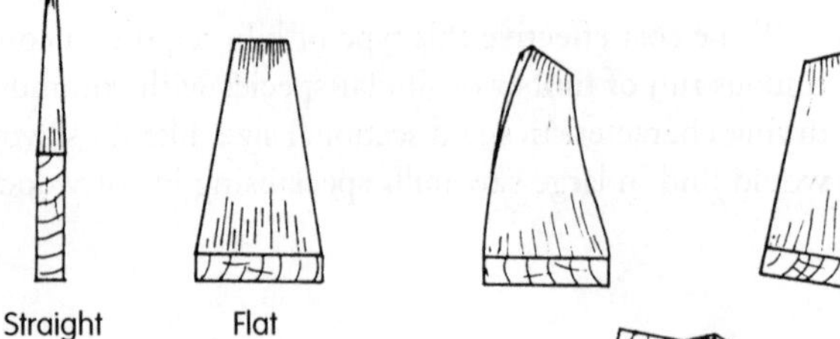

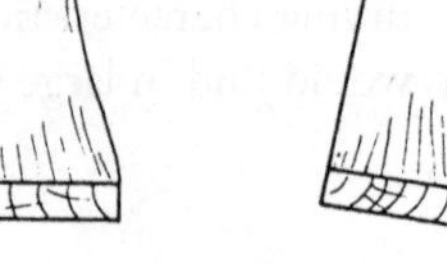

Figure 1.70 Twist (distortion)

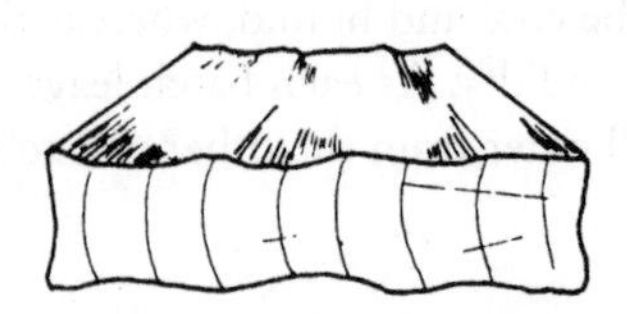

Figure 1.71 Collapse (distortion)

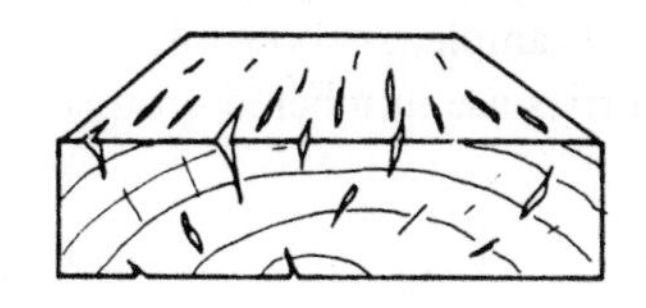

Figure 1.72 Surface and end checking

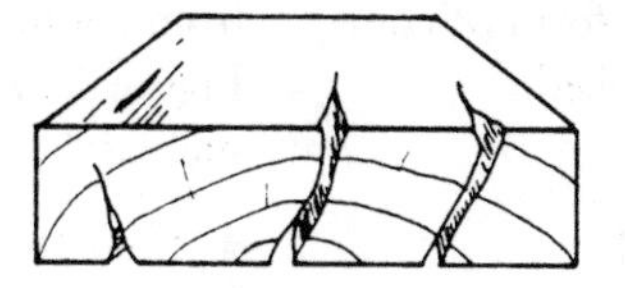

Figure 1.73 End splitting

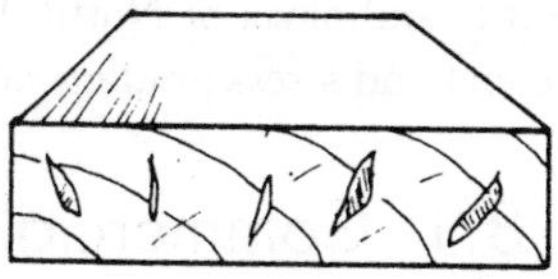

Figure 1.74 Honeycombing

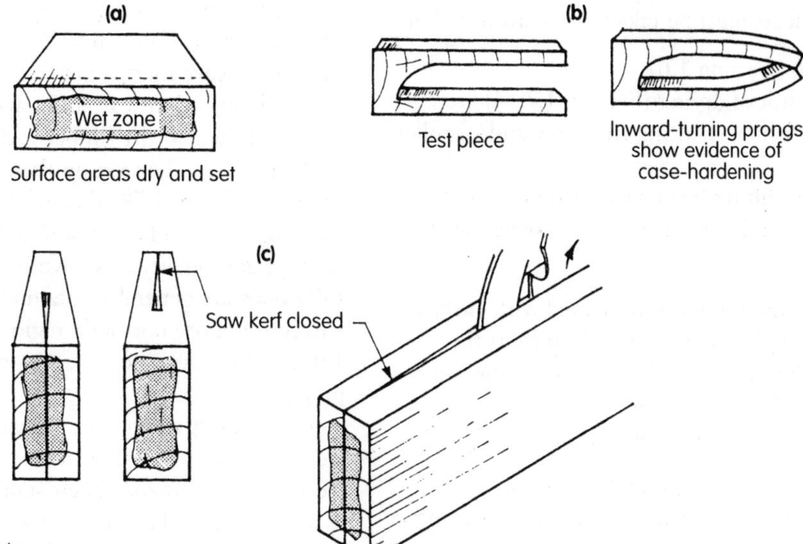

Figure 1.75 Case-hardening

should be able to recognise each of the listed defects. Notice that all the defects have been grouped into 'stains', 'distortions', 'checking and splitting', and 'case hardening'. You should also note that many of these defects are caused by the speed and unevenness with which moisture was removed from the wood. So, don't forget that, as shown in Figure 1.76, when timber is dried, it is the outer layer which dries first, and that, as the moisture is lost through evaporation, it is replaced with that contained within the wood. If this action becomes unbalanced, i.e. the rate at which surface moisture is lost means it cannot be replaced quickly enough, then internal stresses within the wood will be created, resulting in many of the defects listed.

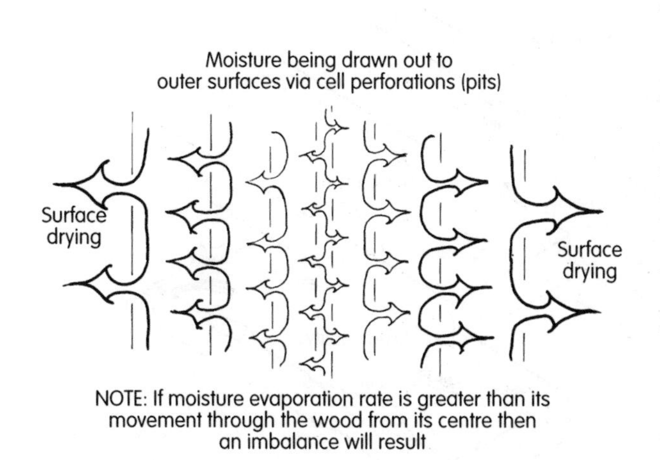

Figure 1.76 Moisture movement through wood

1.8 GRADING TIMBER

When we talk about a grade, we are usually referring to a level of attainment, such as an examination result, or possibly the condition of something. In either case someone has to make some form of judgment as to whether the items in question meet certain standards. These standards will be set by a body of professionals highly qualified in that particular field.

Like many other natural materials with inherent variations, timber has to meet certain standards so that it can be classified as suitable for a particular end use. The two main contributing factors towards end use are appearance and strength, although these are not necessary separate. For the purpose of this chapter we shall call the grades set down due to appearance 'commercial grading', and those grades set down due to strength, 'strength grading'.

No matter from what source the timber is supplied, be it Scandinavia or North America, some form of commercial and stress grading rules will be applied.

1.8.1 Commercial Grading (Softwoods)

The grader takes all the surfaces of the timber into account before making judgments on its grade. Decisions will be based on the 'quality' and 'condition' of the timber.

When judging quality, the type, size and number of the following defects must be taken into consideration:

- natural defects (Section 1.6)
- defects as a result of drying (Section 1.7.14)
- machining defects – resulting from saw and/or roller marks
- other factors resulting from production methods, which could include such things as 'bowing' due to bad stacking, etc.

The condition of the timber is associated with surface staining due to fungi, minerals picked up from the ground, or discoloration due to weathering. The moisture content will also be taken into account.

Three possible stages of grading are:

- while still in log form – grouped according to species, and girth size (distance around its circumference)
- after conversion from green (undried) section size
- after drying.

Rules which put timber into the various categories vary from country to country; for example, as shown in Figure 1.77, European countries use numbered groups such as:

- firsts (for the best quality)
- seconds
- thirds
- fourths
- fifths
- sixths (poorest quality).

These can be sold in mixed batches. Swedish and Finnish mills generally mix first, second, third and fourth qualities and sell them as 'Unsorted'(US), whereas the Russian states may leave fourths out of their US grade.

Grade variations can differ across borders, and the ports from which the timber is dispatched. The identity of the ports can be seen on the cut ends of the timber as shown in Figure 1.78. These marks are known as *'shipping marks'*. They are a useful guide when buying a set type and quality as shown in Table 1.11. Best quality softwoods are generally obtained from trees grown within the more northerly regions of Europe (Figure 1.9) and buyers come to recognise these by their shipping marks.

Canadian grading rules will mainly apply to timber we receive from North America, principally from the west coast of Canada. Their standards use the term *clear* as a basis for timber that is free of any visual defects (best quality), and *common* for timber of the lower standard. We cannot make direct comparison with

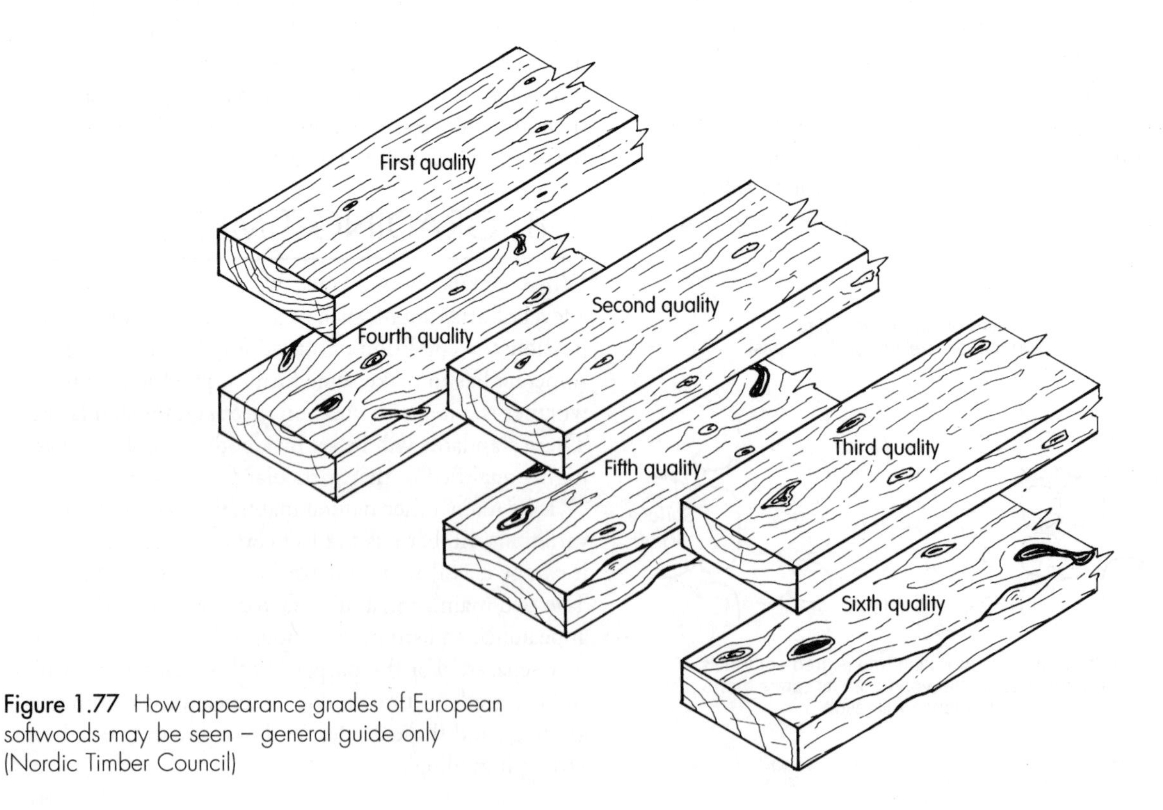

Figure 1.77 How appearance grades of European softwoods may be seen – general guide only (Nordic Timber Council)

Table 1.11 Scandinavian, Polish, Czechoslovakian and Russian softwood grades/marking

		Quality				
		High				Low
Shipper's name	**Port of origin**	**Grade U/S**	**Grade S/F**	**Grade 5**	**Grade 6**	**Grade 7**
		Unsorted made up of grades 1, 2, 3, 4	**Sawfalling usually grades 1, 2, 3, 4, 5**	**Fifths/V**	**Sixths/VI**	**Sevenths/VII specialist low grade end uses**
Sweden						
Bergkvist – Insjon AB (originally Axel-Bergkvist hence 'A&B')	Gothenberg	A ♛ B	A ★ B	A – B	A B	Not produced
Svenska Cellulosa 'SCA Group' This group owns several sawmills with individual marks.	Holmsund (also name of sawmill)	HSUND	–	HMS	HSUS	–
A-TRÃB Skelleftea	Skelleftehamn or Kage	LAPP	–	★ LAPP ★	+LAPP+	
Mellanskog	Gothenberg	♛ GHS ♛	–	★ GHS ★	+GHS+	
Vasterbottens	Kage, Lovóele	B ♛ N	–	B – N	B + N	
Mainly redwood and whitewood						
Finland						
Oy. W. Rosenlew	Manty Luoto Finland	BS ♛ SC	–	BWC	+ R +	
Kemi Oy.	Kemi Finland	KEMI		KME	K – E	
Mainly redwood plus excellent whitewood						
Poland						
Polish timber (state controlled)	Gdansk and Szczecin	L P	L [eagle] P	+L [eagle] P+	WL [eagle] PW	
Redwood and whitewood – generally lower volumes than Scandinavia						
Czechoslovakia						
Czecho-slovakia (state controlled)	Various European ports	LIGNA	–	LGA (in red)	LGA (in green)	
Whitewood generally						

Table 1.11(continued) Shipping marks can be a useful guide to origin and quality. (Compiled by Mr P Kershaw of North Yorkshire Timber Co. Ltd)

Russia

Until the recent upheaval in the former Soviet Union exports of Russian wood goods were controlled by 'Exportles' – the Russian Wood Agency.

Timber was harvested in various forest areas, then milled and graded into the following qualities. A system of marking denoted the port of export; these were originally hammer stamped into the end grain of every piece. They were replaced by red painted end stamps. Buyers in the UK eventually began to prefer various ports' productions as being of superior quality. Some of these marks are gradually being replaced as private enterprises and foreign investment move in. However, some of the more popular end marks still exist.

Port of loading	Grade U/S Unsorted made up of grades, 1, 2, 3	Grade IV Fourths grade 4	Grade V Fifths grade 5	Grade VI Sixths grade 6
Archangel	E**AR	E*AR	E–AR	
Kem	E**K	E*K	E–K	
Kara Sea/Igarka	E**I	E*I	E–I	
Leningrad	E**L	E*L	E–L	
Mesane	E**M	E*M	E–M	
Onega	E**O	E*O	E–O	
Petchora	E**P	E*P	E–P	
	= Scand. U/S	Similar to, but less consistent than Scand. 5ths	Similar to Scand. 6ths	

Mainly redwood and whitewood of generally excellent quality from one of the world's largest forest regions (coniferous).

New marks appearing:

Segezha/Orimi – replaces Kara Sea/Igarka
Maklakov (green bear) – replaces Belamorsk
Taiga

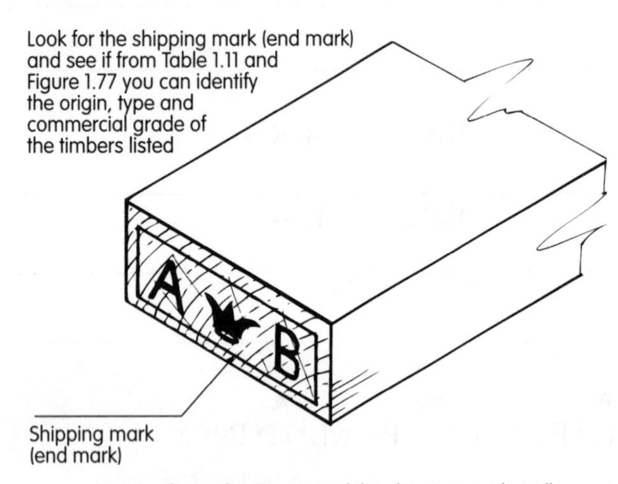

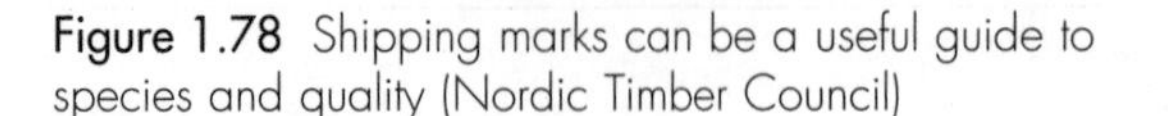

Figure 1.78 Shipping marks can be a useful guide to species and quality (Nordic Timber Council)

European grades, but we can assume with Canadian grades of:

- clears
- select merchantable
- merchantable
- common

that clears will be equal to the best European grades of 'firsts', and likewise 'commons' with say 'sixths'.

Other areas of America, and Brazil use other grading terms.

Figure 1.9 shows the species we would expect to find in the above regions.

1.8.2 Commercial Grading (Hardwoods)

Once again grading rules can vary between countries. The National Hardwood Lumber Association (NHLA) covers the USA, although certain regions may modify these rules. Malaysia has its own grading rules, but apparently uses similar principles when determining the apportionment of 'cuttings' (usable piece of board).

Hardwood is generally, unless otherwise stated, graded from its worst side – edges are not taken into

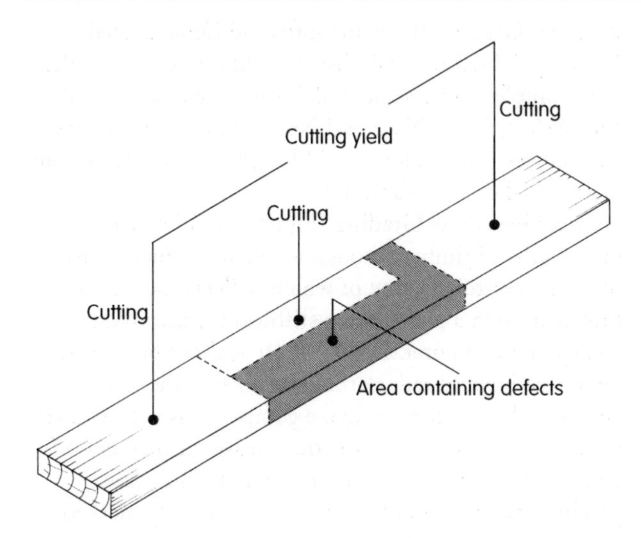

Figure 1.79 Areas of clear yield of a hardwood board

account. Each grade is based on the number, size and condition of cuttings obtainable from each board when cut up. The higher the grade, the greater the area of 'clear' wood (Figure 1.79).

NHLA grades are:

- FAS (firsts and seconds) – which offer long, wide clear cuttings of at least 83.5% of total area of the board. These boards are usually reserved for best quality joinery and mouldings
- FAS 1F and selects – both these grades use both faces to determine the grade. The better one must meet the conditions set for FAS with the opposite side meeting No. 1 common grade. The main differences between the two are their minimum length and width requirements
- No. 1 common grades – these grades will accommodate clear cuttings of medium length and width of at least 66⅔% of the total area
- No. 2 common grades – will contain cuttings of at least 50% which are short and narrow. If clear cuttings are expected then grade No. 2A would be sort.
- No. 3 common grade – the area of clear cuttings can be as low as 33⅓ (3A) or 25% (3B).

Common grades may, like FAS, be mixed with better grades, like say 'No. 2 common and better'.

1.8.3 Timber for joinery use

Classifying timber used in the manufacture of joinery items, such as windows, doors and stairs takes on a different perspective – as the grading assessment must be made upon its potential end use (the finished product).

These assessments will be based on the limitations of knot size over the finished surfaces, and whether these surfaces are to be exposed or concealed. Four classes are used, namely:

- *class CSH* – which represents clear (almost knot free) grades of softwood and hardwood – maximum permitted knot size 6 mm in diameter
- *class 1* – used for high class joinery products and purpose-made items
- *classes 2 & 3* – cover the main joinery use – discretion being given to the better class 2 as required.

1.8.4 Strength Grading (Structural Grading)

We are concerned here with the strength of timber. Grading timber in this way provides the designer of a building or its structural components with guidance on selection, and assurance when having to specify a particular timber section to carry a required loading. Examples could include items such as beams (joists, lintels, purlins, etc.), or roof trusses (rafters, ties and struts). We have seen how commercial grading rules apply when grading for appearance, but in this case the grading rules will be based on the ability of the timber to withstand the internal stresses that could be brought about by varying degrees of external forces.

Figure 1.80 shows the three main areas of concern and how wood tissue has to resist being compressed (compression), stretched (tension), or subjected to forces of shear (scissor action). Therefore every piece of timber used in such situations must be tested beforehand. This

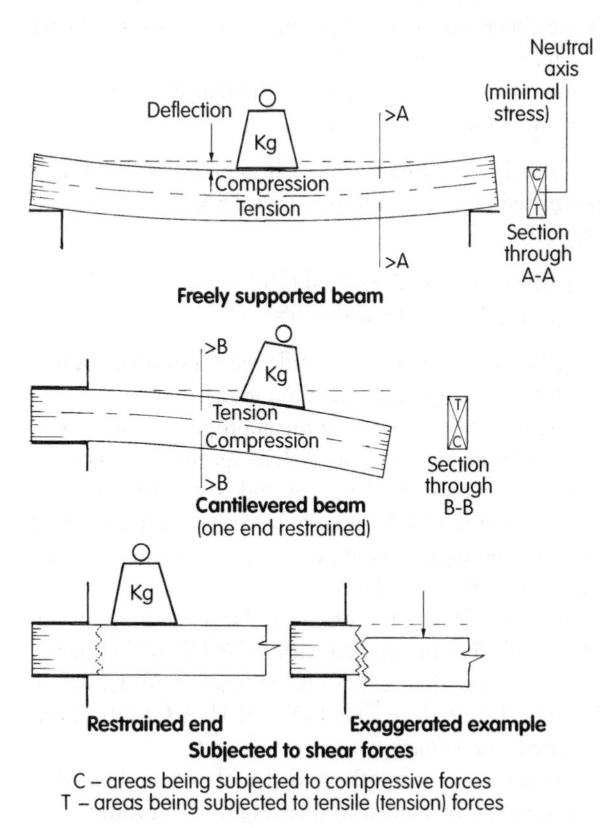

Figure 1.80 Possible effect of loading a timber beam

Table 1.12 Basic softwood sawn sizes available as strength graded timber (adapted from Nordic Timber Council (UK))

Thickness (mm)	Width (mm)						
	75	100	125	150	175	200	225
38	X	X	X	X	X	X	X
47	X	X	X	X	X	X	X
50*	X	X	X	X	X	X	X
63			X	X	X	X	X
75				X	X	X	X

Note: * Check on availability strength graded before specifying

type of testing is termed 'strength grading', which may be carried out at the saw mill. As shown in Table 1.12, basic sawn sizes from Sweden and Finland are made available as strength graded timber. Alternatively, testing may be undertaken in the factory responsible for manufacturing structural components such as trussed rafters.

There are two methods of strength grading timber:

- visual strength grading
- machine strength grading.

With visual strength grading, the grader has to determine the strength of each piece of timber handled by visually inspecting all of its surfaces, and taking into account:

a natural inherent defects – such as size, type, distribution and number of knots, growth rate and pattern of growth (e.g. slope of grain) etc. – see Section 1.6.9
b defects resulting from drying – such as splits, checks, etc. – see Section 1.7.14
c non-visible factors – such as the density of wood, and its moisture content

before rejecting it or placing it into one of two grades permitted under British Standards (BS). The two grades are:

1 general structural – coded GS
2 special structural – coded SS.

NB: The special structural grade requires higher standards than the general structural grade.

Figure 1.81a shows how the appropriate grade, and other relevant information such as species (or species mix), grading company, agency and controlling body (such as the British Standards Institute together with its number and date) may appear stamped on the surface of the timber prior to 1997.

Figure 1.81b shows how a similar grade may appear after 1997. Timber graded within the UK will either display the emblem CATG (Certification And Timber Grading Limited) or TRADA (TRADA Certifications Limited) on its face.

Both visually strength-graded timber from Canada, which would be graded to their National Lumber Grades Authority (NLGA) rules, and the USA to the National Grading Rules for softwood Dimensional Lumber (NGRDL), will display a different grade code. For example, under structural 'joists' or 'plank' you will find, Select, No 1, No 2 and No 3 – under certain circumstances these grades could have been with European grades as shown in Table 1.13.

Machine Stress Grading (Figure 1.82) involves passing lengths of timber through a machine which consists of a series of rollers, one of which deflects (bends) the timber in such a way that its stiffness is measured by means of a computer. Defects such as those previously mentioned would reduce the resistance of the wood, enabling the machine to make comparisons with similar, yet defect-free samples. It is this measured stiffness difference which classifies the timber as a whole as a specific grade; the grade is then automatically marked on to the face of each piece of timber.

Previously, 'stress grading' (as it was known) to BS 4978 provided for the following machine stress grades:

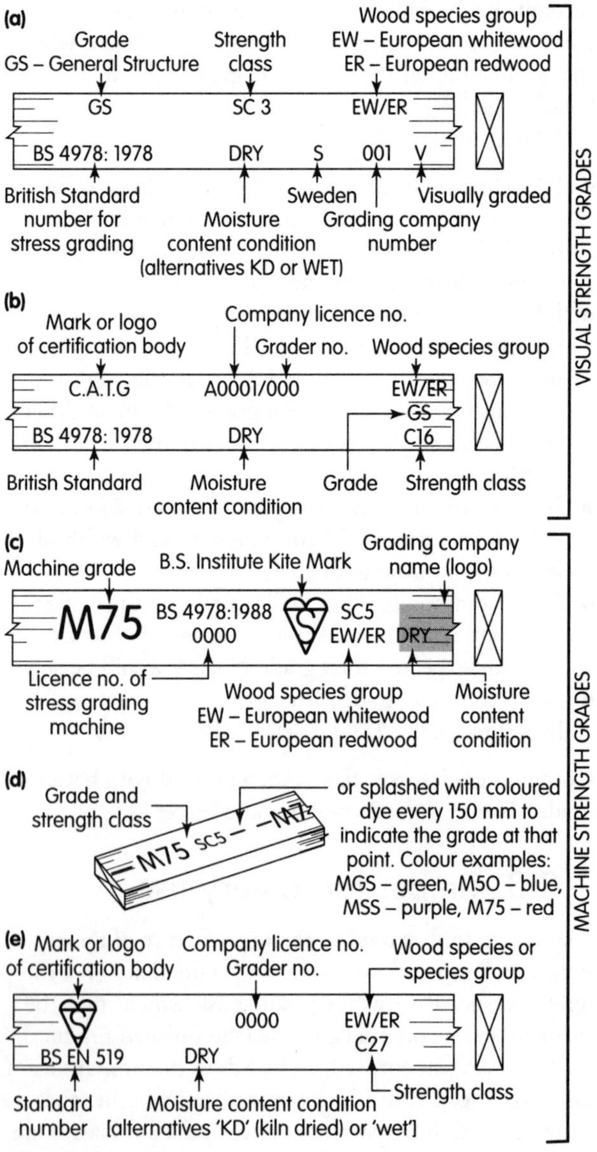

Figure 1.81 Stress graded timber

Table 1.13 Cross referencing softwood species and grades with strength classes 1–5. (Reference TRADA Wood Information Section 1. Sheet 25)

- Timber graded to BS 4978
- Timber graded to Canadian NLGA or American NGRDL Joist and plank grades. (Note: Timber graded to North American Structural light framing and Stud grades are included in BS 5268 Part 2 but not in this table)
- Timber graded to North American machine stress rated grades

			Grades to satisfy strength class				
Species	**Origin**	**Grading rules**	**SC1**	**SC2**	**SC3**	**SC4**	**SC5**
Corsican pine	UK	BS 4978		GS	M50	SS	M75
Douglas fir	UK	BS 4978		GS	M50, SS		M75
Douglas fir	Canada and	BS 4978					
– larch	USA	J & P	No. 3		GS No. 1,	SS	
		Machine	900f-1.0E	1200f-1.2E	No. 2	Select	
					1450f-1.3E	1650f-1.5E	1650f-1.5E
						1800f-1.6E	1800f-1.6E
							1950f-1.7E
							2100f-1.8E
European spruce	UK	BS 4978	GS	M50, SS	Machine graded to strength class		
Hem – fir	Canada	BS 4978					
	USA	J & P	No. 3		GS, M50	SS	M75
	Machine		900f-1.0E	1200f-1.2E	No. 1, No. 2	Select	
					1450f-1.3E	1650f-1.5E	1650f-1.5E
						1800f-1.6E	1800f-1.6E
							1950f-1.7E
							2100f-1.8E
Larch	UK	BS 4978			GS	SS	
Parana pine	Imported	BS 4978			GS	SS	
Pitch pine	Caribbean	BS 4978			GS		SS
Redwood	Imported	BS 4978			GS, M50	SS	M75
Scots pine	UK	BS 4978			GS, M50	SS	M75
Sitka spruce	UK Canada	BS 4978	GS	M50, SS	Machine graded to strength class		
		BS 4978		GS	SS		
		J & P	No. 3	No. 1, No. 2	Select		
Southern pine	USA	BS 4978			GS	SS	
		J & P			No. 1, No. 2,		
					No. 3		
		Machine	900f-1.0E	1200f-1.2E	1450f-1.3E	1650f-1.5E	1650f-1.5E
						1800f-1.6E	1800f-1.6E
							1950f-1.7E
							2100f-1.8E
Spruce-pine							
– fir	Canada	BS 4978			GS, M50	SS, M75	
		J & P	No. 3		No. 1, No. 2	Select	
		Machine	900f-1.0E	1200f-1.2E	1450f-1.3E	1650f-1.5E	1650f-1.5E
						1800f-1.6E	1800f-1.6E
						1950f-1.7E	2100f-1.8E
Western red cedar	Imported	BS 4978	GS	SS			
Western whitewoods	USA	BS 4978	GS		SS		
		J & P	No. 3	No. 1, No. 2	Select		
Whitewood	Imported	BS 4978			GS, M50	SS	M75

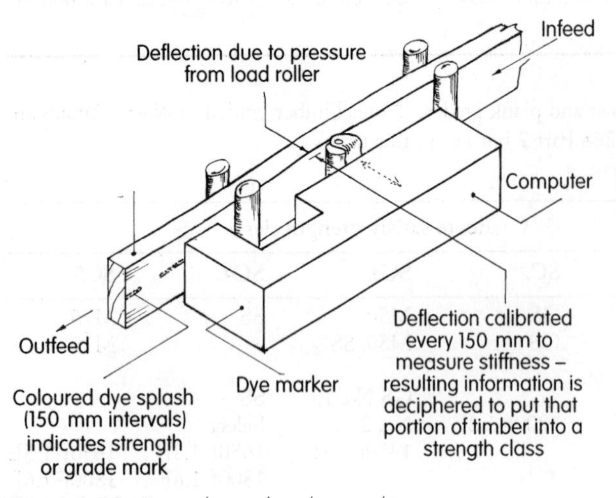

Figure 1.82 Strength grading by machine

1 MGS (Machine General Structural) equivalent to GS
2 MSS (Machine Special Structural) equivalent to SS

and two extra grades:

3 M50 (having 50% strength value of a comparable piece of timber with straight grain and defect free)
4 M75 (having 75% strength value of a comparable piece of timber with straight grain and defect free).

Marks on the timber before 1997 include the grading agency, number and date of BS 4978, company name and licence number of machine, species, MC at the timber's time of grading and grade or strength class, as shown in Figure 1.81c.

These marks have to be visible, at least one on each face, although it is common practice with some agencies to mark the timber at regular intervals along its whole length. The timber may, as shown in Figure 1.81dc also be colour coded with a splash of dye every 150 mm along its length to indicate the strength class at that particular point. Recommended colours were: MGS = green; M50 = blue; MSS = purple; M75 = red.

You will notice that in Figure 1.81c, d a strength class (SC) is shown. There were nine strength classes of timber described in BS 5268 pt 2 – the higher the number, the greater the strength. Softwoods were set within SC1 to 5, whereas hardwoods could extend to SC9.

The introduction of BSEN 338 1995 (standard strength classes) has meant that strength classes are now based on characteristic values for timber strength and stiffness – they are now specifically grouped into softwoods and hardwoods by a prefix 'C' for conifer and 'D' for deciduous. These letters precede a range of numbers (which have no bearing on the previously used strength class (SC)). For example, the softwood 'C' class range is: C14, C16, C18, C22, C27, C30, C35 and C40. There is also an allowance for a special grade to meet the end use of truss rafter material, this is classed as TR26. Whereas the hardwood 'D' class range is: D30, D35, D40, D50, D60 and D70.

It should be noted that the more commonly available grades of softwood will be within strength class C16 or C24 for GS or SS grades (previously SC3 or SC4) – however, true comparisons can only be made by referring to Table 7 within Eurocode 5. An example of face marking to post-1997 grading rules is shown in Figure 1.81e.

Softwoods which are machine stress graded to North American Export Standards will be graded with a number followed by 'f' and a further number followed by 'E', for example: 900f – 1.0E and so on. You may want to check these against Table 1.13 to see how these grades used to compare with the softwood machine grades for the UK and Europe.

Both visual and machine stress-graded structural softwood timber (less than 100 mm thickness) which is intended for internal use, such as walls, floors, or roofs, is at the time of grading required to have an average moisture content of 20% (20% or less, but not exceeding 24%) which reflects the 'DRY' or 'KD' (kiln dried – North American grading) mark alongside the designated structural grade mark.

Conversely, stress-graded softwood timber with an intended exterior use may be graded without these restrictions, but the final grade mark must be accompanied by a 'WET' mark.

1.9 PROCESSING TIMBER

After initial conversion and drying, timber may be used in its rough sawn state (unwrot), recut or planed to a regular section (regularised), resawn to a smaller section size, and/or one or more sides planed smooth (wrot). Square timber may be further reformed by cutting indents, grooves, curves etc., to produce a multitude of shaped profiles (moulded sections).

1.9.1 Regularising

Whenever batches of sawn constructional timber are required to have uniform width (depth), one or both edges are recut either by sawing or planing. Figure 1.83 shows how and why the process is undertaken, together with permitted amounts of wood removal.

1.9.2 Planing Process

Planing provides a means of machining timber smooth, to a uniform thickness and/or width throughout its whole length.

Figure 1.84 shows a common method of planing the surfaces of previously dimensioned timber. First one face is planed flat on a surface planing machine, a process known as *flatting* (Figure 1.84a). Then one face edge is similarly planed flat and squared with the face

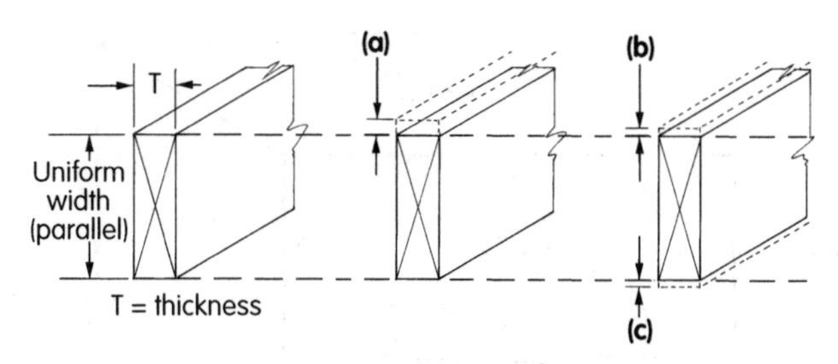

Maximum permitted amount of wood removed off 'a' or 'b+c' is =
3 mm for widths up to 150 mm and
5 mm for widths over 150 mm

Figure 1.83 Regularising – reducing timber to a uniform width

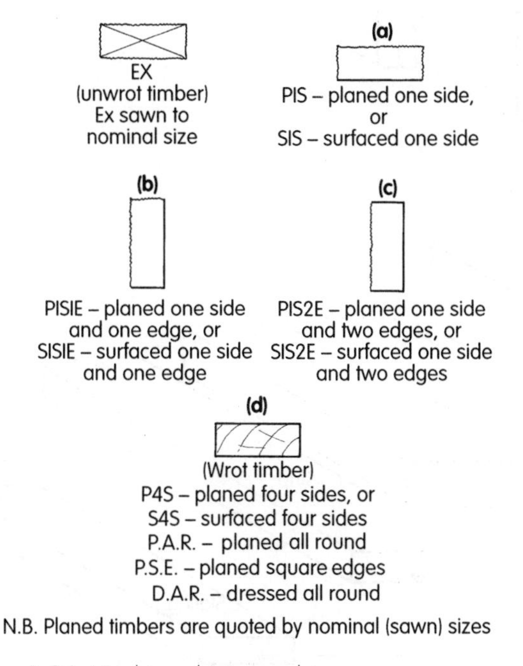

Figure 1.84 Machine planing and terms

side on the same machine (Figure 1.84b). Remaining sides are reduced to their predetermined sizes using a thicknessing machine (Section 9.6.6). However, fully planed timber available on a commercial basis would be processed through a machine capable of planing and reducing all four sides during one pass through a 'four cutter' planing machine. Such timber would be required to meet set tolerances. Table 1.14 shows target sizes for structural timber machined on all four sides.

It is worth noting that processed timber from Canada and the USA (known as CLS and ALS) has rounded corners (arrises) with section sizes of 38 mm thickness and widths ranging from 63 mm to 285 mm, for example, 38 × 63, 38 × 89, 38 × 114, etc.

1.9.3 Methods of Ordering Timber

First let us try to unravel some of the mystique associated with timber trade terminology which is often used to describe a range of timber sectional sizes. Words used include: 'flitch', 'baulk', 'plank', 'deal' (sometimes loosely used to describe 'redwood' as 'red deal', or 'whitewood' as 'white deal'), 'batten', 'board', and 'strip', etc. Figure 1.85 should help you understand these sometimes ambiguous terms more easily. When ordering timber, however, we need only state, among other things, the specific sawn sectional sizes, leaving the terms relating to the range of sawn sizes to the timber merchant.

Figure 1.86 shows a sequence and method of ordering from a supplier which may be adopted.

Large quantities of sawn softwood and some hardwoods are bought and sold by volume. For example, European softwoods could be bought and sold by the cubic metre – an amount which may take some understanding. Try to imagine a cubic metre, or better still an empty box with inside dimensions of 1000 × 1000 × 1000 mm (1 × 1 × 1 m) (Figure 1.87a), and filling it

Table 1.14 Customary target sizes of structural timber (coniferous and poplar) machined on all four sides (Extracted from BSEN 336:1995)

Thickness (to tolerance class 2) in mm	Width (to tolerance class 2) in mm									
	72	97	122	147	170	195	220	245	270	295
19		X	X	X	X	X	X			
22	X	X	X	X	X	X	X			
35	X	X	X	X	X	X	X			
44	X	X	X	X	X	X	X	X		X
60		X	X	X	X	X	X			
72		X	X	X	X	X	X	X	X	X
97		X		X		X	X	X		X
147				X		X				X

Note: * Tolerance class 2: (a) for thicknesses and widths ≤ 100 mm: $\left[\begin{smallmatrix}+1\\-1\end{smallmatrix}\right]$ mm;

(b) for thicknesses and widths > 100 mm: $\left[\begin{smallmatrix}+1.5\\-1.5\end{smallmatrix}\right]$ mm. Target size of 20% moisture content

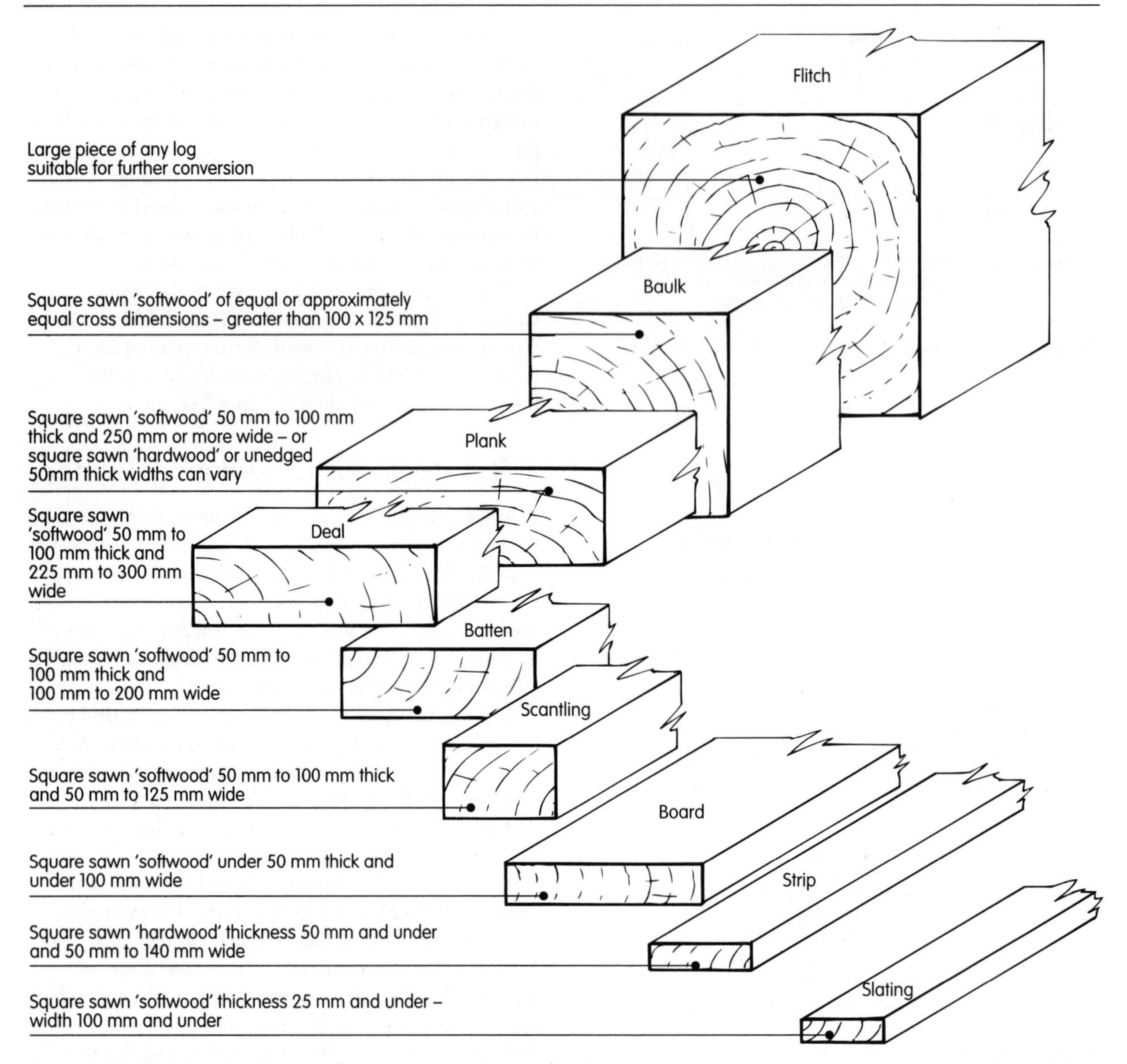

Figure 1.85 Traditional trade terms for cross-sectional sizes of timber

with one metre lengths of timber with a cross section of 200 × 200. How many do you think it would take to fill the box? Figure 1.87b shows that we could get five pieces across its width, and 5 pieces across its breadth; so if we multiply 5 × 5, we have 25 in total. If we were to place all the pieces end to end the total length would be 25 metres! We can now say that 1 cubic metre of 200 × 200 will realise 25 linear metres, (25 m lin), just as:

2 cubic metres (2 m^3) of 200 × 200
= 50 m lin × 200 × 200

or

3 cubic metres (3 m^3) of 200 × 200
= 75 m lin × 200 × 200

and so on.

Let us take another example: what would be the total length of 50 × 100 realised from 3 cubic metres (3 m^3):

$$\textbf{total length} = 3\text{ m} \times \frac{1\text{ m}}{\text{Thickness}} \times \frac{1\text{ m}}{\text{Width}}$$

$$= 3 \times \frac{1000}{50} \times \frac{1000}{100}$$

$$= 3 \times 20 \times 10$$

= 600 linear metres (600 m lin)

Unfortunately, like some other trades there are occasions when we need to be conversant with both metric and imperial measurements. American hardwoods have imperial measurement: in other words measurements in feet and inches will apply. As an example, their board

1 Species – by common name (in some cases, followed by the true botanical name
2 Quality, grade – e.g. commercial grading (joinery or carcassing, etc.) or strength grading (strength classification)
3 Cross-section – quote:

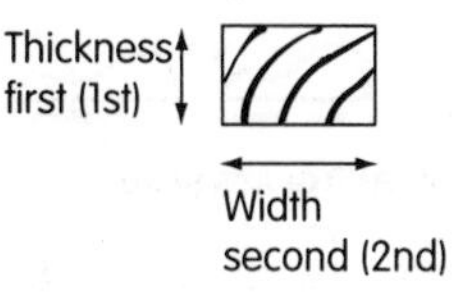

4 Quantity – either by:

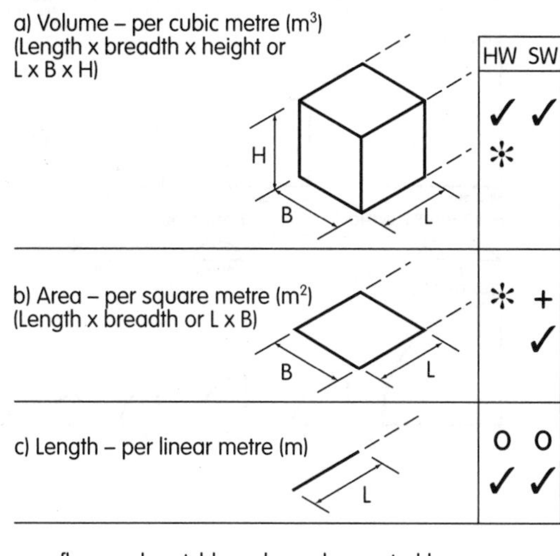

+ – floor and matchboard may be quoted by area
o – small quantities and shaped sections
* – some hardwoods may be sold using imperial measurements; for example a cubic or square foot (see surface measurement and board foot, Figure 1.88)

5 Finish – type of processing
 – regularised (Figure 1.83)
 – planed (Figure 1.84)
 – shaped sections (mouldings, etc.) (Figure 1.89)
6 Moisture content – if appropriate
7 – Preservative (type of treatment) – if required

Figure 1.86 Sequence and method of ordering timber

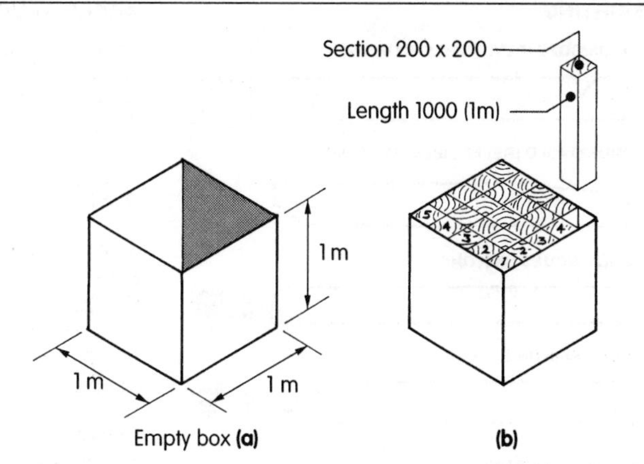

Figure 1.87 One cubic metre of timber – (a 200 mm × 200 mm section)

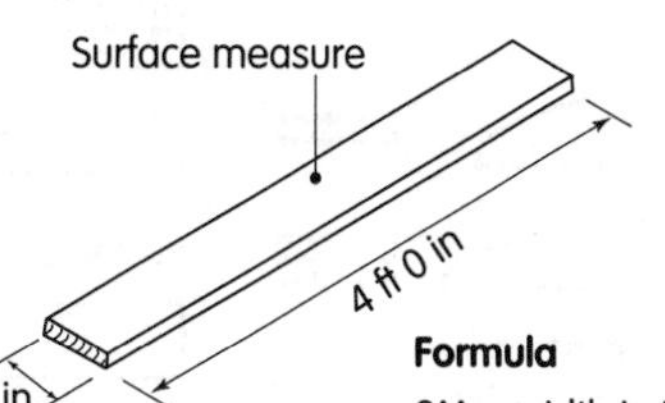

Formula

SM = width in inches × length in feet ÷ 12

SM = W (inch) × L (feet) ÷ 12

$$= \frac{6 \times 4}{12}$$

SM = 2

(a) Surface measure

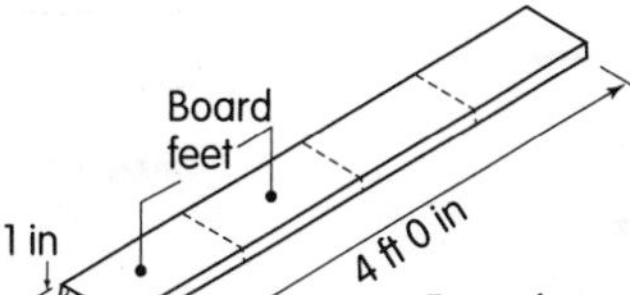

Formula

BF = width in inches × length in feet × thickness in inches ÷ 12

BF = W (inches) × L (feet) × T (inches) ÷ 12

$$= \frac{6 \times 4 \times 1}{12} = \frac{24}{12}$$

BF = 2

NB. Had the board been 2 inches thick, there would have been 4 board feet

(b) Board foot

Figure 1.88 Measuring hardwood (NHLA)

measurement is based on *surface measurement* (square feet – see Figure 1.88a) and a *board foot* (one foot long, one foot wide, and one inch thick – see Figure 1.88b).

It is always worth remembering when ordering timber that by adopting standard sizes wastage can be avoided. This also reduces the build up of *short ends* (NB the term *off-cuts* may sometimes be used, but is usually reserved for waste pieces cut off sheet material, not timber), thereby making the job more cost effective.

It is worth noting that a large part of the industry is made up of small joinery firms with limited machinery and storage facilities. It follows that these firms often rely heavily on the smaller timber merchant to provide them with standard sizes of both sawn and processed timber. The buying of processed timber usually adds considerably to the cost of a job, further stressing the importance of selecting the correct standard sizes to meet the requirement of the job specification.

1.9.4 Shaped Sections (Moulding, etc.)

Many different profiles are commercially available in both hardwood and softwood and a selection of these is shown in Figure 1.89, and Table 1.15. These can be used as a guide to show how named sections can either be connected with an item of joinery, and/or a location where they would most likely be found.

Figure 1.90 shows how mouldings based on the circle can be formed; similarly those shown in Figure 1.91 are based on the ellipse. By combining these, classical mouldings can be formed. Examples are shown in Figure 1.92. You will notice that some of the curves are

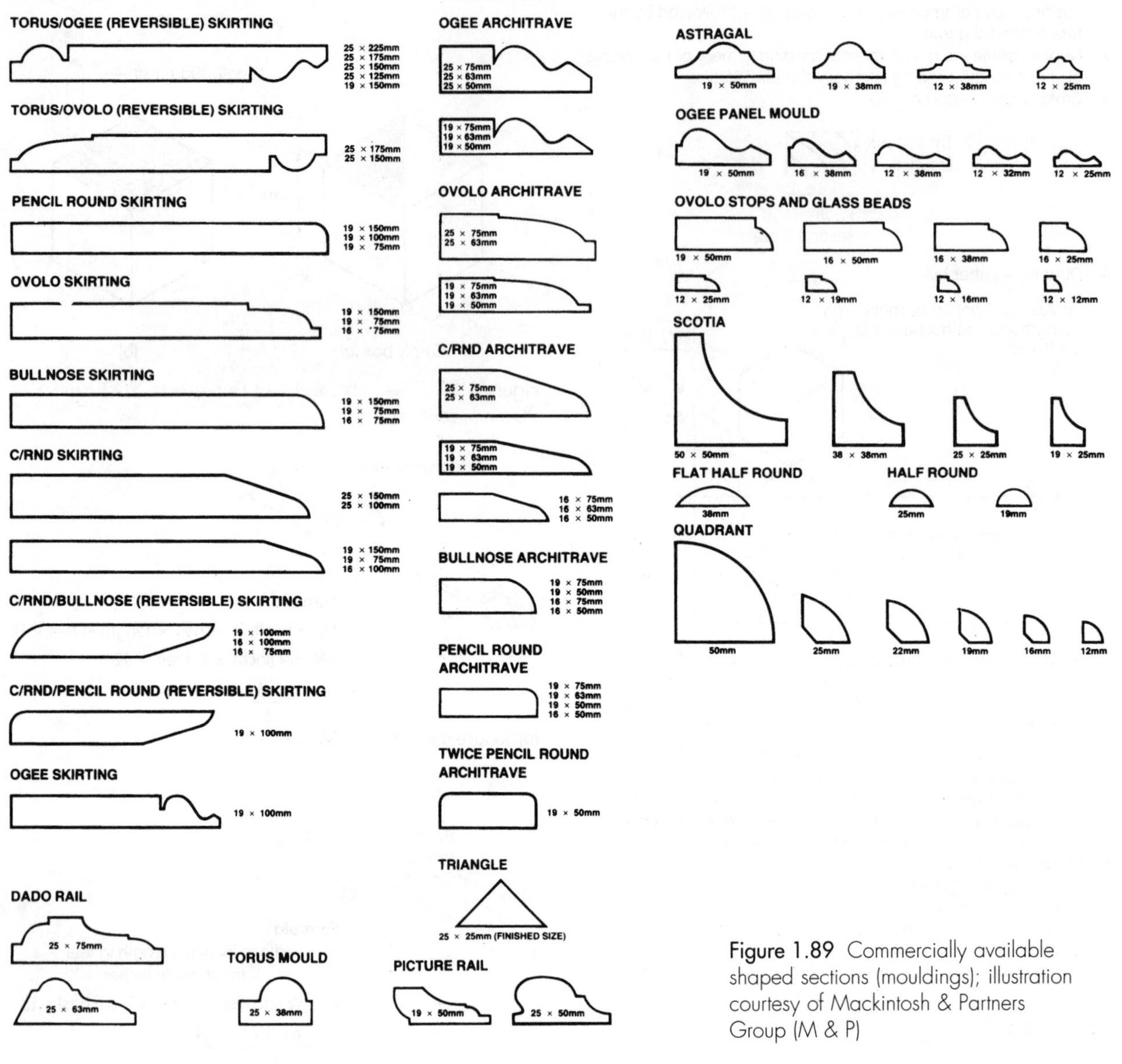

Figure 1.89 Commercially available shaped sections (mouldings); illustration courtesy of Mackintosh & Partners Group (M & P)

Table 1.15 Guide to the use of wood moulded sections

Location (situation)	Moulded section
Roofs	Fascia board, lead roll, triangular fillet
Cladding (exterior)	Weather boarding (shiplap etc.); Match boarding (TG & V)
Flooring	Floor boarding (off-set T & G)
Wall to floor (trim)	Skirting board
Wall to door/window linings (trim)	Architrave
Wall trim	Dado rail, picture rail
Panelling	Panel moulds, bolection moulds, astragals, skirting board, corner beads (scotia & quadrant)
Doors	Panel moulds, bolection moulds, slop laths, glazing beads, weather board, match boarding (TG & V)
Stairs	Hand rails, nosings, cappings, spindles (see stairs), scotia beading/board
Windows	Sash material, glazing beads, sills, drips, window boards

Note: Examples of all the moulded sections are shown in Figure 1.89

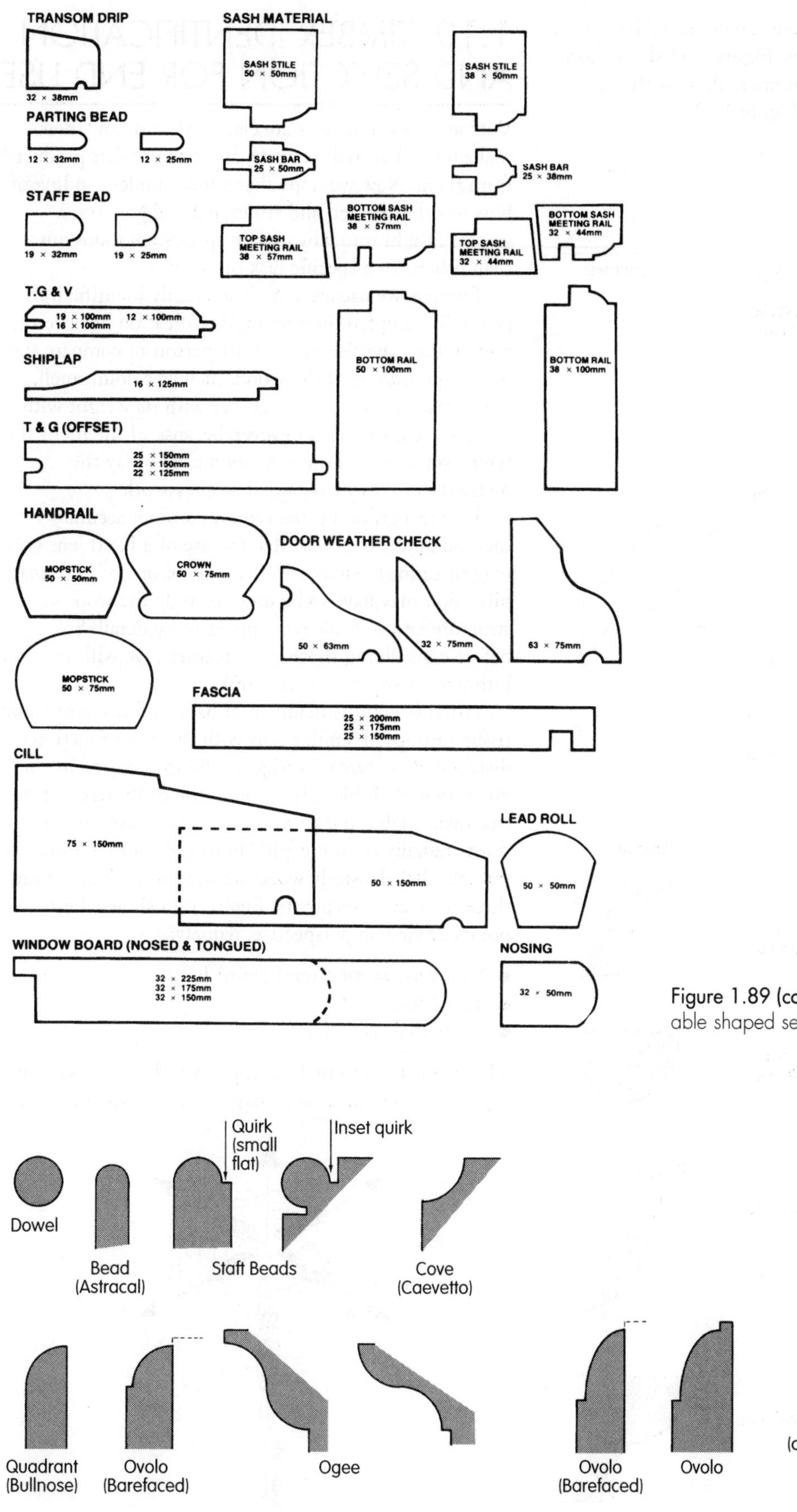

Figure 1.89 (continued) Commercially available shaped sections (mouldings)

Figure 1.90 Mouldings based on the circle

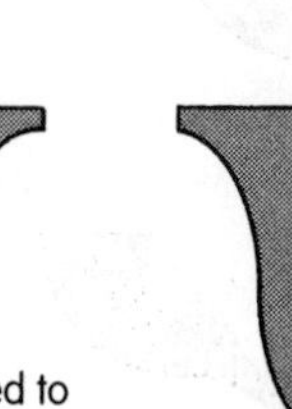

Figure 1.91 Mouldings based on the ellipse

much softer than others – these can be traced back to the Grecian and Roman styles. Figure 1.93 shows how these two styles can differ. Compare these with the build up of shapes shown in Figure 1.92.

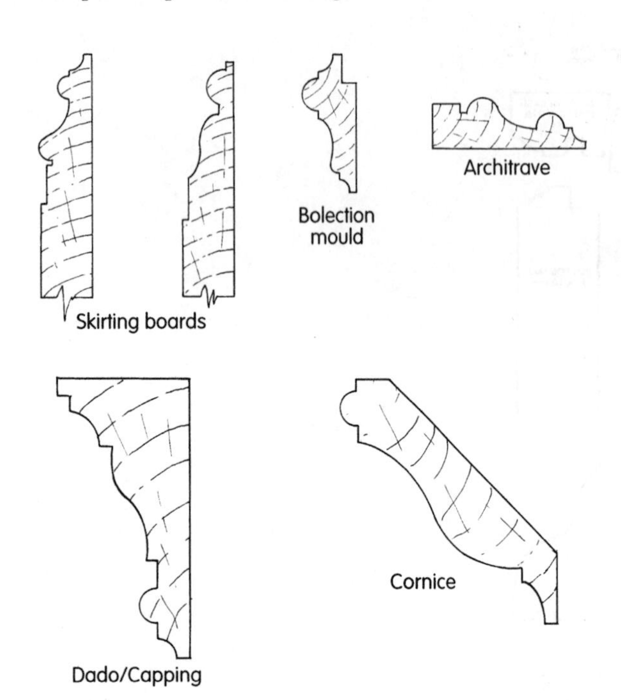

Figure 1.92 Circular and elliptical based moulding to form traditional classical lines

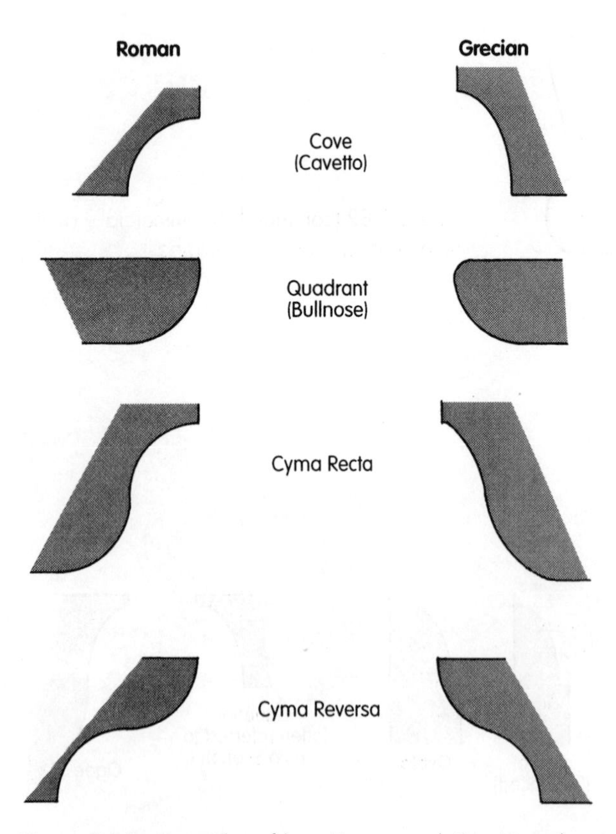

Figure 1.93 Examples of how Roman and Grecian styles of moulding may differ

1.10 TIMBER IDENTIFICATION AND SELECTION FOR END USE

We have seen how trees are classified, and how their make up and growth patterns affect the timber produced from them. Now we hope that a fuller understanding of how wood is formed and structured will give you a greater insight into why certain species are more suitable than others for a specific task or end use.

Timbers we use are not always easily identified: the generally accepted method of identification has been to rely on the expertise of the craftsperson to compare the basic *gross features* of the wood, such as colour, smell, grain, texture and figure together with its weight with other known species. However, because of the many different yet similar species of wood used today this method can only be regarded as guesswork.

True recognition often requires a more accurate method and this can involve the use of a hand lens (×10 magnification). Alternatively a microscope of high magnification may have to be used to study the wood's structure and possible contents in more detail. So, before considering those gross features, we will first look into wood's structure (make up).

Wood (xylem) is made up of bands of different wood tissue (groups of similar cells with the same function) distributed as shown in Figure 1.94 (to be used in conjunction with Table 1.16) around the central axis of the tree (axial cells), and to a lesser extent extending outwards radially from the 'pith' (although not continuous) (radial cells). To study wood we need to look at it from three different viewpoints. Figure 1.95 shows these points of view in perspective, namely:

- transverse section (end grain) E
- radial section R
- tangential section T.

These enable us to make comparisons between known types, distributions, sizes, shapes, etc., of wood cells, and

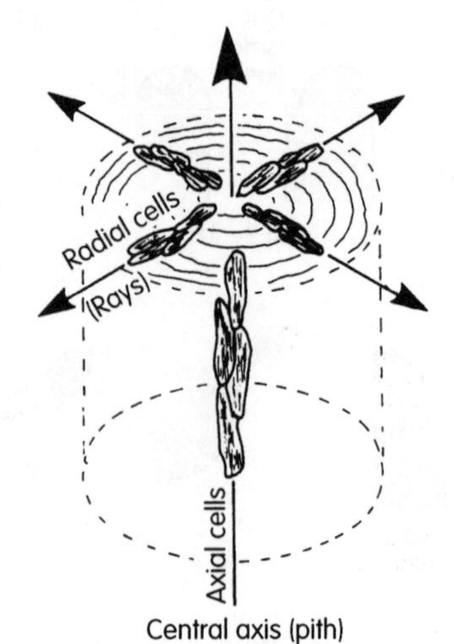

Figure 1.94 Wood cell distribution

Table 1.16 Relationship between wood cell distribution, type and function

Distribution	Tissue type	Location	Function
Axial cells	Tracheids (Figs. 1.96/1.99)	Mainly softwood, some hardwoods (e.g. oak)	Provide strength and conduct sap
Axial and radial cells	Parenchyma (Figs. 1.98 & 1.100)	Softwoods and hardwoods	Conduct and store food
Axial cells	Fibres (Figs. 1.100, 1.102 & 1.103)	Hardwoods	Provide strength
Axial cells	Vessels or pores (Figs. 1.100, 1.102 & 1.103)	Hardwoods	Conduct sap

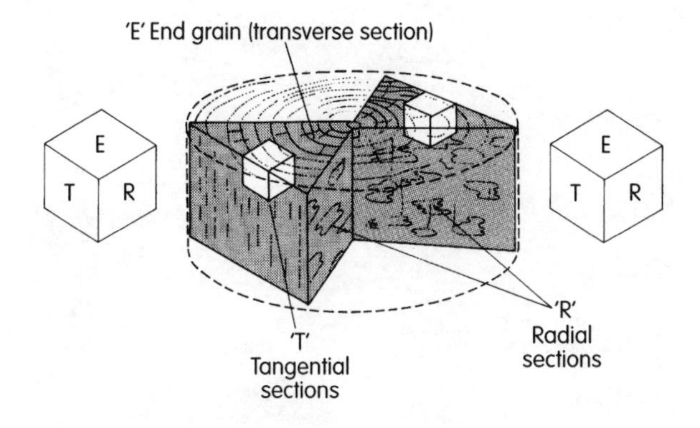

Figure 1.95 Key to viewing wood samples

detect any special features such as the presence of any intrusions into the cell wall or cavity, such as extractives (see Section 1.10.3g).

1.10.1 Structure of Softwood

The bulk of the stem tissue is made up of cells called **tracheids** (see Figure 1.96) which are responsible for conducting sap and providing the timber with its mechanical strength. Tracheids are elongated pointed cells of 2.5 to 5 mm in length, although sometimes longer. Those produced first in the growth cycle and which conduct the bulk of the sap have large cavities with thin walls, while those of the later growth with thick walls provide the strength – the former are known as **earlywood** and the latter as **latewood**. Movement of sap is mainly via holes in the cell walls, which are known as **bordered pits**. Figure 1.97 shows how these pits have the ability to control this flow of sap. Closure of a bordered pit (aspirated pit) may be as a result of injury to the tree, or more likely natural ageing when the sapwood becomes heartwood, which could explain why preservative penetration into the wood is more difficult within heartwood. Tracheids may also be present in the rays.

Rays are responsible for moving sap horizontally and for food storage via tissue known as **parenchyma**, which generally appear as a line of single radial cells stacked in groups one upon another. The cells are brick-shaped, with interconnecting holes in the cell walls (see Figure 1.98) known as **simple pits**. Similar storage cells, known as 'wood-parenchyma', are also present in the vertical axis. In some softwoods, ray parenchyma may also be accompanied by ray tracheids and resin canals. Vertical resin ducts may also be present; a resin duct is shown within

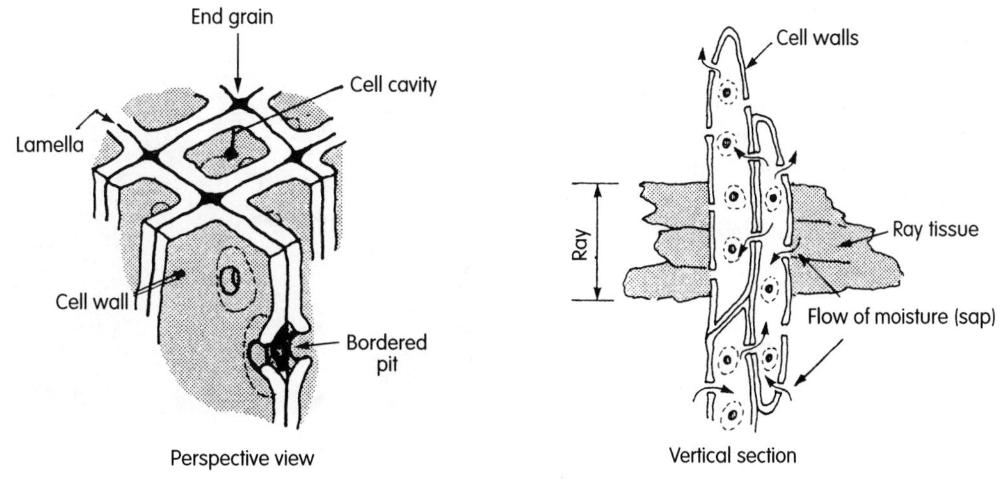

Figure 1.96 Tracheids

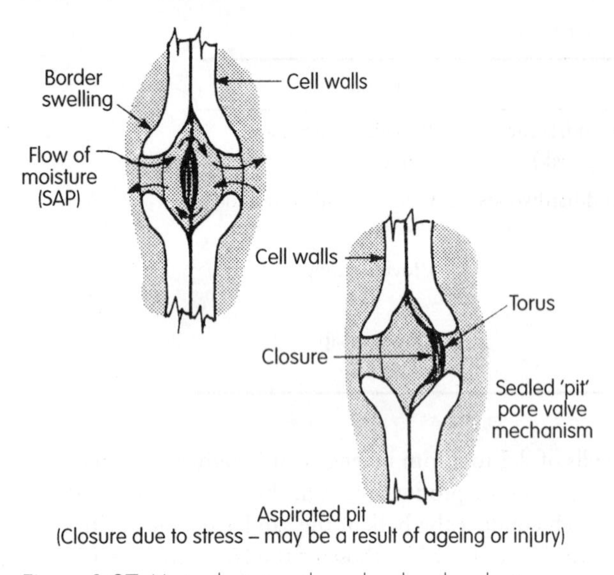

Figure 1.97 Vertical section through a bordered pit

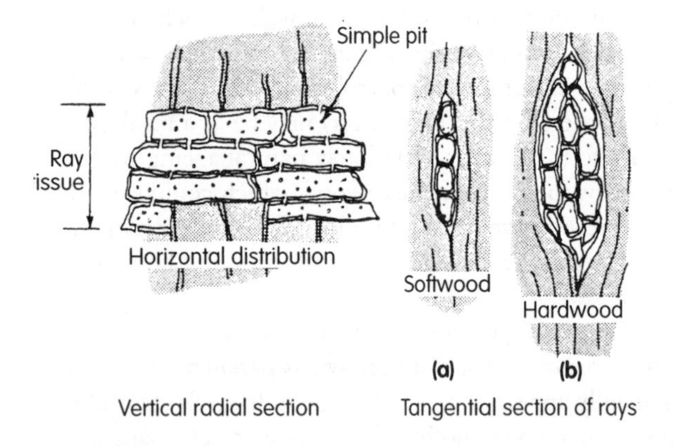

Figure 1.98 Vertical section through parenchyma cells

the transverse section of Figure 1.99. Resin is produced in the parenchyma cells which surround these canals, or ducts, and secrete resin into them.

Resin canals and resin ducts may be regarded as one of the same apertures within the wood structure. Resin canals can be used as a feature for identifying certain species (see Section 1.10.3g), but they may also be formed unnaturally as a result of an injury, just like another feature we call **resin pockets** (pitch pocket). These appear in the tree as vertical saucer shaped cavities, and can follow a line around the growth ring. It is not uncommon to expose these pockets during conversion, when liquid resin may be seen running over the surface of the timber.

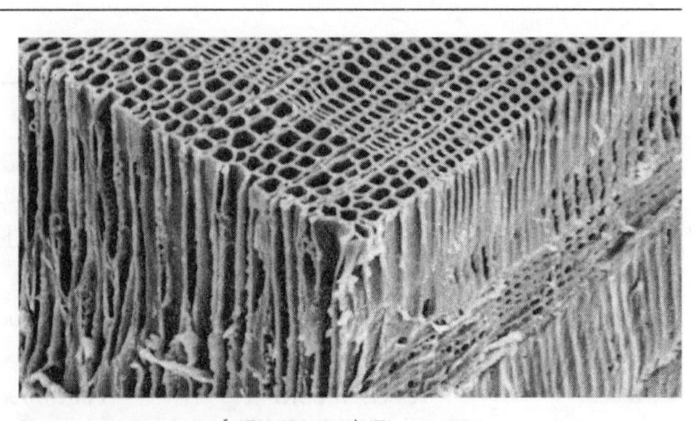
Perspective view of 'E', 'R' and 'T' sections

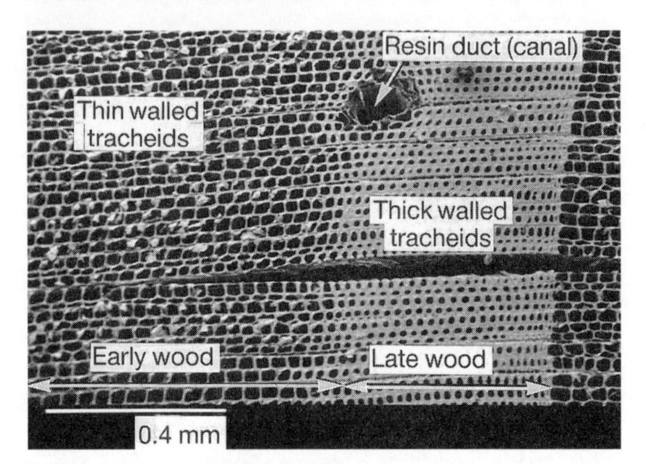

Transverse section (end grain) 'E'

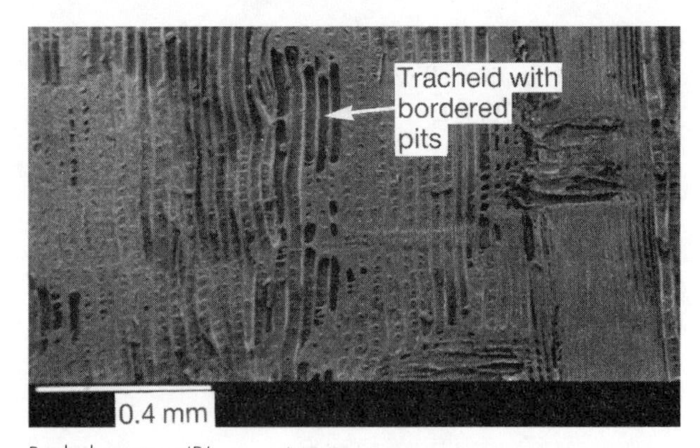

Radial section 'R'

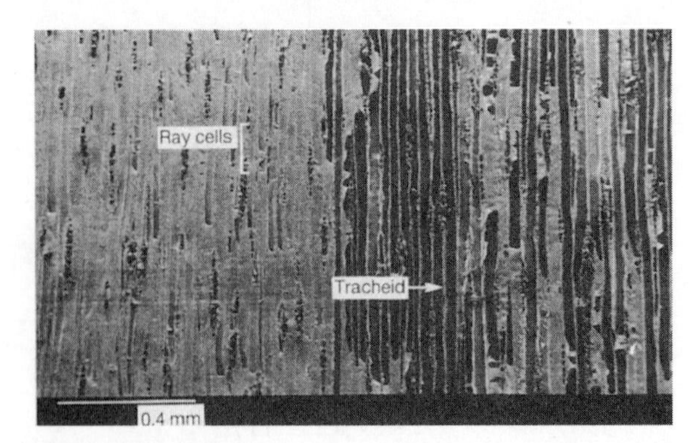

Tangential section 'T'

Figure 1.99 Softwood – Scots pine (pinus sylvestris)

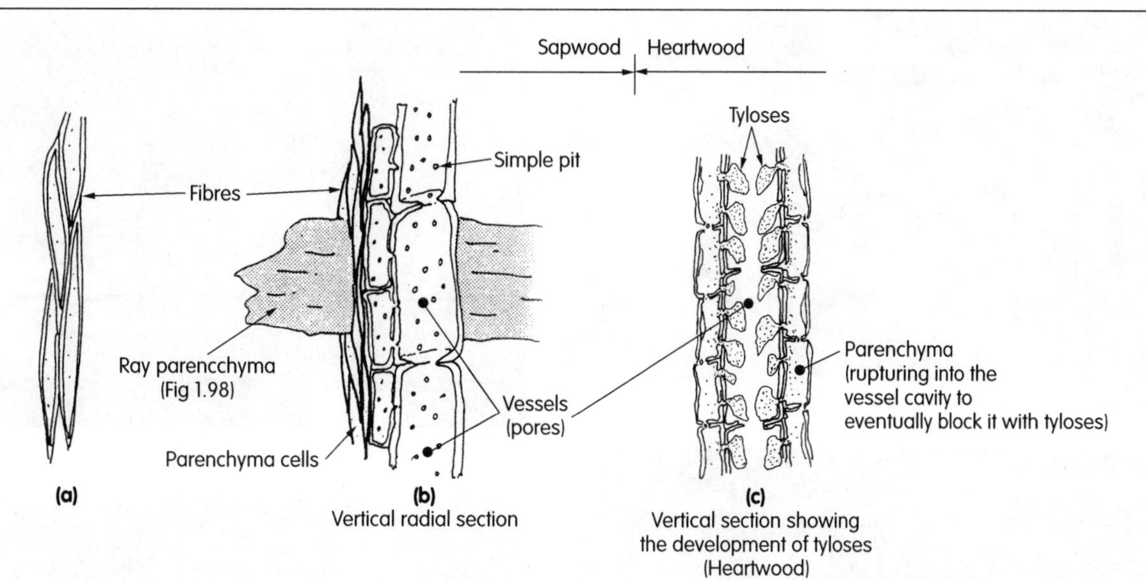

Figure 1.100 Vertical section through hardwood tissue

1.10.2 Structure of Hardwoods

Hardwoods have a much more complicated structure, with a wider range of cell formation (see Table 1.16). **Fibres** (Figures 1.100, 1.102 and 1.103) form the greatest mass of axial tissue. Pitting in the cell walls may be simple (Figure 1.101) or bordered (they would then be called fibre tracheids) to provide conduction. The main task of fibres is to provide wood strength similar to the latewood tracheids of softwood, but with one big difference: the main conducting tissues in hardwood are known as **vessels**, often referred to as 'pores'. For the purpose of this chapter the term 'vessel' has been used when the cells have been exposed longitudinally (radial and tangential sections), and **pores** when exposed transversally (end grain).

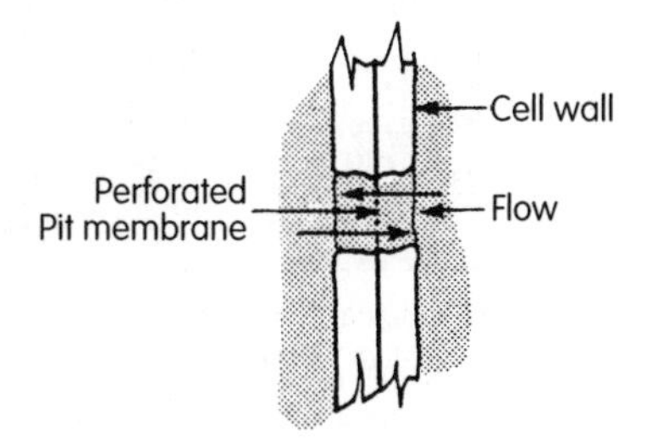

Figure 1.101 Vertical section through a simple pit

These cells start life joined end to end, but eventually, as they swell, the ends split and end up linking each other into a tubular form capable of conducting large quantities of sap as required by all broad leaved hardwoods. Tracheids may also be present in some hardwoods to further help conduction.

The size and distribution of pores (vessels) can, as shown in Table 1.17, vary between species and growth pattern (with or without distinct growth rings). Growing conditions should also be taken into account. When pores (vessels) are more or less the same size and evenly distributed, whether singularly or in groups, the species would be called **diffuse porous**. An example is shown in Figure 1.102 (birch – Betula *spp.*), and diffuse porous types form the biggest group of hardwoods. To a lesser extent, some species have pores (vessels) which diminish in diameter from earlywood (large pores) to latewood (small pores). These species are called **ring-porous** hardwoods, and one of the most common of these is European Oak, as shown in Figure 1.103.

As shown in Table 1.17, this distribution of the pores, together with the presence or absence of a distinct growth ring, can often contribute to identification. We have already established that the presence of a growth ring would generally reflect those species grown within regions with set growing seasons, whereas, conversely, those without distinct growth rings would have grown in regions permitting continuous, or near continuous, growth.

Heartwood, like the heartwood of softwood, is the non-active part of the wood, where conduction has been restricted. In softwood you will recall that this breakdown was in the main due to pit aspiration of the tracheid wall. Some hardwoods on the other hand have their sap flow restricted because, the vessels can become blocked with 'tyloses' (Figure 1.100). This is a membrane that is formed within the vessel as a result of parenchma cell walls rupturing and entering vessel cavities via the pores, due to a difference in pressure between the vessels and their surrounding cells (parenchyma). Hardwood rays have a similar function to those within softwood; the main physical difference is, as shown in Figure 1.98, that they are generally more than one cell wide.

Resin canals, usually referred to as 'gum' canals or ducts, are not very common, and are found in only a few

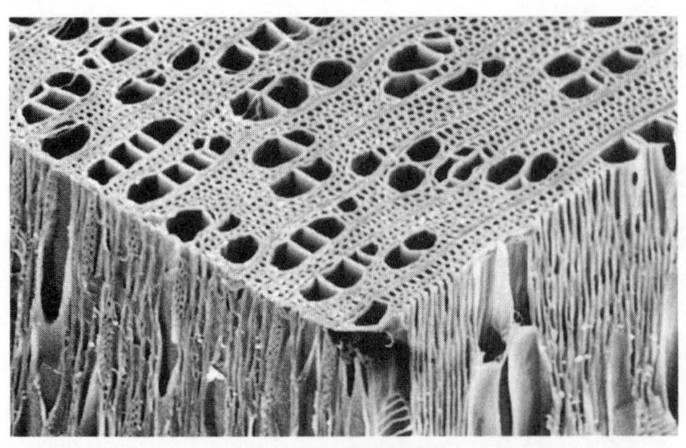

Perspective view of 'E', 'R' and 'T'

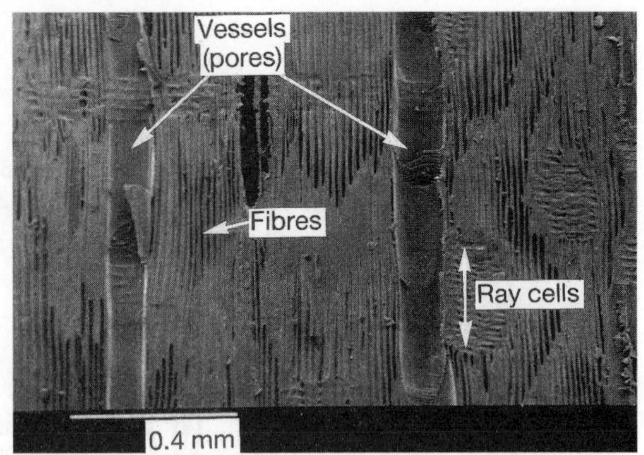

Radial section 'R'

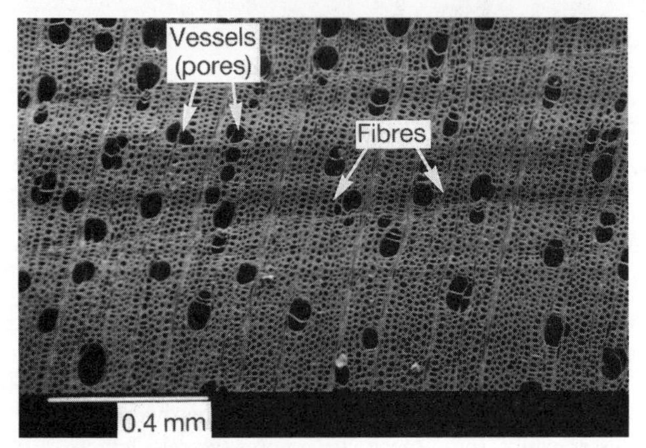

Transverse section (end grain) 'E'

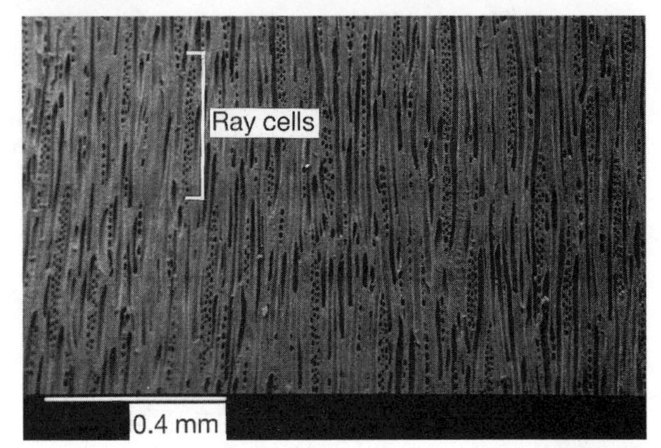

Tangential section 'T'

Figure 1.102 Hardwood: diffuse, porous – Birch (Betula *spp.*)

Table 1.17 Distribution of hardwood pores in relation to species and growth ring

Growth ring	Diffuse – porous species	Ring porous and semi-ring porous species
Easily seen (distinct – conspicuous)	Agba – *Gossweilerodendron balsamiferum* Beech (European) – *Fagus sylvatica* Birch (European) – *Betula spp.* Mahogany (American) – *Swietenia macrophylla* Obeche – *Triplochiton scleroxylon* Sapele – *Entandrophragma cylindricum* Sycamore – *Acer pseudoplatanus* Maple (rock) – *Acer saccharum*	Ash (European) – *Fraxinus excelsior* Elm (English) – *Ulmus procera* Oak (European) – *Quercus robur* Sweet chestnut – *Castanea sativa* Teak – *Tectona grandis* Walnut (European) – *Juglans regia*
Without, or not usually distinct unless using lens	Afrormosia – *Pericopsis elata* Iroko – *Chloraphora excelsa* Jelutong – *Dyera costulata* Keruing – *Dipterocarpus spp.* Mahogany (African) – *Khaya spp.* Ramin – *Gonystylus spp.* Red Meranti – *Shorea spp.*	

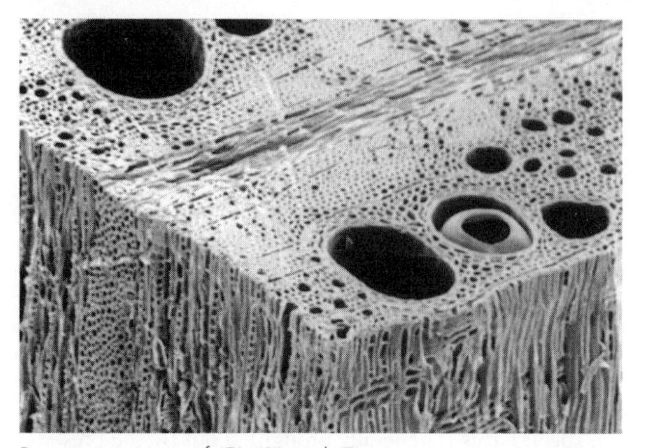

Perspective view of 'E', 'R' and 'T'

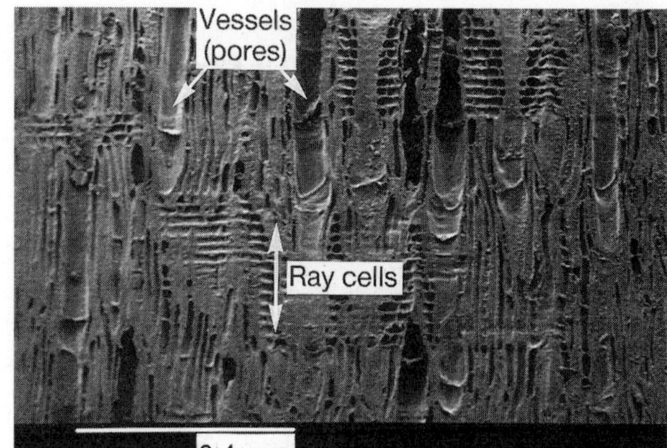

Radial section 'R'

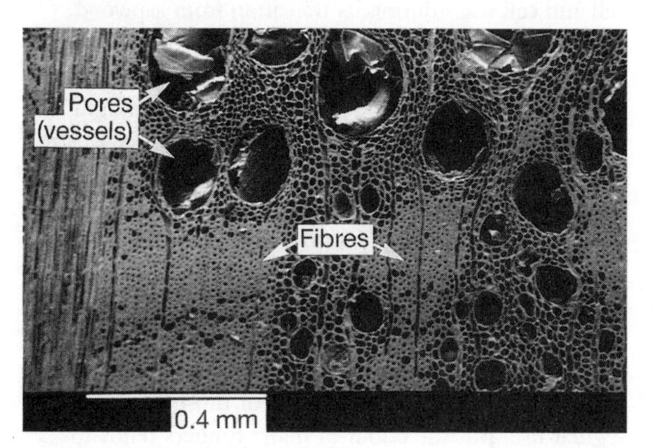

Transverse section (end grain) 'E'

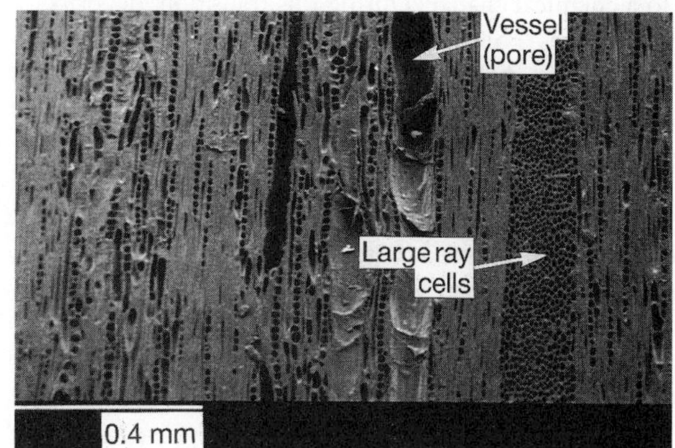

Tangential section 'T'

Figure 1.103 Hardwood: ring, porous – Oak (*Quercus spp.*)

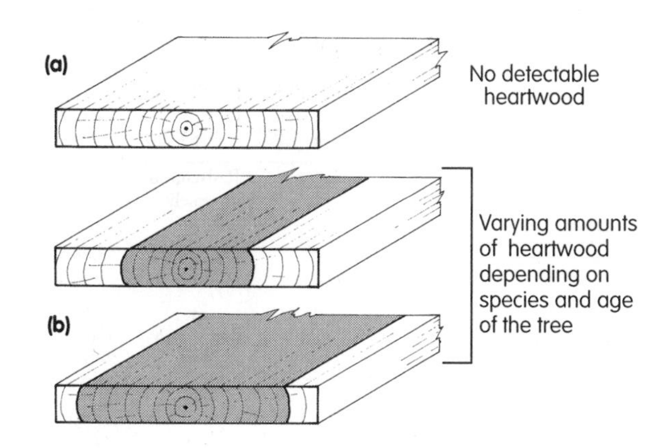

Figure 1.104 Proportions of sapwood

species, one familiar example being Meranti (Lauan). Hardwoods as a whole can be classified as 'porous wood' (NB 'porous' used as a botanical term only). Not only do pores provide a distinguishing feature between the two types (diffuse and ring porous), but the absence of pores would indicate a softwood.

1.10.3 Gross Features of Timber

So far we have looked at the structure of wood as if we were using a hand lens or microscope. What concerns us here are those features that may be apparent simply by using our natural means of detection, such as sight, touch, and possibly smell; these senses become more active once the wood is processed.

What features might we be looking for? Consider the following:

a growth ring
b sapwood and heartwood
c ray
d colour
e lustre
f odour
g cell intrusion (extractives)
h grain
i texture
j figure
k compression wood
l tension wood.

a) Growth Ring

(See also Section 1.2.4.)

With softwood, the darker region of the growth ring is where the cells are small and thick walled, whereas with hardwood, even though the bulk of the cell walls thicken, it is often the size and distribution of different cells which affect the growth ring pattern.

Trees grown in tropical regions may not produce a growth ring annually as we know it, but possibly one which reflects wet or dry seasons. On the other hand, others may sustain continuous growth throughout their lives. As shown in Table 1.17, the growth ring can hold vital clues to the identity of the species. As shown later, the growth ring can also significantly influence the strength of timber.

b) Sapwood and Heartwood

(See also Sections 1.2.5 and 1.2.6.)

Ageing must also be a very important factor, as it may produce an unmistakable feature, that of varying proportions of heartwood and sapwood (Figure 1.104b). Heartwood is often darker in colour than sapwood due to a chemical change. During this transition, deposits (extractives – Section 1.10.3g) are left within the cells, and they can become blocked with inactive dead tissue, all of which can contribute to wood durability (Section 1.10.4e).

If you have the opportunity to see a full radial section of a tree you will usually see the demarcation line between the sapwood (outer bands) and heartwood (inner bands). There are exceptions, e.g. European Whitewood and Beech as shown in Figure 1.104a.

c) Rays

(See also Section 1.2.7.)

Rays can form a recognisable feature, known as 'figure', which will show up on the surface of timber as a result of large and/or broad rayed species being quarter sawn. Oak and beech are good examples.

d) Colour

Colour usually refers to heartwood, as sapwood is generally a much paler colour and ranges from off-white to light brown. Tables 1.18 and 1.19 give a colour for each listed species, but these should only be used as a guide, since the colour can vary between the species, and be affected by exposure to air, light, or heat to the extent that they may darken, or even lighten the colour. Next time you remove a piece of timber from an existing stack of planed redwood, notice the difference in colour where the various pieces overlapped one another.

Where coloration of heartwood is a distinct feature of the species, this is generally brought about by the presence of extractives (Section 1.10.3g) retained within the cell and cell walls during its transition from sapwood.

e) Lustre

If the cut surfaces of wood cells are exposed to light some surfaces take on a bright shiny appearance (lustre), whilst others may appear dull, both of which can be used to contrast one with the other as wood figure. Examples are given in Section 1.10.3j.

f) Odour (Smell)

We all know that wood has its own particular odour, and those who work with wood will quickly become familiar with the particular odour of those species in regular use. But the odour of many timbers tends to lessen with age, whilst others seem, albeit to a lesser extent, to retain their odour.

Table 1.18 A guide to some properties of various softwood species (heartwood)

	Species			(a)	(b)	(c)	(d)	(e)	(f)	
Colour of heartwood	Common name	Latin (botanical) name	Origin	Moisture movement	Approx. density (kg/m³)	Texture (see Table 1.20)	Working qualities (cutting, nailing, etc.)	Durability (See Table 1.22)	Permeability (See Table 1.23)	General usage
Pinkish brown	Douglas fir	*Psuedotsuga menziesii*	Canada USA & UK	Small	530	Fine/medium (straight grain)	Good	Moderate	Resistant to extremely resistant	Internal & external joinery*†
Pinkish brown	European Redwood (Scots pine)	*Pinus sylvestris*	Europe	Medium	510	Fine/medium	Good	Non-durable	Moderately resistant	Internal & external joinery*
Brown to red-dark streaks	Parana pine	*Araucaria angustifolia*	Brazil S. America	Medium	550	Fine (mainly straight-grained)	Medium to good	Slightly durable	Moderately resistant	Internal use only plywood
Pinkish brown	Sita Spruce	*Picea sitchensis*	UK	Small	450	Coarse	Good	Non-durable perishable	Resistant	Construction, pallets
Light brown	Western hemlock	*Tsuga heterophylla*	Canada USA	Small	500	Fine – even (straight grain)	Good	Slightly durable	Resistant	Internal joinery*
Pink to chocolate brown	Western red cedar	*Thuja plicata*	Canada USA	Small	390	Medium to coarse (straight grain)	Good	Durable	Resistant	Internal & external joinery. Cladding
White	Whitewood (European spruce)	*Picea abies*	Europe	Small	470	Fine	Good	Slightly durable	Resistant	Internal joinery*

Note: * Structural – timber suitable for situations where strength is a major factor; † Decorative appearance – timber noted for its decorative properties

Table 1.19 A guide to some properties of various hardwood species (heartwood)

Colour of heartwood	Species: Common name	Species: Latin (botanical) name	Origin	(a) Moisture movement	(b) Approx. density (kg/m³)	(c) Texture (see Table 1.20)	(d) Working qualities (cutting, nailing, etc.)	(e) Durability (See Table 1.22)	(f) Permeability (See Table 1.23)	General usage
Reddish brown	African mahogany	*Khaya spp.*	West Africa	Small	530	Medium – varies (interlocking grain)	Medium	Moderate	Extremely resistant	Internal & external joinery. Plywood veneer
Reddish brown	American mahogany	*Swietenia macrophylla*	Central & S. America	Small	560	Medium	Good	Durable	Extremely resistant	Internal & external joinery
Light brown	Afrormosia	*Pericopsis elata*	West Africa	Small	710	Fine to medium interlocking grain)	Medium	Very durable	Extremely resistant	Internal & external joinery. Veneer
Grey, brown	Ash (American)	*Fraxinus spp*	USA	Medium	670	Coarse	Medium	Perishable	Permeable	Interior joinery Tool handles
White to light brown	Ash (European)	*Fraxinus excelsior*	Europe	Medium	710	Medium/ Coarse	Good	Perishable	Moderately resistant	Interior joinery
Light brown	Beech	*Fagus sylvatica*	Europe	Large	720	Fine (straight grain)	Good	Perishable	Permeable	Internal joinery Plywood – veneer
White to light brown	Birch (European)	*Retula pubescens*	Europe	Large	670	Fine	Good	Perishable	Moderately resistant	Plywood furniture
Yellowish brown	Chestnut (sweet)	*Castanea Sativa*	Europe	Large	560	Medium	Good	Durable	Extremely resistant	Internal & external joinery, fencing
Light brown	Elm	*Ulmus spp.*	Europe USA Japan	Medium (can vary)	580	Medium/ coarse (coarse grain)	Medium	Slightly durable	Moderately resistant	Internal joinery Veneer†
Mid to dark brown	Iroko	*Chlorophora excelsa*	East and West Africa		660	Coarse/ medium (interlocking grain)	Medium to difficult	Very durable	Extremely resistant	Internal & external joinery. Plywood veneer
Reddish brown	Keruing	*Dipterocarpus spp.*	Malaysia (SE Asia)	Large	740	Medium	Difficult	Moderate	Moderate	Internal & external joinery*
Light/ dark	Meranti Lauan	*Shorea spp.*	Malaysia Indonesia Philippines (SE Asia)	Small	550–710	Coarse/ grain) medium (Slight interlocking grain)	Medium	Moderate to durable	Resistant to extremely resistant	Internal & external joinery. Plywood
Light brown	Oak	*Quercus spp.*	Europe America Japan	Medium	670–790	Medium to coarse	Medium to difficult	Durable (not all species)	Extremely to moderately resistant	Internal & external joinery Plywood – veneer†
Yellowish	Ramin	*Gonystylus spp.*	Indonesia (SE Asia)	Large	670	Medium (straight grain)	Medium	Not durable	Permeable	Internal joinery Mouldings and trim
Reddish brown	Sapele	*Entandro-phragma cyndricum*	West Africa	Medium	640	Fine to medium (interlocking grain)	Medium	Moderate	Resistant	Internal & external joinery Plywood – veneer†
Whitish	Sycamore	*Acer Pseudoplatanus*	Europe	Medium	630	Fine (straight grain)	Medium to good	Perishable to not durable	Permeable	Internal joinery Veneer
Light brown	Teak	*Tectona grandis*	Myanmar Thailand Indonesia	Small	660	Medium (straight grain)	Medium	Very durable	Extremely resistant	Internal & external joinery* Veneer†
Reddish brown	Utile	*Entandro-phragma utile*	West Africa	Medium	660	Medium (interlocking grain)	Medium to good	Durable	Extremely resistant	Internal & external joinery. Plywood – veneer

Note: * Structural – timber suitable for situations where strength is a major factor; † Decorative appearance – timber noted for its decorative properties

The odour given off by resinous pines is very familiar to most woodworkers, and similarly the spicy smell of some hardwoods, but I find the odour of individual wood species hard to describe. Therefore it is best for each person to memorise his or her own interpretation of a particular odour.

g) Cell Intrusion

I include here those items termed 'extractives', substances which invade the cells and cell walls but are not normally an intrinsic part of them, and could therefore be chemically removed. Extractives, which are organic compounds such as oils, acids, tannins, latex etc. may be present. Wood from oak, chestnut and western red cedar contains acids, which, under moist conditions, can corrode some metals, particularly ferrous metals.

Extractives are responsible for giving heartwood its darkening effect, and for its durability (see Section 1.10.4e). We should also include here resins and gums already mentioned in Sections 1.10.1 and 1.10.2.

Examples of softwoods where resin canals (ducts) can be found are:

- European redwood (*Pinus sylvestris*)
- European whitewood (*Picea abies*)
- Larch (*Larix spp.*)
- Douglas fir (*Pseudotsuga menziesii*)
- Pitch pine (*Pinus palastris*)
- Yellow pine (*Pinus strobus*)

Examples of hardwoods where resin (gums) canals (ducts) can be found are:

- Red meranti/seraya (*Shorea*)
- Gurjun/keruing (*Diptercarpus*)
- Agba (*Gossweilerodendron balsamiferum*)

Some tropical hardwoods may contain calcium salts or silica grains, for example iroko (calcium carbonate), and keruing (silica). Silica grains can have a serious effect on the cutting edges of tools.

h) Grain

Grain refers to the direction of the main elements of the wood. The manner in which grain appears will depend upon one or more of the following:

- the direction of the cut
- the location of the cut
- the condition of the wood
- the arrangement of the wood cells.

To help qualify this Table 1.20 has been arranged as a quick source of reference to common grain terms and/or conditions, together with a broad explanation of why they are named, etc.

i) Texture

Texture is a surface condition resulting from the size and distribution of wood cells (Figure 1.105), usually associated with touch. Texture can be linked to the conditions shown in Table 1.20 and with the wood properties featured within Tables 1.18 and 1.19. Unless the grain is filled, many high gloss surface finishes can emphasise course texture (Figure 1.105b), just like an uneven texture (Figure 1.105d) will show ripples, even through a painted surface. Of course light direction and intensity will always be important factors in revealing any textured surfaces.

a) Fine texture (open small pores or cells) e.g. Sycamore

b) Coarse texture or open large pores, e.g. Oak

c) Even texture (little or minimum contrast between growth rings)

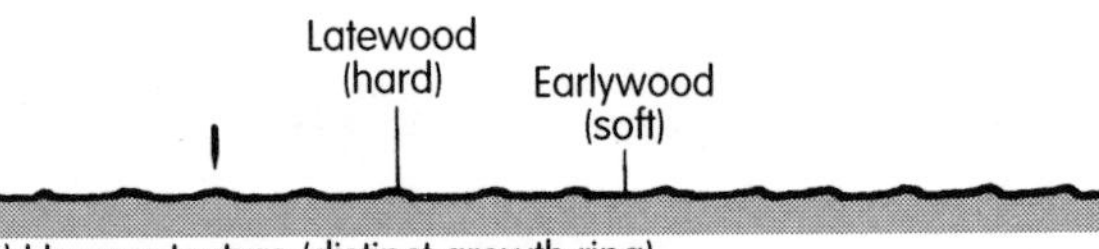

d) Uneven texture (distinct growth ring) e.g. Douglas Fir, Western Hemlock

Figure 1.105 Texture of finished surfaces

j) Figure

Figure is best described as the pattern or markings which are formed on the surface of processed timber as a result of wood tissue being cut through. For example, as shown in Figure 1.106a, quarter sawn oak exposes broad ray tissue to produce what is known as silver figure; whereas, if, as shown in Figure 1.106b, the interlocking grain of African mahogany was quarter sawn it would reveal a stripe or ribbon figure. Tangential (plain) sawn softwood like Douglas fir with distinct growth rings would, as shown in Figure 1.106c, produce a flower or flame figure.

Timber possessing these characteristics can be regarded as having natural inherent decorative properties.

k) Compression Wood (Softwood)

We discussed this condition under structural defects (Section 1.6.1). However as a gross feature it would be fair to say that it may not be apparent, other than by a darkening of the wood to a reddish brown.

Table 1.20 Grain terms and conditions

Grain terms and conditions	Explanation	Texture	Example of species	Remarks
End grain	Cross-cut exposure of axial and radial cell tissue	–	All	See Fig. 1.95
Straight grain	Grain which generally follows a longitudinal axial course	–	Keruing Kapur	Hardwood and softwood
Cross grain	Grain which deviates considerably from being parallel to the edge of the timber	– –	Elm	Hardwood and softwood
Open or coarse grain	Exposed large vessels, wide rays, and very wide growth rings	Coarse	Oak, Ash	Associated with texture
Close or fine grain	Exposed small vessels, narrow rays and/or narrow growth rings	Fine	Sycamore softwood	Associated with texture
Even grain	Generally uniform, with little or no contrast between earlywood and latewood	Even	Spruce	Associated with texture
Uneven grain	Grain elements vary in size and uniformity – distinct contrast in growth zone	Uneven	Douglas fir	Associated with texture
Curly or wavy grain	Direction of grain constantly changing	Uneven	Walnut	Rippled effect
Interlocking grain	Successive growth layers of grain inclined to grow in opposite directions		African mahogany Afrormosia, Sapele	Striped or ribboned figure
Spiral grain	Grain follows a spiral direction around the stem from roots to crown throughout its growth		British species Horse Chestnut	Defect in timber, affecting structural use (often visible as checks in de-barked poles)
Sloping grain (diagonal grain)	A conversion defect resulting from straight-grained wood being cut across its natural axial growth pattern, or a growth defect resulting from an abnormality in an otherwise straight tree		Can occur in any species	Defect in timber, affecting structural use

Note: In addition to the above, 'short grain' may result from timber being cut and may easily split due to the short length of its elements (e.g. within a trench sawn for a housing joint)

Silver figure
(a)

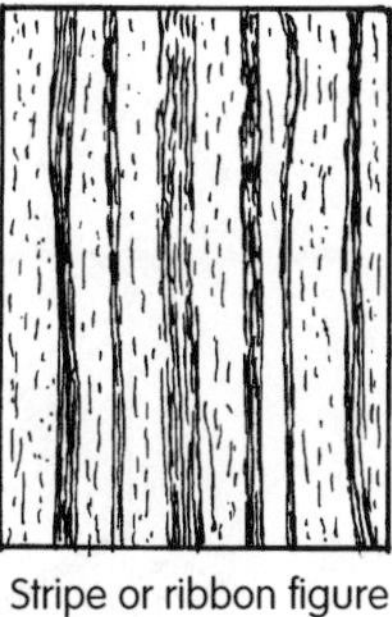

Stripe or ribbon figure
(b)

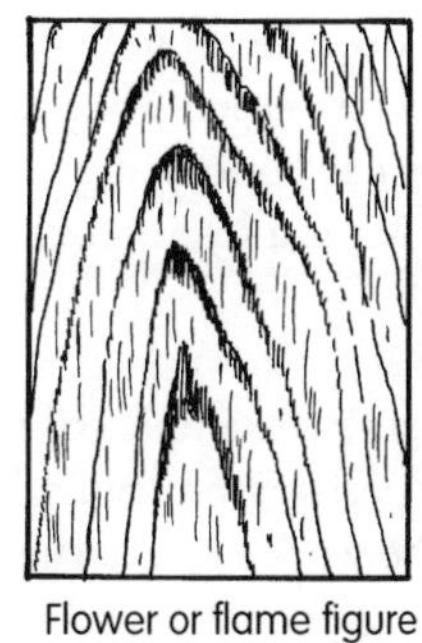

Flower or flame figure
(c)

Figure 1.106 Decorative figure

l) Tension Wood (Hardwood)

We also discussed this condition under structural defects (Section 1.6.1). In contrast to compression wood, tension wood appears lighter in colour. The cutting properties of tension wood are impaired, as the fibres from its cut surface will easily pull away leaving a ragged surface.

NB: There is one other aspect which we have not considered here, that of 'weight'. Weight can be a most important aspect when making comparisons, but it can be just as misleading. What we should really be concerned about is the wood's density, which is dealt with in the next section.

1.10.4 Measurable Properties of Wood (Related to Use) and Selection for End Use

Many of the properties that affect selection will include most, if not all, of the gross features, as they can be identical. For example, colour is one of the most recognisable features of wood, just as it can be a unique property of it. So this section is based around Table 1.18 (Softwoods), and Table 1.19 (Hardwoods), where the following properties are considered:

a moisture movement
b density
c texture
d working qualities
e durability
f permeability.

Other properties (not in the table) that may need be taken into account are:

- strength
- effect of fire
- thermal properties
- electric resistance.

a) Moisture Movement (Column a)

This refers to the amount of movement (small, medium or large) which might affect the dimensional shape of timber when it is subjected to atmospheric conditions liable to change its moisture content after having first been dried to suit its end use, i.e. into a state of equilibrium with its destined environment.

Any increase or decrease in moisture content will result in differential movement (Section 1.7.5) either by expanding (swelling) or contracting (shrinking) – remember that timber is a hygroscopic material (see Section 1.7.4)

Evidence of moisture movement is all around us. Figure 1.107 gives a few examples of this. See if you can recognise them, or perhaps you can relate to one, if not all of them, with the work you are currently undertaking in your practical assignments.

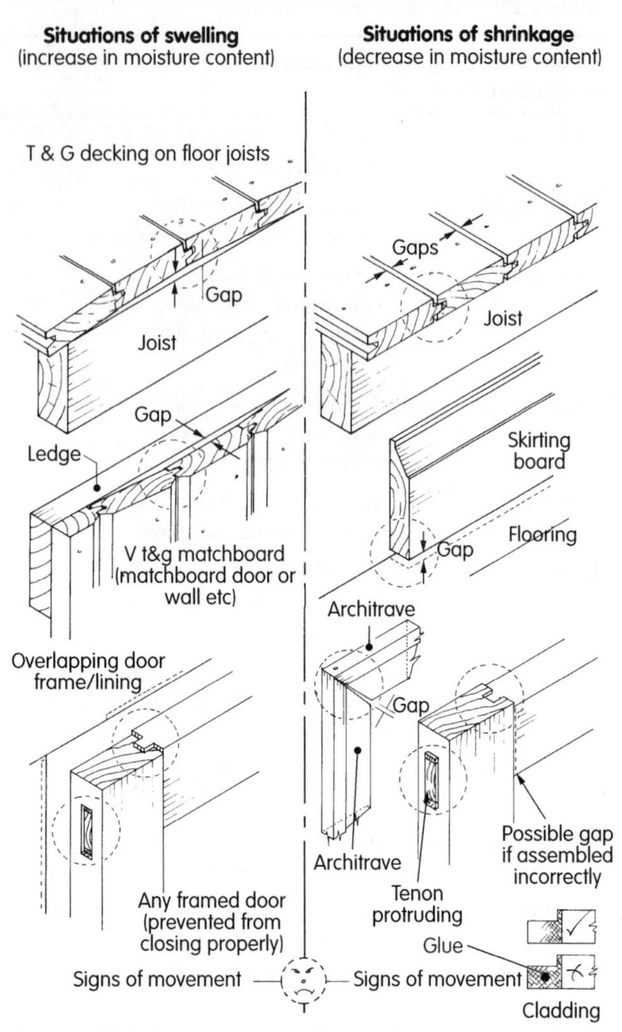

Figure 1.107 Some common effects of moisture movement

b) Density (Column b)

Density refers to the mass of wood tissue, and other substances contained within a unit volume of timber and is usually expressed in kilograms per cubic metre (kg/m^3). Figure 1.108 shows the basis for calculating density. Calculations follow on the next page.

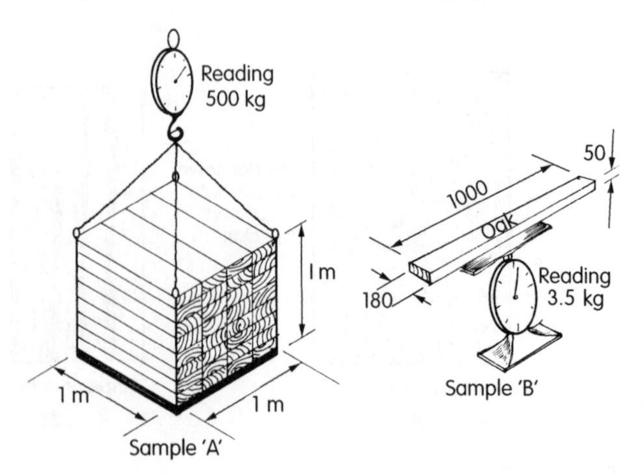

Figure 1.108 Basis by which density of wood is calculated

Calculating density can be derived from Figure 1.108:

Density is expressed as $\dfrac{\text{Weight (kg)}}{\text{Volume}} = \text{kg/m}^3$

or

$$\frac{\text{Sample weight (kg)} = \text{weight}}{\text{Sample volume (m}^3) = \text{length (l)} \times \text{breadth (b)} \times \text{depth (d)}} = \text{kg/m}^3$$

or

$$\frac{\text{Sample weight (kg)} = \text{weight}}{\text{Sample volume (m}^3) = \text{length (l)} \times \text{width (w)} \times \text{thickness (t)}} = \text{kg/m}^3$$

therefore

$\therefore$ **Density of sample A** (Redwood at 15% MC) =

$$\frac{500}{1_m \times 1_m \times 1_m} = 500 \text{ kg per cubic meter (500 kg/m}^3)$$

Or, in a more realistic situation, using a random sample – in this case a piece of oak measuring 1000 × 100 × 50, with a MC of 15% – density can be calculated, as follows.

Density of sample B – as before

$$\frac{\text{Weight (kg)}}{\text{Volume (l} \times \text{w} \times \text{t)m}} = \text{kg/m}^3$$

$$\therefore \frac{3.5}{1.0 \times 0.100 \times 0.050} = 700 \text{ kg/m}^3$$

N.B.: Check these density calculations against those densities shown in Tables 1.18 and 1.19.

Figure 1.108 (continued)

It follows that the moisture content of wood must affect its density, because of the weight of the water contained therein. The amount present in a sample must therefore be a very important factor, as is the presence of any extractive. For a true density to be given, the moisture content must be known and extractives removed. The figures quoted in Tables 1.18 and 1.19 are given as averages for samples at 15% MC.

Figure 1.109 illustrates some of the factors that can influence densities:

- bound water within cell walls (bound moisture)
- free water within cell voids (free moisture)
- the presence or absence of extractives
- the amount of cell tissue in relation to air space.

It is possible to make comparisons between different wood species because we know that the density of solid wood tissue is the same for all (1506 kg/m^3). The difference between the listed species must therefore (providing they are all of equal moisture content with nil extractives) indicate the proportion of wood tissue per unit volume of one cubic metre.

Density of wood is a property closely related to its 'hardness' and 'strength' (see Table 1.21).

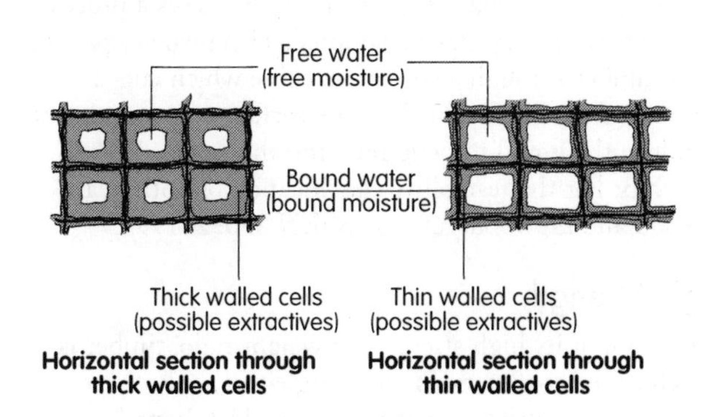

Figure 1.109 Factors that can effect wood density

Table 1.21 Guide to the condition of timber in relation to its general strength characteristics

Condition	Stronger	Weaker
a) Siting – position (stiffness)	kg	kg
b) Density	High (heavy)	Low (light)
c) Moisture content	Low	High
d) Direction of growth rings	Tangential-sawn softwood	Quarter-sawn softwood
e) Fast growth	Ring-porous hardwood	Softwood
f) Slow growth	Softwood	Ring-porous hardwood

c) Texture (Column c)

Texture was first mentioned as a gross feature (Section 1.10.3i) because of its visual nature. As shown in Figure 1.105, texture may be classified as:

- **fine** – surface exposure of small cells, such as pores, tracheids, and rays
- **coarse** – surface exposure of large cells, such as pores
- **even** – surfaces showing little or no contrast between early and latewood, or portraying open cells of similar size across the surface
- **uneven** – the surface shows marked differences between early and latewood, or distinct cell variations across the surface.

Examples of all the above types of texture are given in Table 1.20.

d) Working Qualities (Column d)

This broadly refers to how the timber would respond to being cut by hand, or machine tools, but not necessarily how it would respond to glueing.

Some of the factors influencing working qualities could include grain condition (Table 1.20), and the nature of the wood (e.g. whether it is physically hard, soft, or brittle). Even the presence of intrusions (extractives) which would cause the wood to be abrasive, greasy, or even corrosive could interfere with the working qualities of the wood.

e) Durability (Column e)

Timber may be required to withstand the effects of physical wear and tear, or decay by many different means. But, in this case, we are concerned only with the natural durability (natural resistance) of the heartwood to fungal decay.

This form of durability was classified into five categories of longevity (timespan), as shown in Table 1.22, which was developed by carrying out field tests on several samples of each species of wood. The test, known as a 'graveyard test' consists of driving heartwood stakes of 50 mm × 50 mm section and 600 mm long, 450 mm into the ground (leaving 150 mm above the ground). The stakes are left there until they decay and their condition recorded annually. The time span (in years) taken

Table 1.22 A guide to the classification of the natural durability of heartwood

Classification	Time span (years)
Very durable	More than 25
Durable	25 to 15
Moderately durable	15 to 10
Slightly to not durable	10 to 5
Perishable	Less than 5

Note: For examples of suitable species see Tables 1.18 and 1.19

to decay will determine the final classification of natural durability as shown in Table 1.22. However, the way in which durability is classified has now been changed.

NB: For true durability classification, refer to BSEN 350–2: 1994. For the relationship to hazard classes (Section 4.3) refer to BSEN 460.

f) Permeability (Column f)

We usually think of permeability as a means by which a material allows a liquid to pass through it or be absorbed into it. In this case permeability refers to the ease with which a wood preservative can be impregnated into heartwood when standard pressure methods are used. Table 1.23 shows how wood permeability may be classified. Let us consider how these variations come about.

You may remember that softwoods rely mainly on bordered pits for transference of moisture from cell to cell, both longitudinally and laterally (sideways). You may also remember that these pits can, for various reasons close down (aspirate). For example, natural closure of the heartwood in redwood would contribute to its moderately resistant classification. Whitewood, on the other hand, like all spruces, is generally classified as resistant. This, in part, is due to the very small connections between tracheids and ray tissue resisting the movement of liquid, and possibly to a further closing down of the pits while the timber is being dried for its end use.

Hardwoods, on the other hand, rely in the main on vessels (pores) to move moisture longitudinally, and if these become blocked with 'tyloses' (see Section 1.10.2) on ageing, then preservative penetration is going to be restricted as it would be with European oak. Other considerations are:

- the cell size and distribution
- the moisture content
- the amount of sapwood in relation to heartwood
- knots, their size, number, and distribution
- the presence of any cell intrusions (extractives).

However, there are certain measures that can be taken to increase permeability, including using rough sawn timber instead of planed (the planing process tends to squash the surface of the timber, thereby crushing and closing open ended cells), or in extreme cases a process known as *incising* may be adopted. This involves passing the timber through a special machine which cuts a series of small incisions into the surface of the timber to induce the liquid to penetrate into the wood.

NB: For the testability classification of both heartwood and sapwood, refer to BSEN 350–2: 1994.

g) Strength

Because of its high strength to weight ratio, timber is well suited to situations requiring either, or both, compressive or tensile strength. Figure 1.110 shows how structural timber may be stressed. Posts are being

Table 1.23 A guide to permeability – showing lateral penetration of preservative using standard pressure methods

Classification 'P', 'MR', 'R', 'ER'	Lateral penetration (heartwood)	Species – see also Tables 1.18 and 1.19
Permeable (P) (easy to treat)	No resistance or difficulty	Beech (sapwood of many species)
Moderately resistant (MR)	6–18 mm after 2–3 hours	Ash; Oak (American red); Parana pine; Redwood (European)
Resistant (R) (difficult to treat)	Difficult to penetrate	Douglas fir; Elm (English); Hemlock (western); Oak (American red);Whitewood (European)
Extremely resistant (ER) (extremely difficult to treat)	Minimum absorption	Afrormosia; Douglas fir; Iroko; Lauan (Red); Mahogany; Meranti (Red); Oak (American white); Oak (European); Sweet chestnut; Teak; Western red cedar

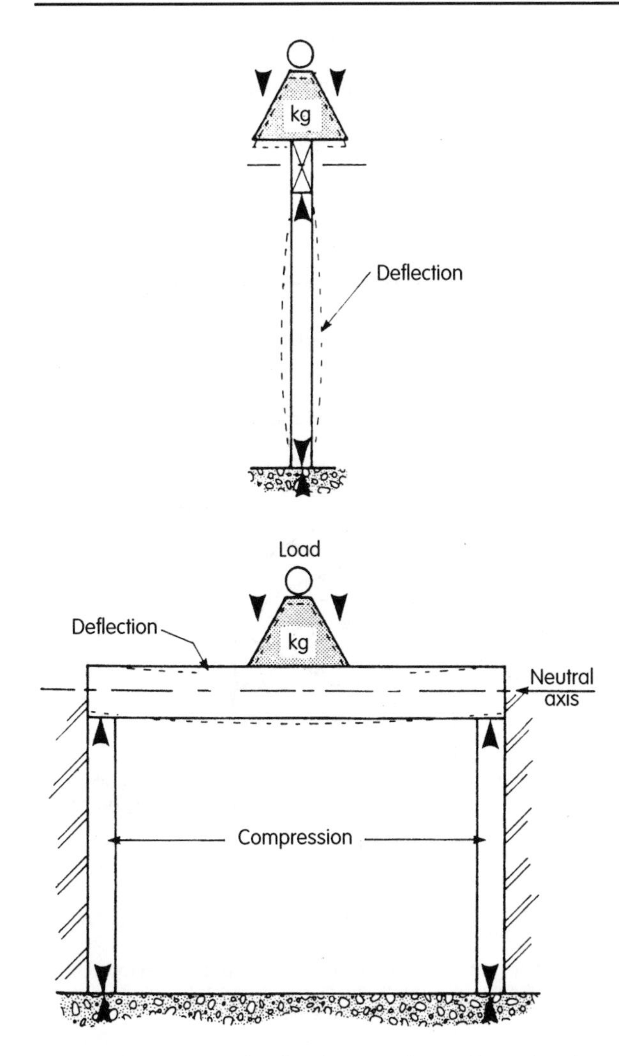

Figure 1.110 Structural timber being subjected to a load

compressed from their ends by the beam plus its load, and also from the subsequent reaction from the ground. The beam is being subjected to stresses of compression (wood tissue being pushed together) within its upper region, and tension (tissue being pulled apart) within its lower region. The neutral axis is a hypothetical dividing line between the two zones where stresses are zero.

If either of these stressed zones, particularly the tension zone, is damaged or distorted by structural (natural) defects such as knots etc.(see Section 1.6.10), or by defects associated with drying (see Section 1.7.14), the beam could be drastically weakened. Cutting notches in stressed timber is strictly controlled.

Structural weakness may also be due to other factors. Table 1.21 gives the following examples:

- timber section in relation to load
- low density of timber (particularly of the same species) at an acceptable moisture content
- high moisture content
- direction of growth ring – wood tissue should be positioned to provide the greatest strength
- fast grown softwoods producing low-density wood because of the large thin-walled cells (tracheids)
- slow grown ring-porous hardwoods producing a large number of large vessels (pores) with a reduced number of strength providing fibres.

h) Effects of Fire

Timber, as we all know, is combustible. The importance of this is not that it burns, but for how long it retains its structural stability whilst it burns. It would be impossible to start a camp fire with just large heavy logs: it

Table 1.24 Examples of timber sections and their possible end use

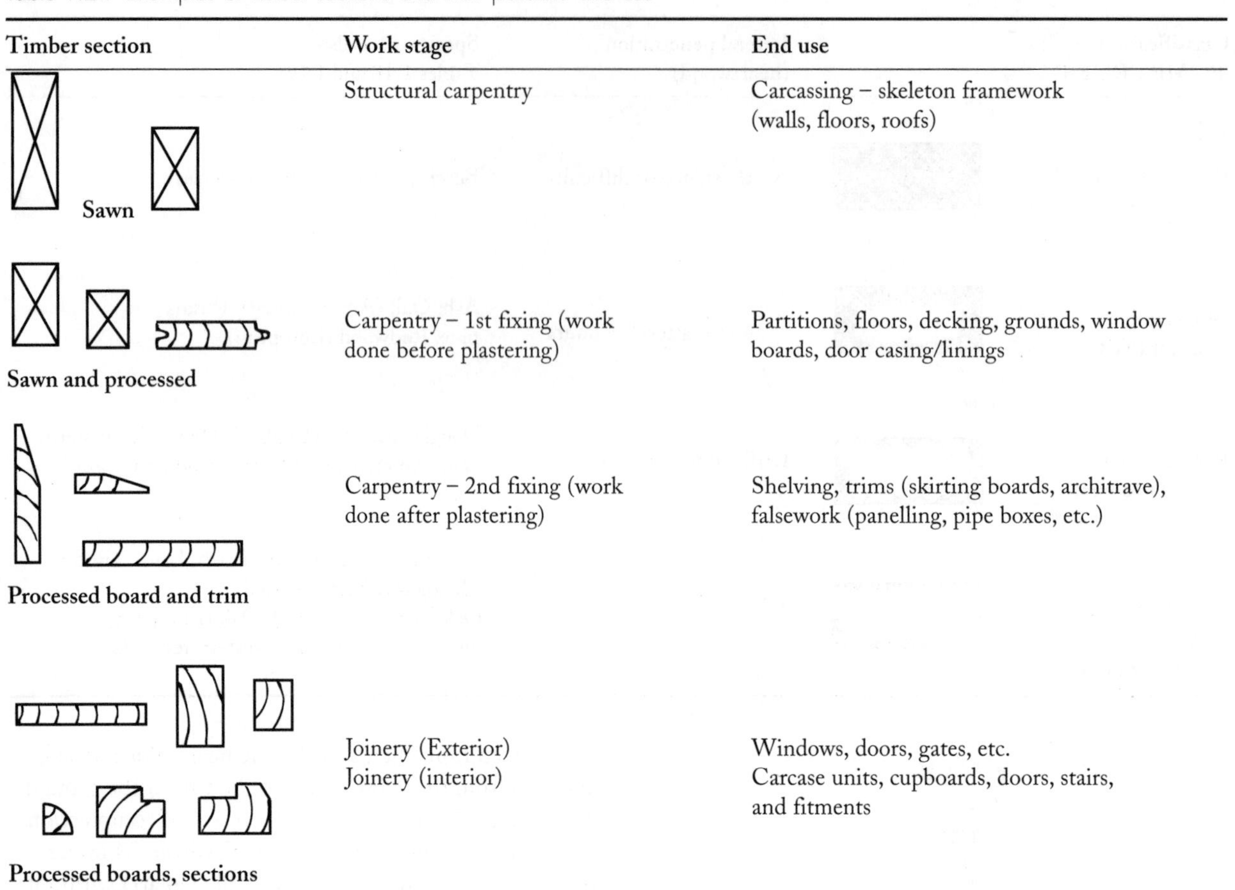

Timber section	Work stage	End use
Sawn	Structural carpentry	Carcassing – skeleton framework (walls, floors, roofs)
Sawn and processed	Carpentry – 1st fixing (work done before plastering)	Partitions, floors, decking, grounds, window boards, door casing/linings
Processed board and trim	Carpentry – 2nd fixing (work done after plastering)	Shelving, trims (skirting boards, architrave), falsework (panelling, pipe boxes, etc.)
Processed boards, sections and trims	Joinery (Exterior) Joinery (interior)	Windows, doors, gates, etc. Carcase units, cupboards, doors, stairs, and fitments

would be more sensible to start with dry twigs and then build up the size of material gradually until the fire has a good hold. Then, and only then, can the logs be added so that the fire will, provided there is enough draught and conditions are right, burn steadily for some time – probably hours without the need for replenishment. It can therefore be said that the rate at which timber burns must be related to:

- its sectional size
- its moisture content
- its density
- air supply.

Provided that an adequate section is used, timber can retain its strength in fire for longer than unprotected steel or aluminium, even though they are non-combustible.

One effect flaming has on timber is to form a charcoal coating over its surface (which is measurable and known as its *charring rate*, as shown in Figure 1.111). This acts as a heat insulator, thus slowing down the rate of combustion. The failure rate therefore depends on the cross-sectional area of the timber.

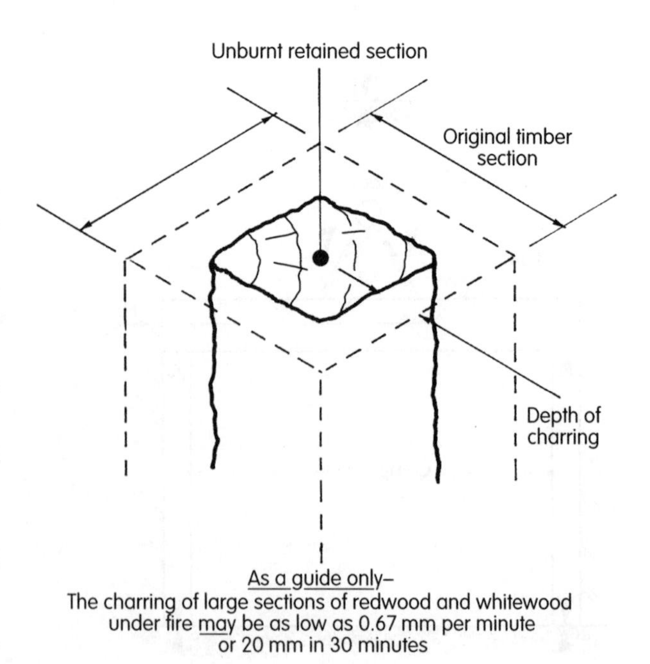

Figure 1.111 Possible rate of charring

j) Thermal Properties

The cellular structure of wood generally provides it with good thermal insulation qualities. Species which are light in weight (low density) can make very effective insulators where structural strength is not important. Examples of good practical thermal insulation and poor conductive properties can be found within wall components of timber framed buildings. These help to prevent heat loss from within and cold intrusion from without, or vice versa depending on the time of year. Another example is where wooden handles are used for metal cooking utensils.

k) Electrical Resistance

Although timber generally has good resisting properties against the flow of an electrical current, this resistance will vary with the timber's moisture content. Moisture is a good electrical conductor, and this is the basis on which moisture content is measured, when using a moisture meter (see Section 1.7.3).

l) General Suitability of Use

The last columns in Tables 1.18 and 1.19 give some guidance as to some end uses, or to situations where each species would possibly be best used, provided they were of the correct quality with regard to strength and/or appearance. There are a few species that can satisfy both these requirements but they are usually very expensive, a good example being English oak.

Table 1.24 should give you some idea of how we often relate a timber section to its end use, and you will notice the use of several trade terms which you may not be familiar with, such as:

- carcassing
- first fixing
- second fixing
- joinery (exterior)
- joinery (interior).

A simple explanation is given below.

Carcassing is a skeletal framework using rough sawn timber (unwrot carcassing timber) in preparation for cladding (covering). Examples include roofs, walls and floors. Carcassing timber is used in the construction of structural components.

First fixing is a stage in the construction of a building when items of carpentry and joinery are fixed in place before any plastering is carried out. Such work may include fixing decking to floors, fixing timber grounds in preparation for panelling etc., window boards, linings to door openings (a narrow lining may be called a 'casing'), or pipe boards onto which a plumber can fix pipework, etc.

Second fixing is work undertaken after plasterwork is complete, and may include such items as wall trims (architrave, skirting boards etc.) hanging doors, false-work such as panelling, and pipe boxes. Shelving and pre-made joinery items, such as cupboards, cabinets etc., would be fixed at this stage. NB Other trades such as plumbers and electricians also undertake their own first and second fixing procedures. Joinery (exterior) includes the making and, in some cases, fixing of items such as exterior doors, gates and windows. Joinery (interior) is the making, and, in some cases, the fixing of interior doors, stairs, cupboards and fitted furniture, etc. How to ensure we use the right timber for the right job will depend on all the factors we have already mentioned. Figure 1.112 shows the many and varied ways in which solid wood, wood veneers, or its lamination can be utilised. Bearing in mind all we have said and shown in Tables 1.18 and 1.19 you should have a clearer view about the many decisions a designer has to make before selecting a suitable species to carry out a specific function.

There are, however, two very important factors I have left until last: they are cost and availability, both of which can change year by year, or even by the month.

1.11 TIMBER STORAGE, HANDLING AND PROTECTION

As shown in Figure 1.113, softwoods will be dispatched from the supplier in one or a combination of three ways. However, as shown in Figure 1.114, hardwoods, with the exception of dimension stock, will be supplied in various widths. This is important when the wood is destined for decorative use, because then the end users can decide for themselves the best method of recutting the wood to suit grain pattern and colour, etc.

Initial handling on delivery may be at the yard or on site. Carcassing qualities of wood may well be stored within an outside shelter sited off the ground, which has been cleared of any ground vegetation and covered with a blinding of concrete, etc. Timber should be stacked off a hard standing with levelled timber bearers, or racked off the ground with side, and/or end access (Figures 1.115a and 1.115b). Racks of this type must be specifically designed to withstand heavy loads and, for reasons of safety, their height restricted to within stated safety limits. The type of stacking will generally be determined by turnover (moving timber on for use), moisture content, and end use.

Timber stacks will require intermediate sticks (stickers) to help stabilise the stacks (sticks help to tie the stacks together), and in some cases to encourage air circulation (see Section 1.7.6 and Figure 1.49).

Joinery quality timber and most hardwoods should be stored within a reasonably controlled environment (under full cover). Stacking in this case would possibly be from the floor via timber bearer, or purpose made racking similar to those shown in Figure 1.115a and

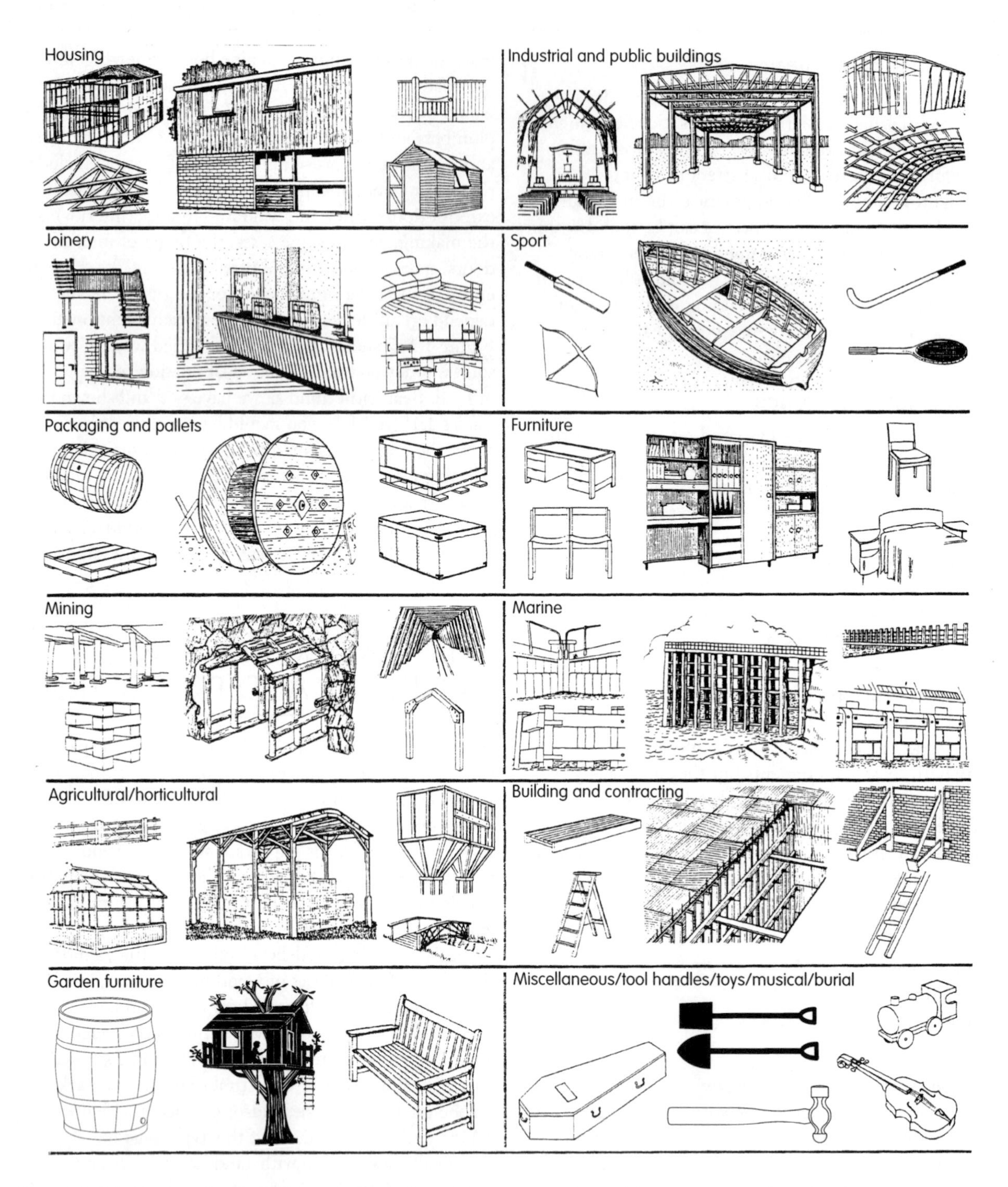

Figure 1.112 Some examples of how solid wood, wood veneers, and wood laminations can be utilised

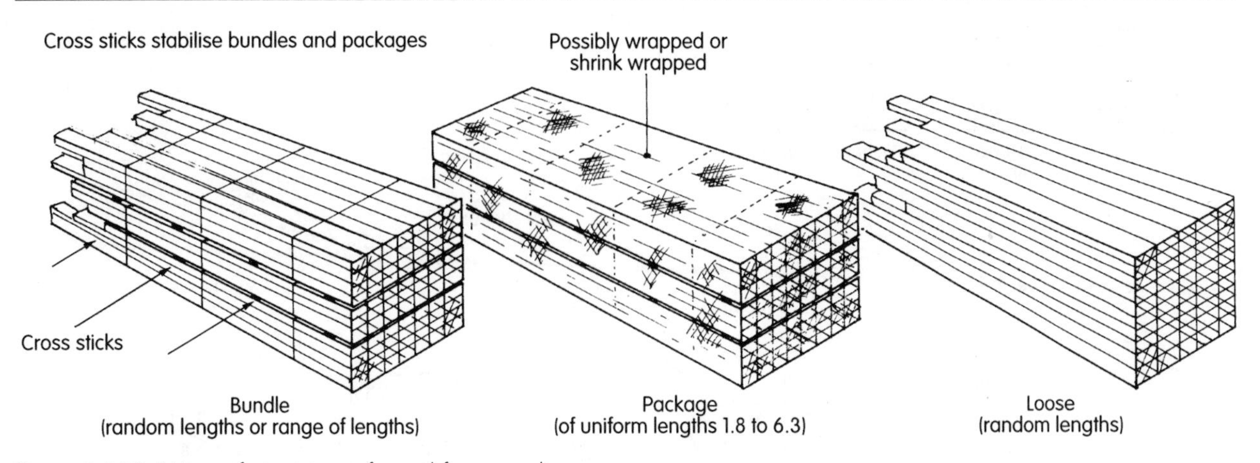

Figure 1.113 Ways of receiving softwood from suppliers

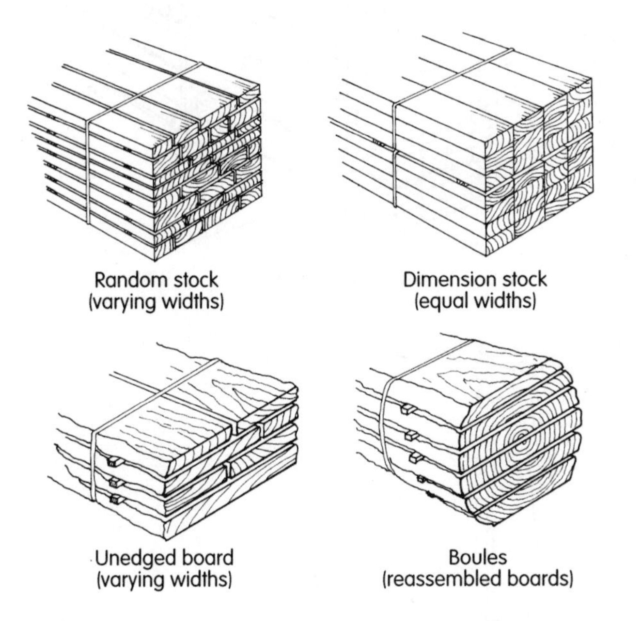

Figure 1.114 Ways of receiving hardwood from suppliers

1.115b. Where random selection of small lots is required, vertical stacking may be considered (Figure 1.115c). However, in this case, base stability and restraint must be established, together with a dedicated detachable guard rail (or chain) to prevent toppling when selection is made.

Before receiving carcassing timber on site (Figure 1.116) a clear space should be prepared well in advance, making sure the area is flat, free draining, cleared of ground vegetation, and with enough timber ground bearers at the ready.

These bearers should be of a section size large enough to sustain the load and of square section, or laid flat if rectangular to prevent toppling over under load or side pressure (Figure 1.116a).

Never use trestles to stack timber (Figure 1.116b) because joiner's trestles can collapse, and like metal trestles have a tendency to topple over when loading and dismantling a stack of timber.

Stacks should be covered to protect them from inclement weather. Special covers are available which reduce the problems of condensation – shower proof breather types may be adequate. Some form of cover anchorage will be needed to resist wind pressure. Tying off points should facilitate access to one end of the stack.

For joinery quality wood (including trims etc.), where possible try to find full under cover protection, such as a garage (Figure 1.116c). Providing good ventilation is very important if moisture pick up is to be avoided as a result of the surrounding wet trades fabric (floors and walls) drying out.

Pay special attention to the following points when receiving or stacking timber:

- always keep stacks clear of the ground and ground vegetation
- ensure that the stack is stable
- keep sticks vertically in line
- provide a marker rod (length of timber marked off at one metre (1 m) intervals) as a quick reference to timber lengths within the stack
- check quantities and qualities against delivery notes – never sign unless you are sure, or state on the note any discrepancies
- check moisture content against any requirements, or as thought necessary
- ensure provision has been made to cover the stack
- when manual lifting don't forget to bend your knees and not your back (keep your back straight)
- always keep well clear of any mechanical lifting operations
- if in any doubt always ask for help and for guidance from the person responsible for such operations.

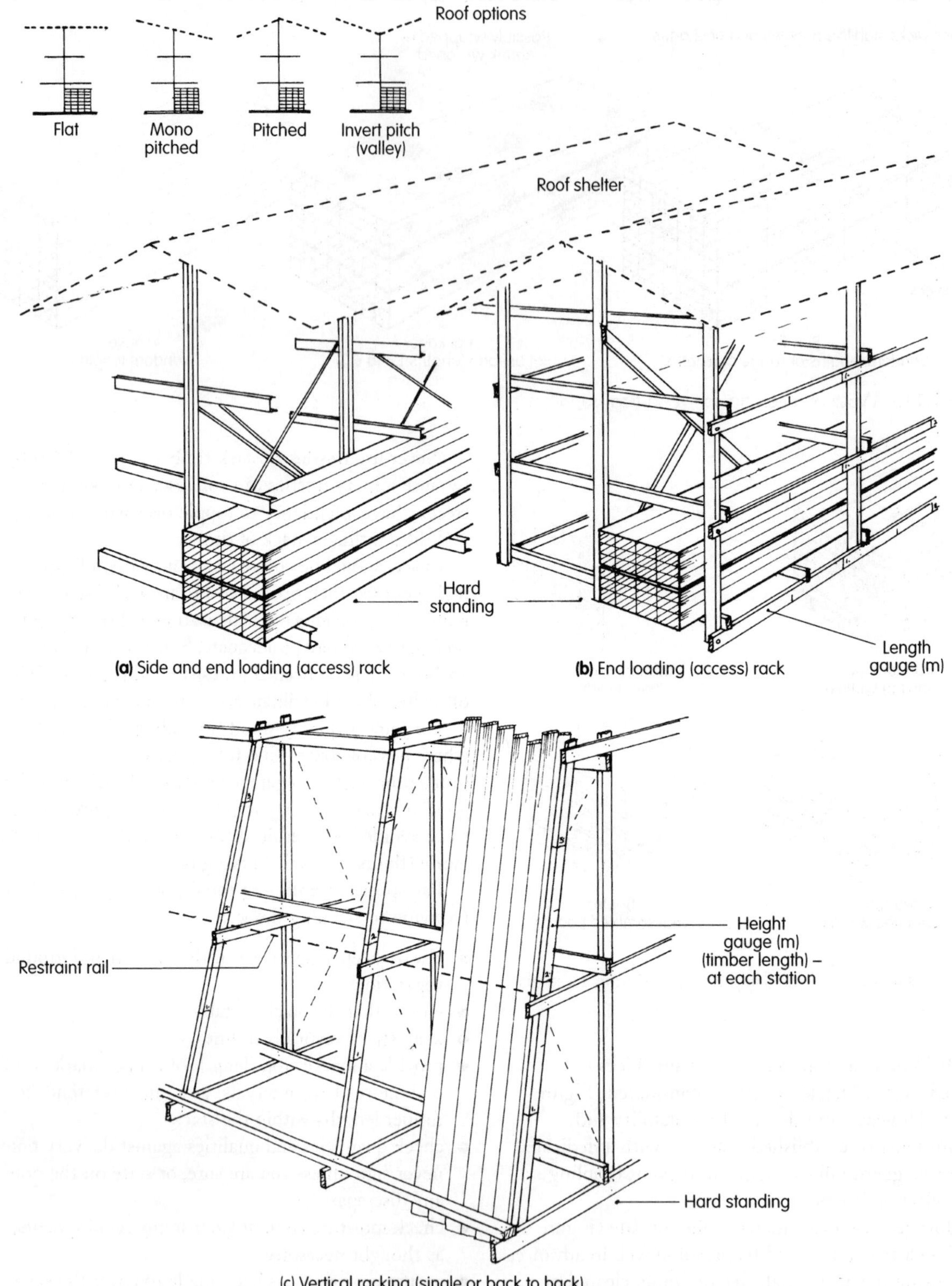

Figure 1.115 Yard racking of timber

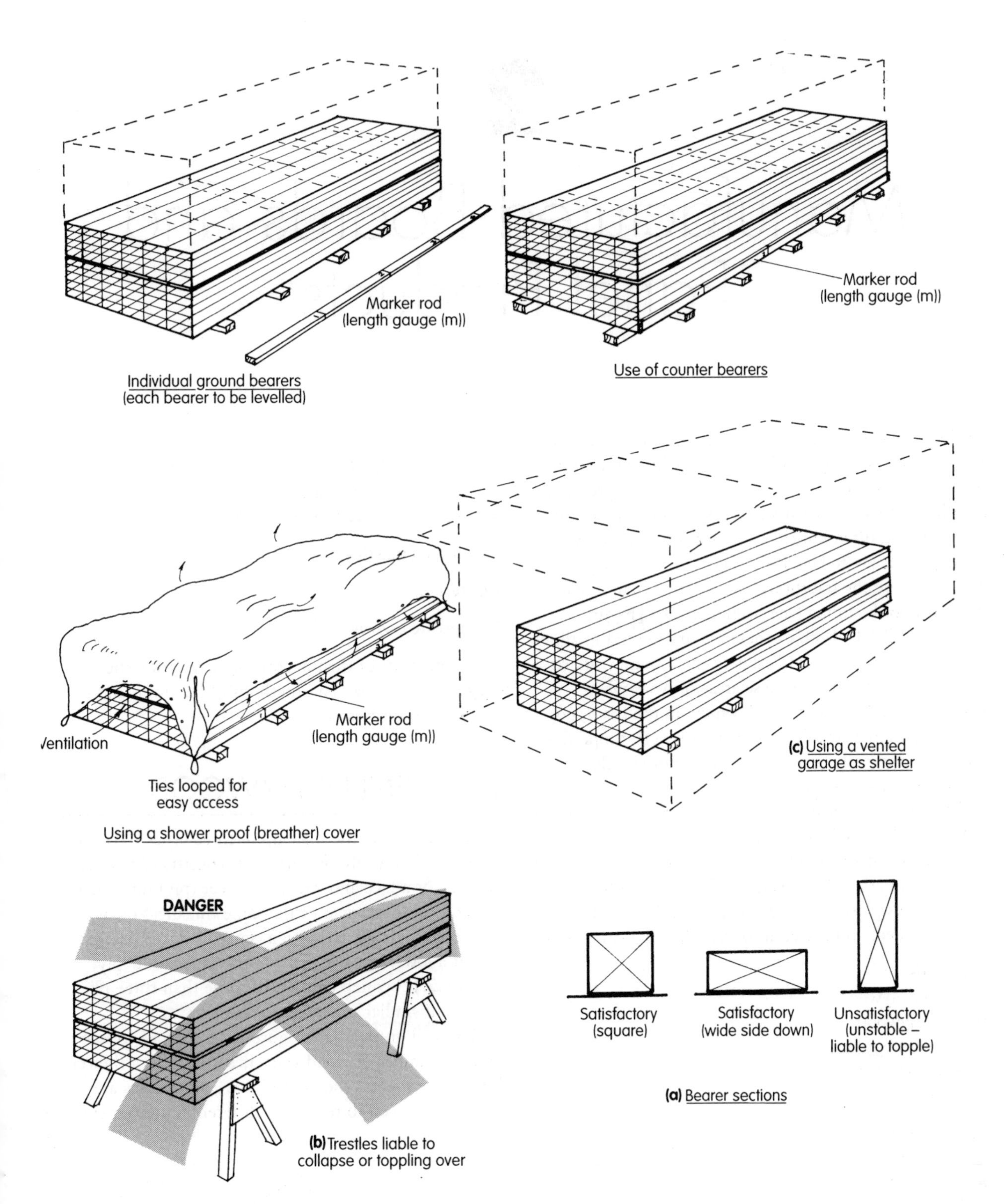

Figure 1.116 Site stacking of timber

2

Manufactured Boards and Panel Products

We have seen in Chapter 1 how timber can be subject to dimensional change and instability when used in its solid form (see Section 1.7.5). These inherent problems, together with solid board width restrictions due to tree size and cost, often limit the use of timber where wide or large areas are to be covered. Two very important exceptions are the traditional use of wood match boarding and floor boarding (see Section 1.10.4 [Figure 1.107] and 7.4.1). Both of these are dry jointed to allow for any movement, or infilled between a framework or moulded margin. It is in these areas of work where manufactured boards are used, which, in the main, possess dimensional stability and equal strength properties both longitudinally (lengthways) and transversely (across the board or sheet).

Manufactured boards used by the carpenter and joiner have traditionally been composed of either wood veneer (thin sheets of solid wood), strips, particles, or their combination. However, in this chapter other materials, such as plastics, gypsum plaster, and cement have been included, and placed into the following categories:

Plywoods
- veneer plywood
- core plywood
  - laminboard
  - blockboard
  - battenboard

Particle boards
- wood chipboard
- flaxboard
- bagasse board
- wood-cement board

Flake boards
- waferboard
- oriented strand board

Fibre boards
- insulation board
- medium board
- hardboard
  - standard hardboard
  - tempered hardboard
- medium density fibreboard (MDF)

Laminated plastics

Plasterboard

Composite boards

Their methods of manufacture, general properties, usage, available size and methods of handling and storage are also discussed.

2.1 VENEER PLYWOOD

Figure 2.1 shows the makeup of three types of veneer plywood. Three-ply (Figure 2.1a) consists of two face veneers, or a face and backing veneer (ply) sandwich of either a central veneer of the same thickness or, as shown in Figure 2.1b, with a central thicker veneer, usually of a lower density. The third type is multi-ply (Figure 2.1c) which consists of more than three plies, usually of similar thickness.

You will notice that (Figure 2.2a) they all consist of an odd number of wood veneers so that, from the centre line of the sheet – the neutral axis – both sides are equally balanced to resist any differential movement (see

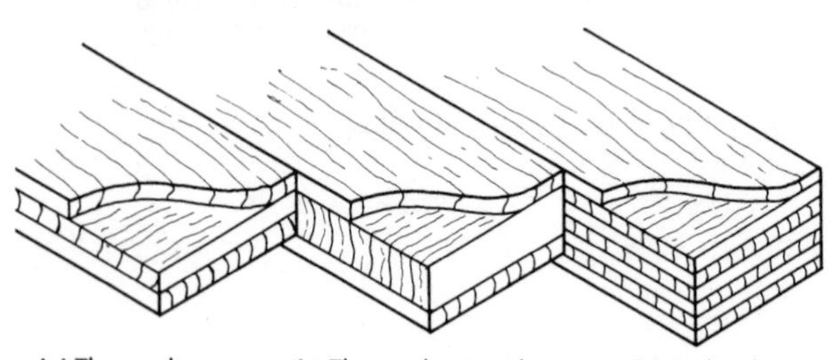

Figure 2.1 Veneer plywood (odd number of veneers)

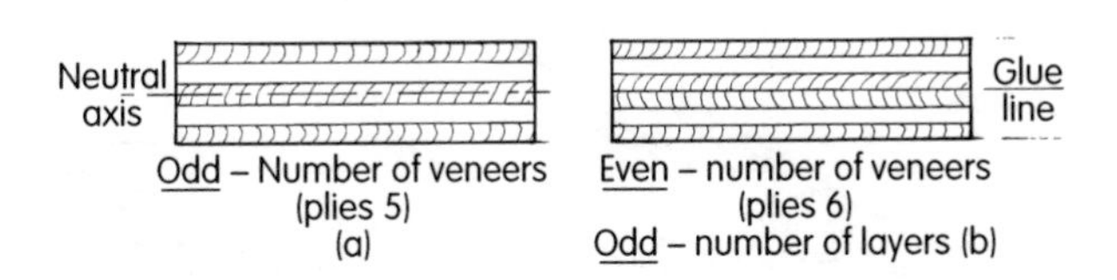

Figure 2.2 Number of plywood layers in relation to wood veneers

Section 1.7.5), which may otherwise occur should an extra layer be added to one of the faces.

However, as shown in Figure 2.2b, you may find that a glue line forms the neutral axis, in which case there would be, in total, an even number of veneers (plies). The important point here is that the veneers either side of the glue line have their grain running parallel to one another, thereby still maintaining a veneer balance to the sheet as a whole, because this central veneer lamination will, as shown, act as a single layer.

2.1.1 Manufacturing and Bonding

Figure 2.3 should give you a good idea of the various operations involved, but the following sequences and procedures may vary between manufacturers.

1 Log Selection

Only certain species of tree are generally used for plywood veneer – those produced by peeling would be required to have straight boles. Softwood species may include: Douglas fir, Western hemlock, true firs, spruces, and pines. Hardwoods used may include: birch, lauan, meranti, and mahogany. The types used usually reflect the country of manufacture, e.g.

- *Finland*: birch, spruce, and pine
- *Canada (COFI)*: Douglas fir, Western hemlock, Western white spruce, Lodgepole pine
- *USA (APA)*: Douglas fir, Southern pine (plus an extensive range of both hardwoods and softwoods)
- *Far East*: tropical hardwoods.

2 Conditioning the Log

After de-barking and cutting to length, the logs are conditioned to a high moisture content either by soaking in hot or cold water, or steaming (depending on species and procedures). This process helps to make the operation of log peeling easier and smoother by softening wood tissue and thereby reducing the risk of surface defects (tearing of grain).

3 Peeling the Log

Pre-cut log lengths are fixed within a log lathe (veneer peeler) and rotated against a full length lathe (knife) to produce a veneer of uniform thickness. This operation could be likened to unrolling a toilet roll.

4 Veneer Cutting (Clipping)

Depending on plywood manufacturing methods, veneers of uniform thickness may be allowed to emerge as a continuous band or cut to a predetermined length, or required size.

5 Veneer Drying

Veneers may be dried in batches of cut veneers or by a continuous process. With either of these, machine drying would achieve about 4–8% moisture content (MC) to facilitate the required adhesive bond and stabilisation of the finished plywood.

6 Veneer Joining/Repairs

Edge to edge joining to desired widths may be by edge glueing, stitching, or the wood may be taped with a perforated adhesive paper strip. Small defects such as knots may be cut out and plugged with good veneer.

7 Veneer Grading

After drying and any restoration work veneers are graded by inspection to classify them into their respective facing categories, and those that are only suited as core veneers.

8 Glueing Stations

Veneers are assembled cross-wise in batches (generally of odd numbers), known as their lay-up.

This composition of veneer lay-up may be either:

- all hardwood (e.g. birch)
- all softwood (e.g. Douglas fir)
- *Combi* * – (e.g. two birch veneers on each face and alternating inner veneers of softwood and birch)
- *Combi-mirror** – (e.g. one birch veneer on the outer faces and alternating inner veneers of softwood and birch)
- *Twin** – (softwood veneers, faced with a birch veneer on each face).
  * These are the terms used by 'Fin Ply'.

The types of synthetic resin adhesive (see Table 2.1) used and the method of application, by roller, spraying or curtain coating, will vary depending on the method of manufacture and type of plywood.

9 Pre-pressing

The lay-up of veneers is pre-pressed to improve bonding and to facilitate easy handling when being inserted into a multi-pressing system.

10 Final Pressing

The pre-pressed lay-up is inserted into a hydraulic multi-open layered press to hold and clamp the sheets for a prescribed curing period. Temperatures and pressure are strictly controlled.

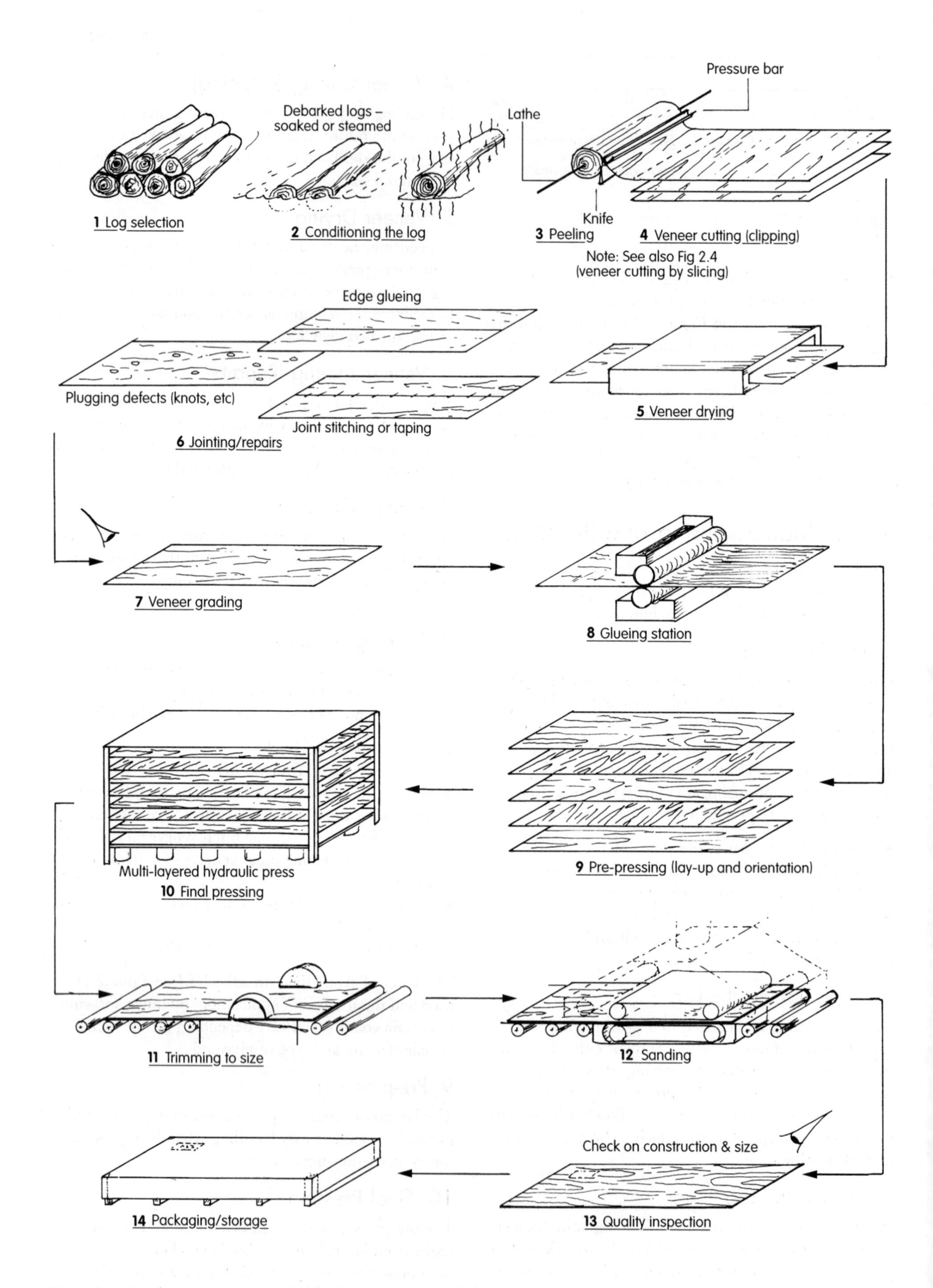

Figure 2.3 Guide to processing wood for the manufacture of plywood

11 Trimming to Size

After a set standing time for cooling the sheets are trimmed on all four edges to their final dimensions.

12 Sanding

Both faces are sanded to produce a very accurate plywood thickness. Final defect filling or patching may be carried out just prior to sanding.

Note: Some grades of plywood are not sanded, and are graded and sold as unsanded.

13 Quality Inspection

The final inspection will check on proper quality (grade) face veneers, general construction, surface dimensions and thickness, before the plywood is labelled and/or edge marked accordingly with its final grade (see Section 2.1.4).

14 Packaging and Storage

The type of packaging will depend on the quality of the product. For example, sheets may be packed on non-returnable pallets, or in crated bundles, where they should remain until required for use. Flat, firm, fully supported bases should be provided with adequate protection against moisture intake.

2.1.2 Surface Finishes

The surface finishes will depend on the end use. Plywood might be used for roofing, walls, formwork or to provide a decorative or hygiene surface.

There are many different finishes available, and a few examples are given below:

1 Natural – this is a sanded, or unsanded self finish of rotary cut face veneer.
2 Natural decorative wood veneer – thin slices of wood are cut from decorative woods by methods chosen to bring out the best features of the wood. Figure 2.4 shows a flat slicing method used to make tangential cuts (flame figure is shown). Wood commonly used as decorative plywood veneers includes oak, ash, and sapele.

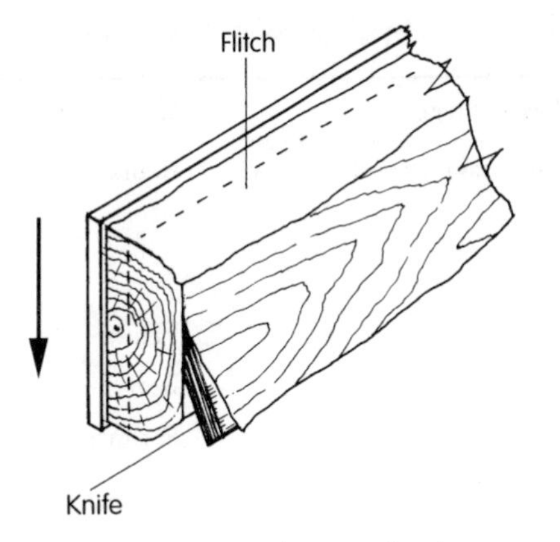

Figure 2.4 Producing a wood veneer by slicing

3 Phenolic resin film faced – the surface consists of a paper impregnated with a phenolic resin. It is suitable as sheeting in formwork.
4 Painted overlay – this is surfaced with phenol impregnated paint based film.
5 Melamine resin faced – this has a coating of melamine resin of various decorative colours.
6 Glass-fibre reinforced surfaces – these produce a smooth or textured decorative surface.
7 Metal faced – here, one or both faces are covered with a veneer of aluminium, steel sheet or foils.
8 Pre-painted – this is treated with either a primer, paint, or varnish.

2.1.3 Plywood by Type

Plywoods can fall into several groups, which usually tend to reflect their end use. Here are a few examples:

1 Marine plywoods – as the name implies these plywoods would be required to have durable wood veneer and bonds (adhesive). Wood species should be classified as at least moderately durable (see Section 1.10.4e) and their bond should be classified to a WBP standard (see Table 2.1).
2 Structural plywood – strength properties are important here, as this plywood could be expected to withstand varying degrees of load and pressure when

Table 2.1 Adhesives used in the manufacture of plywood

Adhesive classification (UK)	Adhesive type (bonding agent)	Plywood use
Interior use only (INT)	Urea formaldehyde (UF)*	Not to be subjected to damp conditions
Moisture resistant (MR)	Urea formaldehyde (UF)†	General interior usage with provision for limited period of exposure to damp conditions
Cyclical boil resistant (CBR)	Melamine urea formaldehyde (MUF)	May withstand extreme weather conditions – exterior with protection
Weather and boil proof (WBP)	Phenol formaldehyde (PF)	Exterior quality giving full protection from all conditions of exposure

Note: * Weak formulation; † strong formulation

Table 2.2 Plywood type against general usage

	Plywood type	General usage	
(a)	Structural plywood	Roofs	Stressed skin panels Sarking (sheathing) to pitched roofs Decking flat roofs Trusses (webs)
		Floors	Decking
		Beams	Box beams and core beams
		Formwork	Decking and sheeting
		Walls	Wall panel sheathing (timber framed buildings) hoardings
(b)	Non-structural plywood	Joinery	Doors, stair risers, panelling, fitments, soffits, fascia boards, shelves
		Shopfitting	Wall cladding, signs, display boards, display cabinets/units
		Furniture	Tables, chairs, cabinets/units
		Toys	

put into situations such as those listed in Table 2.2a. Most of these grades of plywood are manufactured to meet specific requirements and would be graded and classified accordingly.

3 Non-structural plywoods (Table 2.2b) – face veneers are more likely to be of interest here for joinery and furniture, although these plywoods may have some exterior use. Grading of face veneers differs from country to country with no direct comparison as methods of coding, be it by number or letter, frequently differ. However, the range and some scales have been included in Tables 2.3–2.6.

4 Exterior plywoods – unlike interior types, which require only a minimum of non-durable wood veneer, and a bond of a weak formulation of UF adhesive, full exterior grades will require durable wood species (or equivalently treated ones) as shown in Table 2.7, together with a phenol formaldehyde adhesive (Table 2.1).

5 Coated plywoods – as described in Section 2.1.2 – either have a decorative or functional surface, depending on their intended end use, be it a wood veneer, special surface coating, or overlay of fabric, plastics, or metal.

Table 2.3 Examples of how the Council of Forest Industries (Canada) [COFI] grade their exterior plywood for use

General plywood grade (code) or product code	Codes for veneer grades			Finish	Examples of use
	Face	Inner	Back		
Good two sides (G2S)	A	C	A	Sanded	Furniture, shelving, formwork, etc.
Good one side (G1S)	A	C	C	Sanded	Fitments, falsework, formwork, etc.
Select-tight face (SELTF)	B*	C	C	Unsanded	General construction – hoarding
Select (SELECT)	B	C	C	Unsanded	Boards, etc.
Sheathing (SHG)	C	C	C	Unsanded	General construction – sheathing to walls, roofs, and hoarding.
COFI FORM and COFI FORM Plus	A A B C	C C C C	A C C C	Sanded and unsanded	Formwork
COFI: ROOF (TM)	B C	C C	C C	Special edge profile	Roof sheathing and decking
COFI: FLOOR – tongued and grooved (T&G)	B C	C C	C C	Special edge profile	Floors and roof sheathing

Note: * Permissible surface openings filled; A – highest veneer grade; B and C – lowest veneer grades

Table 2.4 Examples of how the American Plywood Association (APA) grades its plywood for use

General plywood grade (code)	Codes for veneer grades			Bond	Finish	Examples of use
	Face	Inner	Back			
A – B Two good surfaces – one less important	A A	C D	B B	Ext Exp 1	Sanded	Furniture, shelving, cabinets signs, etc.
B – B Two good surfaces	B B	C D	B B	Ext Exp 1	Sanded	General purpose – formwork etc.,
C – C Plugged	C*	C	C	Ext	Unsanded	Formwork, decking, etc.
Sturd-i-floor	C* C*	C&D C	D C	Exp 1 Ext	Touch† sanded edges square or T&G	Decking (flooring)
High density overlay (HDO)	A or B	C*	A or B	Ext	Phenolic film faced	Formwork, signs,counter tops, etc.

Note: * Plugged – veneer repaired with small insets (various shapes) of wood or synthetic material;
† Touch sanded – a light surface sanding operation; Ext, exterior – *fully waterproof bond* (WBP);
Exp1, exposure 1 – *highly moisture resistant* (WBP); A – highest veneer grade; D – lowest veneer grade

2.1.4 Grading Plywood

The term 'grade' or 'classification' can be very confusing. So far, we have discussed how the type of adhesive used to bond the veneers together can strengthen plywood against exposure to the elements (weather), and how the durability of the wood veneer can likewise affect plywood exposure classification (Table 2.7). We have also considered the strength (structural) properties of plywood, not to mention the various coatings or films available to make plywood suitable for special purposes. All these factors will contribute to the general classification of the plywood as a whole.

However, plywood grades most commonly refer to the condition and quality of the veneers (plies) used in their construction. Unfortunately, there appears to be no common international coding for these, since quality can be signified by either a letter, a number, or their combination: for example, best quality veneers could be coded 'A' or 'E', or 'I' or '1' whereas worst quality could be 'WG' or 'D', or 'IV' or '3'.

To give you some idea how this operates take a look at Table 2.3 which shows how the Council of Forest Industries (COFI) of Canada grades exterior plywood against veneer grades, and, similarly, Table 2.4 which shows a few examples of how the American Plywood Association (APA) grades its plywood.

Finnish plywood coding for veneers is totally different from the two already mentioned. Table 2.5 lists these and gives a general interpretation of their meaning. There are numerous possible combinations for face and back veneers, as shown in Table 2.6. All Finnish

Table 2.5 Examples of Finnish plywood face veneer coding and possible surface finishes

Wood	Code for face veneer grade	Brief description of veneer surface suitability
Birch	A	Virtually defect free – highest quality veneer for varnishing
	B	Very good surface for paint or varnish finishes
	S	Suitable for a good paint finish
	BB	Used for standard interior paint finishes
	WG	Used where surface appearance is not an important factor
Conifer	E	Virtually defect free – highest quality of veneer for varnishing
	I	Very good surface for paint or varnish finishes
	II	Used where an ordinary painted surface is required
	III	Used as a general painting surface where quality finishes are not so important
	IV	Used where surface appearance is not an important factor

Table 2.6 Finnish plywood face and back veneer combinations

Birch faced plywood				
A/A	B/B	S/S	BB/BB	WG/WG
A/B	B/S	S/BB	BBI/WG	
A/S	B/BB	S/WG		
A/BB	B/WG			
A/WG				
Conifer faced plywood				
E/E	I/I	II/II	III/III	IV/IV
E/I	I/II	II/III	III/IV	
E/II	I/III	II/IV		
E/III	I/IV			
E/IV				

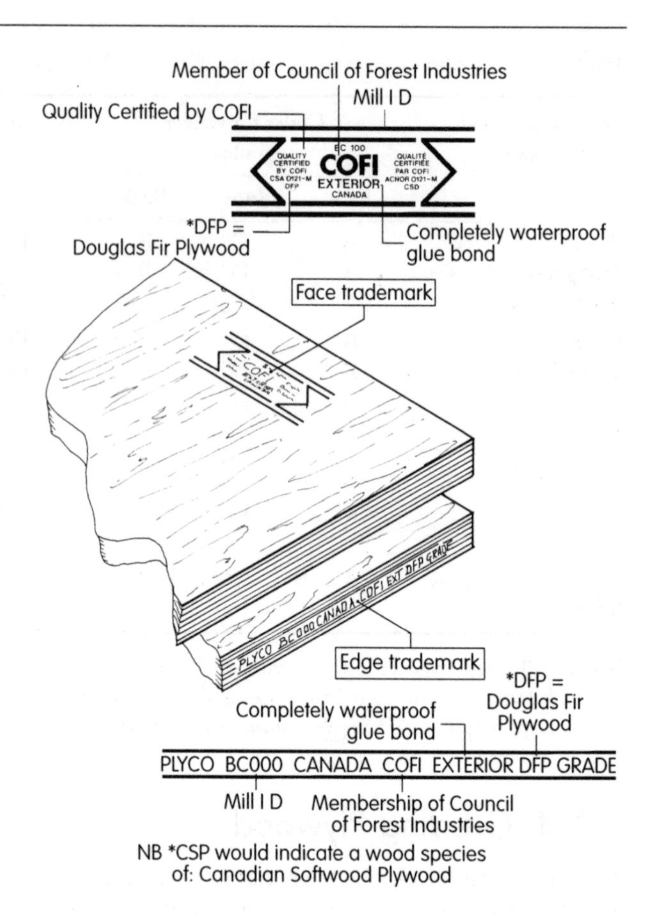

Figure 2.5 Examples of face and edge labelling COFI exterior plywood

exterior grade plywood is bonded with weather and boil-proof adhesive (WBP) in the form of a phenol resin adhesive. It is constructed to meet, depending on the composition of the plywood, various applications, including structural requirements.

NB: Plywood manufactured to BS 5666 Part 6, 1985 for general usage has a standardised comparison classification for surface grades, to which you should refer.

2.1.5 Labelling Plywood

Whether plywood is suitable for a particular task may simply be a question of its surface finish. More frequently it will be a specific requirement of a design or building specification. Fully labelled sheets help you and the supplier to meet these requirements.

Labelling usually consists of a series of coded letters and/or numbers printed on to the surface and/or edge of the sheet. These marks are made by using either a dye coated stamp, an ink-jet spray or a pre-printed stick-on label. Face marks can interfere with the surface finish on sanded sheets, in which case edge marks may be more appropriate.

Figures 2.5–2.7 will give you some idea of the information you can expect to see on these labels.

Plywood labelled under BS 6566: Part: 1,1985 should, as indicated in Figure 2.8, include:

- a British Standards number
- the panel thickness
- an appearance grading
- its bond performance
- its durability classification (if it is treated, details of the method used should be included).

Table 2.7 Plywood durability classification*

Code	Condition
G	Used for general purposes – not requiring a durability classification
E	Highly resistant to wood boring insects (other than termites) – not resistant to decay
M	Moderately resistant or better to decay, but not necessarily wood boring insects such as lyctus and termites etc., unless treated with a suitable wood preservative
H	Highly resistant to decay, but not necessarily wood boring insects such as lyctus, and termites etc., unless treated with a suitable preservative

Note: * Taken from BS 6566: Part: 7/1985

2.1.6 Specifying Veneer Plywood

Table 2.2 will enable you to identify general use against structural and non-suitable locations, and Tables 2.3 and 2.4 set use against general and veneer grades, and surface finishes.

As already mentioned, the uses of veneer plywood are many and varied, including

- formwork
- hoarding

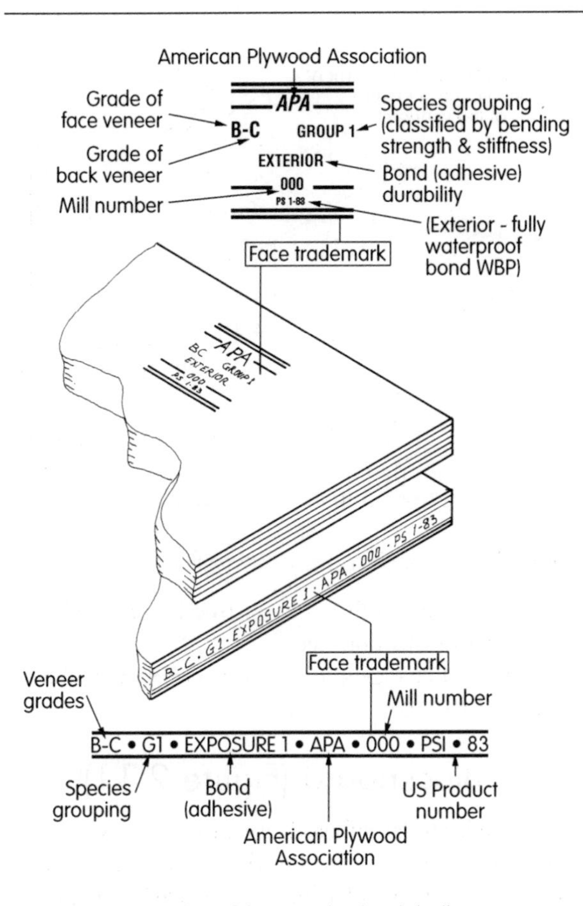

Figure 2.6 Examples of face and edge labelling APA plywood

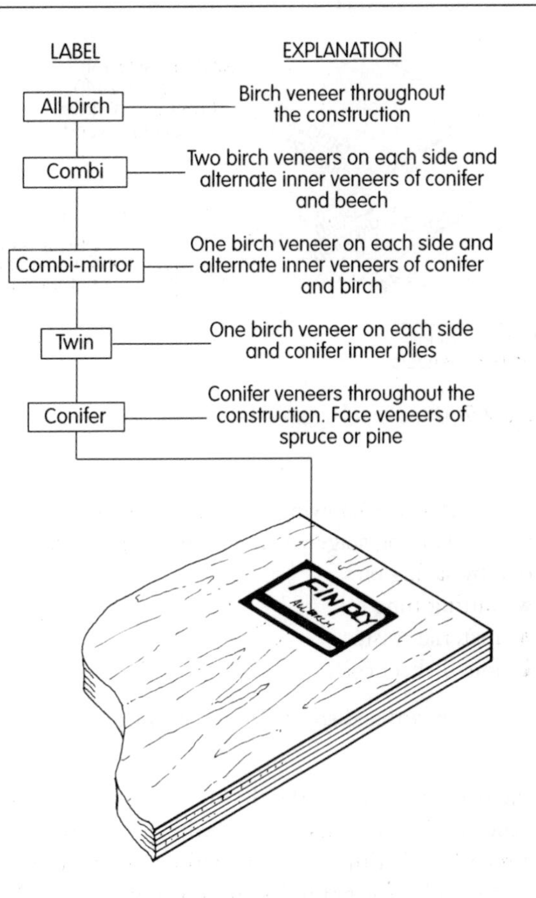

Figure 2.7 Labelling to show veneer composition

- decking roofs
- decking floors
- sheathing (covering timber formwork, such as the wall panels of timber framed houses)
- exterior cladding
- wall panelling
- flush doors
- shelving.

These are just a few of the uses; others include work associated with shopfitting and the furniture industry, as well as materials handling (the construction of boxes, bins, pallets, etc.).

No matter which area of work you are concerned with, the choice of veneer plywood will depend on one or more of the following requirements:

1 its composition, i.e. the species, grade, bond and type (marine, structural, non-structural) (see Section 2.1.3)
2 the veneer appearance (face and back grades) and whether it is sanded or unsanded
3 the surface finish and coating (see Section 2.1.2)
4 the surface grain direction:
   - cross-grain plywood (Figure 2.10a)
   - long-grain plywood (Figure 2.10b).

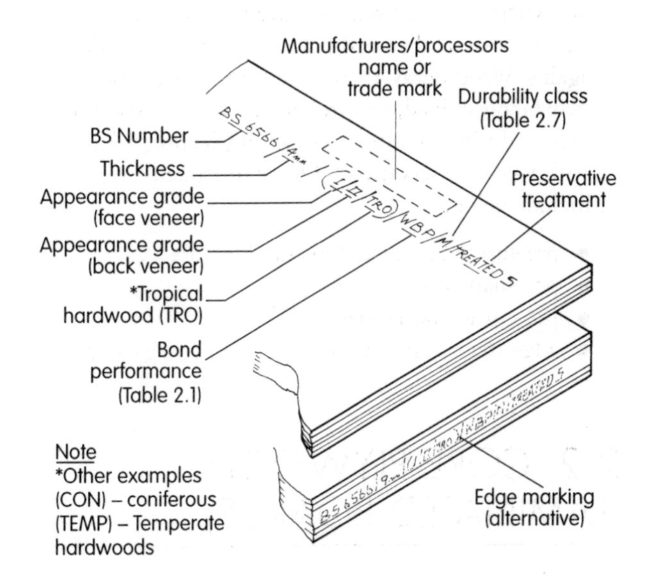

Figure 2.8 Examples of how identification markings appear on the back face or edge when conforming to BS 6566: Part: 1, 1985

(When specifying the face dimensions of plywood it is usual, as shown in Figure 2.9 for the direction of the face veneer grain to run parallel to the direction of the first stated dimension.)

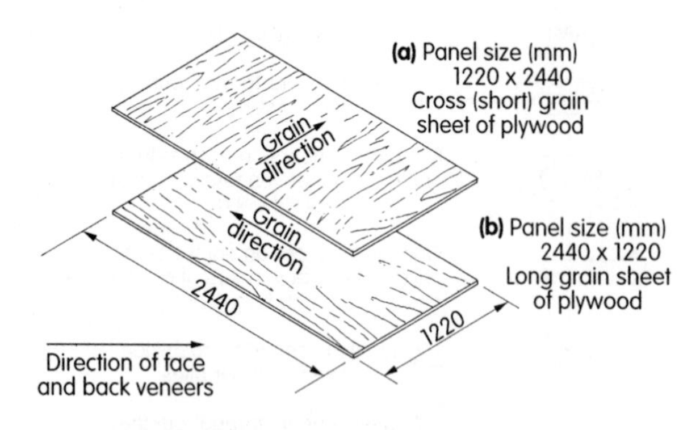

Figure 2.9 Direction of grain in relation to a stated sheet size

5 the sheet size – knowing size variations can help you at the planning stage when designing and setting-out, by saving on:
   - cutting time
   - material waste
   - initial sheet cost.

(Common board and sheet sizes are shown in Section 2.10.)

6 the panel thickness – the thickness and number of veneers will reflect its end use, with regard to strength and stiffness. Combinations are many and varied and a few examples are shown in Section 2.10.
7 the use of special treatments – we have already discussed surface finishes and coatings in Section 2.1.2, but where plywood veneers have to meet certain classes of durability, and fail to meet these standards against wood decay, then a preservative treatment may be prescribed; this treatment can be carried out before, during, or after fabrication (Section 4.3 deals with wood preservatives). Other treatments may include being:
   - pre-formed as an architectural feature
   - pre-shaped by machine
   - pre-cut to special sizes
   - edge treated, for example tongued and grooved.

2.2 CORE PLYWOOD (LAMINATED BOARDS)

Core plywood consists of wood strips contained by one or more veneers on both sides. It is the width of these strips that gives the board its name.

2.2.1 Laminboard (Figure 2.10)

This has a core made up of a glued lamination of narrow wood strips, not exceeding 7 mm in width. This lamination is faced with two wood veneers to each side,

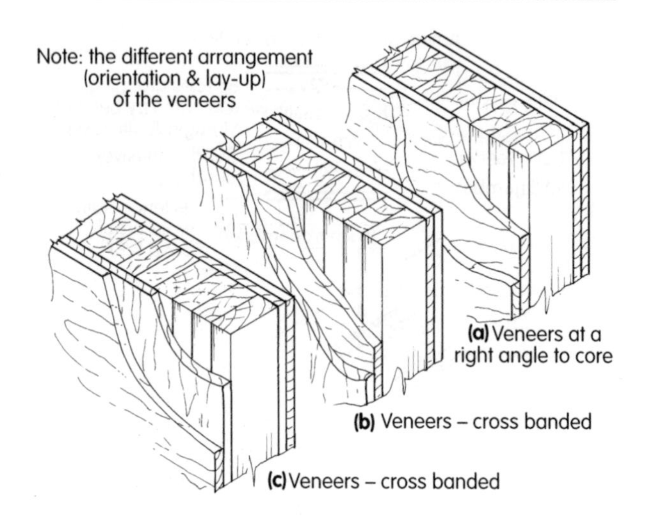

Figure 2.10 Laminboard, 5-ply construction

producing a board of five-ply construction. Laminboard can be almost distortion free.

Methods of laying-up the veneers can vary, as shown in Figure 2.10.

2.2.2 Blockboard (Figure 2.11)

This is similar to laminboard, except that the wood strips which form the core are wider, usually between 19 mm and 30 mm wide. Facing veneers may be singular, as shown in Figure 2.11a, to produce a blockboard of three-ply construction, or double to form five-ply construction as shown in Figure 2.11b, with the face veneers at right angles to the core grain, or parallel to the core grain, as shown in Figure 2.11c.

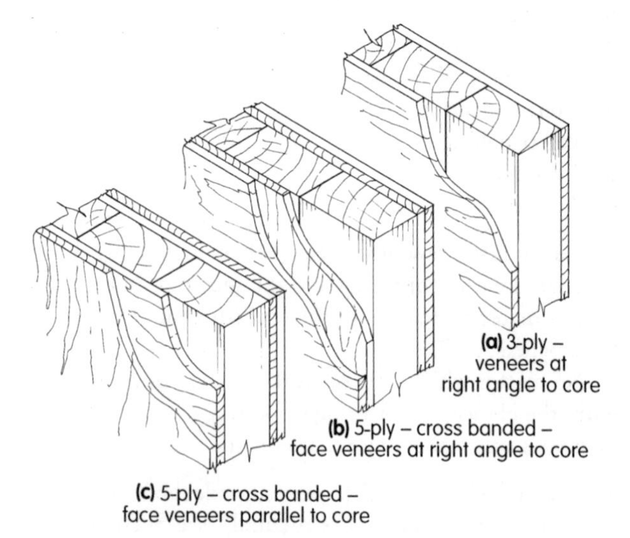

Figure 2.11 Blockboard – 3- and 5-ply construction

2.2.3 Battenboard

Although it is now regarded as obsolete in the UK you may come across battenboard during the course of your work. Battenboard is constructed in a similar manner to blockboard, the main difference being that its wood strips are much wider. Any moisture movement of these, particularly those tangentially sawn (which are more liable to 'cupping') would result in the board having a 'rippled' appearance over its otherwise smooth surface.

2.2.4 Manufacturing and Bonding

The wood strip cores of blockboard are produced from dried boards, previously cut from logs, or redeemed from log off-cuts etc., whereas the cores of laminboard are cut from thick veneers (not thicker than 7 mm).

The strips that form the core of laminboard and some blockboard are first glued together to form a laminated slab. These are then clad on both sides with veneer to make up either three- or five-ply boards.

Laminboard and blockboard are generally bonded with a urea-formaldehyde (UF) synthetic adhesive and classified for interior use – although in some cases an exterior adhesive of phenolic formaldehyde may be used.

2.2.5 General Usage

The uses of laminboard and blockboard as a panel product are similar to those of veneer plywood, the main differences being in their bonding and durability (see Table 2.2b). Face veneers can be of similar appearance and graded for decorative purposes. Board thicknesses do, however, tend to start at about 12 mm and finish at about 25 mm. A range of sizes is given in Section 2.10.

2.3 CHIPBOARD

Chipboard is often referred to as 'particle board', which is a term associated with any of the manufactured boards made from particles of wood, such as :

- chipboard
- flake board
- fibreboard.

The main natural ingredients of chipboard are wood particles derived from softwoods such as spruce, pines, firs and hardwoods, such as birch, in the form of :

- immature trees specially grown as chipping material
- forest thinnings
- residue from saw mills
- wood machining waste (chippings etc.).

Other sources of raw material may be used, such as shives (slivers) from the flax plant (used in the linen fabric industry) to produce a board known as 'flaxboard'. The fibrous nature of sugar cane can also be utilised once the sugar content is removed to produce a particle board known as 'Bagasse board'.

2.3.1 Manufacture (Figure 2.12)

The manufacturing process can differ with the type of board and raw material used. The following list should therefore be used only as a general guide to enable you to understand more fully how the make-up of the finished board type described in Section 2.3.2 can affect its practical application (cutting, machining and finishing etc.) and end use.

1 Raw material – this is received and selected for use.
2 Chip production – the wood is cut and prepared by chipping or grinding machine.
3 Drying process – the moisture content of the woody particles is reduced to about 2–5% MC in special dryers.
4 Grading particles – particles are graded into sizes to suit various types of board or surface finish (see Figure 2.13) and stored.
5 Blending with adhesive (binder) – particles are blended (coated) with a synthetic adhesive; a urea formaldehyde (UF) resin is generally used together with various additives. Improved moisture resistant boards would be bonded with a melamine urea formaldehyde (MUR) resin.

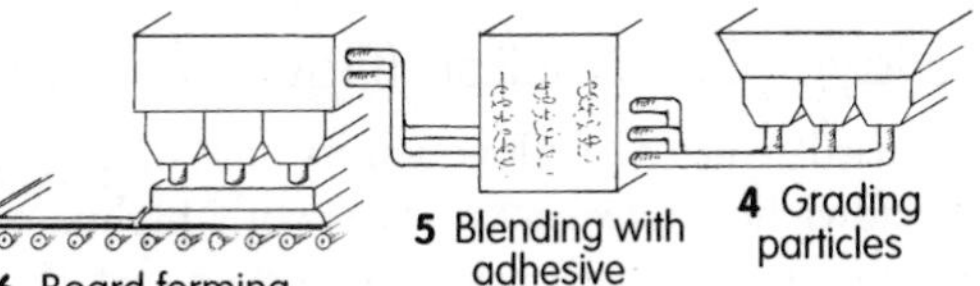

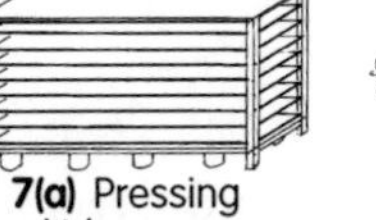

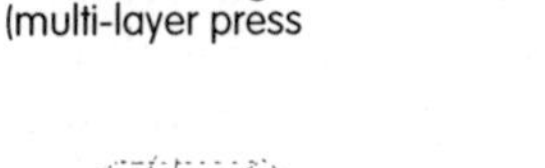

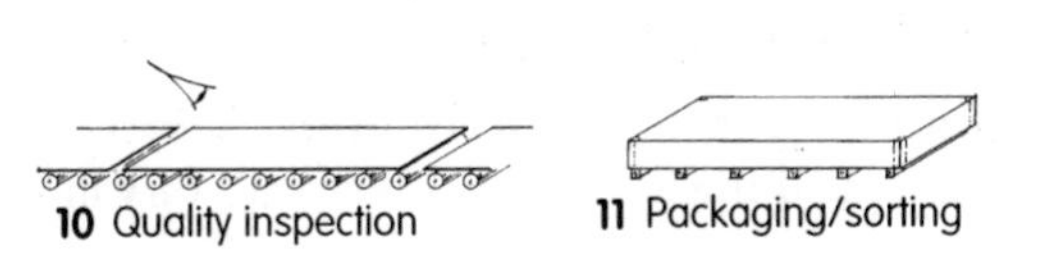

Figure 2.12 Guide to the processing of chipboard

6 Board forming – pre-coated particles are laid up into mats. The distribution of the particles will depend on the required structure of the finished board (Figure 2.13). These mats may be pre-compressed before final pressing.
7 Pressing – mats are compressed to their required thickness and the adhesive cured by heat. Pressing may be a multi-layered operation or a continuous process.
8/9 Trimming and sanding – like plywood, boards will need to be accurately cut square and to size, and finished by sanding.
10/11 Final stages – after checking on quality, various methods of packaging may be used to facilitate ease of transport and storage.

2.3.2 Types of Board Construction

The construction of pressed boards may be of one of four categories:

- single-layer construction
- three-layer construction
- multi-layer construction
- graded density.

Single-layer Construction (Figure 2.13a)

This is a uniform mass of particles of either wood or flax. The type, grade and compacting of these particles will affect the board's strength and working properties. As far as use is concerned, the bond may classify the board as interior structural. Because of their composition (uniformity of particles), single layer boards present very few problems when being cut.

Three-layer Construction (Figure 2.13b)

These boards consist of a low density core of large particles sandwiched between two layers of a higher density. Because of their very smooth even surface they are very suitable for direct painting. They may be classified as a general purpose board for interior non-structural use.

Multi-layer Construction

This is very similar to three-layer board except for the increase in the number of layers, and possibly the inclusion of a high density core layer to improve strength properties. Care is required when cutting all layered boards as, unlike single layer types, they tend split or chip away at the cut edge.

Graded Density (Figure 2.13c)

The structure of these boards is mid-way between single-layer and three-layer types. Particles vary in size, getting smaller from the board centre outwards. They are suitable for furniture production.

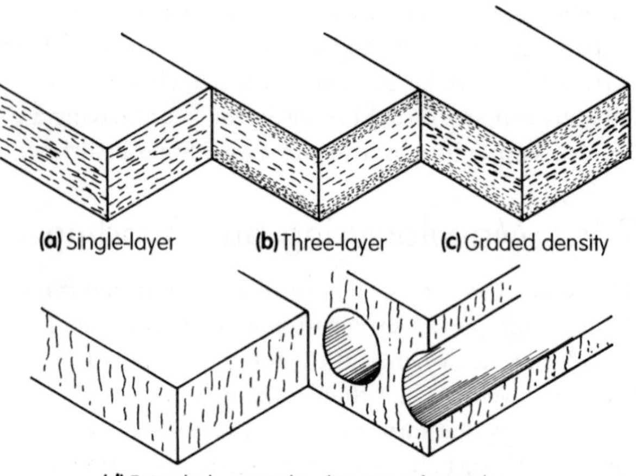

Figure 2.13 Types of chipboard (particle board) construction

Extruded Boards (Figure 2.13d)

In these boards the board forming process differs, in that the blend of chippings (particles) and adhesive is forced through a form of die, resulting in an extruded board of predetermined width and thickness, but of unlimited length. The holes found in some of these boards are the result of metal heating tubes which assist in the curing process, thus enabling much thicker boards to be produced. These holes reduce the overall weight of the board. Chipboard produced in this way will, in the main, have its particles located at right angles to the board's face, thus reducing its strength. However, the main use of these boards is as a core material sandwiched between suitable layers of veneer, be they of wood or other materials, to give the board its required stability.

2.3.3 Types (Grades) in Relation to End Use

Table 2.8 lists the types (grades) of board, together with examples of possible end use based on their mechanical performance levels and reaction to moisture.

Table 2.9 lists the different types against their colour coded edge stripe. Figure 2.14 shows how sheets of chipboard could be surface labelled and edge marked. Surface labelling on one face should show the appropriate British Standard number and date, together with the trade mark or identification mark to indicate the name of the manufacturer.

A range of board sizes is given in Table 2.21.

Table 2.8 Wood chipboard type (grade) in relation to its possible end use (application)

Type of board (grade)	Important properties/features	Typical end use
C1	Standard non-structural board	Lining walls, casing, joinery fitments board (dry situations only)
C1A	Furniture board non-structural	Furniture, as a base for wood veneers, laminated plastics (dry situations only)
C2	Flooring board with good impact resistance	Light duty floors (dry situations only)
C3	Improved moisture resistance	Roofing (decking flat roofs), lining walls, kitchen worktops, casings, joinery fitments, and wall sheathing (timber frame)
C4	Improved moisture resistance plus impact resistance	Domestic and other light duty floors – roofing (decking flat roofs where access is required). Wall sheathing (timber frame). Lining walls, and joinery fitments
C5	Improved moisture resistance plus better mechanical properties. Limited structural use	Heavy duty flooring and racking shelves. Beams ('I' and 'Box' sections), and general roofing

Table 2.9 Wood chipboard edge marks

Type of board, grade (see Table 2.8)	Number of stripes	Identification stripe (Figure 2.14) with colour coding of stripe or stripes (at least 25 mm wide)
C1	One (1)	Black
C1A	Two (2)*	Black
C2	One (1)	Red
C3	One (1)	Green
C4	Two (2)*	One Green – One Red
C5	Two (2)*	One Green – One Yellow

Note: *Gap between stripes shall not be less than 25 mm; coloured stripe/stripes will be visible on either the long or short edge of the board and at near diagonal opposite corners, as shown in Figure 2.14

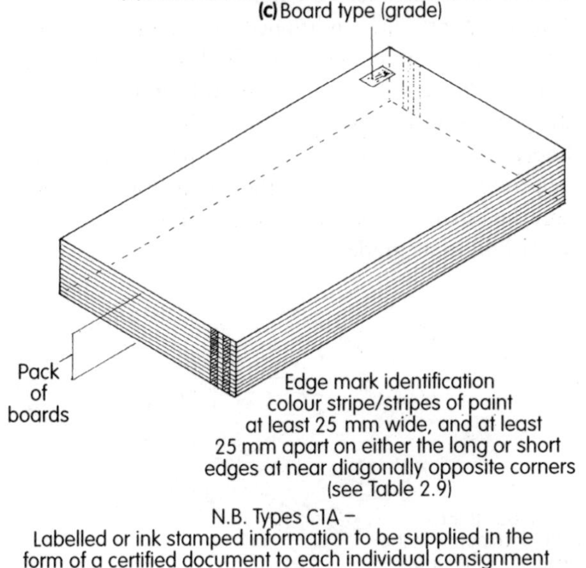

Figure 2.14 Labelling and marking chipboard

2.3.4 Cement-bonded Particle Board

This is a mixture of wood particles bonded with Magnesite or Portland cement to produce two types of high density board, one of which has good moisture resistance, but both of which have high fire resistance. Table 2.10 provides easy cross reference between both types, especially with regard to their end use.

Like other forms of chipboard, identification is very important to end use: methods of marking are very similar. Face markings show the board's compliance with British Standards, the name of the manufacture, and bond type (T1 or T2), and, as stated in Table 2.11, colour edge marks.

A range of available sizes is given in Table 2.21. These boards can be cut satisfactorily by using hand or power tools with tungsten carbide tipped blades.

Table 2.10 Cement-bonded particleboard type (grade) in relation to possible end use

Type of board (grade)	Bonding agent (binder)	Important properties features	Typical end use
T1	Magnesite	Internal use only	Internal linings, walls etc.
	Cement	Low to moderate resistance to moisture	Window boards, casings (pipe boxes, etc.)
		High resistance to fire	
		Liable to attack by wet rot	
		Resistant to insect attack	
T2	Portland cement	Internal and external use	Timber framed buildings (sheathing)
		Good resistance to moisture	External cladding, flooring
		High resistance to fire	Roof decking
		Not liable to attack by wet rot	
		Resistant to insect attack	

Table 2.11 Cement-bonded particleboard edge marks

Type of board grade (see Table 2.10)	Identification stripe (Figure 2.14)	
	Number of stripes	Colour coding of stripes (25 mm wide)
T 1	One	Black
T 2	One	Red

Note: Stripes shall be visible on either the long or short edges of the board and at opposite corners, as shown in Figure 2.14

2.4 FLAKE BOARD

Flake boards are made up of wood flakes. They are classified as either:

- waferboard, or
- oriented strand board (OSB).

Waferboard consists of wood flakes or wafers, cut mainly from hardwood species . These flakes could measure up to 75 mm × 75 mm and be 6 mm thick. Lying flat along the length of the board, the grain direction of the flakes will be randomly orientated. Waferboard has mainly been superseded by oriented strand board (OSB).

2.4.1 Oriented Strand Board (OSB)

As shown in Figure 2.15, large softwood strands or flakes (narrower than waferboard) are collectively layered in such a way that they form a laminated formation on the lines of three-ply plywood.

Manufacturing OSB

The following processes are a simplified overview of the general principles of manufacture.

1. The raw materials (softwoods), mainly pine logs, are debarked and cut to a suitable length for the wafering machine (waferiser).
2. The waferiser receives the cut logs and reduces them to wood strands (flakes) up to about 75 mm long and 35 mm wide. These strands are screeded to remove any fine particles; they are then dried within a tumbler drier prior to being stored.
3. The wood strands are blended with powdered adhesive, either a phenol formaldehyde (PF) resin, or melamine urea formaldehyde (MUF) resin, depending on the type of board, and a proportion of wax.
4. The strands are layered up in a three-layered mat as shown in Figure 2.15.
5. The mats are compressed to their required thickness and adhesive cured by heat, usually within a multi-layered press, then left to cool.
6. The boards are cut, conditioned to take in moisture of 6% MC minimum, then, if required, sanded smooth.
7. Quality checks will be in operation from the start of the process until the finished board is produced. Labelling and certification details (specification) will be printed on the surface of the board (Figure 2.16) and the board will be edge marked (Table 2.13) according to its grade.

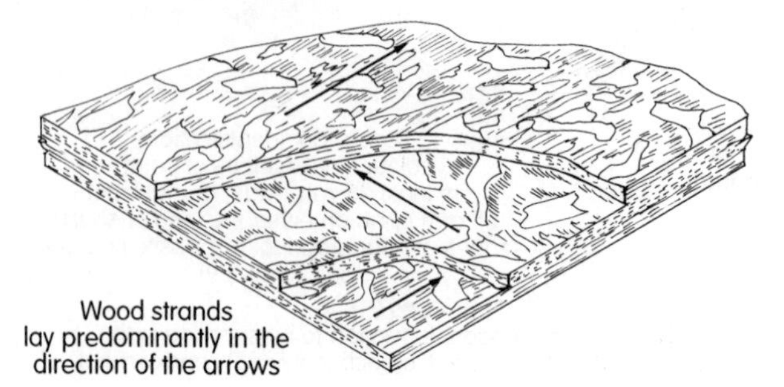

Figure 2.15 Guide to OSB strand orientation

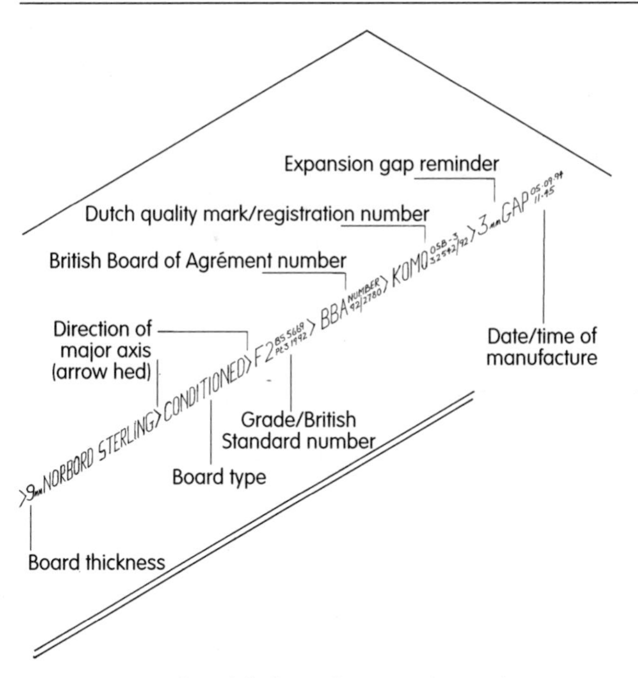

Figure 2.16 Surface labelling of OSB with specification details

Types of OSB

As shown in Tables 2.12 and 2.13, we can recognise two types of OSB: 'standard board' (F1) and 'conditioned board' (F2), together with typical situations which can accommodate their use. OSB will require conditioning to the environment in which it will be used. Section 2.12 explains this in more detail.

A range of available sizes is given in Table 2.21.

2.5 FIBRE BUILDING BOARDS

These boards are produced from wood which has been shredded into a fibrous state and then reconstituted into a uniform (monolithic) board or sheet (thin board) form. The process used to produce these boards is known as either the wet process, or the dry process. The wet process uses the natural inherent adhesive properties of the wood to fuse the wood particles together without the use of any synthetic adhesive. Boards produced in this way include:

Table 2.12 OSB in relation to possible end use

Grade	Type of Board	Bonding agent	Typical end use (examples)
F1	Standard board	Synthetic resin and wax (moisture resistant properties)	Sarking* pitched roofs, garage roofs, site hoarding, cladding, formwork, packing cases, shelving, garden sheds, etc.
F2	Conditioned board	Synthetic resin and wax (moisture resistant than F1, with improved durability)	Sarking* pitched roofs, decking flat roofs, flooring, sheathing (walls), site hoarding, cladding, packing cases, formwork, portable buildings, signs NB: Tongue and groove (T & G) boards available for roof and floor decking

Note: * Exterior covering of roof rafters provide extra stability and resistance to racking etc.

Types according to European Standards will be marked and classified as follows:

Board mark	Definition
OSB 1	General purpose board designed for use internally in dry conditions – non-structural suitable for furniture and fitments, etc.
OSB 2	Load-bearing board can only be used in dry conditions
OSB 3	Load-bearing board can be used in humid conditions
OSB 4	Heavy duty load-bearing board can be used in humid conditions

Table 2.13 OSB edge marks

		Identification stripe (Figure 2.14)	
Grade	Type of board See section 2.12	Number of stripes	Stripe colour (25 mm minimum width)
F1	Standard board	One (1)	Blue
F2	Conditioned board	One (1)	Yellow

- softboard (insulation board)
- mediumboard
- hardboard.

The dry process relies on the introduction of an adhesive made from synthetic resin. The main boards made in this way are known as

- medium density fibreboards (MDF).

The main ingredients for the production of fibreboard are softwoods obtained from:

- immature trees, especially grown for wood pulp
- forest thinnings
- residue from saw mills, such as undersized logs and slabs, etc.

2.5.1 Manufacturing (Wet Process)

Figure 2.17 should help you to understand more fully the following process:

1 The raw material is received and selected for use.
2 The wood is cut to size then passed through a mechanical chipper to produce large wood chippings.
3 After the chippings are steam treated to soften the lignin (natural resin) which binds the fibres together, they are reduced to a fibrous state by passing them through a defibrator machine which contains two grinding discs, one of which rotates.
4 These fibres are then added to water to form a slurry; additives may be added to enhance the resulting properties of the board.

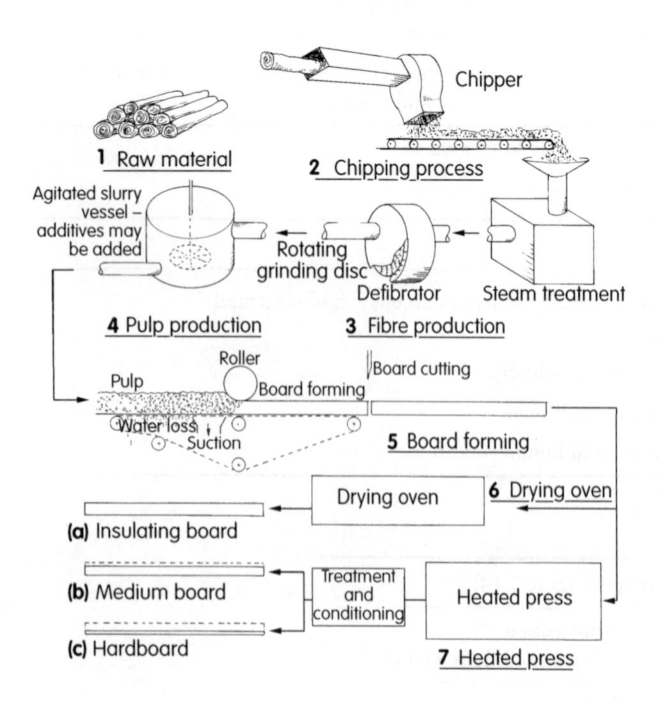

Figure 2.17 Guide to the manufacture of fibreboard using the wet process

5 This slurry or pulp is fed into a slow moving wire mesh conveyer belt where it starts its journey along the board forming machine to become reconstituted. This *wet lap*, as it is known, loses its water in three ways prior to drying: firstly by gravity via its mesh bed, then by under bed suction, and finally by a roller, which squeezes the mass whilst reducing the matt to its uniform thickness. This partly dried wet lap matt is then cut to length ready for drying or pressing. What follows will depend on the required density of the finished board.
6 Drying ovens receive non-compressed boards (softboard).
7 Single or multi-level heated presses reduce the matt on its meshed plates to the required density, either a medium or hardboard. On leaving the presses, boards are conditioned to increase their MC to about 4–8%.

NB: You may see, or indeed need, boards which are smooth on both sides (duo-faced). These may be produced by either bonding two standard boards together back to back by further pressing the wet matt lap between two smooth plates, or by using a dry process of smooth pressing.

2.5.2 Manufacturing (Dry Process)

1 Fibres are produced in the same way as they are with the wet process (stages 1–3).
2 Wood fibres are mixed and coated with the required powdered synthetic resin adhesive and additives (the bonding agent or binder), then dried and stored ready for use.
3 Coated fibres are then formed in mass into a dry mattress, which is pre-pressed to a stabilised form for cutting to size and pressing.
4 After conditioning (cooling), panels are trimmed to size and finished by sanding.

2.5.3 Softboard

Softboard is a low density lightly compressed board dried in an oven. Its cellular make-up is the result of moisture loss (evaporation) during the drying process. It is these voids left by the evaporated moisture that give the board its good thermal properties. This is why these boards are commonly called *insulation board.*

Table 2.14 lists three grades of softboard, together with guidance on their respective density range, appearance, properties, and typical end use.

2.5.4 Mediumboard

This is a mid-density range of compressed boards. Table 2.15 lists the four available grades, two within the low density range and two within the higher density range, together with details of their possible end use.

Table: 2.14 Softboard in relation to possible end use

Grade	Type of board	Density kg/m^3	Bonding agent (binder)	Appearance/properties, etc.	Typical end use
SBN	Softboard (natural)	210 – 350	Natural inherent adhesion of wood fibres	Smooth, dimpled, or mesh patterned to one or both faces coloured finish – in brown, cream, or white A very soft material in sheet, board, or tile form – surfaces may be flat, grooved, patterned, or with stopped holes etc. Specially treated board to resist spread of flame may be available Thermal and acoustic properties	*Internal use only* Lining walls and ceilings floor underlay, core material for doors and partitions*
SBI	Softboard (impregnated)	240 – 400	Wood fibres impregnated with bitumen and/or other additives	Dimpled or mesh patterned – black to brown in colour	Underlay to floor decking
SBS	Softboard (sarking† and sheathing‡)			Water repellent properties	Sarking and sheathing*

Notes: * Treatment required to meet the requirements for 'surface spread of flame'; † Sarking – covering inclined surfaces of pitched roofs, prior to roof covering; ‡ Sheathing – cladding to a structural timber framework to resist racking forces

Table 2.15 Mediumboard in relation to end use

Grade	Type of board	Density kg/m^3	Bonding agent (binder)	Performance* (appearance)	Properties	Typical end use
LMN (Normal)	Mediumboard (Low density)	350 – 560	Natural inherent adhesion of the wood fibres	Normal performance level	Smooth matt surface to one face side, mesh patterned on the other face	*Interior use only* Lining walls and ceilings, panelling, notice boards†
LME (Extra)				Extra (higher) performance level		
HMN (Normal)	Mediumboard (High density)	560 – 800	As above but may also include a binding agent	Normal performance level	Smooth hard shiny surface to one face side, mesh patterned on the other other face	*Interior and limited exterior† use* Lining walls and ceilings, panelling, partitions,‡ exterior cladding, sheathing§, fascia boards and soffits.
HME (Extra)				Extra (higher) performance level		Joinery and shopfitting fitments, signs, chalkboards etc.
				Colour of boards range from grey to brown – embossed pattern and pre-painted boards are available		

Notes: * Performance – levels based on how moisture, humidity, and loading etc., can affect the board; † Enhanced level of durability etc., may be required; ‡ Treatment required to meet the requirements for 'surface spread of flame'; § Sheathing – cladding of a structural timber framework to resist racking forces

2.5.5 Hardboard

As the name implies hardboard can be a very physically hard board. For many years it has been the most common of fibreboards, primarily as a cheap alternative to thin plywood (up to 4 mm thickness), and as a decorative panel product.

Table 2.16 lists the five available grades of hardboard, three within the range of standard hardboard, and two tempered hardboards which have the great advantage of being able to be used externally. The list also includes details of their end use.

2.5.6 Medium Density Fibreboard (MDF)

This group is, with the possible exception of duo-faced hardboard, the odd one out, since it is manufactured using the dry process. These boards can, in many situations, be used as solid wood substitutes because of their ability to respond well to the use of both hand and machine tools. A few examples are given in Table 2.17, together with their respective grades.

2.5.7 Fibreboard Types (Grades) in Relation to their End Use

Typical examples of end usage based on respective properties and reaction to moisture are given in Tables 2.14 (softboards), 2.15 (mediumboards), 2.16 (hardboards), and 2.17 (MDF).

Table 2.18 lists the different fibreboard grades and types against a system of colour coded edge strips. Board identification as a whole should be via both face and edge marking details, as shown in Figure 2.14.

Fibreboards should, as explained in Section 2.12 be conditioned to the environment in which they are to be used. A range of fibreboard sizes is given in Table 2.21.

Table 2.16 Hardboard in relation to possible end use

Grade	Type of board	Density kg/m^3	Bonding agent (binder)	Properties*	Appearance	Typical end use
SHA SHB SHC	Standard Hardboard	800 – 960	Natural inherent adhesion of wood fibres	A' best performance; B' mid-performance; C' lowest performance	One side normally smooth but may be available prepainted plastics, faced textured, embossed perforated (pegboard) or both sides smooth (duo-faced)§	*Interior use only* Lining walls and ceilings, panelling, joinery fitments, under floor covering, door facings, display boards etc.†
				Colour – light to medium brown		
THN (Normal) THE (Extra)	Tempered Hardboard	960 – 1180	Natural inherent adhesion of wood fibres	Normal performance level Extra (higher) performance level	One side normally smooth – other mesh patterned. Boards with both faces smooth (duo faced§) may be available. Perforated and textured boards may be available	*Interior and exterior use* Lining walls and ceilings, external sheathing‡, cladding, formwork, floor covering, roof beam, webs etc.
				Boards are impregnated with oil or resins to improve strength properties and resistance to abrasion and moisture Colour – dark to extra dark brown		

Notes:* Performance – levels based on how moisture, humidity, and loading etc., can effect the board; † Treatment required to meet the requirements for 'surface spread of flame'; ‡ Sheathing – cladding of a structural timber framework to resist racking forces; § Duo-faced – both sides smooth

Table 2.17 Medium Density Fibreboard (MDF) in relation to possible end use

Grade	Type of board	Density kg/m^3	Bonding agent (binder)	Properties	Typical end use
MDF	Medium density fibreboard	600 – 900	Urea-formaldehyde (UF) resin	Both faces sanded smooth; surfaces suitable for painting, veneering etc.	*Interior use only* Furniture and cabinet work, moulded sections, door skins and panels
MDFMR	Medium density fibreboard, moisture resistant		Melamine-urea-formaldehyde (MUF) resin or phenol-formaldehyde (PF) resin	Good machining properties, colour of sandstone	*Interior and exterior use** Joinery sections – staircases treads and risers, window boards, etc; Moulded sections – skirting boards, architraves, cornices, etc.; fascia and soffit boards; Shopfronts – signs and notice boards

Note: * Enhanced level of durability may be required

Table 2.18 Fibreboard edge marks

		Identification stripe (Figure 2.14)	
Grade	Type of board (see Tables 2.14 to 2.17)	Number of stripes	Stripe colour
SBN	Softboard (natural)	One (1)	White
SBI	Softboard (impregnated)	Two (2)	White
SBS	Softwood (sarking and sheathing)	Three (3)	White
LMN (normal)	Mediumboard (low density)	One (1)	Purple
LME (extra)		Two (2)	Purple
HMN (normal)	Mediumboard (high density)	One (1)	Green
HME (extra)		Two (2)	Green
SHA	Standard hardboard	Three (3)	Blue
SHB		Two (2)	Blue
SHC		One (1)	Blue
THN (normal)	Tempered hardboard	One (1)	Red
THE (extra)		Two (2)	Red
MDF	Medium density fibreboard	One (1)	Black
MDFMR	Medium density fibreboard, moisture resistant	Two (2)	Black

Note: * The addition of a yellow stripe would indicate that the surface of the sheet/board has been specially treated for painting etc.

2.6 LAMINATED PLASTICS (DECORATIVE LAMINATES)

These plastics are generally thin synthetic (man-made) plastic veneers of various colours, patterns, and textures, capable of providing both decorative and hygienic finishes to most horizontal and many vertical surfaces.

Figure 2.18 shows an example of how a plastics laminate is built up prior to being bonded together by a combination of heat and pressure. You will find that different grades are available. Table 2.19 lists some of these; notice how the grades intended to be used horizontally (HGS) offer greater wear resistance compared to those intended for vertical use (VGS). Surface finish also influences use. For example, horizontal applications

Table 2.19 Grades of decorative laminated plastics

Grade	Examples of application
Horizontal general purpose standard grade (HGS)	Working surfaces – kitchens, restaurant tables, counters, and vertical surfaces subject to heavy wear
Vertical general purpose standard grade (VGS)	Wall and door panels, occasional use shelving, display cabinets etc.
Horizontal general purpose Post forming grade (HGP)	Heavy duty surfaces parts of which are to be curved by post forming
Horizontal general purpose flame retardant grade (HGF)	Surfaces which have to meet British Standard requirements for class '1' spread of flame, and/or class' 0' statutory requirements of the Building Regulations when suitably bonded to a suitable substrate
Vertical general purpose flame retardant grade (VGF)	Surfaces which have to meet British Standard requirements for class '1' spread of flame, and/or class' 0' statutory requirements of the Building Regulations when suitably bonded to a suitable substrate
Flame retardant grade for post forming (FRP)	As above when post forming is required

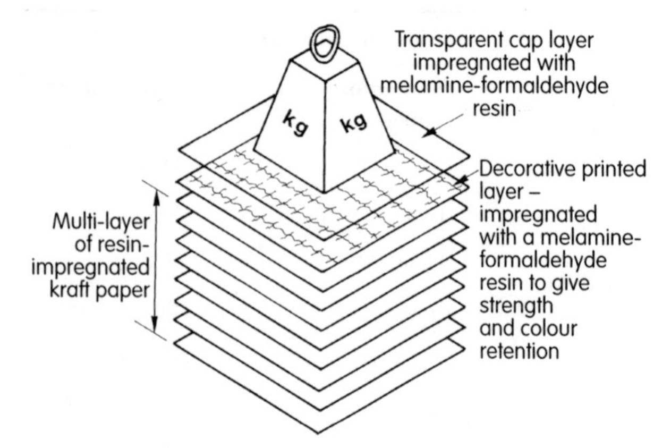

Figure 2.18 Composition of laminated plastics

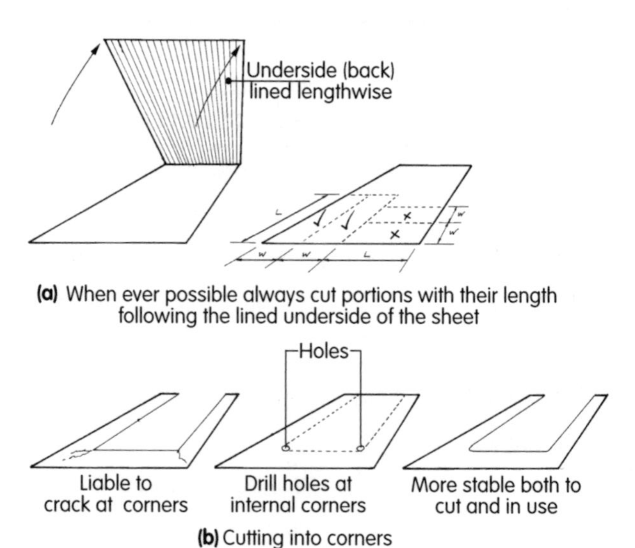

Figure 2.19 Cutting from the sheet

such as worktops are best suited to scratch resistant laminates such as those with patterned and light coloured matt and textured surfaces. In contrast, shiny (glossy) and plain dark colours should be used for vertical application only. Table 2.21 gives a range of sizes.

2.6.1 Cutting, Boring and Trimming

These veneers can be cut successfully from the sheet (see Figure 2.19a) by using hand and/or machine tools. But these thin brittle sheets must always be fully supported on either side of each cut, and, as shown in Figure 2.19b, cuts made into the internal corner should be made towards a rounded (radiused) edge, since this will help avoid corner cracks. More importantly, corner stress cracks can result from restrained differential movement when the laminate is finally in place in high risk areas with varying degrees of humidity, such as an intermittently centrally heated environment. For the same reason, restraining or fixing screws should be used only via movement plates (see Sections 7.3.7 and 11.5.2), or with an over-sized clearance hole as shown in Figure 2.20.

Hand Tools

These veneers can be cut by using either a sharp fine toothed tenon saw, cutting from the decorative face, or by scoring through the decorative face with a purpose-made scoring tool then gently lifting the waste or off-cut side, thus closing the 'V' and allowing the sheet to break along the scored line. A block plane (low angled blade type) and/or file can be used to trim edges.

A special note of caution is needed here. Always remember to keep your hands and fingers away and

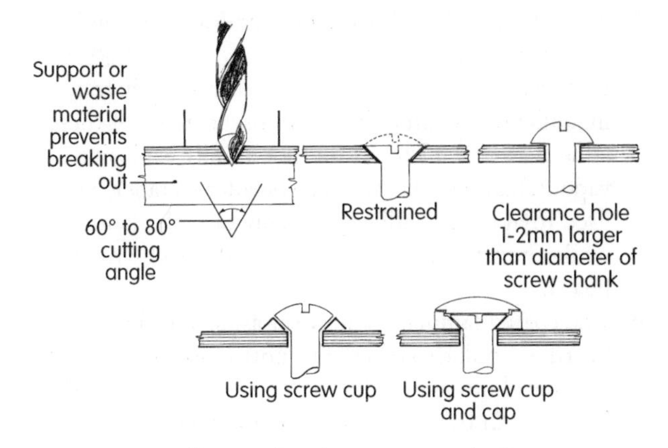

Figure 2.20 Drilling and making provision for fixing screws

clear of these edges whilst this process is being carried out: processed edges can be very sharp!

Machine Tools

Because of its hard and brittle nature, special care must be taken both with methods of holding the laminated plastic while it is being cut and during the machining process. Special blades and cutters (tungsten-carbide tipped TCT) are available; both the manufacturers of laminated plastics and many tool makers offer recommendations about blade and cutter peripheral cutting speeds and techniques.

While processing operations are being carried out there is always the risk of injury to the eyes. It is therefore essential that eye protection is worn at all times, not only by the operator, but also by others in close proximity to the operation.

2.6.2 Veneer Application

Laminated plastics are used as veneers, so their application onto a suitable support (core material) can be dealt with in a similar manner. Support materials include:

- plywood
- blockboard
- laminboard
- chipboard
- hardboard
- MDF
- non-combustible boards (based on calcium silicate).

If the support material is to retain its shape (i.e. flatness), it must, as shown in Figure 2.21, be kept in balance. Therefore any veneer or additional veneers (in the case of plywood) applied to one face (Figure 2.21a) should have an equivalent compensating veneer applied to the opposite face (Figure 2.21b). With laminated plastics, for the best results the balancer (counter veneer) is of the same grade and colour as the face laminate. However standard balancer veneers to a lower standard are available, and where flatness is not essential

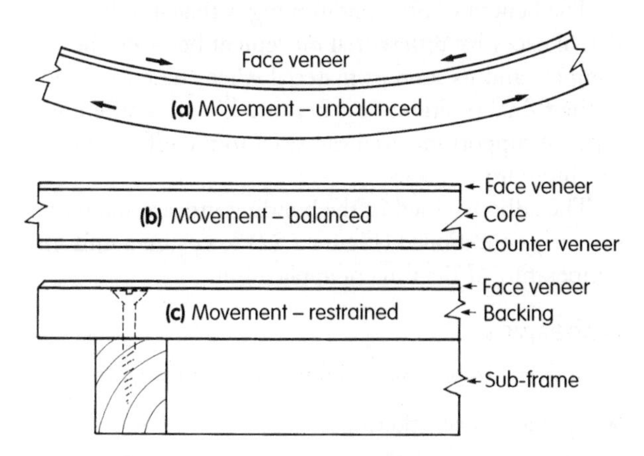

Figure 2.21 Stabilising the possible effect of applying a face veneer

a universal balancer or any backing laminate may be used. In certain circumstances, where an extra thick supported backing is used, as shown in Figure 2.21c, a counter veneer may not be necessary.

Unsuitable backings (core materials) include: plastered and cement-rendered surfaces, gypsum plaster board (see Section 2.8) and solid timber. Narrow (ne. 75 mm wide) quarter sawn boards could be acceptable because of their minimal differential moisture movement (see Section 1.75).

Pre-conditioning

The plastics laminate and its support material will, before they can be stuck together, need to be brought into a reasonably balanced condition: the plastics laminate will need to be at about 5–7% MC and the core material around 9% MC.

These conditions are best achieved within a dry area which can be maintained at about 20° C and at a relative humidity (see Section 7.12) of 50–60%. Techniques used to bring this about vary, but under normal conditions it can take between three and seven days. Figure 2.22 shows one method that may take about three days.

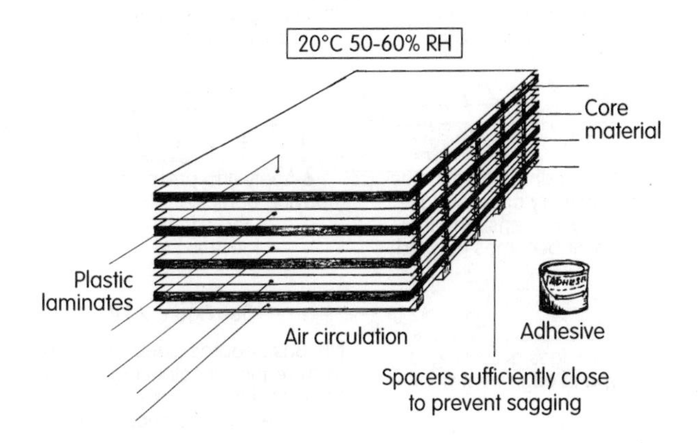

Figure 2.22 Pre-conditioning material prior to bonding

The benefit of pre-conditioning is that it reduces the risk of excessive differential movement between the laminate and its support material when these are fixed in their final position: bear in mind how the different types of support material can react to the effect of relative humidity changes.

The adhesive used in the bonding process should be similarly conditioned (Figure 2.22) to ensure a uniform temperature at the time of application.

Adhesives

Adhesives can be considered within four categories:

- rigid adhesives (thermosetting)
  – urea formaldehyde
  – melamine/urea formaldehyde
- semi-rigid adhesives -PVAc
- flexible adhesive (contact adhesives)
- hot melt adhesives (used only for bonding edge materials).

All should be applied and processed according to their manufacturers' instructions. The most useful of these for on-site work and the small workshop without plate pressing equipment would be contact adhesives. These are either neoprene or natural rubber based, and available as solvent or water-based products.

Figure 2.23 shows the procedure of adhering a one face laminate to a stable base board, as follows.

1 The decorative laminate is cut from the sheet as shown in Figure 2.19.
2 The laminate, base board, and adhesive are conditioned together.
3 In a well ventilated area (you must fully adhere to manufacturers' safety instructions, as many of these adhesives give off toxic and potentially explosive vapours), an evenly serrated spreader (scraper) is used to apply the adhesive evenly over the laminate while the latter is supported (face down) by the base board.
4 The coated laminate is now set aside and the base board is similarly coated, but notice how the spreader (scraper) lines are worked at 90° to those left on the laminate: in this way a better bond between the two will be achieved.
5 After the prescribed time (when the adhesive becomes touch dry) the two surfaces are gradually joined together, all the time ensuring that no air becomes trapped between the surfaces, as once they are in final contact there is no recovery.
6 Moderate yet firm pressure is applied over the whole surface, paying particular attention to the edges.
7 Once the adhesive is set, edge trimming can commence according to the required edge treatment (Figure 2.24).

NB: If a backing veneer is necessary then this is usually applied first, before the face veneer, using the same procedure as above.

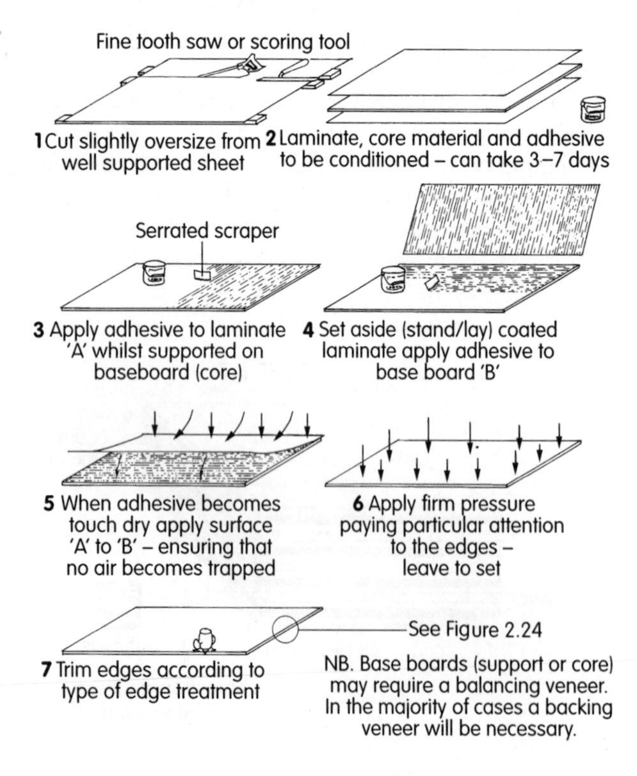

Figure 2.23 Using a contact adhesive

Edge Treatment

This will depend on what is required for the finished item. Figure 2.24 gives several examples of how door and counter/worktop edges may be finished.

In the majority of cases machines are used to carry out this task, tools such as portable electric hand routers and edge trimmers (see Section 8.15 and Figure 8.32), but hand tools must not be ruled out, particularly for finishing on site operations and for fine finishing. Very accurate finishes can be achieved with a block plane, flat and shaped files, and cabinet scrapers.

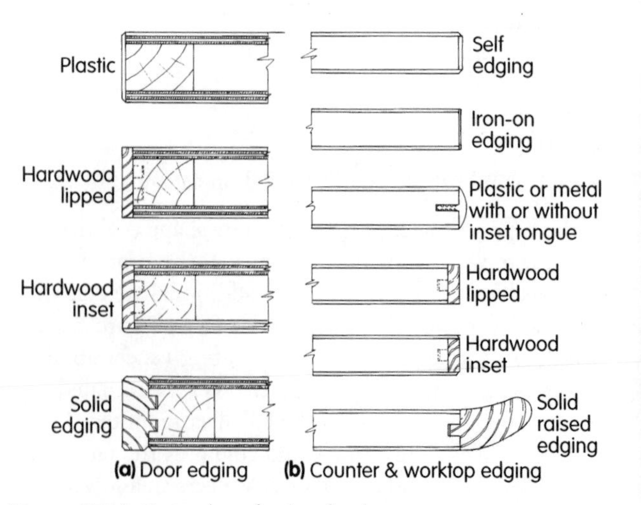

Figure 2.24 Examples of edge finishing

2.6.3 Post-forming

Post-forming is a term used to describe the process involved when bending and bonding specially developed laminates (see Table 2.19) over and into various shapes found on the edges and/or surfaces of worktops and sometimes panels. Figure 2.25 shows a few different examples. Figure 2.26 illustrates a simplified version of post-forming a worktop edge and upstand. Special equipment and jigs are used by firms that specialise in post-forming laminated plastics.

After studying Figure 2.26b you may be wondering how the width of the temporary fillet is determined. If you look at Figure 2.27, you will see that, provided that the radius of the curve (quadrant) is known (which will partly be governed by the thickness of the laminate), all you need do is follow the simple calculation. The cove infill may be made with solid wood or MDF.

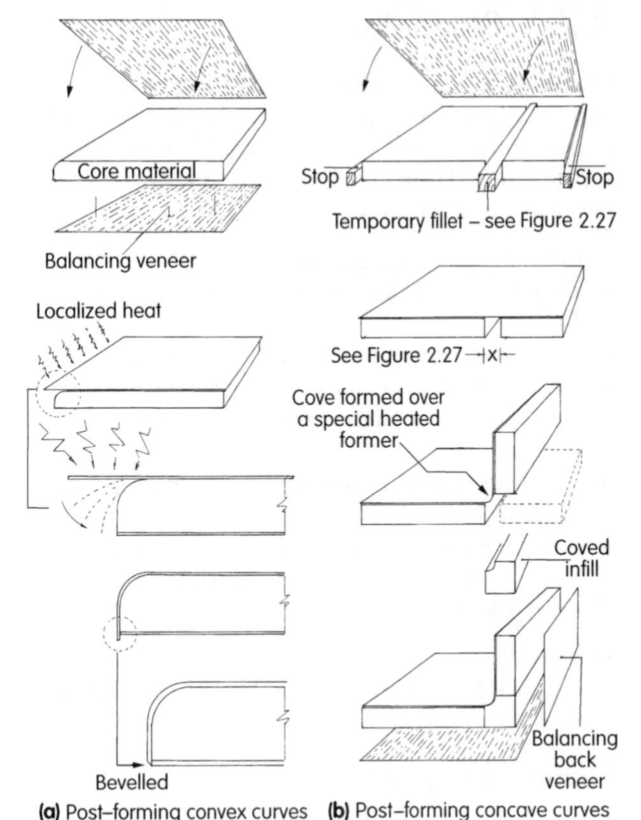

Figure 2.26 A method of post forming curves

2.7 FIBRE CEMENT BUILDING BOARDS

These boards have always been associated with excellent fire resistance, and, until a few years ago, the main component of that resistance was the now obsolete asbestos binding agent. Because of the serious health risks linked to asbestos fibre, it has now been replaced by a calcium silicate binder. The constituents of current boards are largely the result of an interaction of cellulose fibre, lime, cement, silica and a fine protective filler mix during a specialised production process.

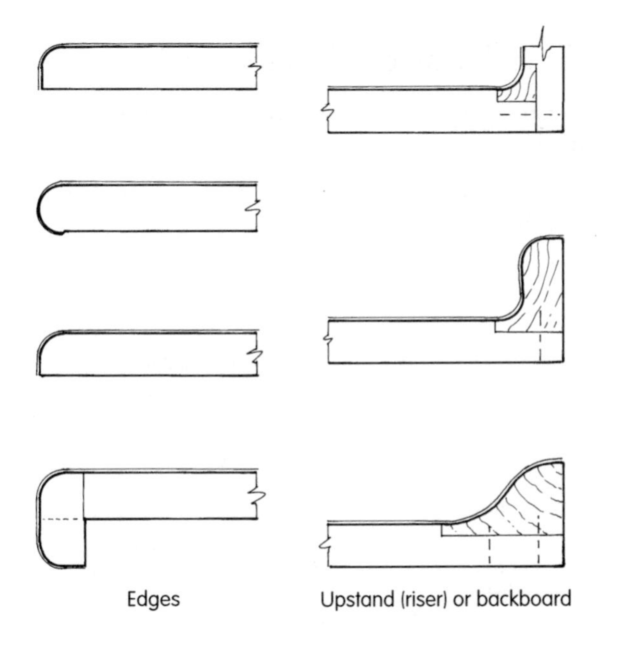

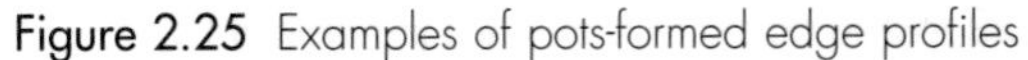

Figure 2.25 Examples of pots-formed edge profiles

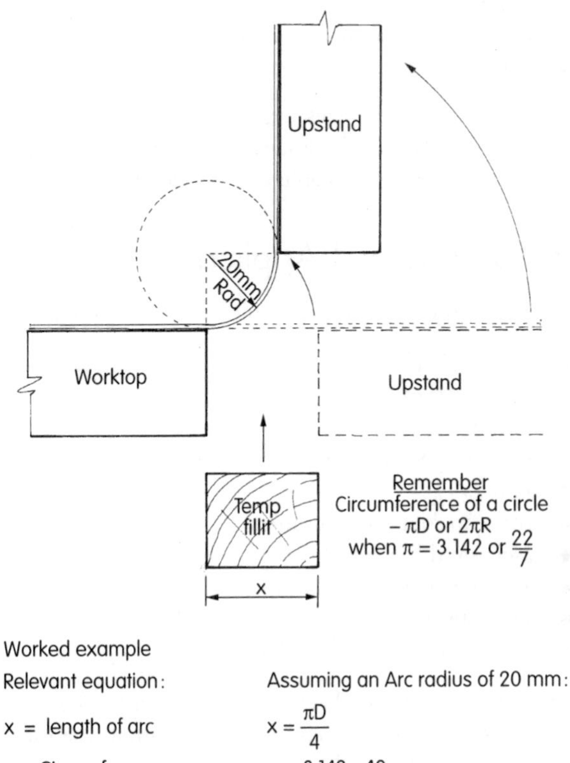

Worked example

Relevant equation:

x = length of arc

$$x = \frac{\text{Circumference}}{4}$$

$$x = \frac{\pi D}{4} \text{ or } x = \frac{2\pi r}{4}$$

Assuming an Arc radius of 20 mm:

$$x = \frac{\pi D}{4}$$

$$x = \frac{3.142 \times 40}{4}$$

$$x = \frac{3.142 \times 10}{1}, \ x = 31.24 \text{ mm}$$

Figure 2.27 Calculating the size of a temporary fillet

2.7.1 Properties

Probably the greatest advantage that these boards have over other panel products, with the exception of wood cement particle board and plasterboard, is their excellent fire resistance. These boards can also compare favourably in other ways too:

- their strength to weight ratio is good
- their impact resistance is comparable with that of MDF
- their resistance to moisture intake is generally good
- weather resistant grades are available
- they have very good dimensional stability (to movement)
- they have reasonable workability (for cutting and shaping)
- they are natural grey/white/brown in colour, with a range usually offering the different shades of natural earth colours.

However, because of their brittle nature, sheets of fibre cement building boards do require careful handling (see Section 2.11).

Form and Appearance

Fibre cement building boards can take the form of:

- flat sheets, either smooth, textured, or faced with a rough aggregate
- flat weather boarding or planking
- corrugated sheets (with various profiles).

A range of flat sheet sizes is given later in Table 2.21.

NB: Roof slates, shingles, eaves guttering, and rainwater fallpipes are also available as fibre cement products.

2.7.2 Use and Application

There are grades of board to suit situations such as :

- vertical cladding to:
  - timber frame walls
  - timber partitions
  - masonry walls via timber grounds
- ceilings
- soffits to roofs and canopies
- fascias and barge boards
- encasing steelwork and pipe work
- substrates for wall tiling (wet areas)
- facings for fire resistant doors.

By corrugating these sheets, extra strength can be obtained. These sheets have been in common use for many years, covering the roofs of domestic garages, outbuildings and large commercial and industrial structures.

A range of flat fibre cement sheets sizes is given in Section 2.10.

Working with Fibre Cement Products

Under properties, workability was classified as reasonably good because most of these sheets and boards can usually be worked satisfactorily with conventional hand tools.

Sheets can be roughly cut to size either by using a special scoring tool to 'V' the surface then snapping over straight edge, or, with the thicker boards, by using a fine toothed hand saw. Extra hard boards will require specially hardened blades, such as those used in sheet saws to cut sheet metal

Power tools will need tungsten carbide tipped blades; see also Sections 2.13 and 8.12.

Fixings

Fixing by nails or screws should only be undertaken after first consulting board manufacturers' literature about the correct type, size and fixing pattern (distances from sheet/board edges and between fixings), since these will vary between situations, sheet/board type and thickness.

2.8 PLASTERBOARDS

As shown in Figure 2.28, plasterboards are building boards composed of an aerated gypsum plaster core, bonded between heavy sheets of paper. The long edges can be formed as wrapped *square*, *tapered*, or *duo*, to meet different finish requirements for being used as a 'dry' lining material for walls, partitions, or ceilings to accept a direct decorative finish, or as a base to receive a traditional plaster finish.

One of the main advantages of plasterboard as a lining material is the fire protection offered by its gypsum core.

2.8.1 Types of Plasterboard

There are many types to suit different requirements, for example:

- gypsum wallboards – a general dry lining used primarily for direct decoration

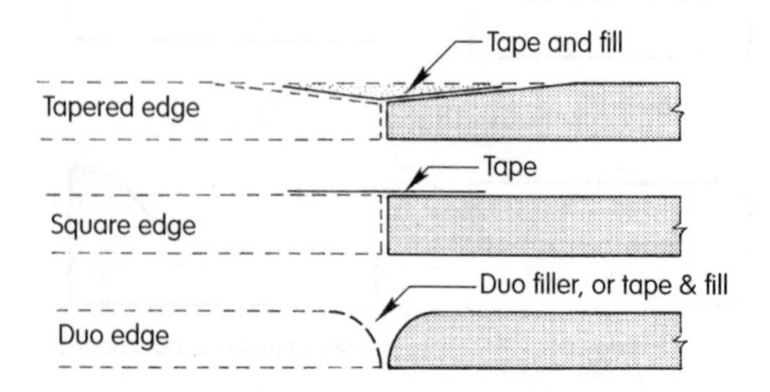

Figure 2.28 Plasterboard and their edge treatment

- gypsum wallboards with metallised polyester backing (Gyproc Duplex) – to provide water vapour resistant and reflective insulator; this backing may be available on 'Gypsum Baseboard' and 'Fireline board'
- gypsum lath (400 mm wide, 9.5 mm and 12.5 mm thick) – a narrow board that provides a base for plastering
- gypsum plank (600 mm wide, 19 mm thickness) – a base for plastering, which is useful for encasing steel beams and columns
- gypsum (thistle) baseboard (900 mm wide, 9.5 mm thick) – a base for plastering
- gypsum plasterboard with glass fibre and additives to the core (Fireline board) – with increased fire resistant properties; it is used to give added protection to metal substrates and timber floors, suspended ceilings, etc.
- gypsum moisture resistant board – plasterboard with added silicon to the core and water repellent paper facings
- gypsum thermal boards – composite board of gypsum and expanded polystyrene to increase thermal insulation.

A range of sizes is shown in Table 2.21.

2.8.2 Use and Application

The plastering trade is usually responsible for fixing plasterboard, but the timber framing to which plasterboard is attached will be fixed by carpenters and joiners.

The sectional sizes and distances between support members, and the size and type of fixings should be set down either within the work specification or working drawings. These details will be determined by the thickness and type of plasterboard. For guidance only, Table 2.20 shows a few examples for fixing gypsum wallboards vertically to wall studs.

All supporting timber should be dried to the required moisture content, and it must be accurately spaced and aligned. Minimum thickness of timber should be as shown in Figure 2.29.

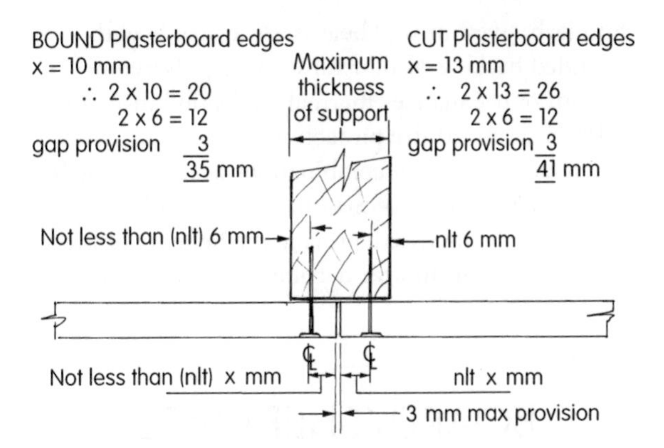

Figure 2.29 Minimum thickness of support material

2.9 COMPOSITE BOARD

Composite board is a special purpose board which consists of one or more dissimilar materials laminated together, enabling it to be used for an individually designed purpose.

For example, by bonding a mineral wool to the underside of a flooring grade chipboard, as shown in Figure 2.30, a resilient floor decking is produced with an inbuilt means of providing enhanced thermal and acoustic properties.

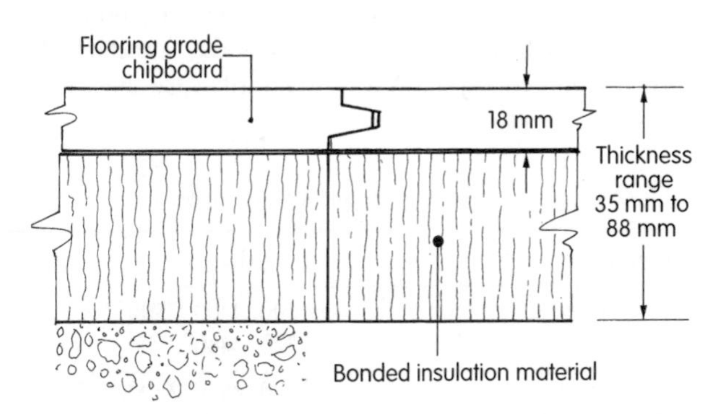

Figure 2.30 Composite flooring material

Table 2.20 Guide for fixing plasterboard vertically to wall studding

Plasterboard	Thickness (mm)	Board width (mm)	Recommended centres (mm)	Fixings – length of Gyproc galvanised nails (mm)
Gypsum	9.5	900	450	30
Plasterboard	9.5	1200	400	30
Gypsum	12.5	900	450	40
Plasterboard	12.5	1200	600	40
Gypsum	15.0	900	450	40
Plasterboard	15.0	1200	600	40

Similarly, gypsum wall boards are available with a pre-bonded backing of mineral wool or polystyrene which not only enhances thermal and sound insulation, but also gives added fire protection.

There are many such combinations available which enhance the properties of manufactured boards, and, as we have already discussed, provide the benefits of pre-bonded surface finishes for decorative and functional reasons.

2.10 BOARD/SHEET SIZES

Table 2.21 gives some indication of the many manufactured board and panel product sheet sizes available at timber merchants. You will notice that the traditional common imperial size of eight feet by four feet is in many cases readily available as 2440 mm × 1220 mm sheets. The nearest available metric size, based on the 600 mm module, would be 2400 mm × 1200 mm.

2.11 STORAGE AND HANDLING

The way in which board and sheet material is kept and handled will affect the end product. If we expect to work with clean undamaged material then we, and all those involved, should ensure that materials are stacked correctly on delivery and protected until every sheet is used, not forgetting that every time one or more sheets are removed from a stack the remaining boards should be left correctly stacked and fully protected.

2.11.1 Storage

Methods of storage will depend on the type of product and location. For example, workshop storage can often be under permanent control both physically and environmentally, whereas site storage can often present difficulties. Protection may be provided by a permanent structure such as a domestic garage, or a temporary cover, which can offer protection from rain and snow, etc., without restricting airflow through the stack. Breather type covers are now available to help reduce undercover condensation, a common problem with standard plastics sheet covers. Undercover condensation can increase the moisture content of wood and wood-based products.

2.11.2 Stacking

Figure 2.31 shows a horizontal stacking arrangement suitable for different types of manufactured boards. Take special note of the set distances between supporting bearers, and see that they are always positioned in line with one another, both horizontally (not twisted) and vertically. Note also the use of a thick base board when the stack is made up of thin sheets. Failure to keep to these rules will almost certainly result in permanently distorted sheet material.

Table 2.21 A guide to the range of sheet sizes

Material	Lengths (mm)	Widths (mm)	Thickness (mm)
Veneer plywood Veneer plywood	1200, 2400, 2500, 3000, 3600 1220, 2440, 2500, 3050, 3660	1200 1200, 1220, others available	4 to 30
Core plywood	1220, 2440, 2500, 3050, 3660	1220	12 to 38
Chipboard (particle board) Chipboard (particle board)	1830, 2440, 2750, 3050, 3660 2400, 2440	1220 600	2.5 to 38
Flakeboard Flakeboard	2400 2440	1200 1220	6 to 25
Hardboard	1220 to 3660	600, 1220, 1372, 1660	1.3 to 9.5
Medium board Medium board	1220 to 5485 Up to 4880	1220 1700	6 to 12.7
Softboard	600, 610, 1200, 1220, 2400, 2440, to 3600, 3660.	600, 610, 1200, 1220	8 to 30
MDF	2440, 2745, 3050, 3660	1220, 1525, 1830	1.6 to 60
Laminated plastics	1830, 2150, 2440, 3050, 3660	610, 1220, 1320, 1525, 1830	0.8 to 1.5
Fibre cement building board	1220, 1830, 2440, 3050	612, 1220	6 to 15
Plasterboard	1800, 2400, 2438, 2700, 3000	900, 1200	9.5 to 15

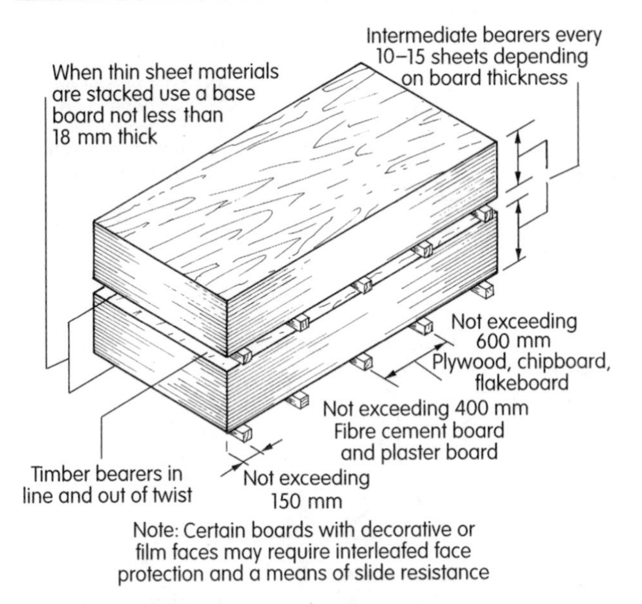

Figure 2.31 Horizontal storage of manufactured boards (except some decorative boards/sheets)

Figure 2.32 shows how, in certain cases, a few sheets may be best suited to stacking on their long edge. However, in general, where space is available, flat stacking should be regarded as being the best choice.

Figure 2.33 shows a compartmentalised arrangement of flat stacking suitable for laminated plastics and other similar products. Decorative sheets should be positioned face to face in pairs; in this way the risk of surface scratching when they are being removed from the stack is reduced. Top sheets should be positioned with their decorative face downwards. The stack should then be covered with a thick cover board to keep all of the under sheets flat. Alternatively, sheets could be stored on inclined racks (Figure 2.32), with a thick cover board to hold them flat and prevent them buckling and sliding forward.

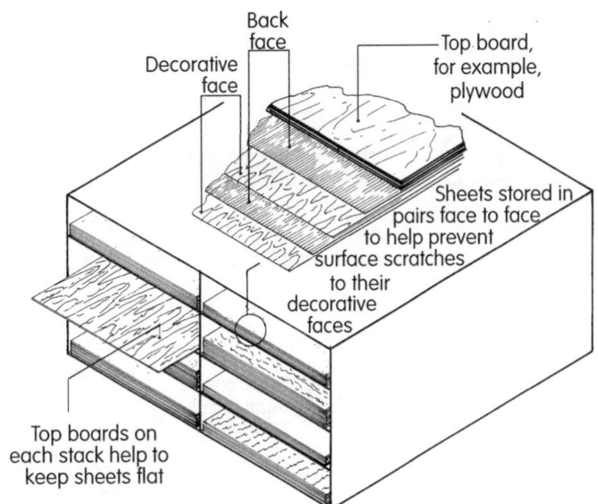

Figure 2.33 Horizontal compartmentalised storage for laminated plastics sheets or similar thin sheet material

2.11.3 Handling

It goes without saying that all sheet materials must be handled with care, not only to prevent damage to the sheet but more importantly to avoid any risk of personal injury to the handler. The correct lifting procedure must be used at all times and protective clothing worn. Of particular importance here is the use of industrial gloves to prevent splinters from wood-based boards entering the hand, and to prevent cuts from sharp edges, especially from laminated plastics. Feet must also be protected with toe protective footwear. This becomes most important when carrying board material edgeways: any board dropped edgeways will inflict damage wherever it lands.

Full rigid and semi-rigid sheets/boards, no matter what their thickness, must always be carried and

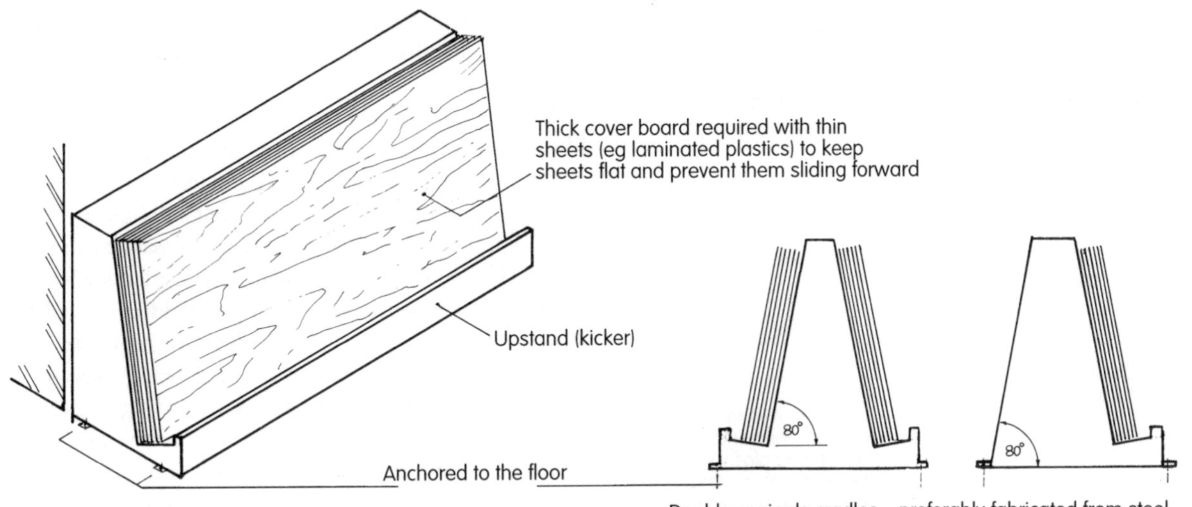

Figure 2.32 Vertical storage using pre-fabricated cradles

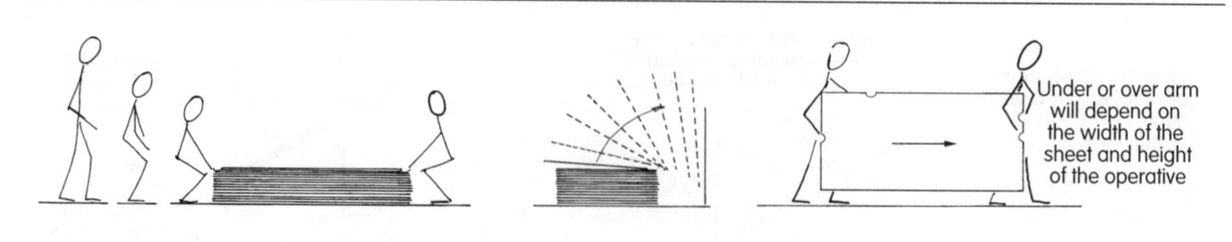

Figure 2.34 Lifting and carrying rigid and semi-rigid sheet material

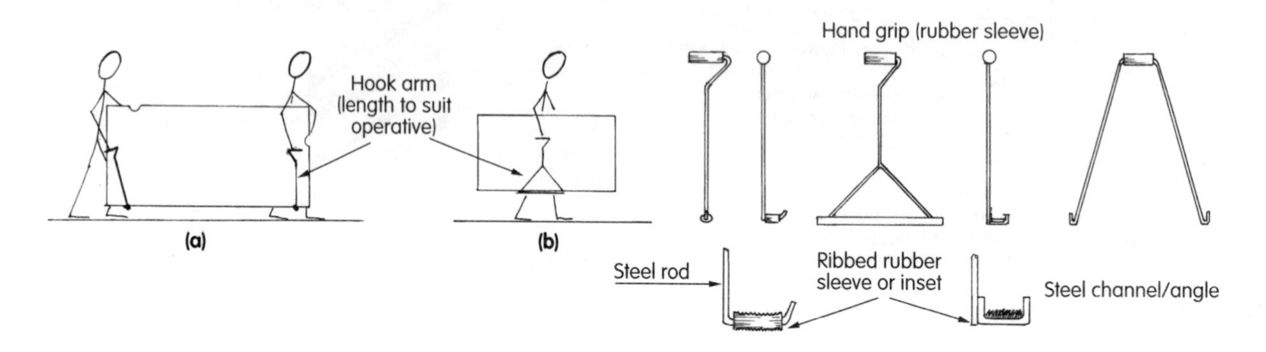

Figure 2.35 The use of purpose made 'Porterhooks'

manoeuvred by two people (Figure 2.34). If the weight of the board, or perhaps its slipperiness, makes edge carrying difficult, or possibly even dangerous, then it may be worth considering using a carrying aid. Figure 2.35 shows how a simple hook arrangement can be improvised to make this task both easier and safer.

Two single hooks, one at each end, as shown in Figure 2.35a, can be used to carry long sheets with relative ease. A single handed double hook could be used to handle smaller yet awkward sheets, as shown in Figure 2.35b.

Flimsy sheet material such as laminated plastics must be handled differently. As Figure 2.36 shows, if the sheets are bowed slightly, they become more rigid, enabling them to be moved more easily. Very thin sheets, such as the thinner grades of laminated plastics, may in some cases be rolled into a cylinder about 600 mm in diameter, and, with their decorative face held inwards with strong cord or a band. Special care must be taken when releasing the cord, since the laminate will tend to spring apart, and the edges of thin plastic laminates can be very sharp.

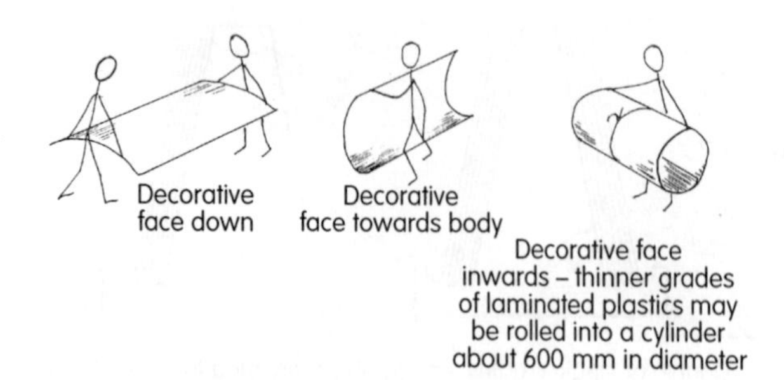

Figure 2.36 Handling laminated plastics

2.12 CONDITIONING WOOD-BASED BOARDS AND SHEET MATERIAL

Nearly all natural wood products and wood derivatives such as wood-based boards and sheet material will, because of their hygroscopic nature (inherent ability to take up and shed moisture), require conditioning if moisture movement is to be limited.

Conditioning will mean open stacking the material with air exposure to both faces and as many edges as possible for prescribed periods of time (a minimum of 48 hours). The air temperature and relative humidity must be similar to those where the material will eventually be placed. In this way, moisture movement will be allowed to stabilise, thus avoiding excessive expansion or contraction.

The increase or loss of moisture will vary until equilibrium with the environment is reached, but as a rough guide, in a fully centrally heated environment, one could expect to find moisture content levels of about 6–9%. Within unheated new buildings, an increase to about 16% MC would not be unusual. This explains why, when you fix wood-based boards as cladding to walls, gaps are left between boards to allow for moisture movement. Similarly when decking is being fixed to floors, even though board joints are fixed tight, gaps are left around the floor edges at the perimeter walls.

Standard and tempered hardboard and type LM mediumboard are the only types of manufactured board that may be conditioned with water to accelerate their moisture intake. This method of conditioning involves

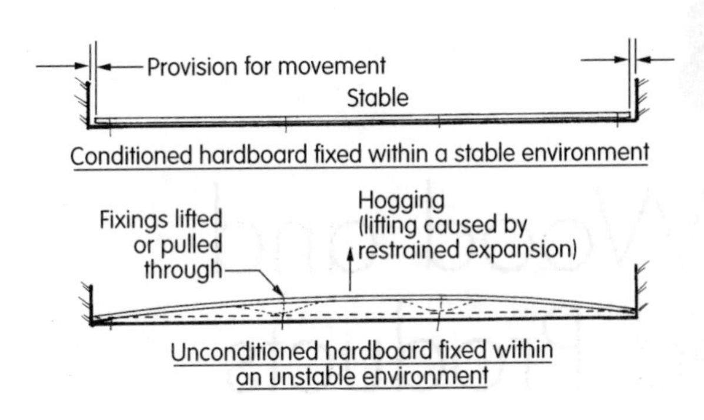

Figure 2.37 Possible effect of using 'conditioned' and 'unconditioned' hardboard

rubbing clean water into the back of the sheet or board (mesh face) with a brush or mop. Rubbing starts from the centre and the whole surface should be covered. Then, after open stacking (wet face to wet face) the material should be allowed to stand for at least 48 hours (tempered hardboard for 72 hours).

Figure 2.37 shows how unconditioned hardboard may react if subjected to dampness after being fixed in position.

It is important to note, however, that sheets/boards treated with flame retardant must never be conditioned with water, or, for that matter, used or fixed where they may become damp.

2.13 HEALTH AND SAFETY

Cutting, shaping, forming and drilling any of these products will result in the generation of dust particles which can be harmful to health. Under the Control Of Substances Hazardous to Health (COSHH) Regulations, exposure to dust should be prevented or controlled. Limits are set under the Occupational Exposure Limits laid down within Health & Safety at Work Act documentation.

Dust is usually controlled by a means of extraction from individual machines.

Operators employed on site work will be required to use dust masks and goggles for all operations involving cutting, shaping, forming and drilling these materials.

Skin irritation can be an added problem caused by wood, cement or plaster dusts. Therefore, whenever you are handling newly cut boards, gloves should be considered where appropriate.

3

Enemies of Wood and Wood-based Products

When a tree dies naturally, or prematurely due to drought or storm, the wood will eventually settle to the forest floor, where it will decay by being broken down by fungi and insects into humus (decomposed organic matter) and returned to the soil to help sustain new plant life. It is this natural wood decay that we interfere with when we wish to retain wood as timber.

Because we mainly cultivate trees specifically for timber and wood-based products, we tend (for economic reasons) to cut down trees prematurely. We then convert them into timber etc., with useful purposes in mind – unfortunately nature does not see it that way, as it has no means of discriminating between dead wood and timber. Therefore, provided that favourable conditions exist, wood is constantly being sought out for destruction by the most natural of means: fungal and/or insect attack. This process could, with some species, be achieved within a few years. However, conditions could mean that wood could survive in its natural dead state for over 100 years.

Information like this allows us to proceed with confidence, knowing that natural decay can be avoided and that timber can have a useful life for many years.

There are at least four common factors that can be responsible for the destruction, or degradation, of otherwise defect-free wood:

- weathering
- non-rotting fungi
- wood-rotting fungi
- fire (see Section 1.10.4(h)).

3.1 WEATHERING

We are often told that, if we leave timber unprotected (unpainted etc.) outside, exposed to the elements such as wind, rain and snow it will rot; it can if the conditions are right (see Section 3.3). However, provided that the timber is not in contact with the ground, or sited where moisture from rain, etc. can become trapped or retained as shown in Figure 3.1, then perhaps the worst of the weathering problems would be:

- gradual surface erosion due to constant or a periodic bombardment of rain and abrasive wind borne particles over the exposed surfaces, which can take on a corrugated appearance (depending on wood species) with a downy texture due to breaking up of woody tissue and constant surface 'checking' (Table 1.10); it may take 50 years to erode 3 mm of the surface of the wood
- discoloration due to exposure to sunlight and the removal of the wood's natural colouring by leaching induced by rain. This will eventually result in the wood taking on a silvery discoloration often accompanied by, and associated with, local atmospheric conditions or pollutants.

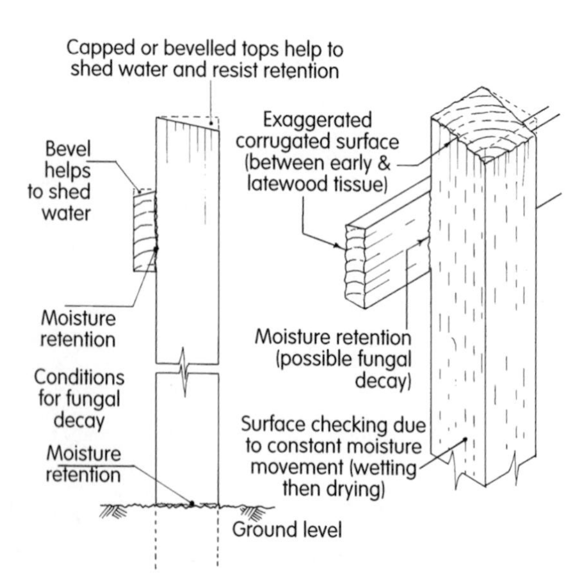

Figure 3.1 Possible effect of weathering on fence posts, rails, and points of moisture retention

3.1.1 Preventative Measures

If it is simply a question of preventing surface erosion, then surface treatment with a water repellent should serve that purpose, but if the wood's natural colour has to be retained, then a proprietary clear varnish could be the answer.

It is worth mentioning here that if timber is to be left to weather naturally it would be advisable to choose a wood species with a durability classification of at least 'moderately durable' (see Tables 1.18 and 1.19). Special attention should be paid to any joint, intersections and end grain, as these are the most vulnerable areas where moisture pick up is most likely to occur.

In this case the term 'weathering' also refers to provision made when preparing (processing) certain externally used timber sections such as:

- fence and gate posts
- fence and gate rails
- window sills
- door weather boards, etc.

With these and other sections, bevelled edges or capping, and/or profiled shapes (weather boards) enable water to be quickly shed from their surfaces.

3.2 NON-ROTTING FUNGI

This type of fungus does not significantly damage the timber other than degrade it by appearance. Surface staining as a result of an attack may appear blue to black in colour.

3.2.1 Sap Stain (Blue Stain) Fungi

As the name implies these fungi take their nourishment from the tree's food reserves left within the sapwood of timber produced from softwood species and light coloured hardwoods. These fungi may attack:

- softwoods; those with distinct heartwood such as pines and larches are particularly susceptible to staining, unlike spruces (without distinct heartwood) such as whitewood, which is less prone to attack
- temperate hardwoods; those most liable to attack are poplar and ash
- tropical hardwoods; species most liable to attack are obeche, jelutong, ramin and balsa wood.

Attacks from sap stain can start soon after the tree is felled and continue until the wood achieves a moisture content below 25%.

3.3 WOOD-ROTTING FUNGI

There are many types and species of fungi, all belonging to the plant kingdom, but they are unlike plants that contain 'chlorophyll' (see Section 1.2.11) a necessary ingredient for producing plant food. Fungi are parasites in that they prey on and live off other living and dead plant material (usually when the plant is either sick or dying) for their source of food.

We have already discussed how the sap stain types of fungi operate. We are concerned here with those types that are responsible for wood decay, which causes the wood to slowly decompose and eventually collapse.

Whenever the following conditions exist, one or more of these types of fungi will eventually become established. The type of fungus and its characteristic life style will in the main be determined by the amount of moisture present in the wood. Fungi need the following to become established:

- food, in the form of cellulose from the woody tissue of sapwood and non-durable heartwood (see Section 1.10.4(e))
- moisture – in the first instance, wood will have had to attain a moisture content in excess of 20%
- temperature – temperatures between 30°C and 37°C; low temperatures may reduce growth; high temperatures will kill the fungi.
- air, which is an essential requirement for growth and respiration of fungi.

Once fungi become established, it is only a matter of time before the wood substance starts to decompose and structural breakdown occurs with the result that the wood:

- loses its strength
- becomes lighter in weight (has a lower density)
- changes its colour, either by becoming darker (brown rot), or lighter (white rot), depending on the type of fungus
- loses its natural smell and takes on a musty or fusty smell
- becomes more prone to insect attack.

All wood-destroying fungi have a similar life cycle, and a typical example is shown in Figure 3.2. The spores have been transported from the parent plant (fruiting body or sporophore) by wind, insects, animals or an unsuspecting human, to a suitable piece of fertile wood where germination can take place. Once established, the fungus spreads its roots (hyphae) into and along the wood in search of food, eventually becoming a mass of tubular threads which collectively are called 'mycelium'. These will eventually produce a *fruiting body* (sporophore).

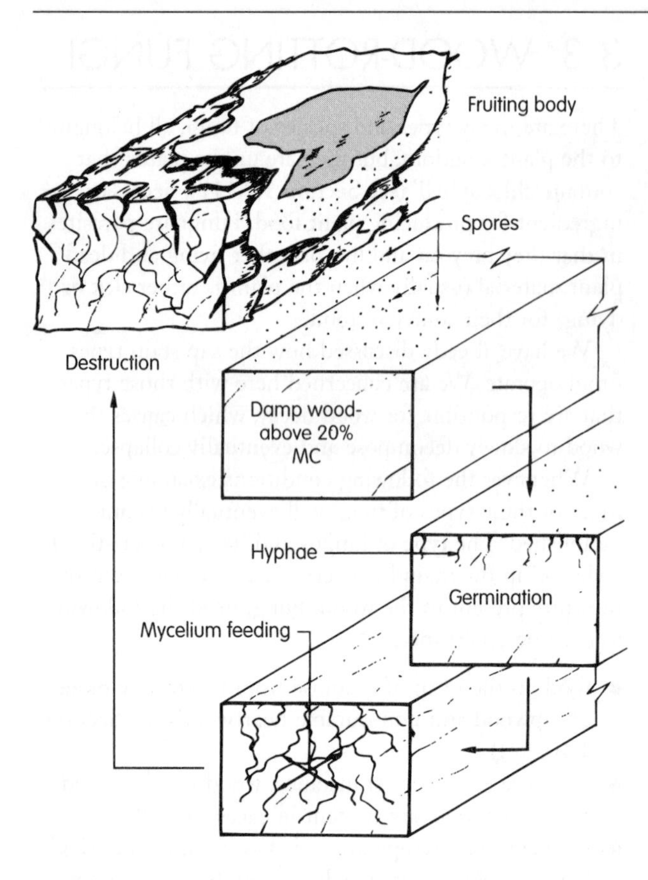

Figure 3.2 Life cycle of a wood-destroying fungi

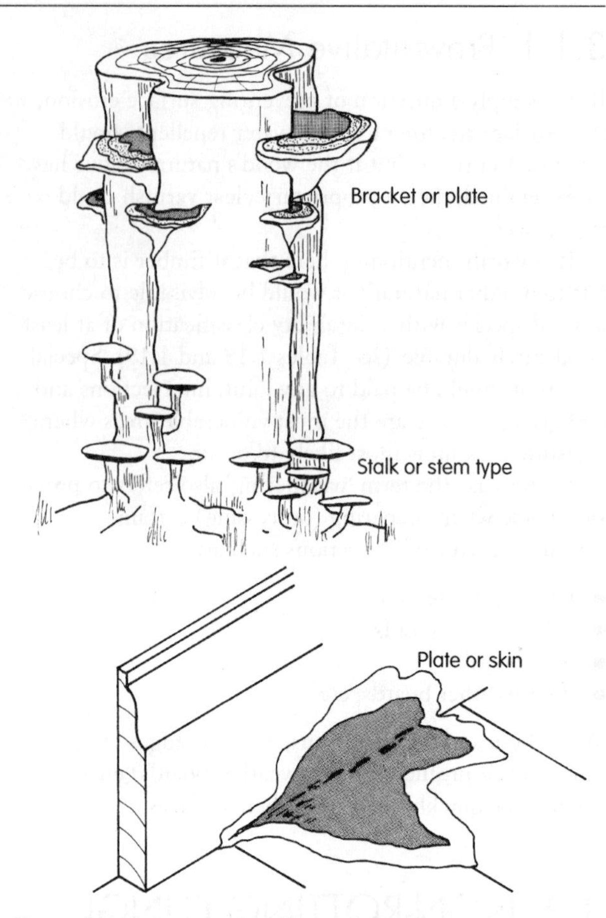

Figure 3.3 Fruiting bodies (Sporophores)

As shown in Figure 3.3 a fruiting body can, depending on the species, take the shape of a 'stalk', 'bracket' or 'plate'. Each fruiting body is capable of producing and shedding millions of minute spores, of which only a very small proportion will germinate.

Wood-rotting fungi can be classified as being either a *brown rot* or a *white rot*. The distinction is quite easy to remember, because:

- brown rots darken the colour of the wood under attack; once dry the residue will be brittle and have a crazed appearance (these include both the 'dry rot fungus' and some 'wet rot fungi')
- white rots, lighten the colour of the wood they attack; they belong within the group of 'wet rots' and are often associated with the decay of external joinery.

This chapter is, however, concerned only with two types of wood destroying fungi – both of which are brown rots – i.e. the dry rot fungus (Serpula lacrymans), and possibly the most common of wet rots, cellar rot (Coniophora puteana). The following sections should be read in conjunction with Table 3.1.

3.3.1 Dry Rot Fungus (Serpula lacrymans)

Timber becomes liable to dry rot attack when its moisture content exceeds 20% and it is sited in positions of poor ventilation. Good ventilation is one of the prime factors in keeping timber below 20% MC. Probably the most ideal conditions for the development of dry rot are situations of high humidity with little or no ventilation and a moisture content between 30–40%. Under these conditions dry rot can spread to distant parts of a building, provided that a source of moisture is available; nothing seems to stand in the way of the rot reaching fresh supplies of wood.

This fungus has moisture conducting strands which help it to sustain growth as it travels behind plaster, or through walls in its search for food.

Once established, its initial starting point (often its main source of moisture) becomes more difficult to find: the fungus may have travelled from room to room, from floor to floor, even into or from the roof, and have re-established itself with another moisture supply. Places which are most at risk will have evolved from bad design, bad workmanship, or lack of building maintenance. Some of the locations most at risk are shown later in Figure 3.7. A typical outbreak of dry rot within

Table 3.1 General characteristics of the wood-rotting fungi: dry rot and cellar rot

	Name of fungus	
	Dry rot (serpula lacrymans)	**Cellar rot (coniophora putaena)**
Appearance		
fruiting body (sporophore)	Plate or bracket, white-edged with rust-red centre (red spore dust)	Plate – not often found in buildings – olive green or brown
mycelium	Fluffy or matted, white to grey, sometimes with tinges of lilac and yellow	Rare
strands	Thick grey strands – can conduct moisture from and through masonry	Thin brown or black strands, visible on the surface of timber and masonry
How wood is affected	Becomes dry and brittle, breaking up into large and small cubes, brown in colour	Exposed surfaces may initially remain intact. Internal cracking, with the grain and to a lesser extent across it. Dark brown in colour
Occurrence	Within buildings, originating from damp locations	Very damp parts of a building, e.g. affected by permanent rising damp, leaking water pipes, etc.
Other remarks	Once established it can spread to other, drier, parts of the building	Fungal attack ceases once dampness is permanently removed

Note: Both the above fungi can render structural timbers unsafe

a suspended timber floor is shown in Figures 3.4 and 3.5. Tell-tale signs of an outbreak of dry rot could be one or more of the following:

- the smell, which is a distinct mushroom-like odour (damp and musty)
- a distorted wood surface, which is warped, sunken (concaved), and/or with shrinkage cracks; lightly tapping the surface with a hammer will often produce a hollow sound, and the wood will offer little, if any resistance to being pierced with a knife or bradawl
- the appearance of a fruiting body, either in the form of a 'plate' (skin), or 'bracket'
- the presence of fine red dust, which is the spores from a fruiting body.

Exposed untreated timber recently under attack may reveal a covering of a soft carpet of mycelium. Established mycelium may appear as a greyish skin with tinges of yellow and lilac. Underlying wood will have contracted into cuboidal sections (Figure 3.6).

The fruiting body of the fungus, which is capable of producing millions of spores, can appear as an irregular shaped rusty-red fleshy overlay with white edges.

Figure 3.4 Surface effect of a dry rot attack (with kind permission from Rentokil)

Figure 3.5 Severe attack of dry rot exposed (with kind permission from Rentokil)

Figure 3.6 Dry-rot damage (with kind permission from Rentokil)

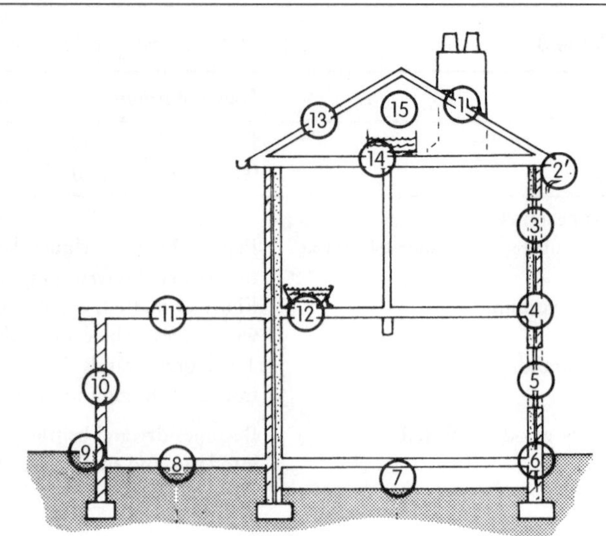

Figure 3.7 Possible causes of dampness. **1** – defective step flashing; **2** – defective or blocked gutter/fallpipe; **3** – window condensation, insufficient external weathering; **4** – bridged cavity wall (mortar droppings, etc.); **5** – defective or omitted vertical damp proof course (DPC) to doors and/or windows; **6** – defective or omitted horizontal DPC, blocked air brick/grate, ground above DPC; **7** – no ventilation to under-floor space; **8** – defective or omitted DP membrane; **9** – defective or omitted DPC, ground above DPC; **10** – solid wall of porous masonry; **11** – defective roof covering, unvented void (cold-deck construction), defective or omitted vapour barrier; **12** – defective plumbing/water spillage; **13** – defective roof covering; **14** – defective plumbing; **15** – unvented roof space

3.3.2 Eradication of Dry Rot Fungus

Treatment of major infestations should only be undertaken by companies who specialise in this area of work. However, the following stages will normally be considered necessary even for the smallest of outbreaks:

a Eliminate all obvious sources of dampness (some possible causes are shown in Figure 3.7).
b Investigate the extent of the outbreak, removing woodwork and plasterwork as necessary.
c Search for further causes of dampness both within and outside the area of attack.
d Remove *all* affected woodwork and fungus from the building (timber should be cut back as shown in Figure 3.8). Take the material to where it can be safely incinerated or disposed of. *In both these cases the method and type of disposal will generally require local authority approval.*
e Surrounding walls and concrete floors etc. may require sterilising and treatment with a fungicide.
f All replacement timber should have a moisture content of less than 20%.
g In areas where there is a risk of re-infestation (those with high humidity) replacement timber will require preservative treatment (with special attention being given to end grain).

See also Section 4.3 on wood preservatives.

3.3.3 Cellar Rot (Coniophora puteana)

This wet rot fungus is responsible for much of the damage caused to timber in and around our buildings when allowed to sustain levels of high humidity and as a result wood is allowed to attain a high level of moisture content.

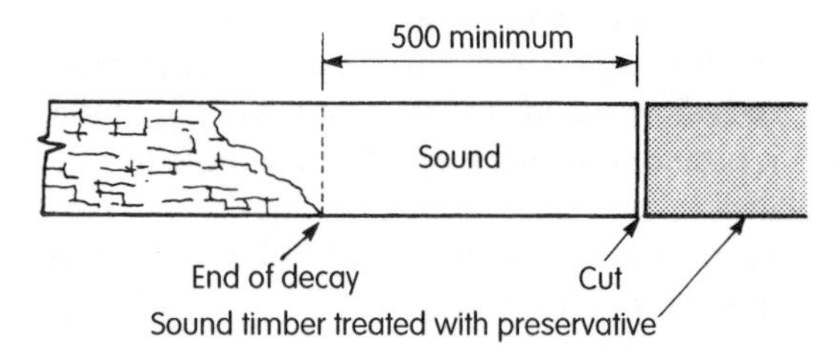

Figure 3.8 Cutting away timber affected by rot

The most favourable locations for this rot are those that can permanently provide wood with a moisture content of about 40–50%. Examples of suitable conditions are:

- in damp cellars or other rooms below ground level
- beneath leaking water pipes and radiators
- behind and under sinks and baths, etc., due to persistent overspill and splashing
- in areas of heavy condensation (check windows, walls, roofs)
- in areas where water can creep (by capillarity) and remain free from evaporation, e.g. behind damaged paintwork, under window sills (Figure 3.9) and thresholds, etc.

Figure 3.9 A case of a wet-rot attacking a window sill (with kind permission from Rentokil)

- above a defective damp-proof course (DPC) or membrane (DPM)
- in timber which is in permanent contact with the ground, or sited below a DPC or DPM.

See also Figure 3.7.

However, unlike dry rot, once the source of moisture has been removed and the wood has dried out, cellar rot becomes inactive. The effect of wet rot damage on end grain is shown in Figure 3.10.

Figure 3.10 Wet rot damage to end grain (with kind permission from Rentokil)

The extent of the decay may not at first be obvious, because the outer surfaces of the timber often appear sound. It is not until these surfaces are tested with the point of a knife or bradawl that the full extent of the damage is known. This rot is, however, reasonably confined to the area of dampness.

One of the biggest dangers associated with cellar rot is that the area of confinement may, in certain circumstances, be taken over by dry rot (see dry rot requirements), in which case a major problem may well have developed.

There are several other species of wet rot, all of which require reasonably high percentages of moisture content. This can make formal identification difficult.

3.3.4 Eradication of Cellar Rot

Unlike dry rot, the one controlling factor here is dampness. Wet rots should be treated in the following manner:

a remedy as necessary any sources or defects responsible for the dampness (Figure 3.7)
b dry out the building
c remove and safely dispose of affected timber, cutting back well into sound wood (Figure 3.8)
d where there is risk that the timber may not be kept dry, for example, in contact with the ground, then suitable wood preservative should be applied to timbers at risk (paying attention to end grain treatment); alternatively, at risk replacement timber may be acquired as pre-treated timber (see Section 4.4)
e all replacement timber must be dry with a moisture content below 20%.

Note that, because the strands of cellar rot do not penetrate masonry, treatment is more localised.

3.3.5 Preventing Wood Rot

In theory, rot should not occur in a building which has been correctly designed, constructed and maintained. For prevention, all that is required is that the moisture content of wood should never be allowed to exceed 20%.

The only obvious places where wood is subjected to conditions likely to achieve high moisture levels are those external situations with full exposure to the elements, such as garden fences and gates, particularly the fence and gate posts at ground level. Items of external joinery, such as doors and windows etc. have partial exposure to the elements. Each of the items will either require:

- permanent protection against the entry of moisture
- siting where high moisture levels cannot be maintained
- natural durability classification (see Section 1.10.4(e) and Table 1.22)
- full pressure treatment with a suitable wood preservative
- a combination of all of the above.

3.4 WOOD BORING INSECTS

There are many species of wood boring insects, each with their own special lifestyle. For example there are those that prefer to prey on living, or recently felled trees, and those with a preference for different types and conditions of timber.

This section is concerned with those insects that attack wood and timber in use within the UK construction industry. Table 3.2 lists these, together with their general characteristics, which should be referred to as you study each insect separately.

Table 3.2 General characteristics of some wood-destroying insects

Name of insect	Powderpost beetle (*Lyctus spp.*), Figure 3.12	Common furniture beetle (*Anobium punctatum*), Figure 3.13	Death-watch beetle (*Xestobium rufovillosum*), Figure 3.14	House longhorn beetle (*Hylotrupes bajulus*), Figure 3.15	Wood-boring weevils (*Pentathrum* or *Euophryum spp.*), Figure 3.16
Adult characteristics					
size (length)	5 mm	5 mm	6–8 mm	10–20 mm	3–5 mm
colour	Reddish brown to black	Reddish to blackish brown	Chocolate brown	Grey/black/brown	Reddish brown to black
flight	May to September	May to August	March to June	July to September	Any time
Where eggs laid on or in wood	Vessels of large-pored hardwoods	Crevices, cracks, flight holes, etc.	Fissures* in decayed wood, flight holes	Fissures* in softwood, sapwood	Usually on decayed wood
Size of larvae	Up to 6 mm	Up to 6 mm	Up to 8 mm	Up to 30 mm	3–5 mm
Bore dust (frass)	Very fine powder soft and silky	Slightly gritty ellipsoidal pellets	Bun-shaped pellets	Cylindrical pellets	Very small pellets
Diameter of flight (exit) hole	0.75–1.5 mm	2 mm	3 mm	6–9 mm ellipsoidal (oval)	Up to 1.5 mm
Life cycle	10 months to 2 years	2 years or more	3–15 years	3–11 years	7–9 months
Wood attacked					
type	HW, e.g. Oak, Elm, Obeche	SW and HW plywood with natural adhesive	HW, especially oak	SW	SW and HW, plywood with natural adhesive
condition	Sapwood fungus	Mainly sapwood fungus	Previously decayed by heartwood later	Sapwood first	Usually damp or decayed
location	Timber and plywood while drying or in storage	Furniture, structural timbers, etc.	Roofs of old buildings, e.g. churches	Roofs and structural timbers	Area affected by fungus (usually wet rot) – cellars etc.
Other remarks	Softwood and heartwood immune	Accounts for about 75% of woodworm damage in UK Resin-bonded plywood and fibreboard immune	Attacks started in HW can spread to SW	Restricted to areas of Surrey and Berkshire	Attack continues after wood has dried out in UK

Note: * Fissures – cracks, narrow openings and crevices

General identification can often be made by noting one or more of the following characteristics:

- the habitat
- the size and shape of beetles
- the size and shape of larva
- the size and shape of bore dust (frass)
- the size and shape of exit (flight) holes
- the sound (death watch beetle).

A typical life cycle for these insects (with the exception of wood boring weevils) is as follows (Figure 3.11).

1 Eggs will be laid by the female beetle in small cavities below the surface of the wood.
2 After a short period (usually a few weeks) the eggs hatch into larvae (grubs) and enter the wood, where they progressively gnaw their way further in, leaving excreted wood bore dust (frass) in the tunnel as they move along.
3 After one or more years of tunnelling (depending on the species of the insect) the larva undermines a small chamber just below the surface of the wood, where it pupates (turns into a chrysalis).

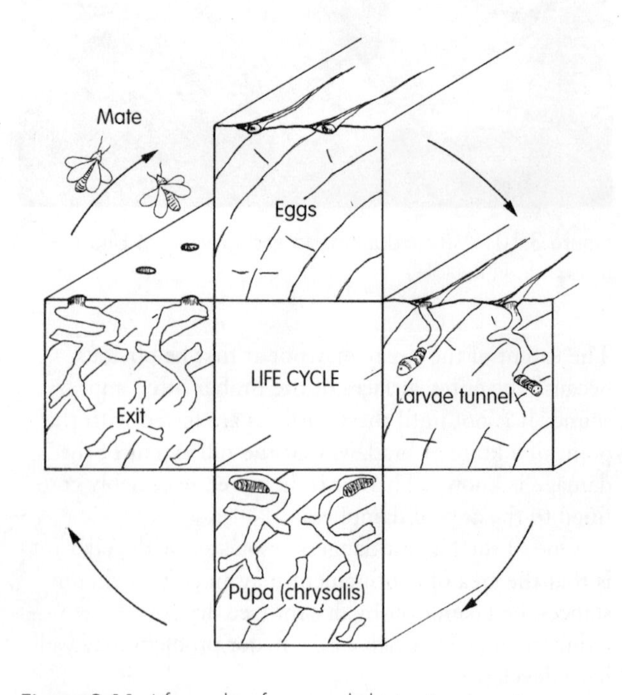

Figure 3.11 Life cycle of a wood destroying insect

4 The pupa then takes the form of a beetle and emerges from its chamber by biting its way out, leaving a hole in the surface of the wood. This hole is known as an exit or flight hole. (Collectively these holes are usually the first sign of any attack). The insect is now free to travel or fly at will, to mate and complete the life cycle.

Two insects not mentioned in Table 3.2 are ambrosia beetles (including the pinhole borer) and the wood wasp, which are usually associated only with attack on a standing tree or green wood. However, we often see the results of their work on or within the timber we use.

3.4.1 Ambrosia Beetle (Pinhole Borers)

With the ambrosia beetle, damage is noticeable through the black stained edges of their bore holes (tunnels), which generally run across the grain of the wood.

Tunnels become lined with a fungus (ambrosia fungus) which is introduced by the beetles and thereby provides food for both the beetle and its larvae. But as soon as the moisture content of the wood is reduced to about 30%, the fungus dies, and, as a consequence, so do the beetles as their food supply ends, leaving behind those unmistakable non-active black lined bore holes which will be clearly visible when exposed during conversion to timber, etc.

3.4.2 Wood Wasp

Damage from wood wasps can result in quite large holes, up to 9 mm in diameter, being bored through the wood. Trees most susceptible to an attack by wood wasps are usually unhealthy softwood trees, and as a result timber derived from them may contain active larvae. However, any emerging wood wasp would be unable to reinfest the wood once the wood had been dried.

The adult wood wasp cannot sting, and is harmless to humans, although at first sight, because of its size (10–50 mm long), yellow and black/blue markings, long spine and loud buzzing sound, it is often mistaken for a hornet or similar stinging species of wasp.

3.4.3 Powderpost Beetle (*Lyctus spp.*) (Figure 3.12)

This species is generally associated with partially dried hardwoods. There are many species of 'Lyctus' denoted by '*spp.*', the most common of which is Lyctus brunneus, to which this text applies. 'Powderpost' refers to the very fine powdered bore dust (frass) left by the lava.

This beetle attacks the sapwood of hardwoods, especially those with large 'pores (vessels)', e.g. oaks, elms, ash, and some tropical species such as obeche and ramin. When they are in a partly dried state, for example when stacked in a timber yard, these hardwoods are very vulnerable. Infestation can be passed into the home via hardwood furniture, flooring, or panelling, etc.

Hardwood species with small pores (which restrict egg laying) or insufficient starch content (which is essential for larval growth) are generally safe from attack. We can therefore say that some hardwoods, softwoods and heartwoods are generally immune.

3.4.4 Common Furniture Beetle (Anobium punctatum) (Figure 3.13)

In Britain, the larvae (woodworm) of this beetle are responsible for most of the damage caused by insects to property and contents, attacking both softwoods and temperate hardwoods, but showing a preference for sapwood. Adults range from reddish to blackish brown in colour, 3–5 mm in length, and live for about 30 days. Eggs, up to about 80 in number, are laid by the female in small fissures or old flight holes.

Actual size about 5 to 6 mm long — Severe internal damage and apparently superficial external damage — Bore dust

Figure 3.12 Powderpost beetle (*Lyctus spp.*) (with kind permission from Rentokil)

Actual size approximately 3 to 5 mm long

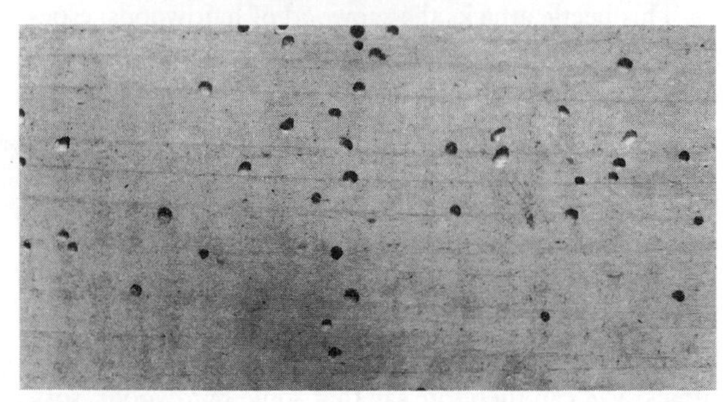

Flight holes

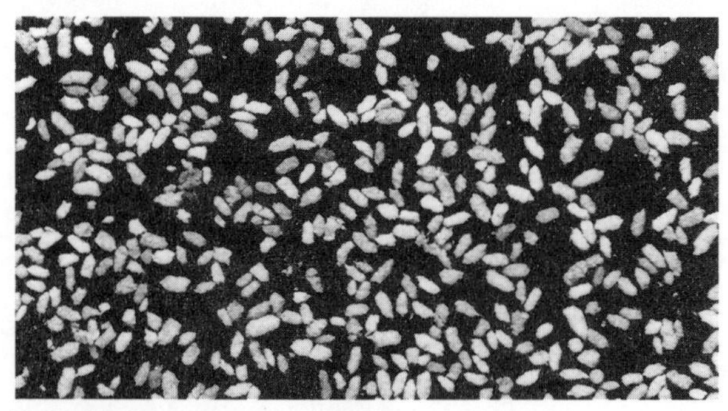

Bore dust

Figure 3.13 Common furniture beetle (with kind permission from Rentokil)

The resulting larvae (up to 6 mm long) may then burrow through the wood for over two years, leaving in their trail a bore dust (frass) of minute ellipsoidal pellets resembling fine sand. The larvae finally come to rest in small chambers just below the surface of the timber, were they pupate (change into chrysalides).

Once the transformation to beetle form is complete (between May and August) they bite their way out, leaving 'flight' (or 'exit') holes about 2 mm in diameter.

Because of the beetles' ability to fly, it seems that very few timbers are exempt. However, some of the tropical hardwoods seem to be either immune or very resistant to attack, e.g. Abura, Afrormosia, Idigbo, Iroko, Sapele, and African Walnut.

Evidence of an attack is provided by the flight holes and, in many cases, the bore dust ejected from them. Old and neglected parts of property, including outbuildings, provide an ideal habitat for these insects. Attics and cellars are of particular interest to them, as they may remain undisturbed there for many years.

3.4.5 Death-watch Beetle (*Xestobium rufovillosum*) (Figure 3.14)

The Death-watch beetle is aptly named, because of the ticking or tapping noise it makes during its mating season from March to June and its often eerie presence in the rafters of old building and churches.

Its appearance is similar to that of the common furniture beetle, but, not only can it be twice as long, it generally attacks only hardwoods which have been previously affected by fungi. It seems to have a preference for oak. The duration of the attack may be as short as one year or as long as ten (or more). Much depends on the condition of the wood, the amount of fungal decay, and environmental conditions.

Because the life cycle can be lengthy, infestations over wide areas are limited.

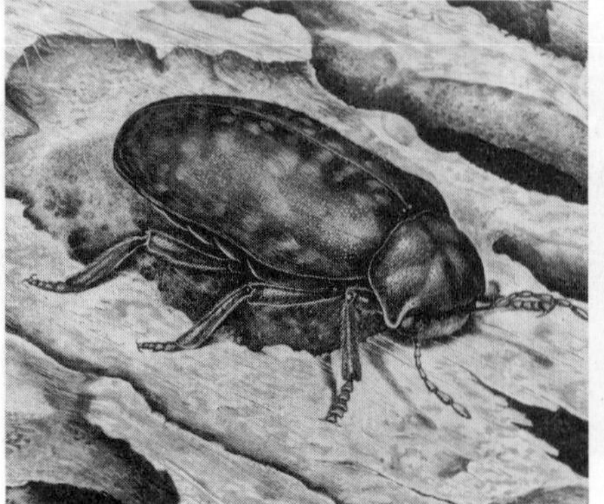

Actual size approximately 6 mm long

Typical damage

Bore dust

Figure 3.14 Death-watch beetle (*Xestobium rufovillosum*) (with kind permission from Rentokil)

Actual size approximately 25 mm long

Typical damage to rafters

Bore dust

Figure 3.15 House longhorn beetle (*Hylotrupes bajulus*) (with kind permission from Rentokil)

3.4.6 House Longhorn Beetle (*Hylotrupes bajulus*) (Figure 3.15)

This large beetle takes its name from its long feelers. It is a very serious problem on the mainland of Europe and in parts of Britain, mainly in Surrey and Berkshire, where it attacks the sapwood of dried softwood.

Because of the size of its larvae and the subsequent bore holes left by extensive tunnelling over a number of years, the extent of the damage caused to structural timbers has led to much concern. It is mandatory under the building regulations to treat with a wood preservative all softwoods used for the purpose of constructing a roof or ceiling within those geographical areas stated in the regulations (the areas are mainly Surrey and Berkshire).

3.4.7 Wood-boring Weevils (*Pentarthrum huttoni* and *Euophryum confine*) (Figure 3.16)

Beetles in this group have distinct protruding snouts from which their feelers project. Their attacks are usually confined to damp and/or decayed (usually wet rot) sapwood and occasionally heartwood of both softwood and hardwood.

Unlike the other wood-boring beetles mentioned, both the adult and the larvae of these insects carry out 'boring' activities. Tunnels tend to follow the pattern of the wood's grain.

3.4.8 General Eradication of Wood-boring Insects

Both obvious and suspected areas of activity should be fully investigated to determine:

- the extent and nature of the attack
- the size and shape of flight holes
- the amount and nature of bore dust (frass)
- the moisture content of the wood

Actual size approximately 3 to 5 mm long

Typical internal damage

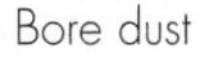

Bore dust

Figure 3.16 Wood-boring weevils (*Pentarthrum huttoni*) (with kind permission from Rentokil)

- if fungal attack is in evidence
- the species and nature of the wood under attack.

It should then be possible to establish which beetle or beetles are responsible for the attack.

False alarms can occur if flight holes are the only symptom, as an attack by the ambrosia beetle (see Section 3.4.1) may have taken place before use, when the timber was 'green' (not dried), in which case the culprit would have been killed during timber drying.

However, assuming that the outbreak being investigated is active, then the following measures can be taken:

a Where possible, open up the affected area. Cut away any wood which is badly attacked, carefully remove it from the site and dispose of it in a safe manner (see also Section 3.3.2(d) and 3.3.4(c)).
b If load bearing timbers are affected, professional advice should be sought about their structural stability. Repair or replacement of such items usually involves propping and shoring, to safeguard against structural movement or collapse.
c Thoroughly clear the whole area of debris. Dust removal may involve the use of an industrial vacuum cleaner.
d If the attack is localised to a small area, then treat the replacement and remaining timbers just beyond that area with an approved preservative (insecticide). On the other hand, if the attack is more extensive, then the work should be undertaken by a firm specialising in insect infestations. In either case the application of wood preservatives must be carried out according to the manufacturer's instructions and the Health and Safety at Work Act guidelines. The method of preservative application will depend on the amount of treatment required. With a small outbreak brush application may be employed, in which case special attention should be given to treating wood end grain, fissures, and joints.

Operatives must wear an approved face mask, for protection against inhaling fumes and/or vapours given off. Protective clothing, goggles, and gloves must be used, to stop eyes and skin from coming into contact with preservative or treated materials. Precautions must always be taken against the possibility of fire, since, at the time of application, organic solvent types of preservative give off a vapour which is a fire risk (see Section 4.7).

3.4.9 Preventative Measures

Probably the best means of preventing insect attack is, where practicable, to keep the moisture content of wood below 10%, for at this level these insects will be discouraged from breeding. Unfortunately many structural timbers during the winter months may well have moisture contents that exceed this figure. However, most situations do not under normal circumstances warrant any special treatment, other than general hygiene and good housekeeping. There are instances where buildings have a history of insect attack, or are sited within a geographical area where infestation by certain species is common (see Section 3.4.6). In these cases preservative/preventative treatment may have to be considered.

In general terms, it would be fair to say that the use of chemical (synthetic) wood preservatives (insecticides) should be considered only as a last resort.

4

Wood Preservation and Protection

Except where timber and wood products are used in their natural state, either for environmental, practical, or economical reasons, they are generally treated with either paint, water-repellent stain, wood preservative, or special solutions to reduce or retard the effects shown in Figure 4.1, and itemised below:

a weathering
b exposure to sunlight
c moisture movement
d fire

and to reduce or eliminate the risk of:

e fungal attack
f insect attack.

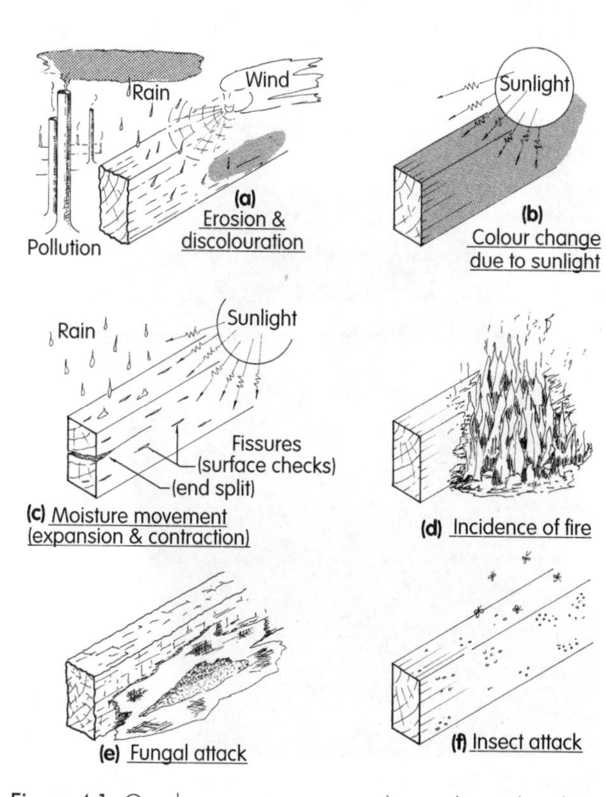

Figure 4.1 Good reasons to protect timber and woodproducts

4.1 PAINTS AND VARNISHES

Here we are looking at exterior-quality paints (opaque coatings) and varnishes (transparent coatings) which also offer decorative properties as well as surface protection from:

- entry of water (moisture) into the wood
- abrasive particles driven by wind and rain
- exposure to sunlight.

The weathering effect of wind, rain and sunlight will gradually discolour and degrade the surface of unpainted timber by breaking down its surface.

Water penetration of non-durable wood is the greatest problem – not only will this result in variable dimensional change (expansion on wetting, contraction on drying) and possibly splitting or checking (see Table 1.10), it also increases the risk of fungal attack. Sunlight causes the wood to change colour and contributes to its degradation.

Correctly painted surfaces should give relatively good protection for up to five years, prior to repainting, provided that:

- the timber is thoroughly dry (i.e. it has the appropriate moisture content) before paints are applied
- the surfaces have been prepared correctly
- the paint manufacturer's recommended number of coats is given
- the end grain has received particular attention with regard to coverage
- inspections are carried out regularly.

If a painted surface becomes damaged, moisture may enter the wood and become trapped behind the remaining film of paint, and further paint failure and wood decay could quickly follow.

This failure may well have been the result of joint movement which produced a hairline fracture of a non-flexible paint film. This could then allow moisture ingress by inducing capillary attraction.

Two ways of resolving moisture ingress are as follows:

1 to use a paint which, when dry, leaves a flexible film over the surface of the wood. This film would be able to stretch when required, when a small amount of moisture movement occurred
2 to use a 'breather' or 'microporous' paint which allows any buildup of moisture vapour behind the protective coating to escape via minute pores in the film. It is, however, important to note that, because of the possible movement of water vapour and the different formulations used (emulsions), any fittings or fixings used and associated with these paints should be 'rust proofed'.

4.2 EXTERIOR STAINS

Stains of this type can provide a clear or coloured, matt or semi-gloss, water repellent surface with fungicidal and microporous properties; but unlike paint these wood stains are translucent, thus allowing wood grain to show through. These stains usually give protection from both weathering and fungal staining (if a fungicide is included within the formulation) for up to four years or possibly more, depending on the stain quality.

Again, because these stain types are usually microporous, only rust proofed fixings and fittings should be allowed to come in contact with them.

4.3 WOOD PRESERVATIVES

Wood, after drying to 20% MC or below, may be put into situations (usually exterior) which allow it to pick up moisture (remember wood is hygroscopic; see Section 1.7.4) and bring the moisture content above 20%. Then those portions of sapwood and non-durable heartwood will be liable to attack by fungi, and in some cases liable to forms of insect infestation. (For classification of wood durability, see Section 1.10.4(e) and Table 1.22). The important aspect here is that the wood retains this high moisture content, since unprotected wood will acclimatise itself to its surroundings until it reaches an equilibrium moisture content.

Timber can, however, be classified according to its potential end use by placing it into a 'hazard class' (HC). For example:

HC(1), being the least hazardous, is intended when timber is sited above ground level in a dry situation with a moisture content (MC) below 18%
HC(2) is when timber is at risk of occasionally getting wet and consequently exceeding 20% MC
HC(3) is when timber is left uncovered and frequently exceeds 20% MC.
HC(4) is when timber is permanently in contact with water and thereby constantly exceeds 20% MC
HC(5) is as HC(4) but applies to salt water.

We nearly always offer the wood some form of surface protection against moisture pick up, such as coating it with paint. The trouble here is that, as we have already discussed, if moisture becomes trapped behind the coating, it can be retained in the wood, and, provided that the other necessary requirements are met, give ideal conditions for fungi to germinate. Probably the best example of how permanent moisture retention can cause or contribute to fungal decay is the traditional garden fence. If the fence is exposed to drying wind, you will usually find wood decay only in areas where moisture can be retained even on a dry day, for example, where posts come into contact with the ground, or where tops of posts are left level with unprotected end grain exposure (see Figure 4.2). You may also find decay where members overlap one another, such as rails over posts, or palings over rails, etc. The rest of the fence can last in its unprotected weathered state for many years, although it must be said that this would be totally impractical unless a means could be found to protect those areas vulnerable to fungal attack.

In this and similar cases we can interfere with nature and make those non-durable parts of wood durable by introducing wood preservatives (fungicide) into the cell structure of the wood to make it toxic to fungi. Such treatment can also have the same effect on wood-boring insects when an insecticide is used. A combination of fungicide and insecticide could also be used.

Figure 4.2 The effect of weather on unprotected flat-topped timber fence posts

Wood preservation may be required in certain vulnerable situations as insurance against possible incidences of fungal and/or insect attack. Or it may be specified as a precaution against bad workmanship or constructional detailing. Either way it should be realised that wood preservatives have a vital part to play in our industry, and are a contributing factor in assuring the longevity of our modern buildings.

Wood preservatives generally form three main groups:

1 organic-solvent types (OS)
2 water-borne types (WB)
3 tar-oil (TO).

4.3.1 Organic-solvent Preservatives

These use a medium of organic solvents to transmit the toxic chemicals into the wood. After application the solvents evaporate, leaving the wood toxic to fungi and insects.

Methods of application for new timber are normally by double vacuum impregnation and immersion, and, where remedial treatment is required, brush or spray. The solvents used in the preparation, such as white spirit, are generally volatile (quick drying) and flammable. Extreme care must therefore be taken at the time of application and in storing containers.

These preservatives do not affect the dimensions of the timber (swelling) or have a corrosive effect on metals. The ability to glue or paint timber is unaltered after treatment once the preservative has dried.

Table 4.1 Treatment and general usage of organic-solvent preservatives

Method of treatment	General use
Low pressure (double vacuum)	Joinery timber (e.g. windows)
	Carcassing timber
	Roof trusses
Immersion	Remedial treatment
Brush	Localised treatment
Spray	

4.3.2 Water-borne Preservatives

These use water to convey the chemicals into the wood. The most common of these are the copper, chrome, and arsenic (CCA) type. Copper is the fungicide, arsenic the insecticide, and once impregnated into the wood the chrome acts as a fixing agent to make the compound non-soluble, in other words fixed within the wood. Water-borne preservatives do have to be applied by a pressure process (usually an empty cell process, see Section 4.4.2 and Table 4.2) to ensure thorough penetration: they are unsuitable for non-pressure application. These preservatives have no smell and are usually non-flammable. When the timber has been re-dried after treatment it can be painted or glued.

Table 4.2 Treatment and use of water-borne preservatives

Method of treatment	General use
Pressure	Structural building timbers
	Agricultural timbers
	Horticultural timbers
	General fencing posts (stakes) and rails
	Railway sleepers
	Marine work

Other types of water-borne preservatives are available, one of which is based on boron; this is applied to green timber by a method known as diffusion (see Section 4.4.1(f)).

4.3.3 Tar-oil Preservatives

These are derived from coal tar and are ideal for preserving exterior work which is not to be painted. They do not usually have any corrosive effect on metals, but they will stain most of the porous materials they contact.

The most common form of tar-oil preservative is 'creosote' (Table 4.3), which can be light to dark brown in colour and can be applied by various processes (pressure and non-pressure), including brushing and spraying. It is flammable and has a strong odour for some time after application.

Table 4.3 Treatment and use of tar-oil type (creosote) preservatives

Method of treatment	General use
Pressure	General fencing posts (stakes) and rails
Immersion	Telegraph posts
Steeping	Railway sleepers
Hot and cold tank	Farming timber
Brush	Fence timber

4.4 METHODS OF APPLYING WOOD PRESERVATIVES

Preservatives are applied by one of three general methods:

1 non-pressure
2 pressure
3 low-pressure.

Figure 4.3 gives some indication of how the method used can affect the depth of penetration, which will also be affected by the permeability of the wood species (see Section 1.10.4(f)).

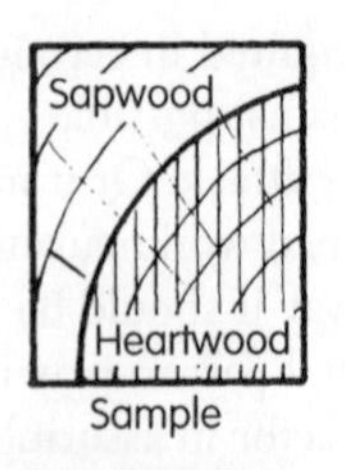

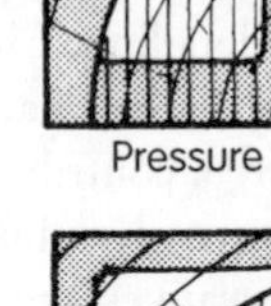

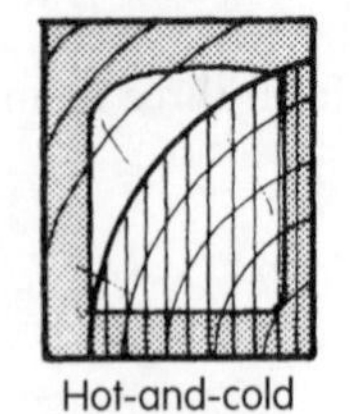

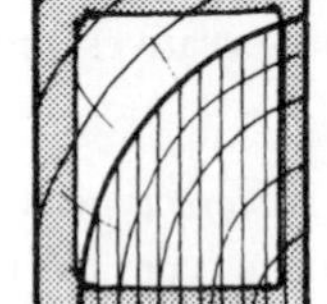

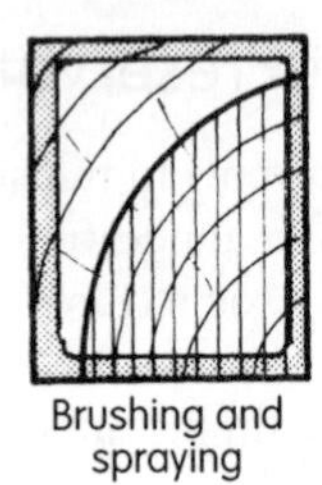

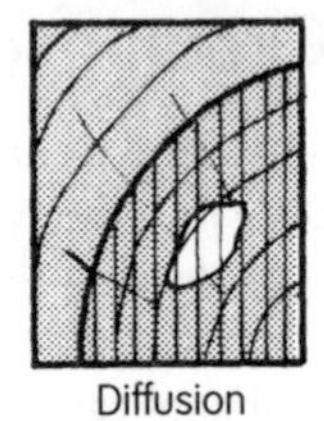

Figure 4.3 A guide to preservative penetration (exaggerated) after various methods of application to a permeable species

4.4.1 Non-pressure Methods (Figure 4.4)

These include:

- brushing
- spraying
- deluging
- immersion
- hot and cold open tank treatment
- diffusion.

Brushing (Figure 4.4a)

Brushing can be used for applying creosote and organic types of preservatives but, because of low penetration, it is not a suitable method for timber which comes in contact with the ground. As a rule, re-treatment is advisable every three to four years.

Spraying (Figure 4.4b)

Spraying achieves similar levels of penetration as brushing, and, because of the health risks associated with applying wood preservatives, the following precautions should always be taken, particularly when spraying, to ensure that:

- only coarse sprays are used, to avoid atomisation
- work areas are well ventilated
- operatives are suitably clothed
- hands are protected by gloves
- mouth and nose are protected by an approved face mask
- eyes are protected with snug fitting goggles – not glasses
- manufacturer's instructions are followed.

Deluging (Figure 4.4c)

In deluging, the timber is passed through a tunnel of jets which spray it with preservative. In the main, organic-solvent preservatives are used in this process, but creosote may be used.

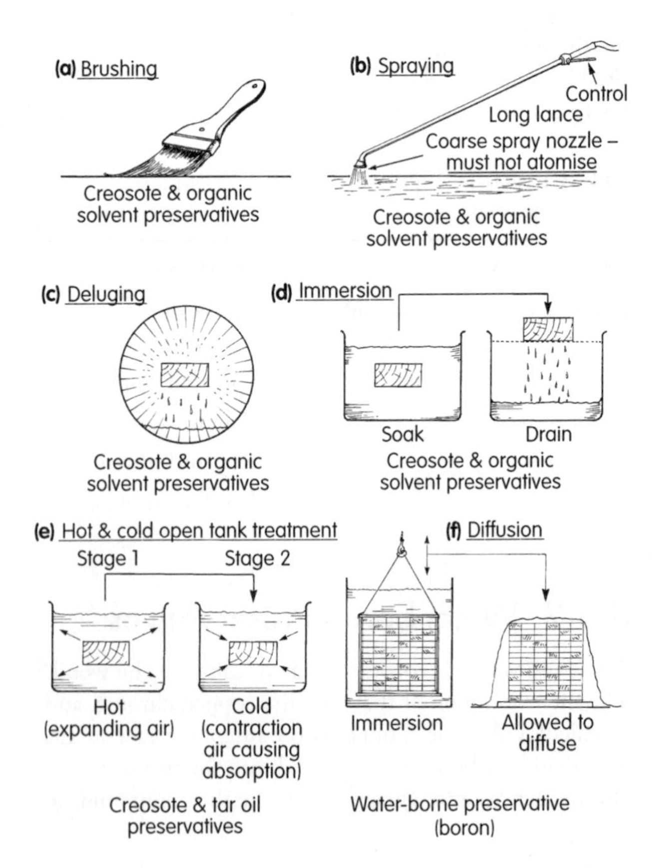

Figure 4.4 Illustrative guide to different non-pressure methods of applying wood preservatives

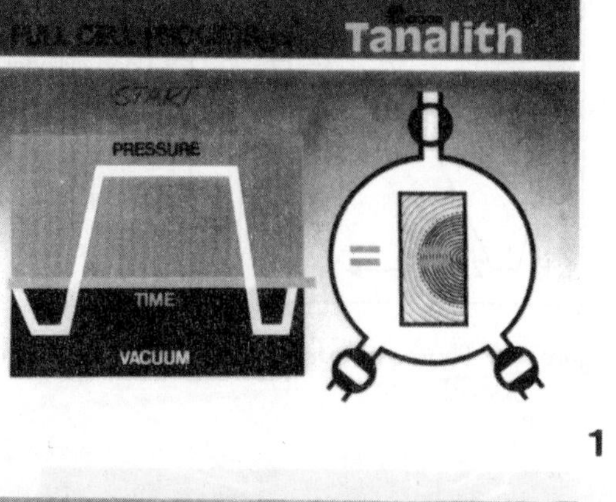

1

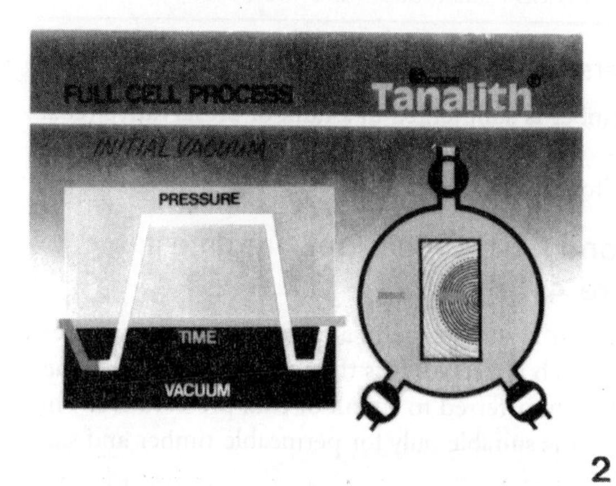

2

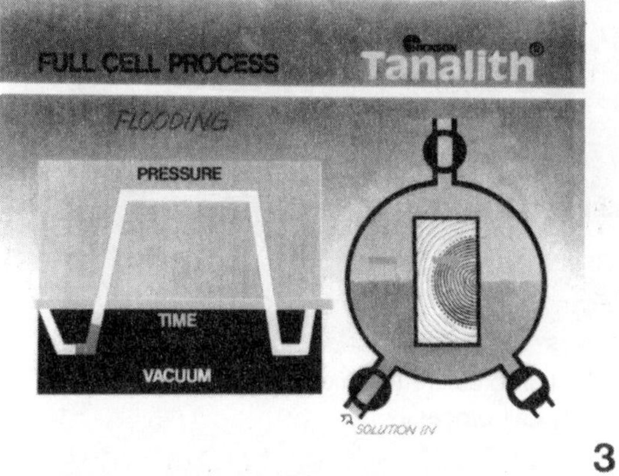

3

FULL CELL PROCESS Tanalith PRESSURE STAGE PRESSURE TIME VACUUM VESSEL FULL

4

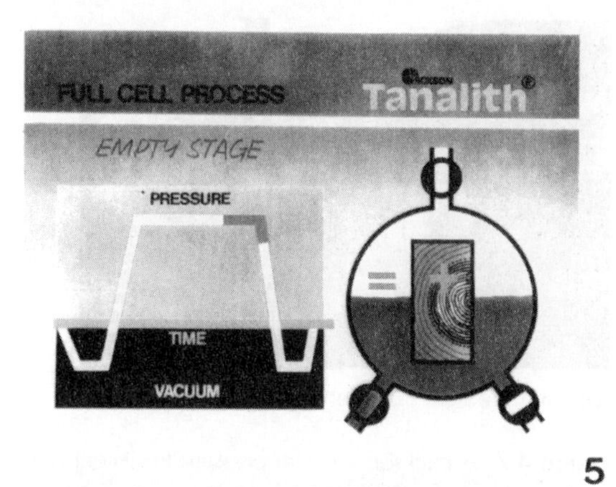

5

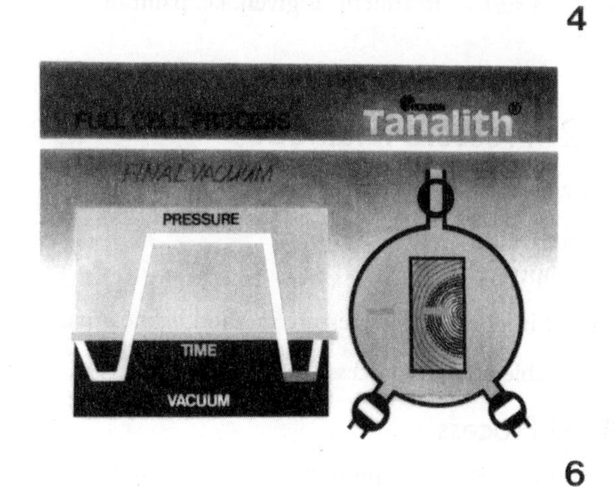

6

FULL CELL PROCESS Tanalith FINISH PRESSURE TIME VACUUM

7

Figure 4.5 Full-cell process (with kind permission from Hicksons)

Immersion (Dipping) (Figure 4.4d)

The timber is submerged in a tank of preservative (coal-tar oils or organic-solvent types) for a prescribed period, then allowed to drain.

Hot and Cold Open Tank Treatment (Figure 4.4e)

The timber is submerged in a tank of preservative (creosote) which is heated. It is then allowed to cool in the tank or is transferred to a tank of cold preservative. This treatment is suitable only for permeable timber and sapwood.

Coal-tar oils (creosote) are flammable, and therefore extra care is necessary with the heat source.

Diffusion (Figure 4.4f)

This method of treatment is only associated with freshly felled green timber which, at the saw mills, is immersed in a water-borne preservative (usually boron salts) and then close-piled and placed under cover until the preservative has diffused into the wood.

The type of water-borne preservative used in this process is liable to leach out from the wood, which makes the timber unsuitable in wet locations unless an impervious surface treatment is given, i.e. paint or varnish.

4.4.2 Pressure Methods

Pressure methods include:

- the full-cell process
- the empty-cell process.

The most important low-pressure method is known as:

- the double vacuum process.

Full-cell Process

As we will see later, this process achieves maximum preservative retention within the wood. It differs from the empty-cell process by using an initial vacuum to remove air which surrounds the wood and which would provide resistance to preservative penetration. Figure 4.5 shows in diagrammatic form the process from start to finish. Figure 4.6 shows how the five operational stages can be linked together on the one graph.

Figure 4.7(a) shows a typical treatment plant suitable for a full-cell process.

Empty-cell Process

The empty-cell process is unlike the full-cell process, where, after treatment, penetrated wood cell cavities supposedly retain preservatives. In the empty-cell process it is assumed that, after treatment, the cell cavities are left empty of preservative but the cell walls are left fully coated.

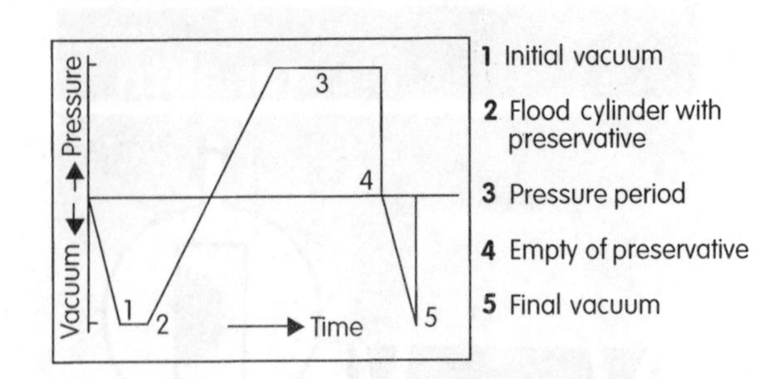

Figure 4.6 Phases of a full-cell cycle (with kind permission from Hicksons)

(a) High pressure treatment plant (full cell and empty cell process)

(b) Low pressure treatment (vac-vac)

Figure 4.7 A Hickson vacuum pressure treatment plant (with kind permission from Hicksons)

As can be seen from Figures 4.8 and 4.9, this process is similar to the full-cell process except that there is no initial vacuum at the start. Air is therefore present before and during the introduction of the preservative. After the pressure period, preservative is pushed out of the wood cell cavities by the compressed air which would have been trapped therein. The final vacuum draws off any surplus solution.

A typical treatment plant is shown in Figure 4.7(a).

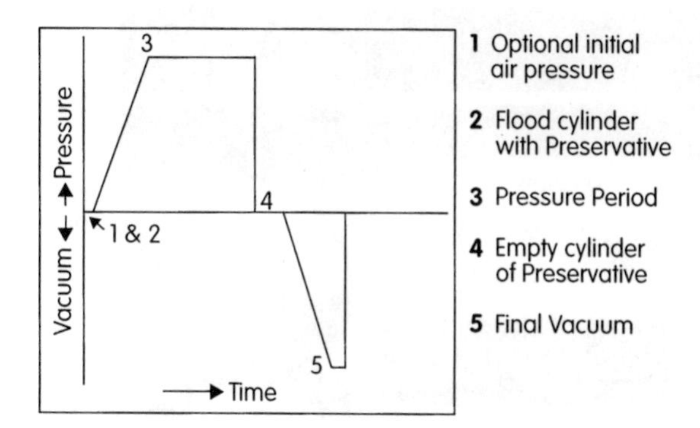

Figure 4.8 Phases of an empty-cell cycle (with kind permission from Hicksons)

Double-vacuum Process

The principles behind this process are the same as those of the full-cell process; but, as you will see from Figures 4.10 and 4.11, the pressures used are only at, or about, atmospheric, so the absorption factor of preservative penetration will be much less. As shown in Table 4.1, the main function is to treat each item of joinery and interior carcassing timber above DPC with organic solvent preservatives.

A typical treatment plant is shown in Figure 4.12a and a loading operation in Figure 4.12b and 4.7(b).

4.5 FLAME RETARDANT TREATMENTS

There are two types of flame retardant treatment:

1 impregnation
2 surface coating.

By impregnating timber using a vacuum pressure treatment with various chemical solutions, it is possible to reduce the rate at which flame would normally spread over the timber's surface. Some panel products, such as chipboards and other particleboards and fibreboards, may have flame retardants incorporated during their manufacture.

Timber impregnated with flame-retardant salts is not normally suitable for exterior use because the salts are liable to leach out – the strength properties of the timber may be reduced as a result of this treatment. However, by using other formulations, leaching can be avoided and strength properties of timber remain unaltered.

Some paints and varnishes are 'intumescent' (they swell when subjected to high heat) and protect the surface of the wood by forming an insulating layer over its surface. Others give off a gas which protects against flaming.

NB: Flame retardants do not increase fire resistance of timber.

4.6 MOULD OILS AND RELEASE AGENTS

These solutions are applied to formwork to prevent concrete sticking to the surface of the forms (parts of the formwork that come in contact with the concrete). A release agent (sometimes called a parting agent) may be in the form of an oil (mould oil), emulsion, or a synthetic resin or plastics compound to give the form's surface a hard, abrasive resistant protective coating. The latter may also require mould oil treatment.

Treatment as a whole increases the life of the forms (reusability) and helps reduce the number of surface blemishes and 'blow holes' (holes left by pockets of air) appearing on the surface of the concrete.

Some treatments are not suitable for metal forms, as they tend to encourage rusting. The manufacturers' recommendations should always be observed with regard to treatment use and application. Methods of application may include spraying, but mould oils are more usually applied by swab or brush.

NB: Care should be taken not to get mould oil (release agents) on any steel reinforcement or areas where adhesion is important.

4.7 HEALTH AND SAFETY

As with all chemical products, care must be taken by operatives during their application, and when handling treated products. Each product must be dealt with according to the manufacturer's or supplier's instructions and safety guidelines issued with each product.

At all times the requirements of the Health & Safety at Work Act and the Control of Substances Hazardous to Health Regulations (COSHH) must be implemented. Particular attention should be given to the application of insecticides, fungicides and flame-retardant chemicals and products treated with them.

Remedial Treatment

Some general safety guidelines when using wood preservatives during remedial (repair) work are listed below.

- Protective clothing must be worn, and include a full overall and PVC gauntlets.
- Protective footwear must be resistant to the chemicals used, and with non-slip soles.
- A dust/fume face mask (of the appropriate approved type) must be used when cleaning prior to treatment.
- A respirator (of the appropriate approved type) must be used when spraying insecticide and/or fungicide solutions.
- Goggles or a full face visor must be worn during the application of a preservative.

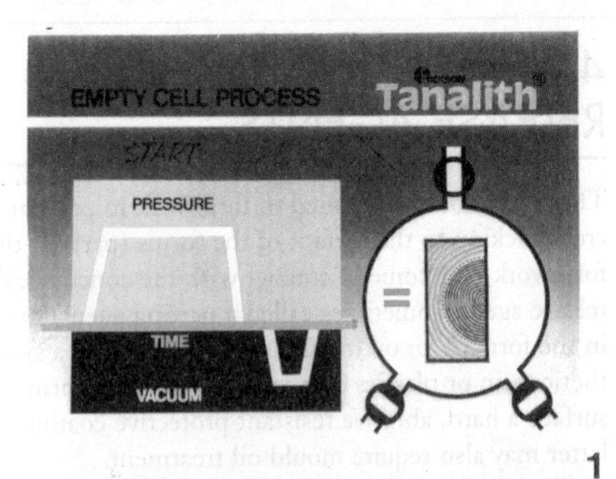

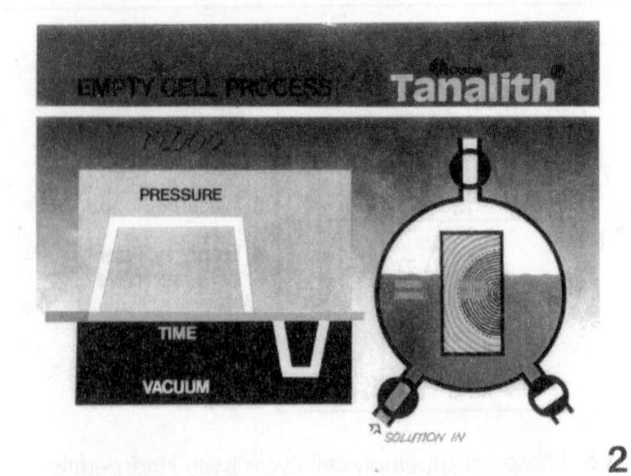

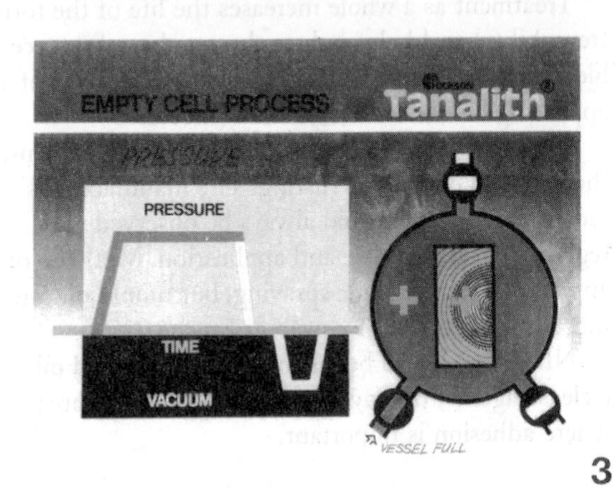

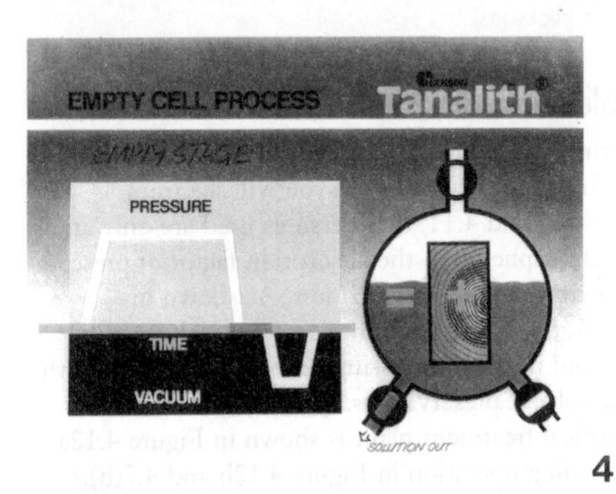

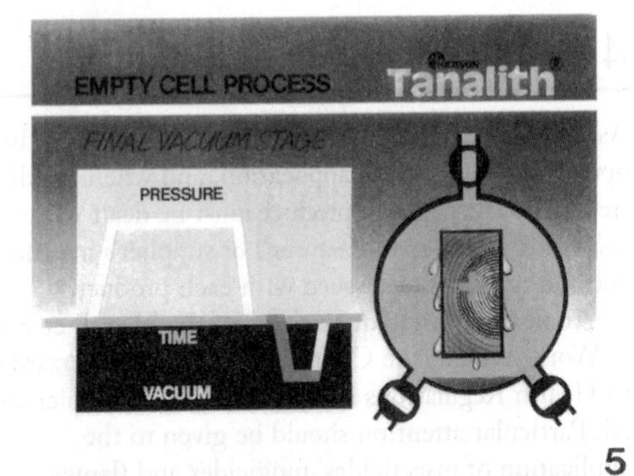

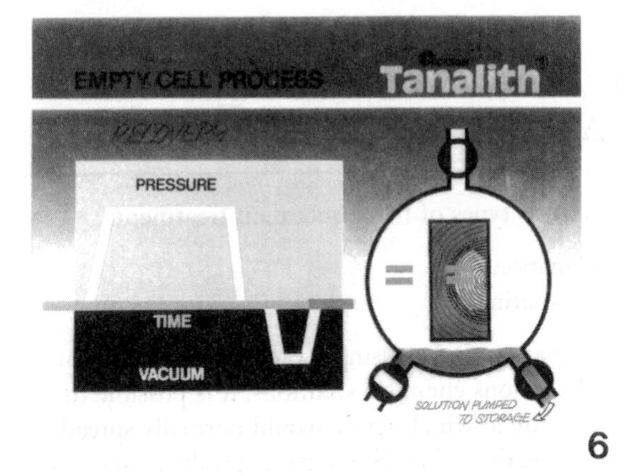

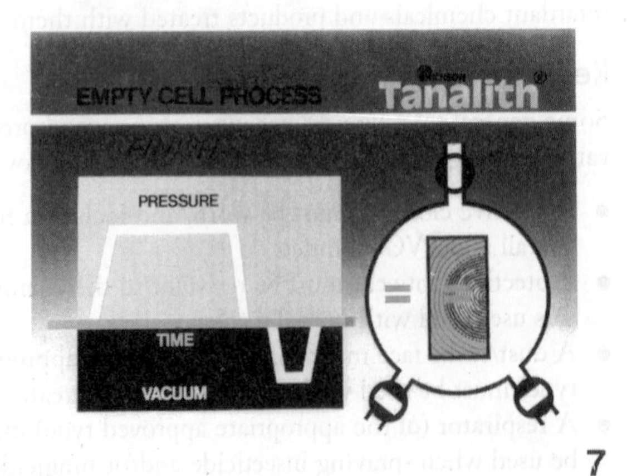

Figure 4.9 Empty-cell process (with kind permission from Hicksons)

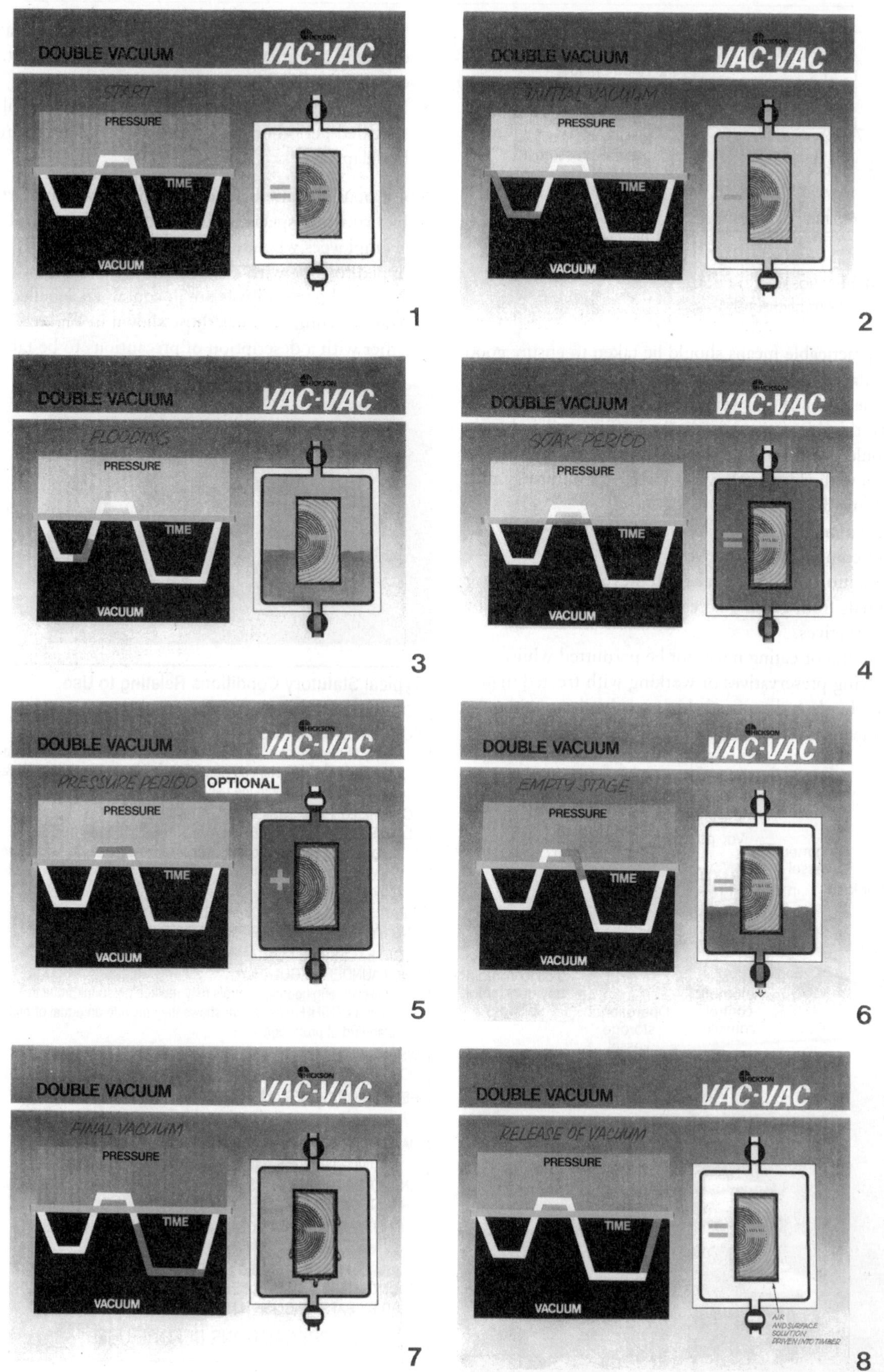

Figure 4.10 Double vacuum process (with kind permission from Hicksons)

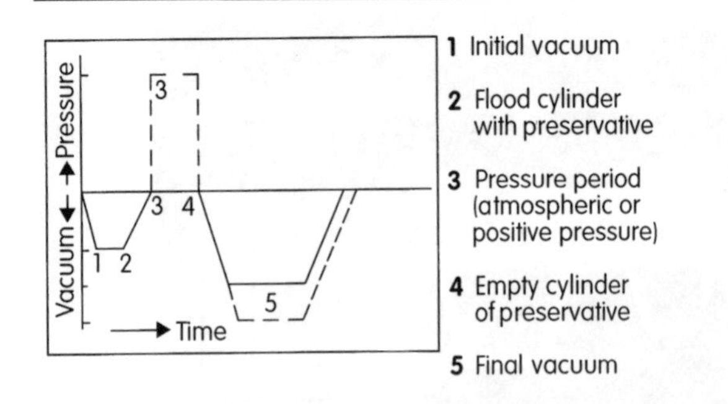

Figure 4.11 Phases of a double vacuum process (with kind permission from Hicksons)

- All practicable means should be taken to ensure good ventilation when applying preservatives.
- In the event of a preservative coming in to contact with the skin, it must be washed off immediately. Should preservatives come in contact with eyes, wash them with clean fresh water for 10–15 minutes and seek medical advice.
- Have a normal bath or shower of the whole body after completion of the work.
- Precautions should be taken to prevent people and/or animals from gaining access to areas contaminated by preservatives.
- Drinking or eating must not be permitted whilst applying preservatives or working with treated material. Hands and exposed skin must be thoroughly washed after work and before eating or smoking.

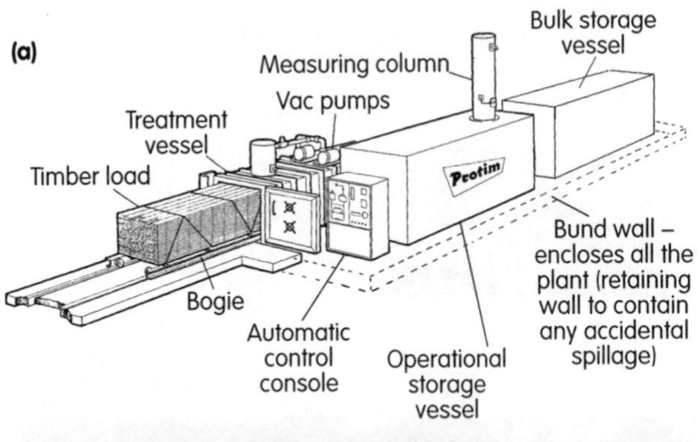

Figure 4.12 Protim Prevac double vacuum treatment (Courtesy of Protim Ltd)

- Smoking is strictly prohibited in areas being treated or where treatment has been recently carried out, or even whilst handling treated material.
- Newly treated areas must not be reoccupied until the specified period of time given by the manufacturer has elapsed.

NB: Extensive remedial work should only be undertaken by firms that specialise in this type of work, and who have employees who are specially trained and fully equipped to cope with all eventualities.

All wood preservatives are in containers labelled with a hazard warning, such as those shown in Figure 4.13, together with a description of precautions to be taken.

Typical Statutory Conditions Relating to Use

FOR USE ONLY AS A WOOD PRESERVATIVE

FOR PROFESSIONAL USE

Apply at the rate of 1 litre of product per 2.5 metres of wood surface. The Control of Substances Hazardous to Health Regulations 1988 (COSHH) may apply to the use of this product at work

FOR USE BY PROFESSIONAL OPERATORS

FLAMMABLE, AVOID NAKED FLAMES AND HOT SURFACES.

Engineering control of operator exposure must be used where reasonably practicable in addition to the following items of personal protective equipment:

WEAR SUITABLE PROTECTIVE CLOTHING (COVERALLS) AND SYNTHETIC RUBBER/PVC GLOVES when using

AVOID EXCESSIVE CONTAMINATION OF COVERALLS AND LAUNDER REGULARLY

However, engineering controls may replace personal protective clothing if a COSHH assessment shows they provide an equal or higher standard of protection

WHEN USING DO NOT EAT, DRINK OR SMOKE

DO NOT APPLY TO SURFACES on which food is stored, prepared or eaten

REMOVE OR COVER ALL FOODSTUFFS before application

AVOID ALL CONTACT WITH PLANT LIFE

DANGEROUS TO FISH AND OTHER AQUATIC LIFE

Do not contaminate watercourses or ground.

UNPROTECTED PERSONS AND ANIMALS SHOULD BE KEPT AWAY FROM TREATED AREAS FOR 48 HOURS, UNTIL SURFACES ARE DRY

THIS MATERIAL AND ITS CONTAINER must be disposed of in a safe way

ALL BATS ARE PROTECTED UNDER THE WILDLIFE AND COUNTRYSIDE ACT 1981. BEFORE TREATING ANY STRUCTURE USED BY BATS, ENGLISH NATURE SHOULD BE CONSULTED

READ ALL PRECAUTIONS BEFORE USE

Figure 4.13 Hazard warning signs shown on the containers of wood preservatives (with kind permission from Hicksons) along with a typical statutory conditions display

5

Hand Tools and Workshop/Site Procedures

Tool skills will, together with good procedures, usually be reflected in the quality of the end product.

Two important ingredients are to make the right tool selection and to be able to maintain the tool's optimum cutting edge and/or functional operation. This chapter has been designed to help operatives make these decisions with confidence and to provide some direction towards the upkeep or maintenance of tools.

5.1 MEASURING TOOLS

Measuring tools are used either to transfer measurements from one item to another or for checking pre-stated sizes.

5.1.1 Scale Rule

At some stage in your career you will have to take sizes from, or enter sizes onto, a drawing. You must therefore familiarise yourself with methods of enlarging or reducing measurements accordingly. It is essential to remember that all sizes will often have to be proportionally reduced to suit the following scales: 1:2 (half full size), 1:5, 1:10, 1:20, 1:50, 1:100, 1:200, 1:1250, 1:2500.

Figure 5.1 illustrates the use of a scale rule, which enables lengths measured on a drawing to be converted to full size measurements and vice versa.

You should also study Section 13.1 which deals with metric and imperial measurement.

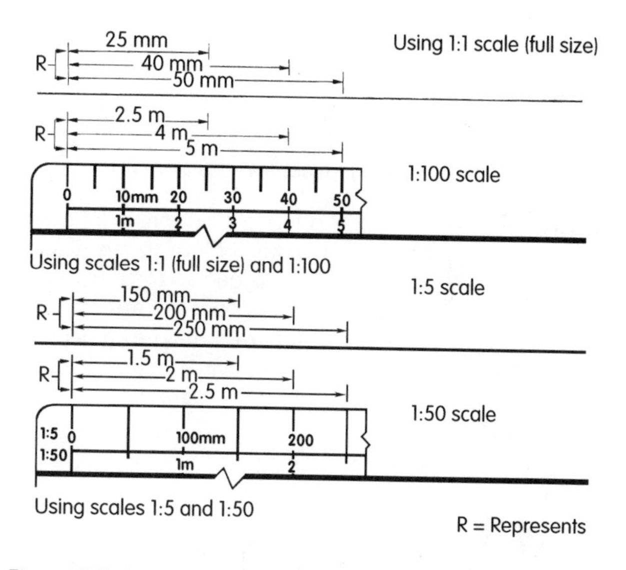

Figure 5.1 Using a scale rule

5.1.2 Four-fold Metre Rule

This rule should have top priority on your list of tools. Not only is it capable of accurate measurement, it is also very adaptable (see Figure 5.2). It is available in both plastic and wood and calibrated in both imperial and metric units. Some models (clinometer rules) also incorporate in their design a spirit level and a circle of degrees from 0° to 180°.

With care, these rules will last for many years so it is important when choosing one to find a type and make that suits your hand. Ideally, the rule should be kept about your person while at work. The most suitable place while working at the bench or on site at ground level is usually in a rule slide pocket sewn to the trouser leg of a bib and brace or overall, etc. The use of a seat or back pocket is not a good idea.

5.1.3 Flexible Steel Measuring Tapes (Figure 5.3)

These tapes retract on to a small enclosed springloaded drum and are pulled out and either pushed back in or have an automatic return which can be stopped at any distance within the limit of the tape's length. Their overall length can vary from 2 to 8 m and they usually remain semi-rigid for the first 500 mm of their length. This type of tool is an invaluable asset, particularly when involved in site work, as it fits easily in the pocket or clips over the belt.

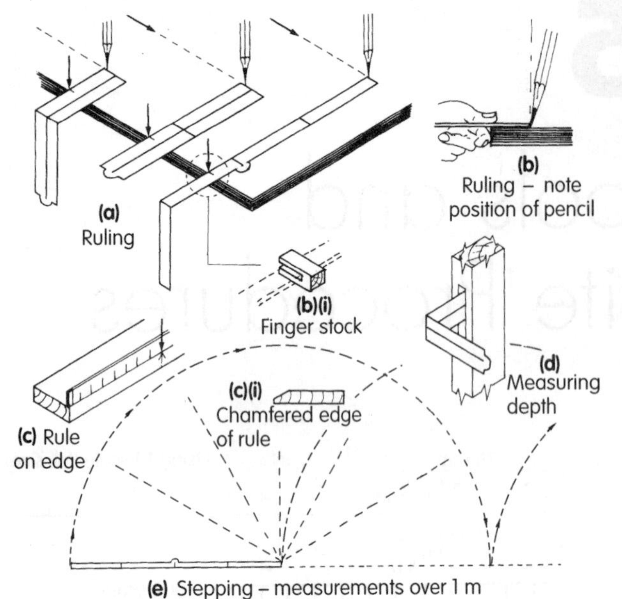

Figure 5.2 The versatility of a four-fold metre rule. (a) ruling different widths; (b) standard method of holding rule at (a); (b)(i) ruling aid – prevents cut fingers and splinters; (c) transferring measurements; (c)(i) chamfered edge of rule – provide accurate flat reading; (d) rule as used as a depth gauge; (e) stepping – measurements over 1 m (not an accurate means of measurement, only used to give an approximate reading)

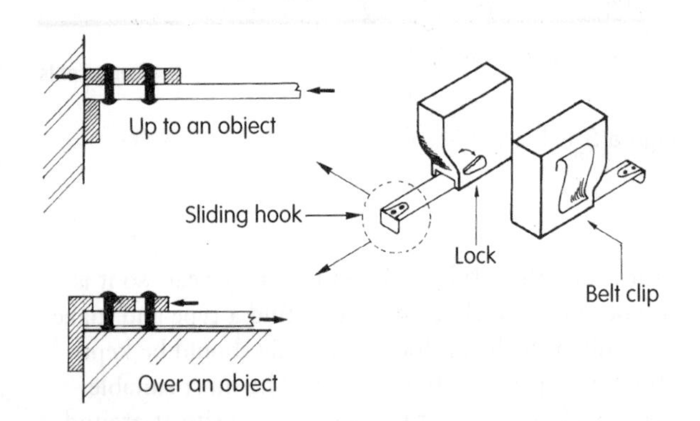

Figure 5.3 Flexible steel tape

It seems to have become common practice of late to use a tape as a substitute for a metre folding rule, though it is better used to complement it.

5.2 SETTING-OUT AND MARKING-OUT (MARKING-OFF) TOOLS

The drawings of a piece of work produced by the designer are usually reduced to an appropriate scale (Figure 5.1) so that an overall picture may be presented to the client. Once approval has been given, the setting-out programme can begin. This will involve redrawing various full-size sections through all the components necessary for the construction, to enable the joiner to visualise all the joint details, etc. and make any adjustments to section sizes.

5.2.1 Setting-out Bench Equipment

Setting-out is done on what is known as a rod. A rod may be a sheet of paper, hardboard or plywood or a board of timber. By adopting a standard setting-out procedure, it is possible to simplify this process. For example (see Figure 5.4):

- draw all sections with their face side towards you (Figure 5.4a)
- draw vertical sections (VS) first with their tops to your left (Figure 5.4b)
- draw horizontal sections (HS) above VS, keeping members with identical sections in line on HS and VS, e.g. the top rail with stile in Figure 5.4c
- allow a minimum of 20 mm between sections (Figure 5.4d)
- dimension only overall heights, widths and depths.

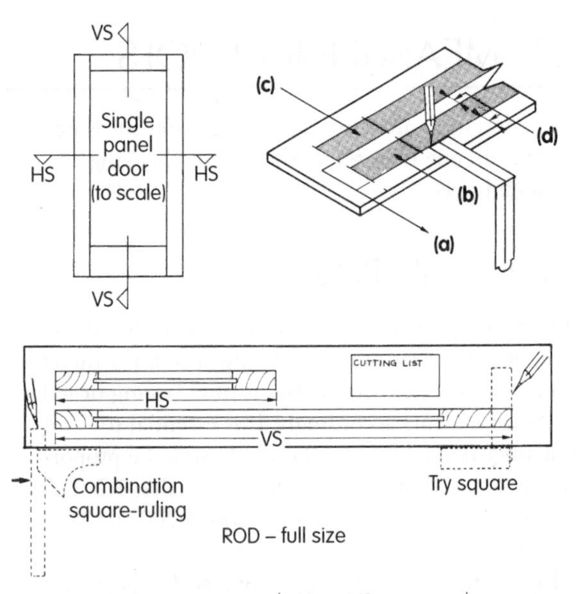

Figure 5.4 Setting out a rod. Key: VS – vertical section; HS – horizontal section

Setting-out will involve the use of some, if not all, of the following tools and equipment:

- a scale rule (Figure 5.1)
- a straight-edge
- a four-fold metre rule (Figure 5.2)
- drafting tape, drawing-board clips or drawing pins
- an HB pencil
- a try-square (Figure 5.4 and see Figure 5.9)
- a combination square (Figure 5.4 and see Figure 5.9)
- dividers (Figure 5.5)
- compasses (Figure 5.5)
- a trammel (Figure 5.6).

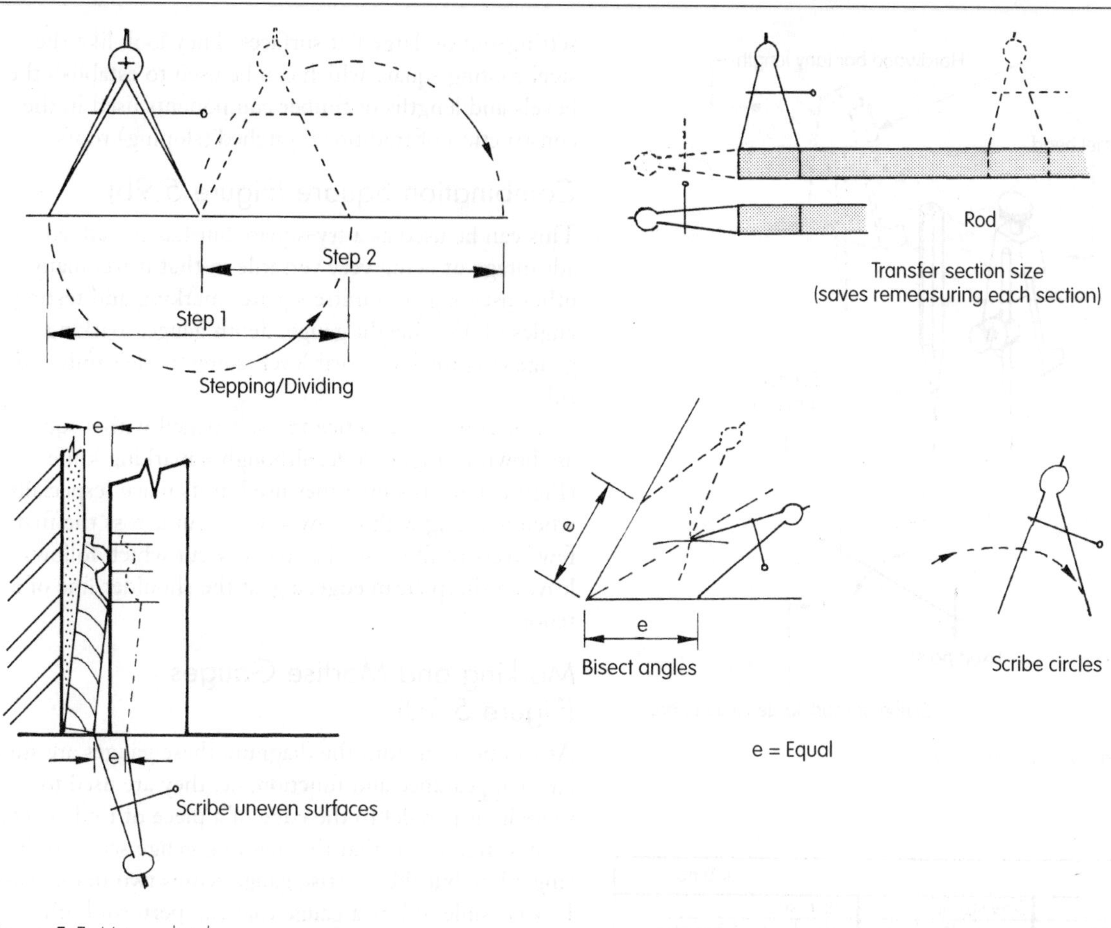

Figure 5.5 Using dividers or compasses

Figure 5.4 shows a space left on the rod for a cutting list. This is an itemised list of all the material sizes required to complete a piece of work. A typical cutting list is shown in Figure 5.7, together with provision for items of hardware (ironmongery), etc. Information about timber sizes and quantity will be required by the wood machinist (Chapter 9).

Marking off involves the transfer of rod dimensions to the pieces of timber and/or other materials needed. Provided that the rod is correct (double check), its use (see Figure 5.8) reduces the risk of duplicating errors, especially when more than one item is required.

Once all the material has been reduced to size (as in the cutting list) and checked to see that its face side is not twisted and that all the face edges are square with their respective face sides, the marking-off process can begin. This is then followed by marking-out.

Figure 5.8 illustrates a typical marking-off and marking-out procedure for a simple mortise-and-tenoned frame.

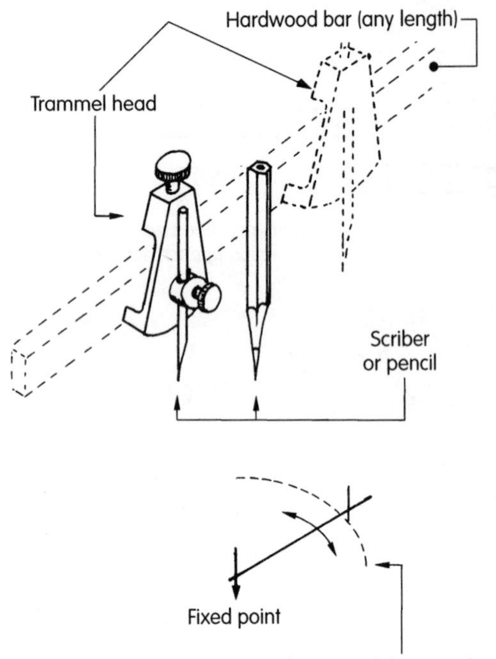

Figure 5.6 Using a trammel

Job title							Job no.
Quantity		Saw size (Ex)			Finish size		
No.	Item	L	W	T	W	T	Remarks

Sundries – Nails, screws, hardware, etc.	
Quantity	Description

Figure 5.7 Cutting and ironmongery list

5.2.2 Marking-out (Marking-off) Tools

Try-squares (Figure 5.9)

As their name suggests, these test pieces of timber for squareness or are used for marking lines at right angles from either a face side or a face edge.

It is advisable to test your try-square periodically for squareness (see Figure 5.9a). Misalignment could be due to misuse or accidentally dropping it on the floor.

Large all-steel graduated try-squares are available without a stock which makes them very useful when setting-out on large flat surfaces. They look like the steel roofing square which can be used to establish the bevels and lengths of timber components used in the construction of traditional pitched (sloping) roofs.

Combination Square (Figure 5.9b)

This can be used as a try-square but has the added advantage of being very versatile in that it has many other uses, e.g. as a mitre square (marking and testing angles of 45°), height gauge, depth gauge, marking gauge (Figure 5.4), spirit level (some models only) and rule.

It is common practice to use a pencil with a square, as shown in Figure 5.9c, although a marking knife (Figure 5.9d) is sometimes used in its place (especially when working with hardwoods) to cut across the first few layers of fibres so that the saw cut which follows leaves a sharp clean edge, e.g. at the shoulder line of a tenon.

Marking and Mortise Gauges (Figure 5.10)

As can be seen from the diagram, these gauges are similar in appearance and function, i.e. they are used to score lines parallel to the edge of a piece of timber. The main difference is that the marking gauge scores only a single line but the mortise gauge scores two in one pass. It is possible to buy a gauge that can perform both operations simply by being turned over (see Figure 5.10b).

Cutting Gauge (Figure 5.10d)

This is used to cut across the fibres of timber. It therefore has a similar function to that of a marking knife.

5.2.3 Marking Angles and Bevels

The combination square (shown in Figure 5.9) has probably superseded the original mitre square, which looked like a try square but had its blade fixed at 45° and 135° instead of 90°. However, two of the most useful pieces of bench equipment for dealing with mitres are a mitre template and a square and mitre template. Examples of their use are shown in Figure 5.11.

Angles other than 45° will have to be transferred with the aid of either a template pre-marked from the rod or the site situation, or by using a sliding bevel. This has a blade which can slide within the stock and be locked to any angle (Figure 5.12). The bevel as a whole can sometimes be a little cumbersome for marking dovetail joints on narrow boards. This can be overcome quite easily by using a purpose-made dovetail template (see Figure 5.12).

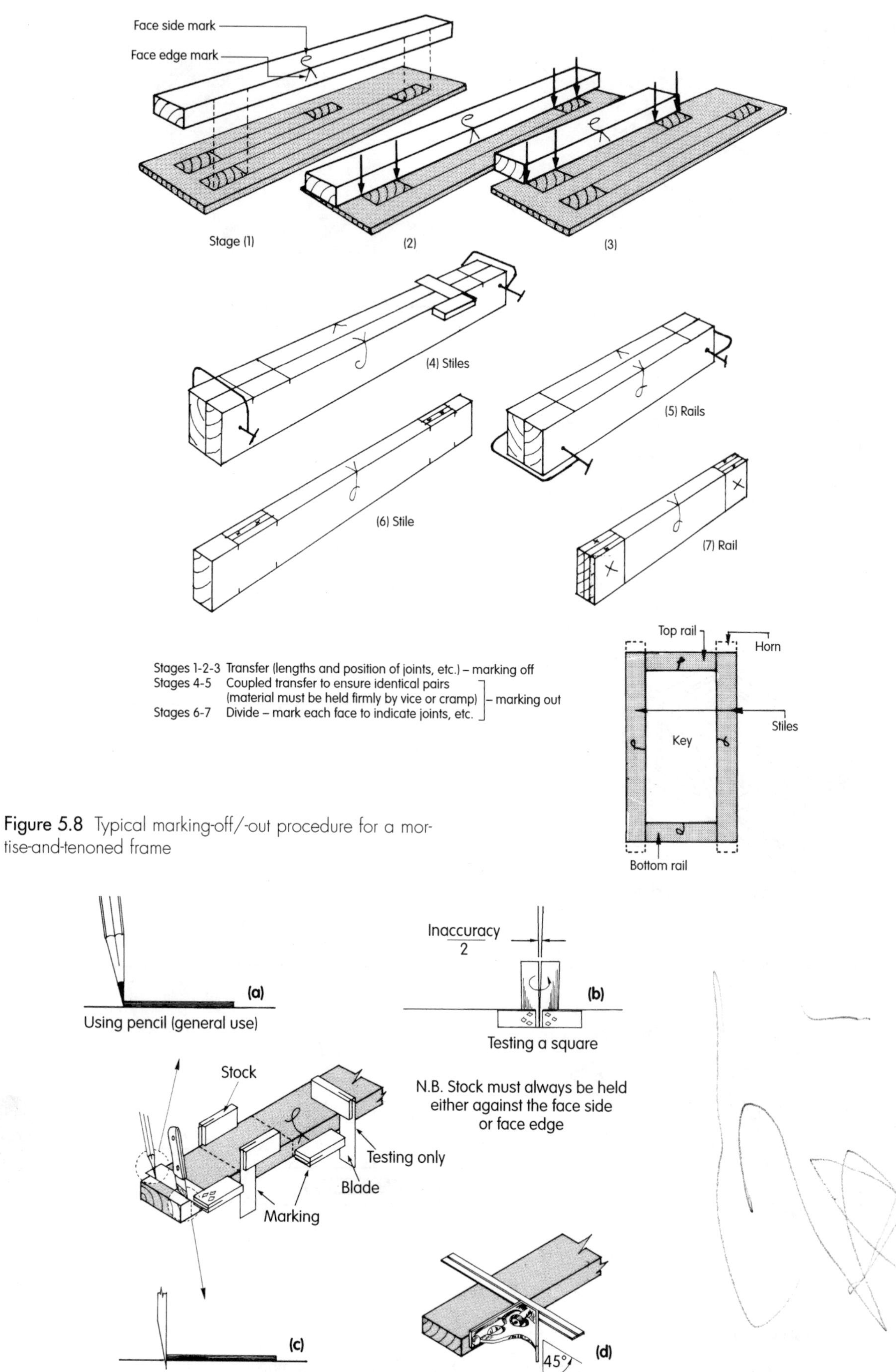

Figure 5.8 Typical marking-off/-out procedure for a mortise-and-tenoned frame

Figure 5.9 Using try-squares

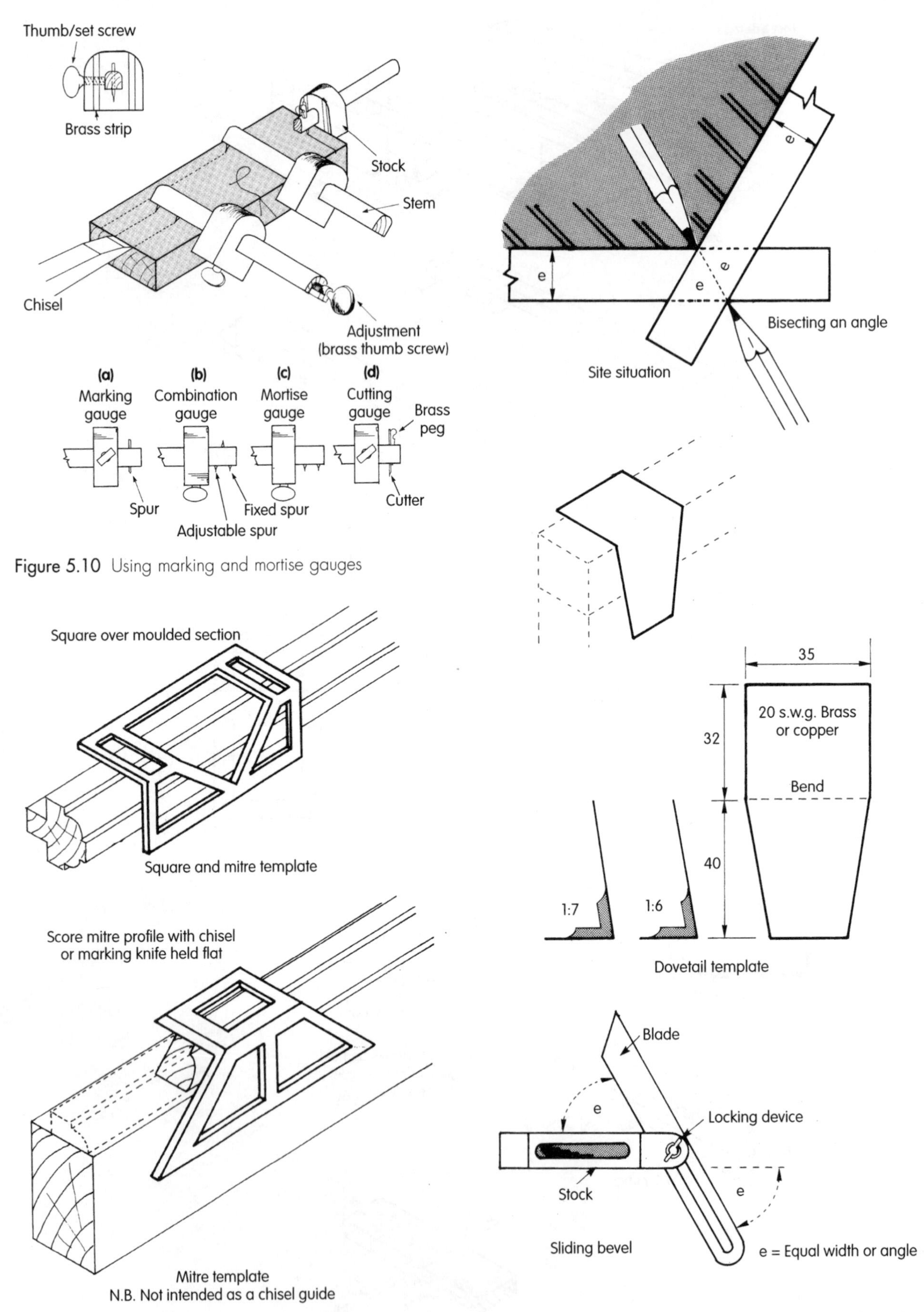

Figure 5.10 Using marking and mortise gauges

Figure 5.11 Marking 45° onto and 90° around moulded sections prior to cutting

Figure 5.12 Aids for marking and transferring bevels and angles

5.3 SAWS

Saws are designed to cut both along and across the grain of wood (except the rip saw – see Table 5.1) and the saw's efficiency will be determined by:

- the type and choice of saw
- the saw's condition
- the application
- the materials being cut.

5.3.1 Choice of Saw

Broadly speaking, saws can be categorised into four groups;

1 handsaws
2 backed saws
3 framed saws
4 narrow-blade saws.

The last two are used for cutting curves. As can be seen from Table 5.1, each type/group can be further broken down into two or three specifically named saws, which are available in a variety of sizes and shapes to suit particular functions.

5.3.2 Condition of Saw

It is important that saws are kept clean (free from rust) and sharp at all times (see Section 5.13). Dull or blunt teeth not only reduce the efficiency of the saw, but also render it potentially dangerous. For example, insufficient set could cause the saw to jam in its own kerf and then buckle or even break (see Figure 5.23).

5.3.3 Method of Use

The way the saw is used will depend on the following factors:

- the type and condition of the wood being cut
- the direction of cut – ripping or crosscutting
- the location – bench work or site work.

Practical illustrated examples are shown in Figures 5.13 to 5.22. Note the emphasis on safety, i.e. the position of hands and blades and body balance, etc.

5.3.4 Material Being Cut

A vast variety of wood species and manufactured boards are used in the building industry today and many, if not all, will at some time be sawn by hand. The modern saw is ideally suited to meet most of the demands made upon it, although there are instances where it will be necessary to modify general sawing techniques, for example when dealing with wood that is:

- very hard
- of very high moisture content
- extremely resinous
- case hardened.

Figure 5.13 Ripping with a handsaw

Figure 5.14 Crosscutting with a hand saw

Figure 5.15 Sawing down the grain (vertically) with a panel saw

Figure 5.16 Sawing down the grain with the material angled – two saw lines are visible

Table 5.1 Saw fact sheet

Type or group	Saw	Function	Blade length (mm)	Teeth shape	Teeth per 25 mm	Handle type	Remarks
Handsaws	Rip (Fig. 5.13)	Cutting wood with the grain (ripping)	650	60° 3°	4–6		Seldom used below six teeth per 25 mm
	Cross-cut (Fig. 5.14)	Cutting wood across the grain	600 to 650		7–8		Can also be used for rip sawing
	Panel (Figs. 5.15/ 5.16)	Cross-cutting thinner wood and manufactured board (MB)	500 to 550	14° 60°	10		Easy to use handle
Backed saws	Tenon (Figs. 5.17/5.18)	Tenons and general work	300 to 450		12–14		Depth of cut restricted by back strip (blade stiffener)
	and dovetail (Fig. 5.19)	Cutting dovetails and fine work	200 to 250		18–20		
Framed saws	Bow	Cutting curves in heavy sectioned timber and M/B	200 to 300		12 ±		Radius of cut restricted blade width
	Coping (Fig. 5.20)	Cutting curves in timber and M/B	160		14		Thin narrow blade
	Hacksaw (Fig. 5.21)	Cutting hard and soft metals	250 to 300		14–32		Small teeth – to cut thin materials. The larger the teeth, the less liable to clog – small frame hacksaw (Fig. 5.22)
Narrow-blade saws	Compass	Cutting slow curves in heavy, large work	300 to 450		10 ±		Interchangeable blades of various widths – unrestricted by a frame
	Pad or key hole	Enclosed cuts – piercing panels, etc	200 to 300		10 ±		Narrow blade partly housed in a handle, therefore length adjusts

Note: M/B – manufactured board

Figure 5.17(a) Tenon saw – starting a cut, vice held

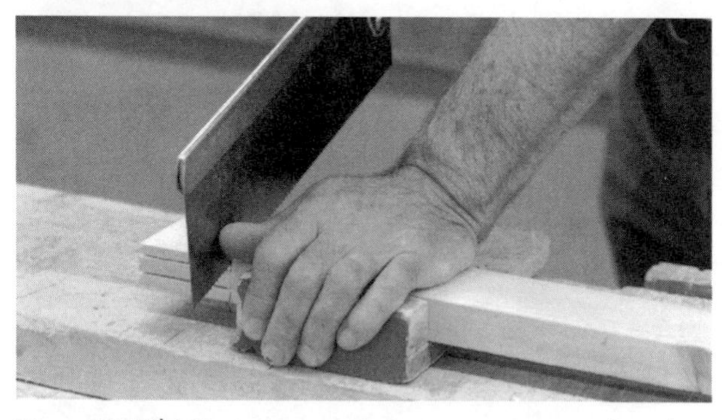

Figure 5.17(b) Tenon saw – starting a cut, using a bench hook

Figure 5.18 Sawing down the grain with a tenon saw

Figure 5.19 Dovetail saw – starting a cut

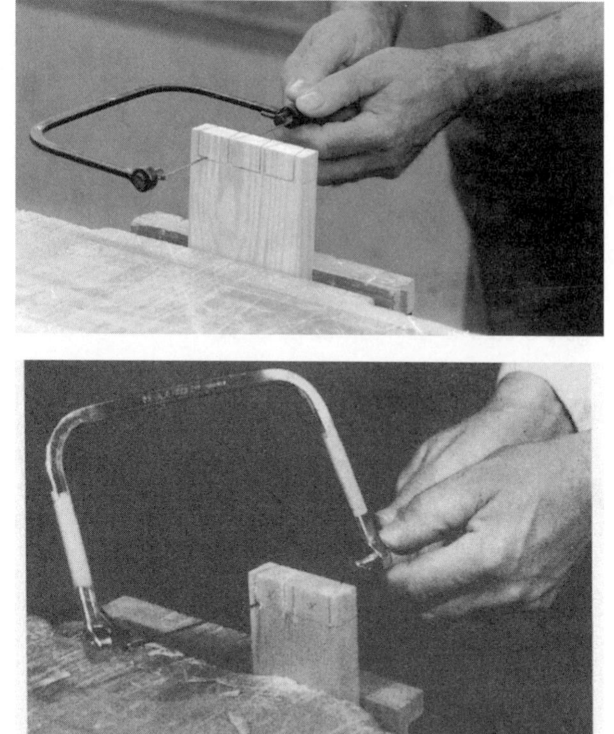

Figure 5.20 Using a coping saw

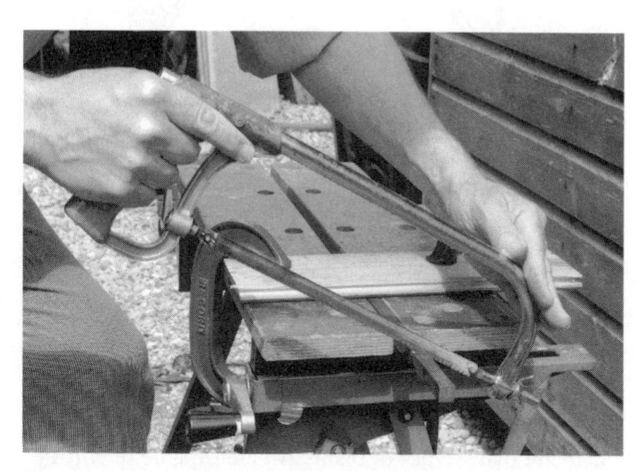

Figure 5.21 Using a hacksaw

Figure 5.22 Using a small hacksaw

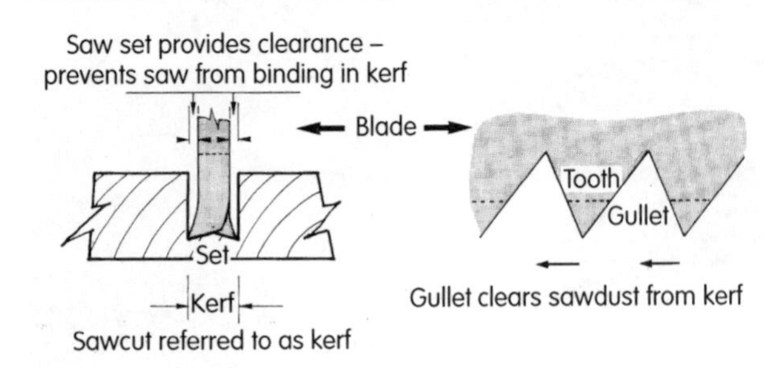

Figure 5.23 Providing saw blade clearance

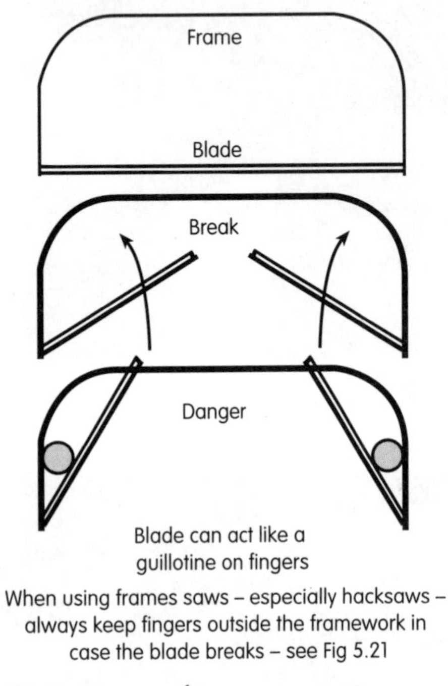

Figure 5.24 Frame saw safety

5.4 PLANES

There are many types of plane. All are capable of cutting wood by producing shavings, but not all are designed to produce plane flat surfaces, as the name implies. However, as can be seen from Table 5.2, each plane has its own function and, for the sake of convenience, planes have been placed within one of two groups:

1 bench planes
2 special purpose planes.

5.4.1 Bench Planes (Figures 5.25–28)

Wooden-bodied bench planes have been superseded by the all-metal (with the exception of the handle and front knob) plane, although the wooden jack plane is still regarded by some joiners as the ideal site plane, as it is light and less liable to break if dropped. Probably the greatest asset of the wooden bench plane is its ability to remove waste wood rapidly – where accuracy is not too important.

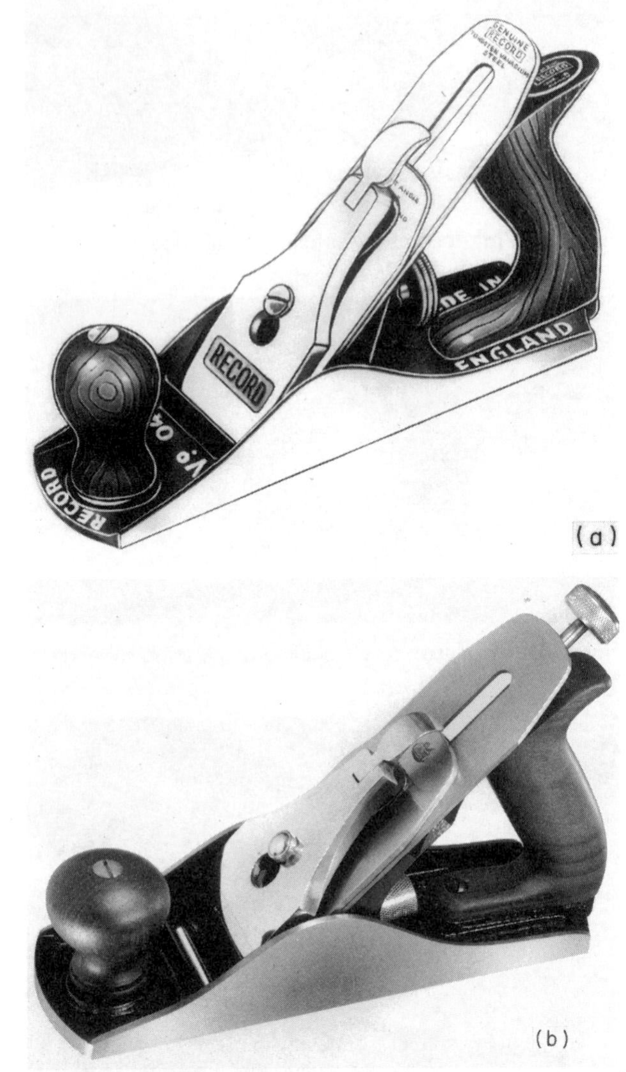

Figure 5.25 Smoothing planes (with kind permission from Record Tools Ltd)

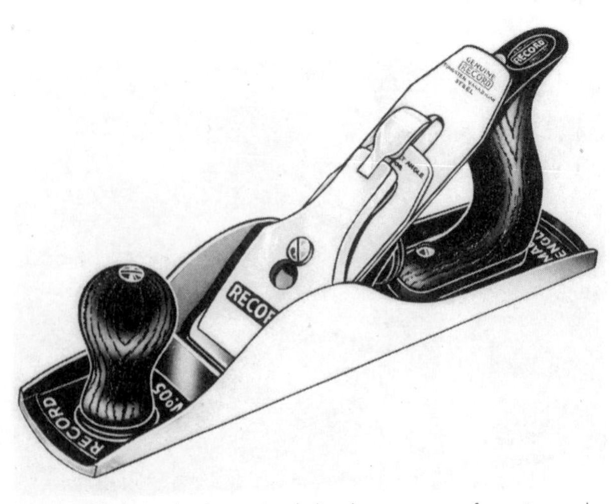

Figure 5.26 Jack planes (with kind permission from Record Tools Ltd)

Table 5.2 Plane fact sheet

Group	Plane	Function	Length (mm)	Blade width (mm)	Remarks
Bench planes	Smoothing (Fig. 5.25a)	Finishing for flat surfaces	*240, *245, 260	45, 50, 60	50 mm the most common blade width
	Record CSBB (Fig. 5.25b)	As above	245	60	'Norris' type cutter adjustment – combining depth of cut with blade adjustment
	Jack (Fig. 5.26)	Processing saw timber	*355, 380 mm	50, 60	60 mm the most common blade width
	Fore jointer (Fig. 5.27) (try plane)	Planing long edges (not wider than the plane's	*455 *560, 610	60	The longer the sole, the greater the degree of accuracy
	Bench rebate (carriage or badger) (Fig. 5.28)	Finishing large rebates	235, 330	54	Its blade is exposed across full width of sole
	Note: * available with corrugated solestors (Fig. 5.29) which are better when planing resinous timber.				
Special planes	Block (Fig.5.30)	Trimming – end grain	140, 180, 205	42	Cutter seats at 20° or 12° (suitable for trimming laminated plastics, depending on type
	Circular (compass plane) (Fig. 5.31)	Planing convex or concave surfaces	235, 330	54	Spring-steel sole adjusts from flat to either concave or convex
	Rebate (Fig. 5.32)	Cutting rebates with or across the grain	215	38	Both the width and depth of rebate are adjustable
	Shoulder/rebate (Fig. 5.33)	Fine cuts across grain, and general fine work	152, 204	18, 25, 29, 32	Some makes adapt to chisel planes
	Bulnose/shoulder rebate (Fig.5.34)	As above, plus working into confined corners	100	25, 29	
	Side rebate (Fig. 5.35)	Widening rebates or grooves – with or across	140		Removable nose – works into corners. Double-bladed – right and left hand. Fitted with depth gauge and/or fence
	Plough (Fig. 5.36)	Cutting grooves of various widths and depths – with and across grain	248	3 to 12	Both width and depth of groove adjustable
	Combination	As above, plus rebates, beading, tongues, etc.	254	18 cutters, various shapes and sizes	Not to be confused with a multi-plane, which has a range of 24 cutters
	Open-throat router (Fig. 5.37)	Levelling bottoms of grooves, trenches, etc.		6, 12 and V	A fence attachment allows it to follow straight or round edges
	Spokeshave (Fig. 5.38) (a) flat bottom (b) round bottom	Shaving convex or concave surfaces – depending on type (a) and (b) see Fig. 5.51	250	54	Available with or without 'micro' blade depth adjustment

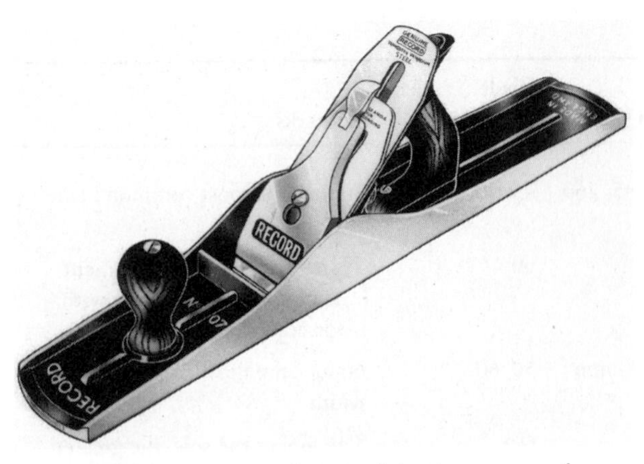

Figure 5.27 Jointer or try-plane (with kind permission from Record Tools Ltd)

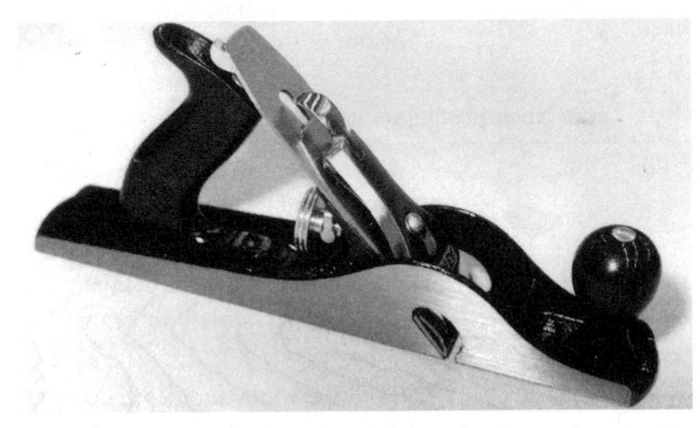

Figure 5.28 Bench rebate (carriage or badger) plane (with kind permission from Stanley Tools Ltd)

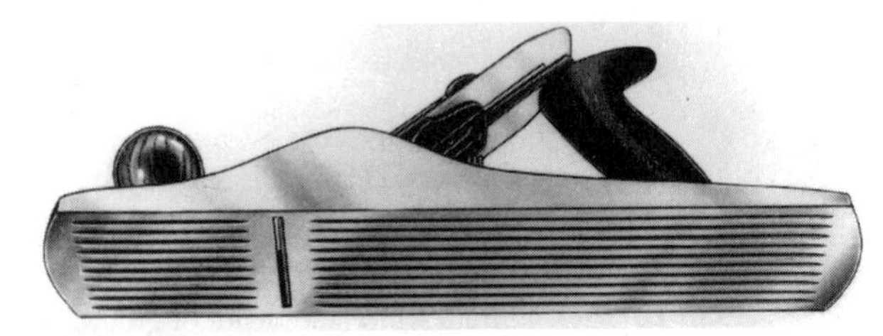

Figure 5.29 Corrugated sole (with kind permission from Record Tools Ltd)

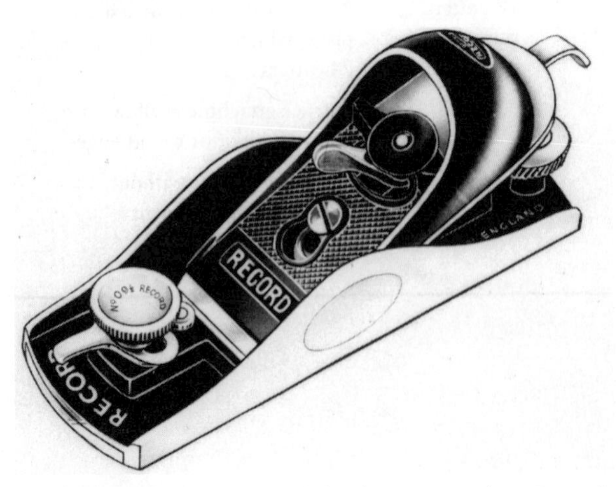

Figure 5.30 Block plane (with kind permission from Record Tools Ltd)

Figure 5.31 Circular (compass) plane (with kind permission from Record Tools Ltd)

Figure 5.32 Rebate plane (with kind permission from Record Tools Ltd)

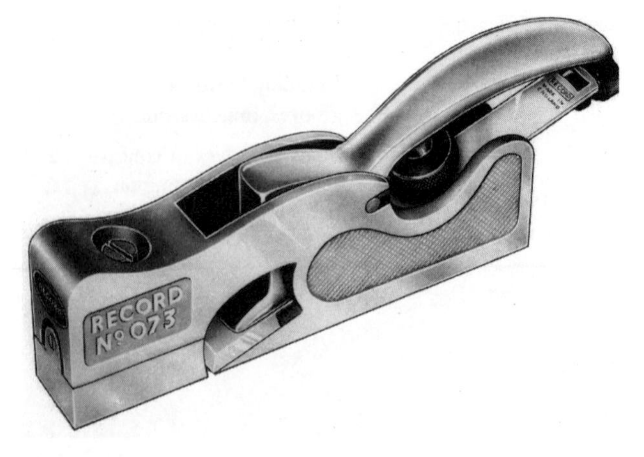

Figure 5.33 Shoulder/rebate plane (with kind permission from Record Tools Ltd)

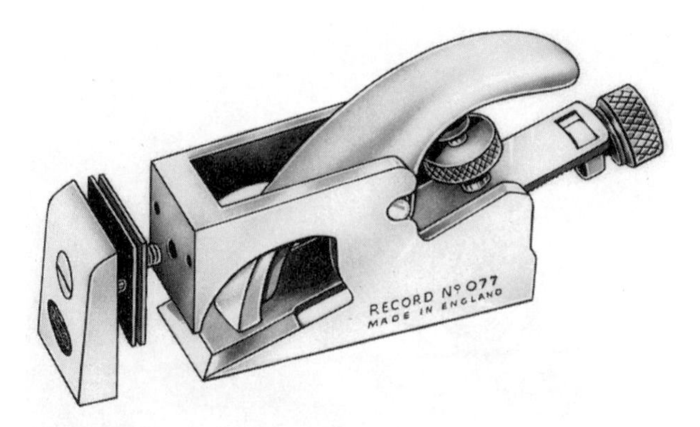

Figure 5.34 Bullnose/shoulder rebate plane (with kind permission from Record Tools Ltd)

Figure 5.35 Side rebate plane in use (with kind permission from Stanley Tools)

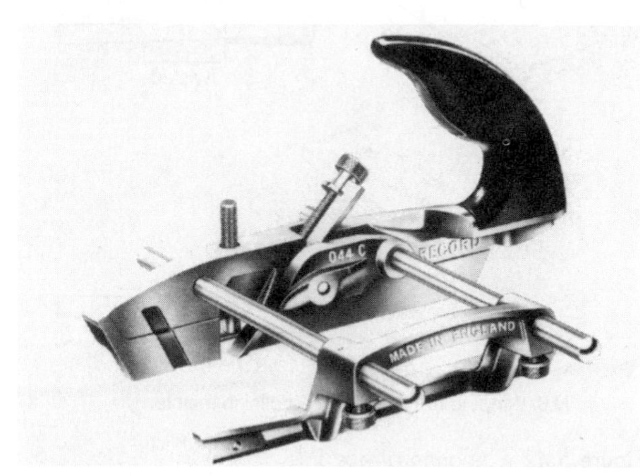

Figure 5.36 Plough plane (with kind permission from Record Tools Ltd)

All metal bench planes are similarly constructed with regard to blade angle (45°), adjustment and alignment (Figure 5.39); variations are primarily due to the size or design of the plane sole, which determines the function (Table 5.2).

Figures 5.40 and 5.41 show a jack plane being used for *flatting* and *edging* a short piece of timber. Notice particularly the position of the hands in relation to the operation being carried out.

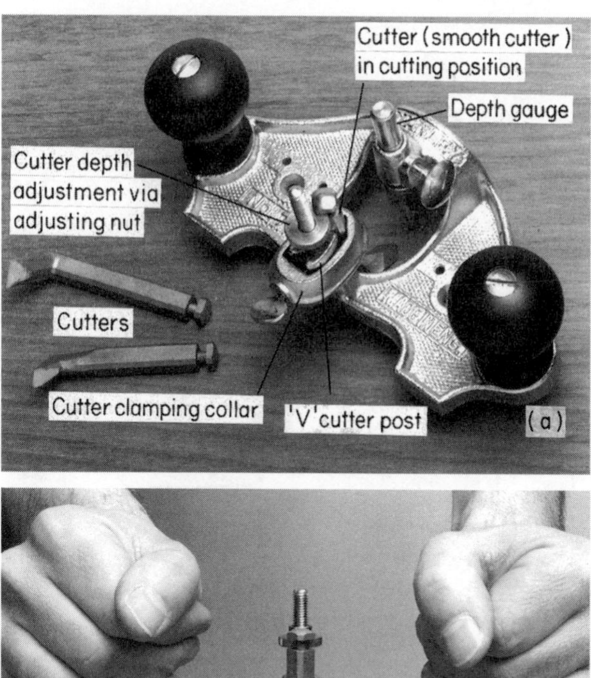

Figure 5.37 Open-throat router and its use (with kind permission from Stanley Tools)

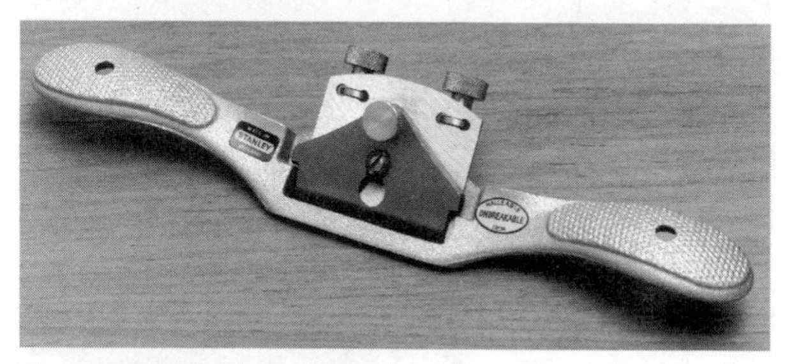

Figure 5.38 Spokeshave (with kind permission from Stanley Tools)

Processing a piece of sawn timber by hand is carried out as follows:

1 Select and, using a jack plane, plane a face side (best side) straight (Figure 5.42a) and out of twist (Figure 5.42b). (Winding laths are used to accentuate the degree of twist.) Label the side with a face-side mark (Figure 5.8).
2 Plane a face edge straight and square to the face side (keep checking with a try-square), using either a jack plane or a try-plane (Figure 5.43), depending on the length of the timber being processed. As can be seen from Figure 5.44, the shorter the sole the more

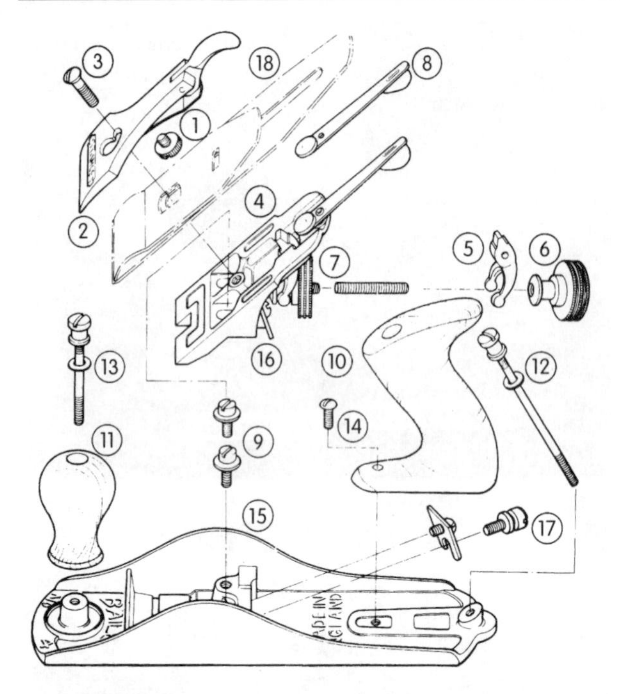

Figure 5.39 Exploded view of a Stanley bench plane (with kind permission from Stanley Tools). 1 – cap iron screw; 2 – lever cap; 3 – lever screw; 4 – frog, complete; 5 – 'Y' adjusting lever; 6 – adjusting nut; 7 – adjusting nut screw; 8 – LA (lateral adjustment) lever; 9 – frog screw and washer; 10 – plane handle; 11 – plane knob; 12 – handle screw and nut; 13 – knob screw and nut; 14 – handle toe screw; 15 – plane bottom, sole; 16 – frog clip and screw; 17 – frog adjusting screw; 18 – cutting iron (blade) and cap iron

Figure 5.40 Flatting – planing the face of a piece of timber placed against a bench stop

Figure 5.41 Edging – planing the edge of a piece of timber held in a vice with finished face positioned towards the operator. The forefinger acts as a guide to centralise the plane during the operation, *not as a fence*

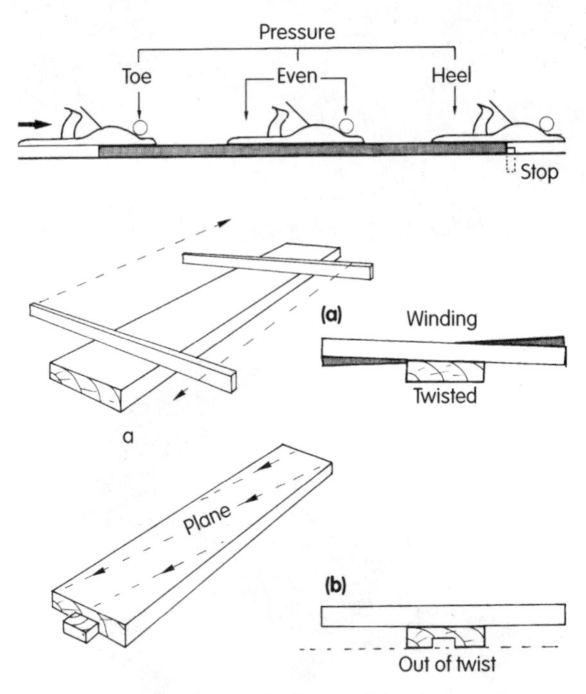

Figure 5.42 Preparing a face side

difficult it is to produce long straight edges. Long lengths will require end support to prevent tipping and this can be achieved by positioning a peg in one of a series of pre-bored holes in the face or leg of the bench (see Figure 5.45). On completion, identify the edge with a face-edge mark (Figure 5.8).

3 Gauge to width (Figure 5.10), ensuring that the stock of the gauge is held firmly against the face edge at all times, then plane down to the gauge line.
4 Gauge to thickness – as in step 3 but this time using the face side as a guide.

Note: When preparing more than one piece of timber for the same job, each operation should be carried out on all pieces before proceeding to the next operation, i.e. face side all pieces, face edge all pieces and so on.

The smallest and most often used of all the bench planes is the smoothing plane. This is very easy to handle and, although designed as a fine finishing plane for dressing joints and surfaces, it is used as a general-

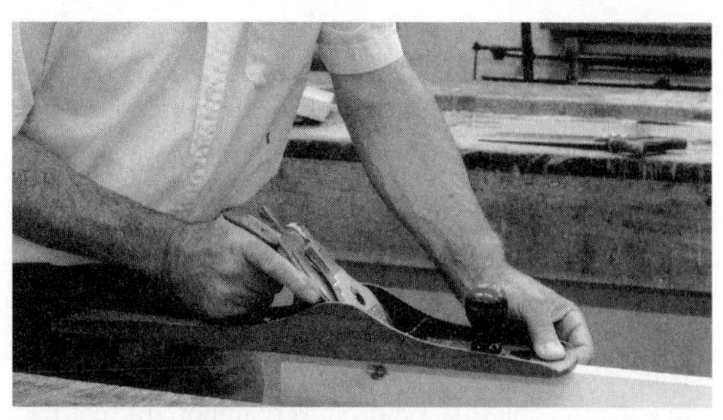

(a) Hold down to middle

(b) Even pressure – self-weight of plane

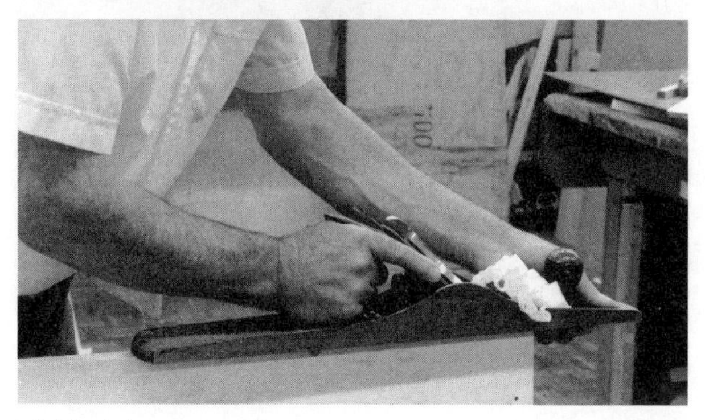

(c) Heel down to end

Figure 5.43 Using a try-plane (jointer)

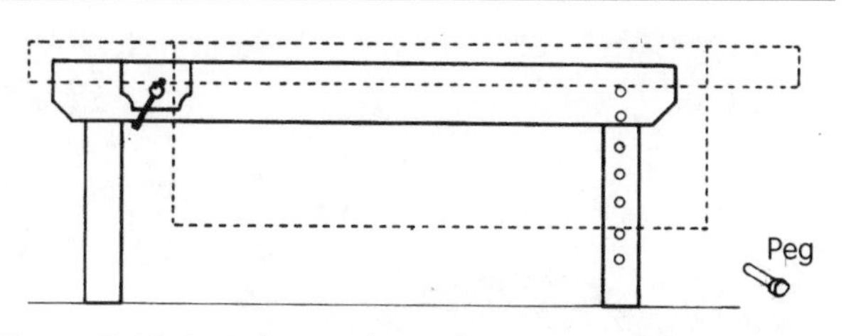

Figure 5.45 Providing end support

purpose plane for both bench and site work. Figure 5.46 shows a smoothing plane being used to dress (smooth and flat) a panelled door and how, by tilting the plane, it is possible to test for flatness (this applies to all bench planes). The amount and position of the light showing under its edge will determine whether the surface is round or hollow.

NB: The direction of the plane on the turn at corners or rail junctions – for example, working from stile to rail or rail to stile – will be determined by the direction of the wood grain.

5.4.2 Special Planes (Figures 5.30–38)

It should be noted that there are more special planes than the ten listed in Table 5.2 and there are also variations in both style and size.

The types of planes selected for your tool kit will depend on the type of work you are doing. There are, however, four planes which, even if not essential in your work, you should find very useful. They are:

1 a block plane (Figure 5.30)
2 a rebate plane (Figure 5.32)
3 a plough plane (Figure 5.36)
4 spokeshaves (Figure 5.38).

a) Block Planes (Figure 5.30)

These are capable of tackling the most awkward cross-grains of both hardwood and softwood, not to mention

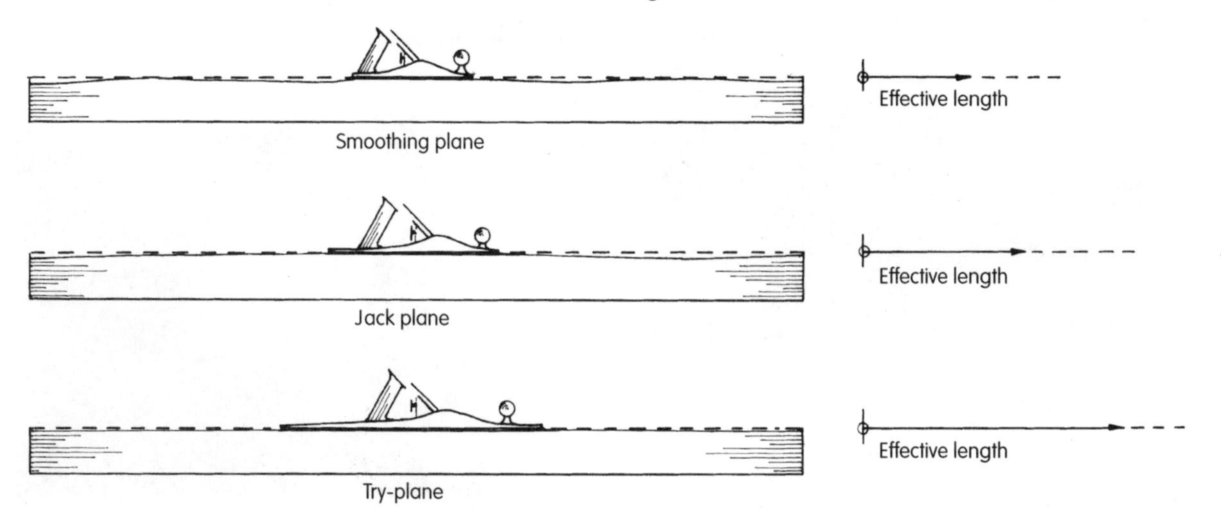

Figure 5.44 The longer the grain side, the greater the accuracy for a straight edge

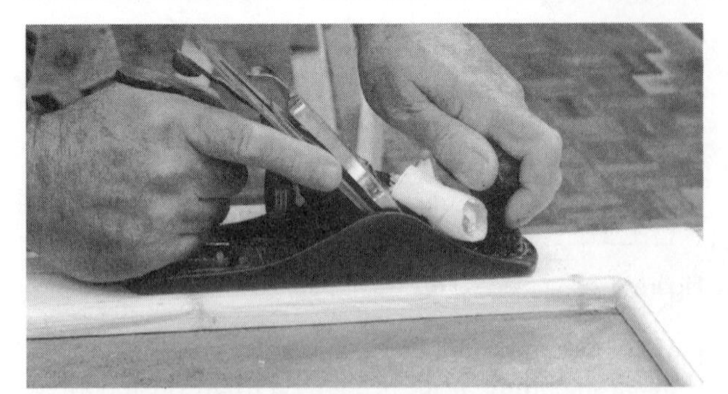

(a) Smoothing a door stile

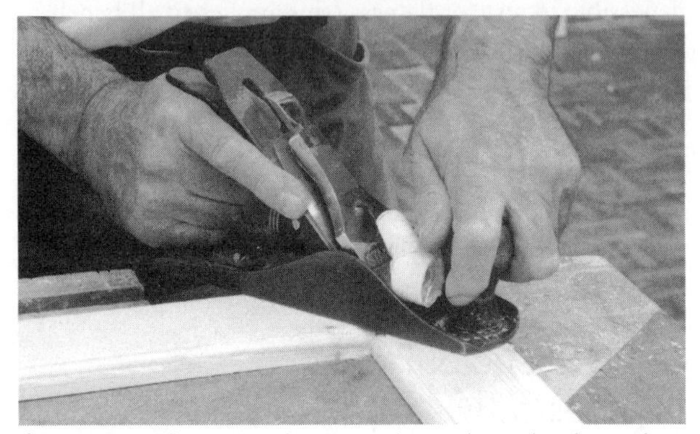

(b) Dressing flat a joint between door stile and rail – a slow turning action towards the joint helps to avoid unnecessary tearing of the grain

(c) Smoothing a door rail

(d) Testing a corner joint for flatness by tilting the plane – notice the light showing between the edge of the plane sole and the surface of the timber, indicating that, at this position, the surface is hollow

Figure 5.46 Using a smoothing plane to dress the faces of fully assembled, glued, cramped and set panelled door

the edges of manufactured boards and laminated plastics. Some models have the advantage of an adjustable mouth (Figure 5.135) and/or have their blade set to an extra low angle of 12°. Such a combination can increase cutting efficiency. Block planes are designed to be used both single- and double-handed.

b) Rebate Plane (Figure 5.32)

Figure 5.47 shows a rebate plane being used to cut a rebate of controlled size, by using a width-guide fence and a depth stop. It is, however, very important that the cutting face of the plane is held firmly and square to the face or the edge of the timber be held firmly throughout the whole operation. Because the blade of a rebate plane has to extend across the whole width of its sole, a cut finger can easily result from careless handling. Particular care should therefore be taken to keep fingers away from the blade during its use and especially when making the desired depth and/or width adjustments.

Figure 5.47 Planing a rebate

c) Plough plane (Figure 5.36)

Figure 5.48 shows a plough plane cutting a groove in the edge of a piece of wood. It is also capable of cutting rebates to the cutter widths provided and by the method shown in Figure 5.49.

Figure 5.48 Plough plane cutting a groove

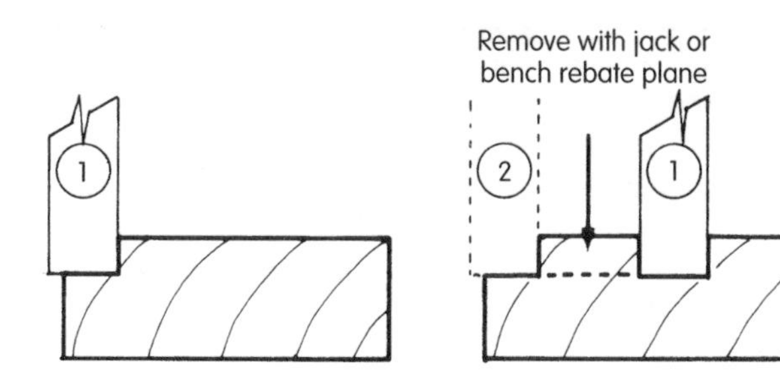

Figure 5.49 Method of forming a rebate with a plough plane

The method of applying the plane to the wood is common to both rebate and plough planes, in that the cut should be started at the forward end and gradually moved back until the process is complete (see Figure 5.50).

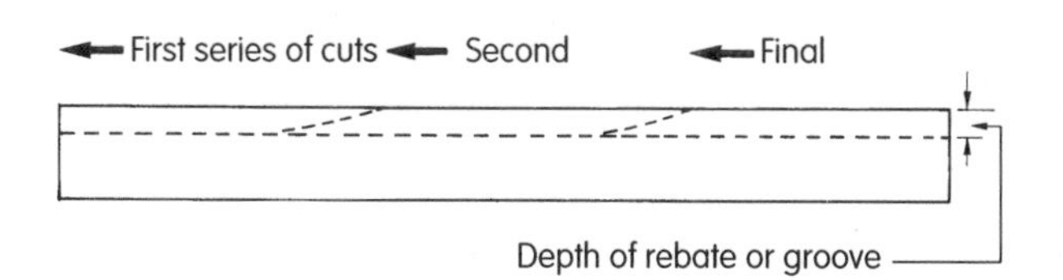

Figure 5.50 Application of a rebate or plough plane

d) Spokeshaves (Figure 5.38)

Joiners often prefer to use larger flat-bottomed planes for shaping convex curves, but the efficiency and ease of operation of a flat-bottomed spokeshave can only be realised when the technique of using this tool has been mastered. However, planing concave surfaces (Figure 5.51) is another matter. As shown in Figure 5.52, a spokeshave with a rounded (convexed) sole is being used – note particularly the position of the thumbs and forefingers, giving good control over the position of the blade in relation to its direction of cut (always with the grain).

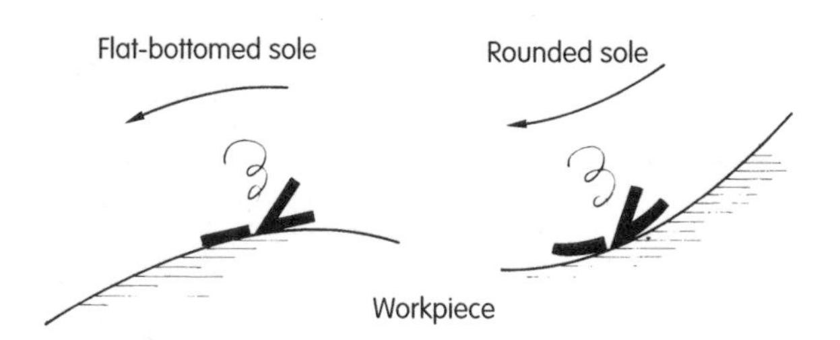

Figure 5.51 Spokeshave use

Figure 5.52 Spokeshave in use

5.5 BORING TOOLS

Boring tools are used for cutting or scraping circular holes of a predetermined size into and below the surface of a given material. They can be divided into three groups:

1 standard bits
2 special bits
3 drills.

Table 5.3 classifies the above under the headings of:

- type
- function
- motive power
- hole size
- shank section.

Table 5.3 should be used in conjunction with the illustrations of tools shown in Figures 5.53 to 5.66.

Table 5.3 Characteristics of bits and drills in common use. (Not all types are available in metric sizes.)

Group	Type of bit/drill	Function	Motive power	Range of common hole sizes	Shank section	Remarks
Bits (standard)	Centre bit (Fig. 5.53)	Cutting shallow holes in wood*	CB	$\frac{1}{4}$–$2\frac{1}{4}$ in	❑	
	Irwin-pattern solid-centre auger bit (Fig. 5.54)	Boring straight holes in wood*	CB	$\frac{1}{4}$–$1\frac{1}{2}$ in 6–38 mm	❑	General-purpose bit
	Jennings-pattern auger bit (Fig. 5.55)	Boring straight, accurate, smooth holes in wood*	CB	$\frac{1}{4}$–$1\frac{1}{2}$ in	❑	
	Jennings-pattern dowel bit (Fig. 5.56)	As above, only shorter	CB	$\frac{3}{8}$ and $\frac{1}{2}$ in only	❑	Used in conjunction with wood dowel
	Combination auger bits (Fig. 5.57)	Cutting very clean holes in wood	CB ED	$\frac{1}{4}$–$1\frac{1}{4}$ in 6–32 mm	⬡	Must only be used with slow cutting speeds
	Countersink (Fig. 5.58)	Enlarging sides of holes	CB ED	$\frac{3}{8}$, $\frac{1}{2}$, $\frac{5}{8}$ in	❑❍	Rose, shell (snail) heads available depending on material being cut and speed
Bits (specified)	Expansive (expansion) bit (Fig. 5.59)	Cutting large shallow holes in wood*	CB	$\frac{7}{8}$–3 in	❑	Adjustable to any diameter within its range
	Scotch-eyed auger bit (hand) (Fig. 5.60)	Boring deep holes in wood	T	$\frac{1}{4}$–1 in		Turned via 'T' bar. Length determined by diameter of auger
	Forstner bit (Fig. 5.61)	Cutting shallow flat-bottomed holes in wood*	CB ED	$\frac{3}{8}$–2 in	❑❍	Ideal for starting a stopped housing
	Dowel-sharpener bit	Chamfering end of dowel	CB		❑	Pointed dowel – aids entry into dowel holes
	Turn-screw bit (Fig. 5.62)	Driving large screws	CB	$\frac{1}{4}$, $\frac{3}{16}$, $\frac{3}{8}$, $\frac{7}{16}$ in	❑	Very powerful screwdriver
	Flat bit (Fig. 5.63)	Bores holes in all forms of wood quickly, cleanly	ED	$\frac{1}{4}$–$1\frac{1}{2}$ in 6–38 mm	❍	Deep holes may wander Extension shank available
	Screw bit (Screwmate) (Fig. 5.64)	Drills pilot, clearance and countersink in one operator	ED	Screw size 1 in × 6 to $1\frac{1}{2}$ in × 10 25 mm × 6 to 38 mm × 10	❍	Saves time changing different bits
	Screw sink (Fig. 5.65)	Combination counterbore all-in-one boring of screw hole and plug hole	ED	$\frac{3}{4}$ in × 6 to 2 in × 12	❍	Depth of counterbore can be varied. Use with Stanley plug cutter
	Plug cutter (Fig. 5.66)	Cuts plugs to fill counterbored holes	ED	To suit screw gauges 6 to 12 and hole sizes $\frac{3}{8}$, $\frac{1}{2}$, $\frac{5}{8}$ in	❍	Use only with a drill fixed into a bench drill stand, and with the wood cramped down firmly (see manufacturer's instructions for maximum safe speed.)
Drills	Twist drills	Boring wood*, metals, plastics	HD ED CA	$\frac{1}{16}$ to $\frac{1}{2}$ in 1–13 mm	❍	A few sizes available as ❑
	Masonry drills (tungsten-carbide	Boring masonry, brickwork, concrete	HD ED	Numbers 6–20 $\frac{5}{16}$ to $\frac{3}{8}$ in†	❍	Available for both rotary and impact (percussion-action) drills

Note: † Larger diameter drills are available; CB – carpenter's brace; HD – hand drill (wheel-brace); ED – electric drill (N.B. Bits used in ED must never have screw point other than with a very slow cutting speed); CA – compressed-air drill; ❑ – square-tapered shank; ❍ – straight rounded shank; * wood and all wood-based products; ⬡ – hexagonal end shank

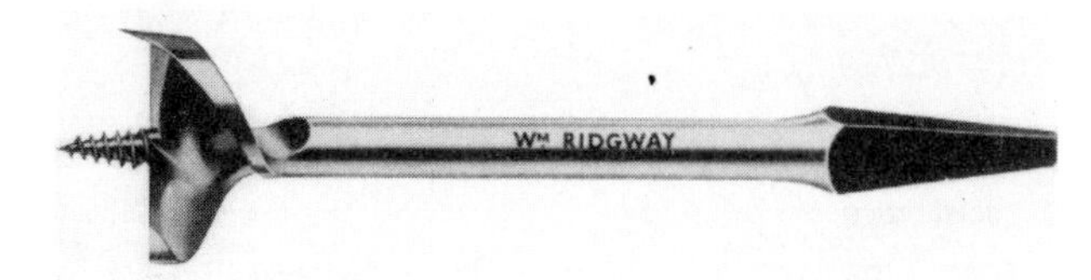

Figure 5.53 Centre bit

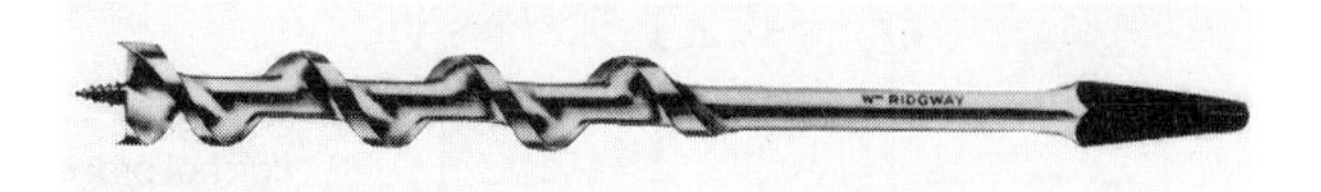

Figure 5.54 Irwin-pattern solid centre bit (with kind permission from Record Tools Ltd)

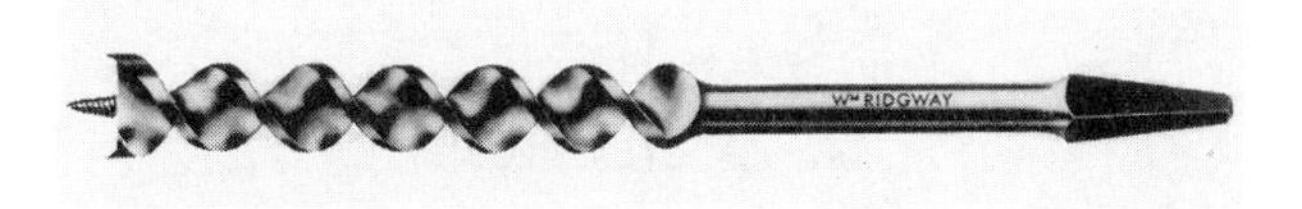

Figure 5.55 Jennings-pattern auger bit (with kind permission from Record Tools Ltd)

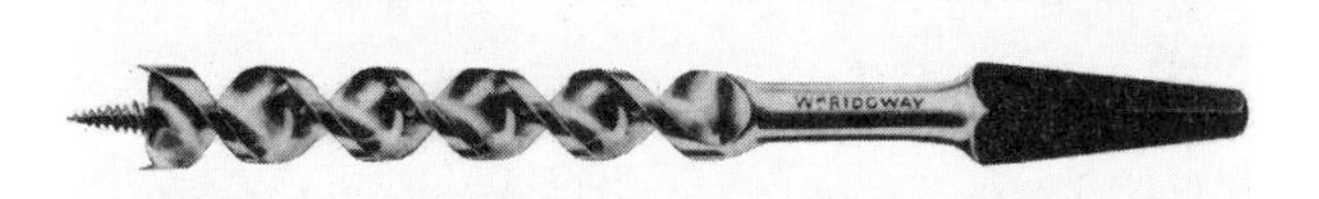

Figure 5.56 Jennings-pattern dowel bit (with kind permission from Record Tools Ltd)

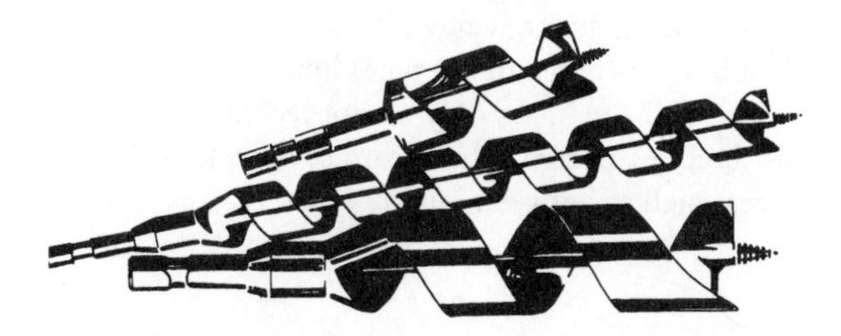

Figure 5.57 Combination auger bit

Figure 5.58 Rosehead countersink bit (with kind permission from Record Tools Ltd)

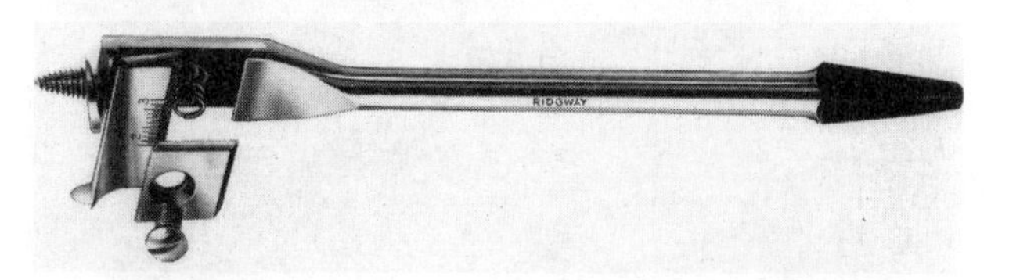

Figure 5.59 Firmgrip expansive bit (with kind permission from Record Tools Ltd)

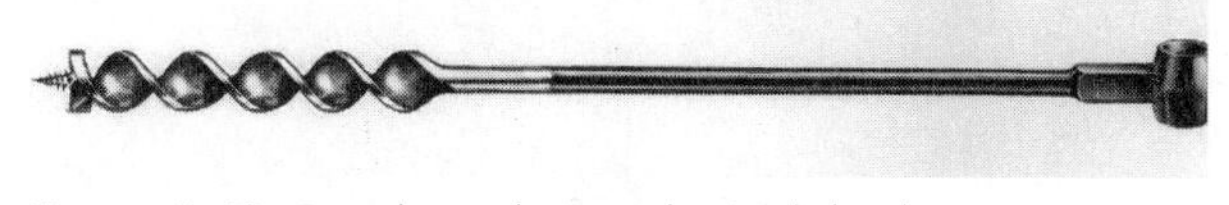

Figure 5.60 Scotch-eyed auger bit (with kind permission from Record Tools Ltd)

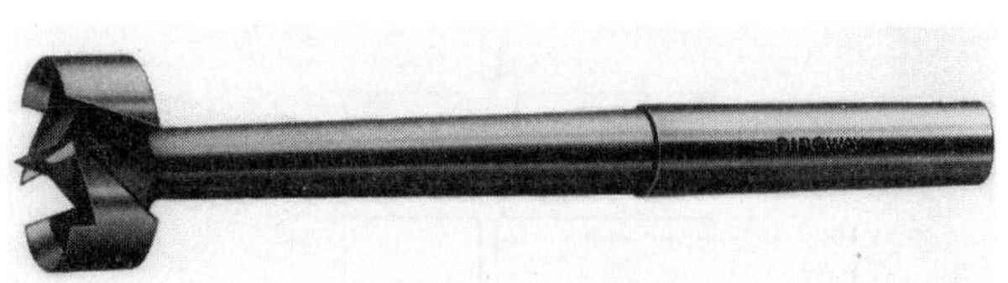

Figure 5.61 Forstner bit (with kind permission from Record Tools Ltd)

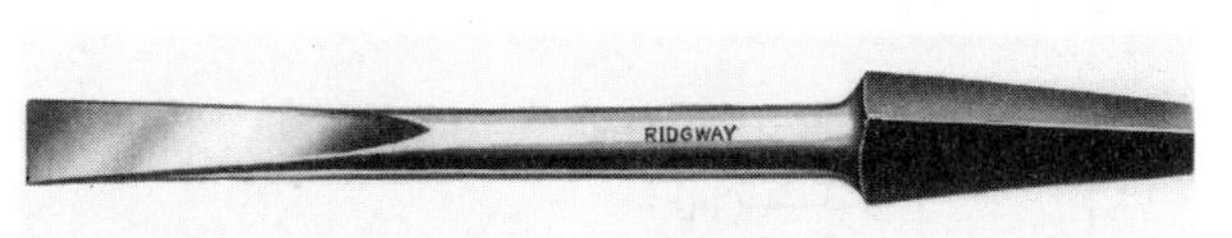

Figure 5.62 Bright cabinet screwdriver bit (turn-screw bit) (with kind permission from Record Tools Ltd)

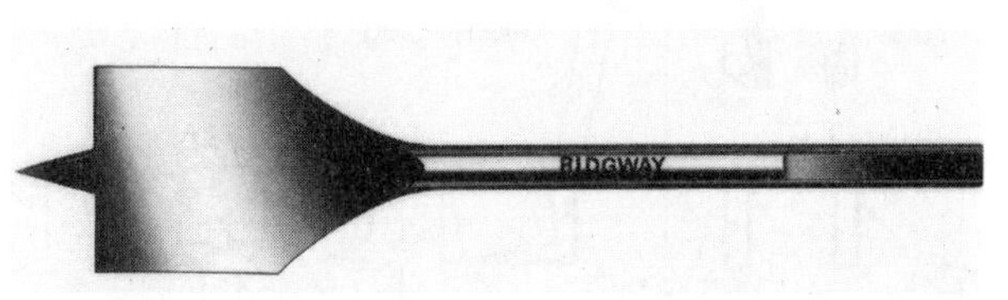

Figure 5.63 Flat bit (with kind permission from Record Tools Ltd)

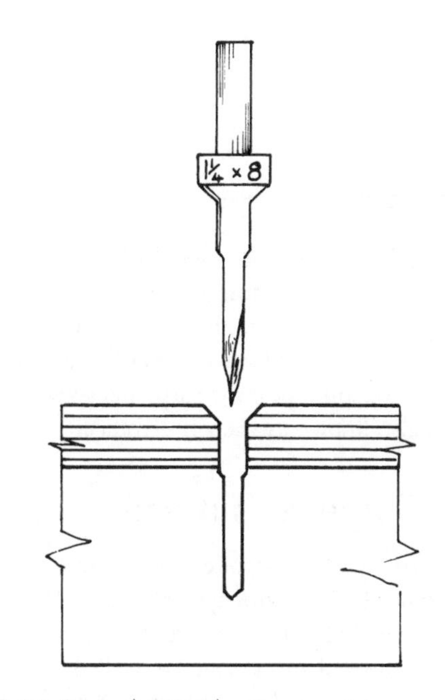

Figure 5.64 Screw bit (Stanley Screwmate)

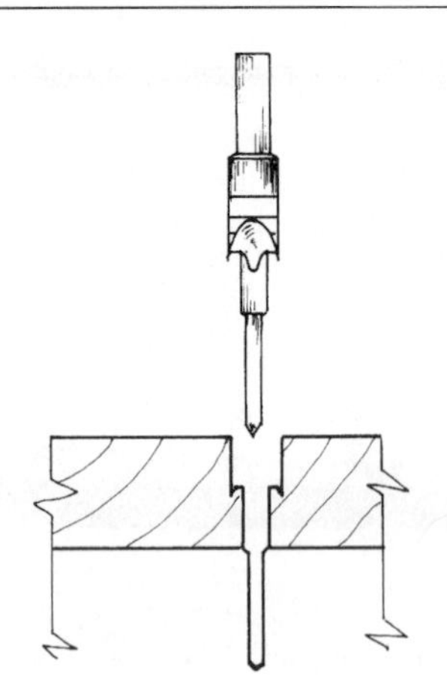

Figure 5.65 Stanley Screwsink

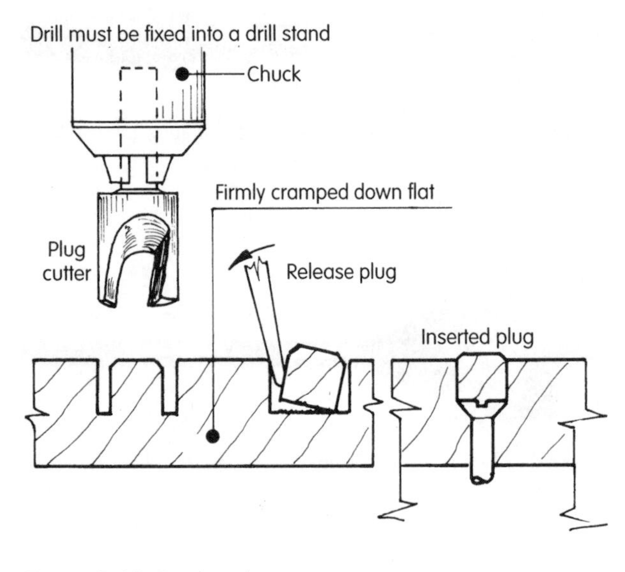

Figure 5.66 Stanley plug cutter

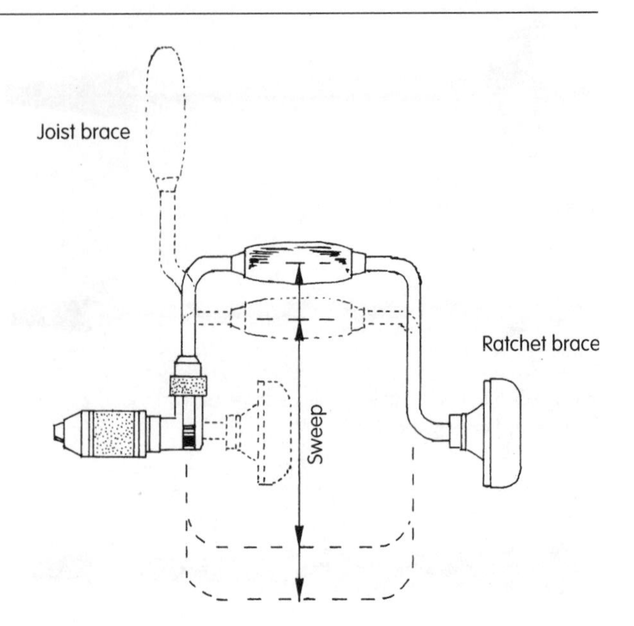

Figure 5.67 Types of carpenter's brace

The driving force to operate such tools is provided by hand, by electricity or by compressed air.

Hand operations involve the use of a bit- or drill-holder (chuck) operated by a system of levers and gears. There are two main types: the carpenter's brace and the hand drill, also known as the wheel-brace.

5.5.1 Carpenter's Brace

This has a two-jaw chuck, of either the alligator or the universal type. The alligator type has been designed to take square-tapered shanks, whereas the universal type takes round, tapered and straight as well as square-tapered shanks. The amount of force applied to the bit or drill will depend largely on the sweep of the brace. Figure 5.67 shows how the style and sweep can vary.

There are three main types of brace:

1 Ratchet brace – the ratchet mechanism allows the brace to be used where full sweeps are restricted (an example is shown in Figure 5.68). It also provides extra turning power (by eliminating overhand movement), so often needed when boring large-diameter holes or using a turn-screw bit.
2 Plain brace – (non-ratchet type) limited to use in unrestricted situations only, so not recommended.
3 Joist brace – (upright brace) for use in awkward spaces, such as between joists.

Figure 5.68 Use of the ratchet mechanism when the full sweep of the brace is restricted

Figure 5.69 Vertical boring using hand signal guidance from an assistant taking the vertical viewpoint

5.5.2 Method of Use

Probably the most difficult part of the whole process of boring a hole is keeping the brace either vertical or horizontal to the workpiece throughout the whole operation. Accuracy depends on all-round vision so, until you have mastered the art of accurately assessing horizontal and vertical practice, seek assistance.

Figures 5.69 and 5.70 show situations where the assistant directs the operative by simple hand signals. Note also that for vertical boring the operative's head is kept well away from the brace, giving good vision and unrestricted movement – only light pressure should be needed if the bit is kept sharp. Horizontal boring support is given to the brace by arm over leg as shown; in this way good balance can be maintained throughout the process. The stomach should not be used as a form of support.

The use of an upturned try-square, etc. placed on the bench as a vertical guide is potentially dangerous since any sudden downward movement, due to the bit breaking through the workpiece, could result in an accident.

Figure 5.70 Horizontal boring using hand signal guidance from an assistant taking the horizontal viewpoint

Figure 5.71 shows the use and limitations of some of the bits mentioned, together with two methods of preventing breaking through the opposite side of the wood without splitting it, i.e. reversing the direction of the bit or temporarily securing a piece of waste wood to the breakthrough point.

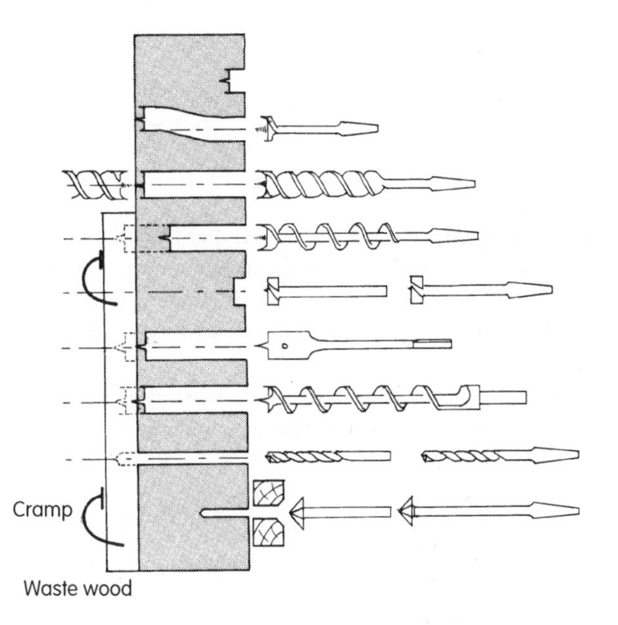

Figure 5.71 Application of various bits and drills

5.5.3 Hand Drill (Wheel-brace) (Figure 5.72)

This has a three-jaw self-centring chuck, designed specifically to take straight-sided drills. It is used in conjunction with twist drills or masonry drills.

5.5.4 Boring Devices and Aids

Probably the most common of all boring devices is the bradawl (Figure 5.73). This is a steel blade fixed into a

Figure 5.72 Hand drill (wheel-brace) (with kind permission from Stanley Tools)

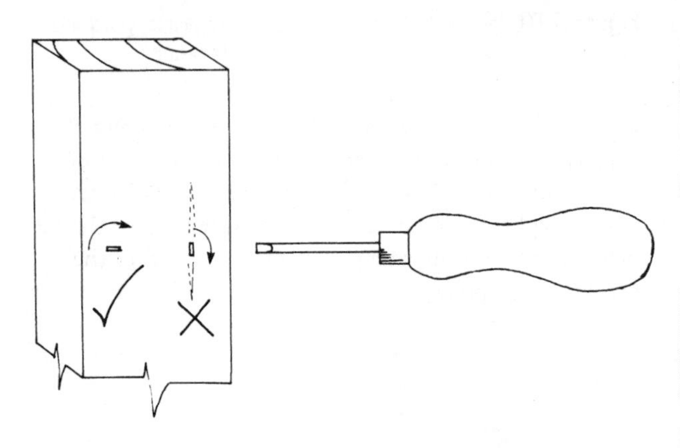

Figure 5.73 Bradawl (pricker)

wooden or plastic handle and used mainly to bore pilot holes for screw threads. The spiral ratchet screwdriver, featured in Section 5.8.4(c), may be adapted to drill small short holes and to carry out countersinking operations.

For gauging the depth of a hole, use either a proprietary purpose-made depth gauge or make your own, as in Figure 5.74.

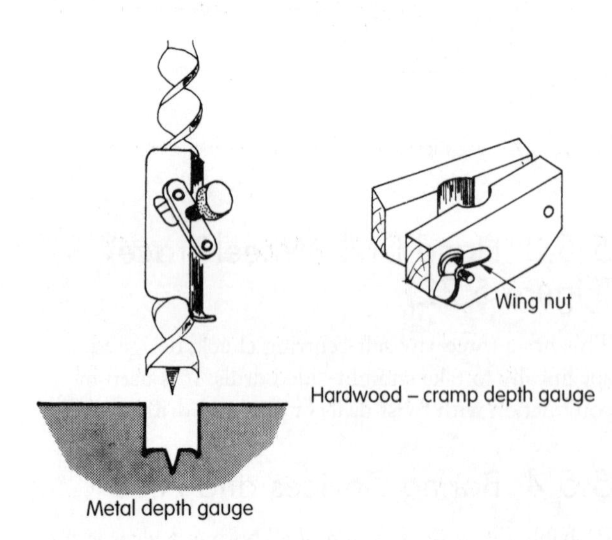

Figure 5.74 Bit depth gauge

5.6 CHISELS

Woodcutting chisels are designed to meet either general or specific cutting requirements. Table 5.4 lists those chisels in common use, together with their characteristics. Figures 5.75 to 5.80 illustrate some popular examples.

Chisels can be divided into two groups – those which cut by one of two ways:

1 paring
2 chopping.

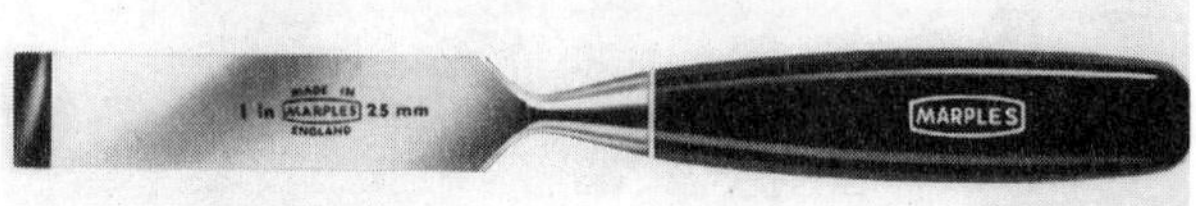

Figure 5.75 Firmer chisel

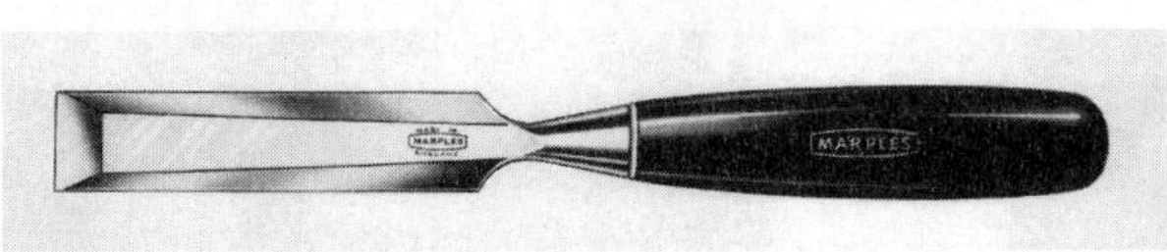

Figure 5.76 Bevel-edged chisel

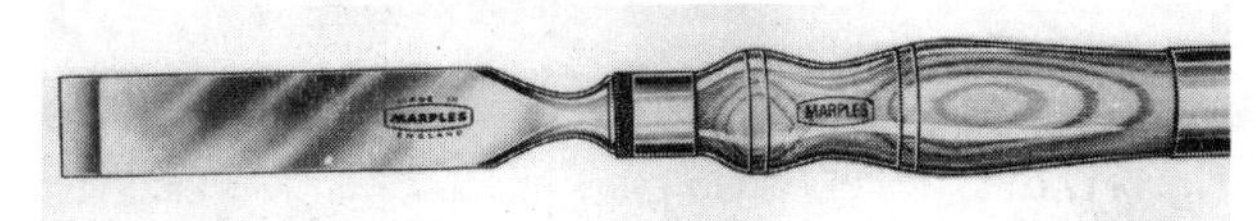

Figure 5.77 Registered chisel

Figure 5.78 Mortise chisel

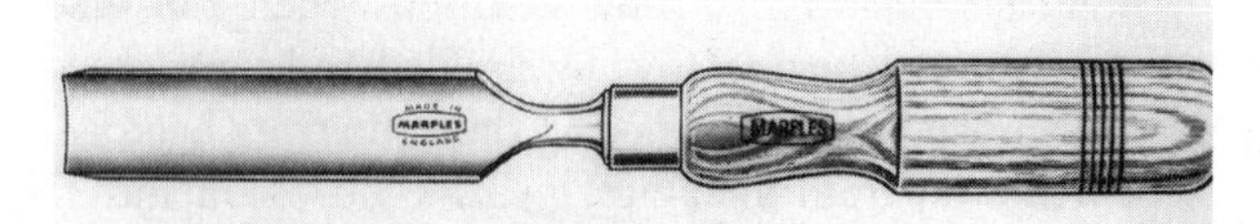

Figure 5.79 Out-cannel (firmer) gouge

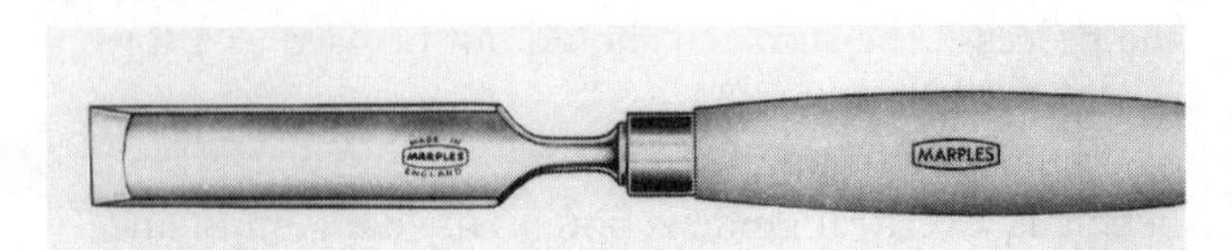

Figure 5.80 In-cannel (scribing) gouge

Table 5.4 Characteristics of woodcutting chisels (not all types are available in metric sizes)

Chisel type	Function	Handle material	Blade widths	Blade section	Remarks
Firmer (Fig. 5.75)	Paring and light chopping	HW plastics	1/4–1 1/4 in 6–32 mm		General bench work etc. Plastics-handle types can be lightly struck
Bevel-edged (Fig. 5.76)	As above	HW plastics	1/8–1 1/2 in 3–38 mm		As above, plus ability to cut into acute corners
Paring	Paring long or deep trenches	HW	1/4–1 1/2 in		Extra-long blade
Registered (Fig. 5.77)	Chopping and light mortising	HW	1/4–1 1/2 in 6–38 mm		Steel ferrule prevents handle splitting
Mortise (Fig. 5.78)	Chopping and heavy mortising	HW plastics	1/4–1/2 in		Designed for heavy impact
Gouge – firmer (out-cannel)* (Fig. 5.79)	Hollowing into the wood's surface	HW	1/4–1 in 6–25 mm		Size measured across the arc
Gouge – scribing (in-cannel)* (Fig. 5.80)	Hollowing an outside surface or edge				Extra-long blades available (paring gouge)

Note: * Out-cannel (firmer) gouges have their cutting bevel ground on the outside; in-cannel (scribing) gouges have it on the inside

5.6.1 Paring

This simply means cutting thin slices of wood, either across endgrain (Figure 5.81) or across the grain's length (Figure 5.82). Chisels used for this purpose are slender and designed for easy handling.

Figures 5.81 and 5.82 show two examples of paring. Note the method of support given to the body, workpiece and chisel (both hands behind the cutting edge).

Figure 5.81 Vertical paring

5.6.2 Chopping

These chisels need to be robust, so as to withstand being struck by a mallet. Their main function is to cut (chop) through endgrain, usually to form an opening or mortise hole to receive a tenon – hence the common name, mortise chisels.

Figure 5.83 shows a mortise chisel and mallet being used to chop out a mortise hole. Notice the firm support given to the workpiece and how the 'L' cramp (a

Figure 5.82 Horizontal paring

Figure 5.83 Chopping a mortise hole. Note the use of scrap wood to protect the bench and workpiece, and how the cramp has been laid flat with protruding bar and handle positioned towards the well of the bench, out of harm's way

'G' cramp could have been used) has been laid over on its side so as not to obscure the operative's vision.

5.7 SHAPING TOOLS

All cutting tools can also be regarded as shaping tools so the previous tools should also be included under the heading of shaping tools.

5.7.1 Axe (Blocker)

Provided an axe is used correctly, i.e. always keeping fingers and body behind its cutting edge, it is a highly efficient tool and invaluable to the site worker for quick removal of waste wood or for cutting wedges, etc. However, it must be kept sharp and at the finish of each operation the blade must be protected with a thick leather sheath. Figure 5.84 shows an axe being used to cut a wedged-shaped plug (see Section 11.6.1). Note the piece of waste wood on the floor, to protect both the floor and the axe cutting edge.

5.7.2 Surform Tools

This is a very versatile range of shaping tools, capable of tackling most materials, depending on the blade (Table 5.5). They are a valuable asset for joiners involved in house maintenance, where conventional tools are often impractical and the bench-hand will also find these files and shapers very useful.

Figure 5.85 shows just two of these tools and Table 5.5 shows their blade capabilities.

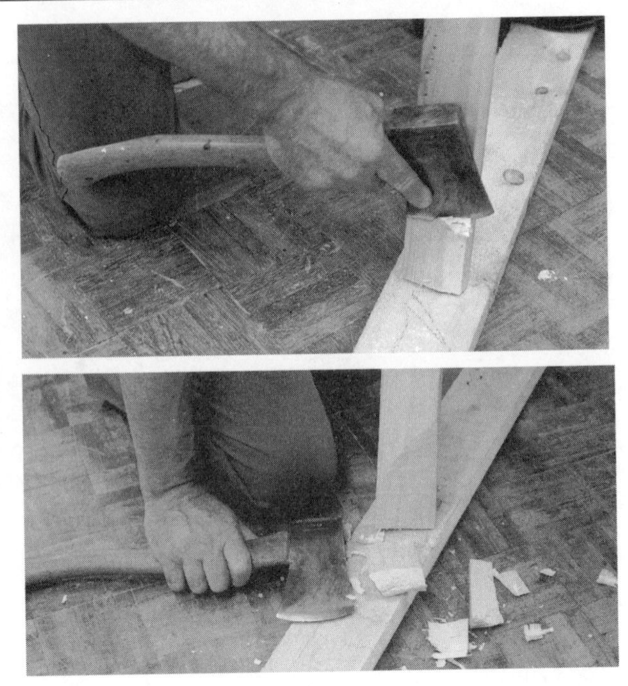

Figure 5.84 Using an axe to cut a wedge-shaped plug. NB: For removing larger amounts of wood, the hand would be moved up the shaft to increase striking power

Table 5.5 Stanley Surform blades

	Standard cut 21–505 Plane Planerfile Flat file	Fine cut 21–506 Plane Planerfile Flat file	Half round 21–507 Plane Planerfile Flat file	Metals & plastics 21–508 Plane Planerfile Flat file	Round 21–558 Round file	Fine cut 21–520 Block plane Ripping plane	Curved 21–515 Shaver tool
Hardwoods	■	■	■		■	■	■
Softwoods	■	■	■		■	■	■
End grain		■				■	
Chipboard	■		■		■		■
Plywood	■		■		■		■
Blockboard	■		■		■		■
Vinyl	■	■	■		■	■	■
Rubber	■		■		■		■
Plaster	■	■	■		■	■	■
Thermalite	■		■		■		■
Chalk	■		■		■		■
Glass fibres	■		■	■	■		■
Brass		■		■		■	
Lead		■		■		■	■
Aluminium		■		■		■	
Copper		■		■		■	
Mild steel				■			
Plastic laminates				■			■
Plastic fillers			■	■		■	■
Nylon	■		■		■		
Linoleum	■		■		■		
Ceramics		■					

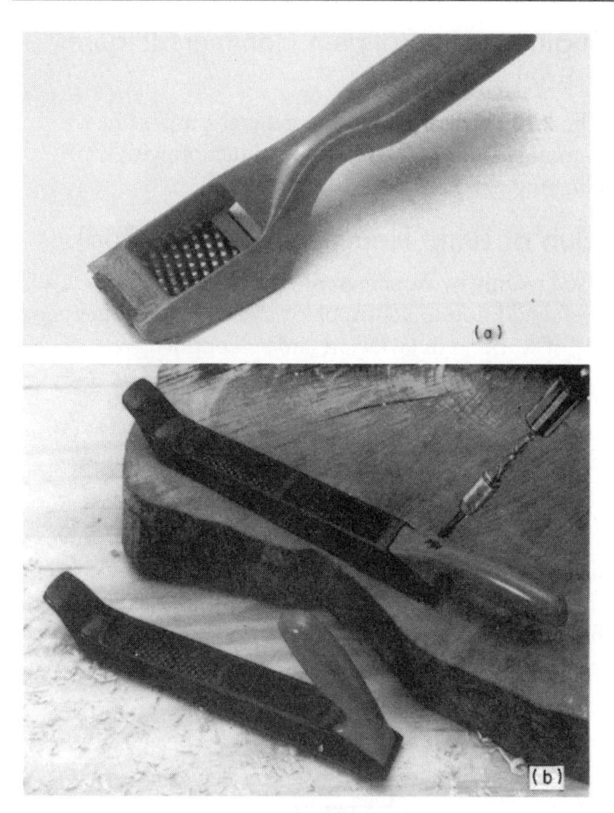

Figure 5.85 A selection of Stanley Surform tools (with kind permission from Stanley Tools)

5.8 DRIVING (IMPELLING) TOOLS

These are tools which have been designed to apply a striking or turning force. For example:

- hammers – striking
- mallets – striking
- screwdrivers – turning
- carpenter's braces – turning
- hand drills (wheel-braces) – turning.

5.8.1 Hammers

There are four types of hammer you should become familiar with:

a claw hammers
b Warrington or crosspein hammers
c engineer's or ballpein hammers
d club or lump hammers.

Claw Hammers (Figure 5.86a)

These are steel-headed with a shaft of wood or steel and a handgrip of rubber or leather. They are used for driving medium to large nails and are capable of withdrawing them with the claw. Figure 5.87 shows a claw hammer being used to withdraw a nail. Notice the

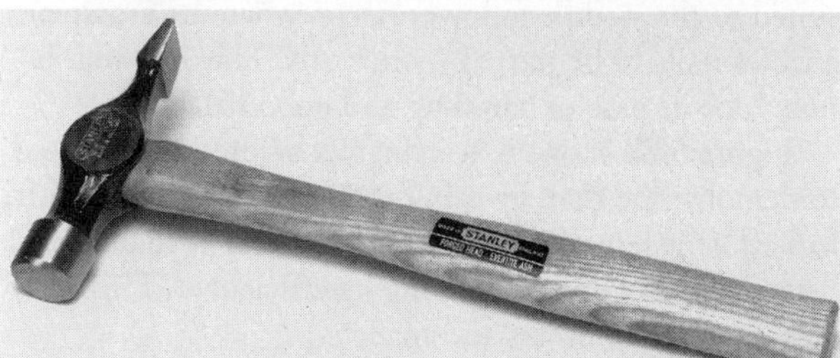

Figure 5.86 Hammers (with kind permission from Stanley Tools)

Figure 5.87 Claw hammer withdrawing a nail. Leverage should be terminated and hammer re-positioned by using a packing (as shown) before the hammer shaft reaches the upright position, or an angle of 90° to the face of the work-piece – otherwise undue stress could damage the hammer shaft

waste wood used both to protect the workpiece and to increase leverage.This type of hammer is the obvious choice for site workers involved with medium to heavy construction work. It can, however, prove cumbersome as the size of nail decreases.

Warrington or Crosspein Hammers (Figure 5.86b)

Although capable of driving large nails, this is better suited to the middle to lower range, where its crosspein enables nails to be started more easily. This hammer is noted for its ease of handling and good balance.

Figure 5.88 shows a Warrington hammer being used to demonstrate that, by using the full length of its shaft, less effort is required and greater accuracy is maintained between blows, thus increasing its efficiency. (This applies to all hammers and mallets.)

Tools associated with this hammer are the nail punch (see Section 5.8.2(e)) and pincers. Pincers (Section 5.9.2) can withdraw the smallest of nails.

(a) Right

(b) Wrong

Figure 5.88 Holding a hammer

Engineer's or Ballpein Hammers (Figure 5.86c)

The larger sizes are useful as general-purpose heavy hammers and can be used in conjunction with wall-plugging chisels, etc.

Club or Lump Hammers (Figure 5.86d)

Used mainly by stonemasons and bricklayers, this is a useful addition to your tool kit as a heavy hammer capable of working in awkward and/or confined spaces.

Warning: Hammer heads should never be struck against one another or any hardened metal surface, as this action could result in the head either splitting or splintering – particles could damage your eyes.

If a hammer face becomes greasy or sticky, your fingers or workpieces may suffer a glancing blow. Always keep the hammer's striking face clean by drawing it across a fine abrasive paper several times.

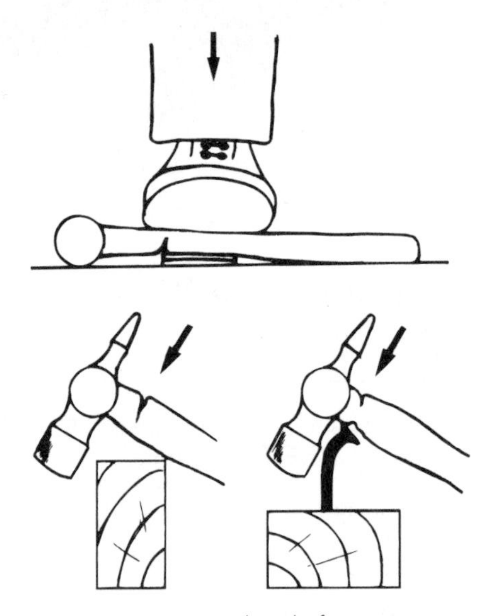

Figure 5.89 Damage to wooden shaft

Wooden shafts are still preferred by many craftsmen, probably because of their light weight and good shock-absorbing qualities. Users of wooden-shafted hammers must, however, make periodic checks to ensure that the head is secure and that there are no hairline fractures in the shaft. Figure 5.89 illustrates three examples of how a shaft could become damaged.

5.8.2 Punches

Punches can be classified as:

- a pin and nail punches (Figure 5.90)
- b pin and nail drivers (Figure 5.91)
- c centre punches (Figure 5.92)
- d name punches (Figure 5.93).

a) Pin and Nail Punches (Figure 5.90)

With the help of a hammer, these are used to sink a pin or nail head (depending on the tip diameter of the punch) just below the wood surface. This will enable the fixing to be concealed with the appropriate wood filler.

The point of the punch is cupped to help prevent it sliding off the nail head whilst being struck.

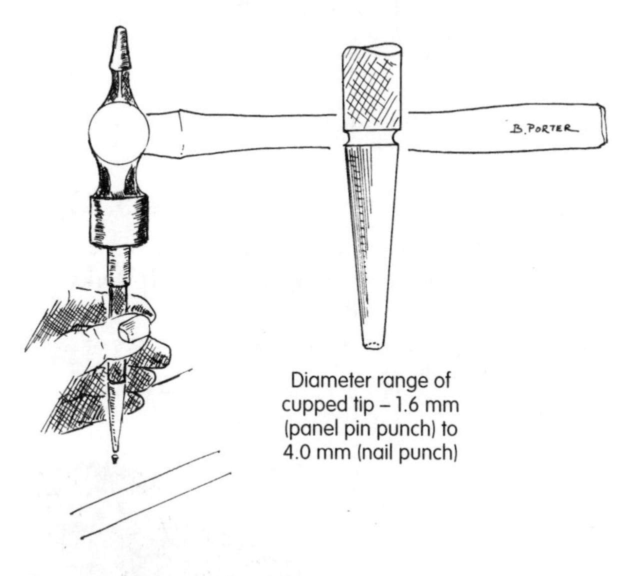

Figure 5.90 Pin/nail punches

b) Pin and Nail Drivers (Figure 5.91)

Although not a true nail punch, when used in conjunction with panel pins or small nails this can achieve the same effect as when a nail punch is used as a 'drifter' (between the pin/nail and hammer head). A driver is used when a hammer would be impracticable or liable to cause damage to the area surrounding the fixing. Drivers use a built-in impeller which is activated by a one-handed push operation.

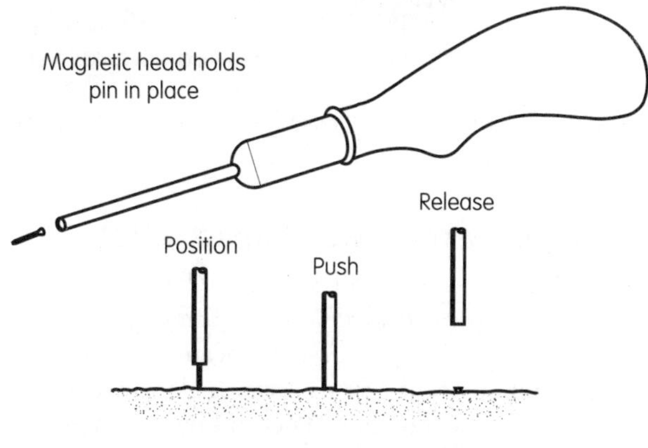

Figure 5.91 Pin and nail drivers

c) Centre Punches (Figure 5.92)

These are generally used by the metal worker, but because the joiner is often called upon to drill holes in metalwork this punch has been included. The main function of a centre punch is to accurately indent the surface of the metal (Figure 5.92a) at the point of drilling, thereby providing the drill bit with an initial pilot. Pictured with other examples of punches, the lowest two punches in Figure 5.92b are automatic centre punches, which indent the metal by simply pushing to activate the built-in impellers.

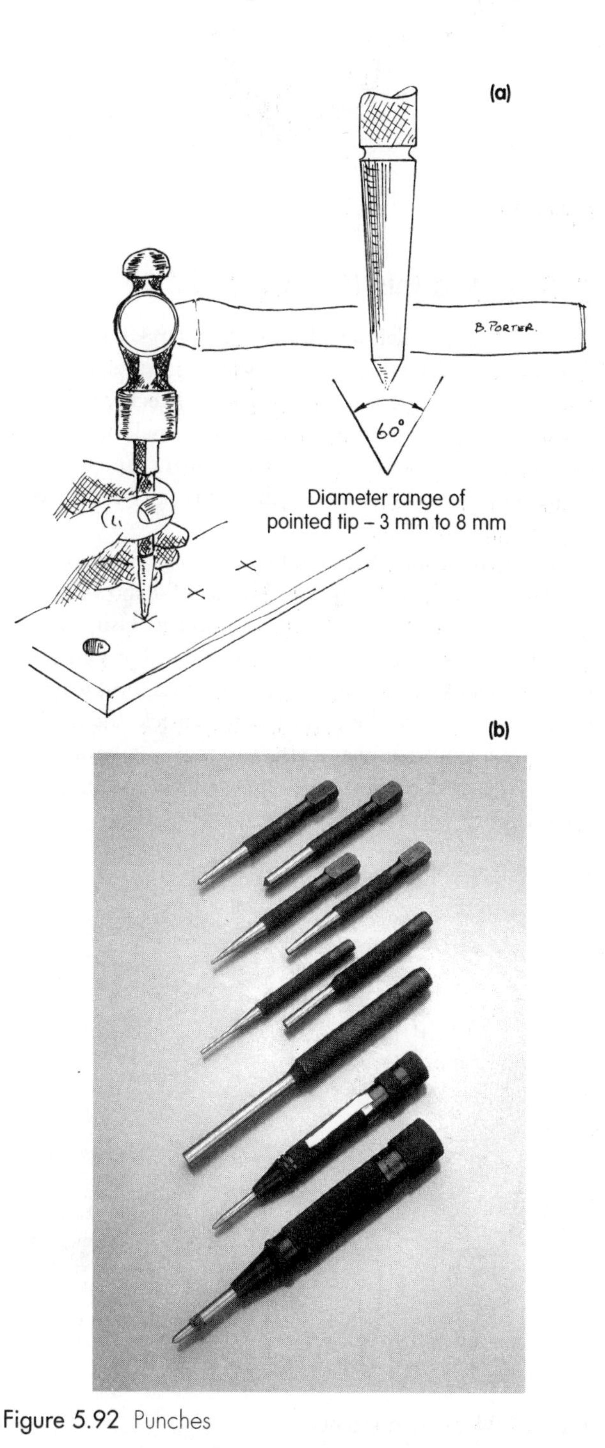

Figure 5.92 Punches

h) Name Punches (Figure 5.93)

Traditionally joiners' tools were nearly always personalised by embossing the joiner's name or code on them. Name punches are available to stamp wood, plastics and some soft metals. Today the means of identification may be a concealed post code.

Figure 5.93 Name punch

5.8.3 Mallets (Figure 5.94)

The head of a mallet, which provides a large striking face and its shaft, which is self-tightening (tapered from head to handle), are usually made from beechwood and weigh between 0.4 kg and 0.6 kg; choice will depend on the mallet's use and the user. Many joiners prefer to make their own mallet, in which case it can be shaped to suit their own hand.

The joiner's mallet should be used solely to strike the handle of woodcutting chisels. Figure 5.94 shows its correct use and Figure 5.83 shows it in use. Using a mallet for knocking together timber frames or joints should be regarded as bad practice because, unless protection is offered to the surface being struck, the mallet will have a similar bruising effect to that of a hammer. However, soft-headed mallets are available specifically for this purpose.

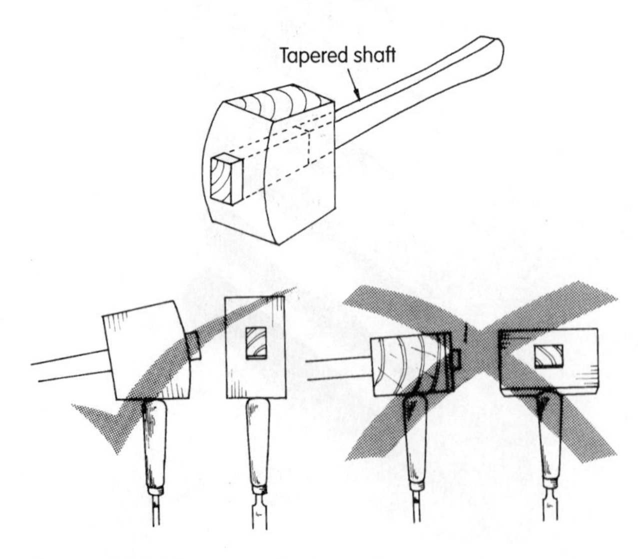

Figure 5.94 Using a mallet correctly

5.8.4 Screwdrivers

The type and size of screwdriver used should relate not only to the type and size of screw but also to the speed of application and the location and quality of the work.

There are three basic types of screwdriver used by the carpenter and joiner:

a the fixed or rigid-blade screwdriver
b the ratchet screwdriver
c the spiral ratchet or pump screwdriver.

Each is capable of tackling most, if not all, of the screws described in Section 11.2.

a) Rigid-blade Screwdriver (Figure 5.95)

This is available in many different styles and blade lengths, with points to suit any screw head. It works directly on the screw head (screw eye) to give positive driving control. Figure 5.95 illustrates two types of rigid-blade screwdriver.

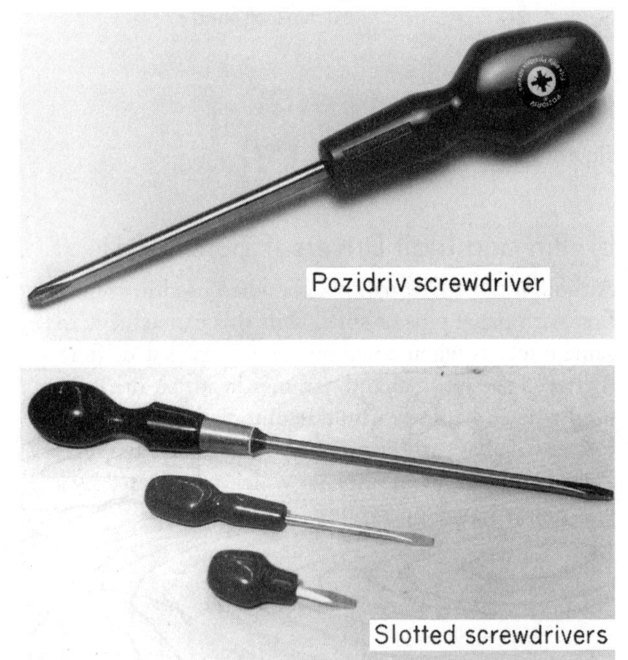

Figure 5.95 Rigid-blade screwdrivers (with kind permission from Stanley Tools)

b) Ratchet Screwdriver (Figure 5.96)

This can handle slotted and Superdriv (Posidriv) headed screws and is operated by rotating its firmly gripped handle through 90° and back and repeating this action for the duration of the screw's drive. A clockwise or counter-clockwise motion will depend on the ratchet setting. A small sliding button is used to preselect any of the following three operations:

1 forward position – clockwise motion
2 central position – rigid blade

3 backward position – counter-clockwise motion.

Because the driving hand retains its original grip on the screwdriver throughout its operation, this ratchet facility speeds up the process.

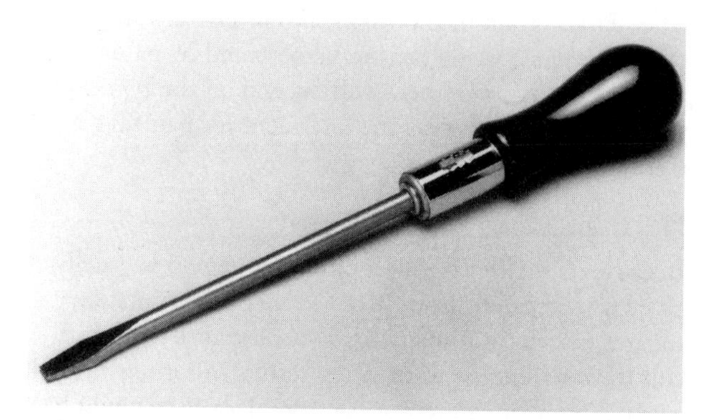

Figure 5.96 Ratchet screwdrivers (with kind permission from Stanley Tools)

c) Spiral Ratchet Screwdriver (Figure 5.97)

This is often termed a pump screwdriver because of its pump action and is by far the quickest hand method of driving screws. Not only can it handle all types of screws, it can also be adapted to drill and countersink holes.

Figure 5.97 shows a Stanley Yankee spiral ratchet screwdriver and Figure 5.98 indicates some of the many accessories available. Its ratchet control mechanism is similar to that of the standard ratchet screwdriver.

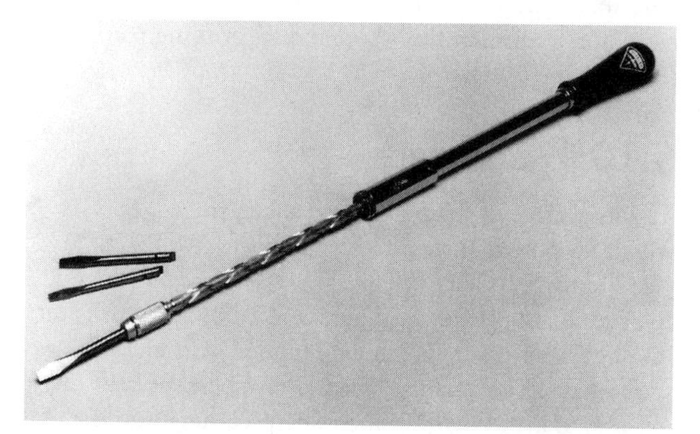

Figure 5.97 Spiral ratchet screwdriver and bits (with kind permission from Stanley Tools)

However, its driving action is produced by pushing (compressing) its spring-loaded barrel over a spiral drive shaft, thus rotating the chuck (bit-holder) every time this action is repeated.

Warning: By turning its knurled locking collar, the spiral drive shaft can be fully retained in the barrel, enabling it to be used as a short rigid or ratchet screwdriver. However, while the shaft is springloaded in this

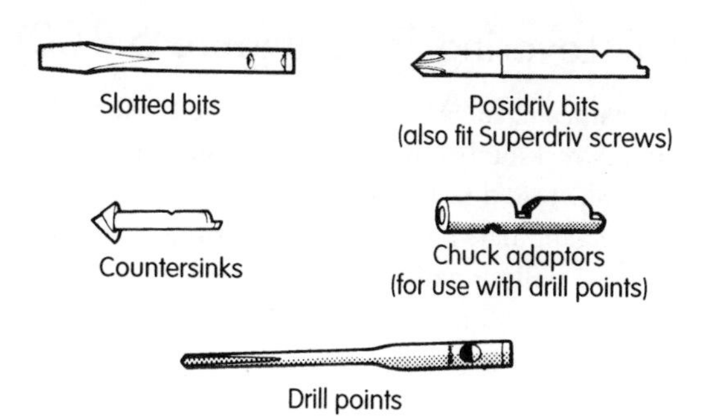

Figure 5.98 Some spiral ratchet screwdriver accessories

position, its point must always be directed away from the operator, as it is possible for the locking device to become disengaged, in which case the shaft will lunge forward at an alarming rate and could result in serious damage or injury.

NB: *After use, always leave this screwdriver with its spiral shaft fully extended. The spring should never be left in compression.*

Screwdriver Efficiency

With the exception of the stub (short-blade) screwdriver, the length of blade will correspond to the point size. If driving is to be both effective and efficient, it is important that the point (blade) must fit the screw eye correctly. Figure 5.99 illustrates how face contact with a slotted eye can affect the driving efficiency – (a) and (b) are inclined to come out of the slot so (c) should be maintained at all times.

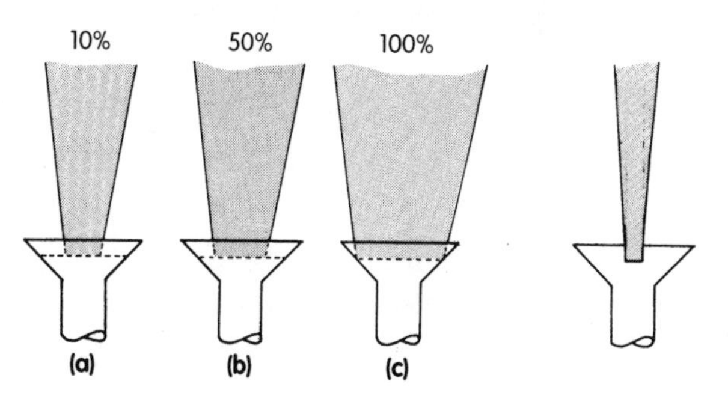

Figure 5.99 A guide to slot screwdriver efficiency

5.9 LEVER AND WITHDRAWING TOOLS

We have already seen how the claw hammer can be very effective in drawing small to medium-sized nails. But where a heavy application is required (large nails, etc.), extra leverage is often needed. This can be provided by using wrecking bars. Conversely, the drawing of small nails or panel pins will usually require the use of pincers.

5.9.1 Levering Tools (Figure 5.100)

Various names have been used to describe these tools, including wrecking bar, nail bar or Tommy bar, and two types are illustrated in Figure 5.100.

All wrecking bars are suitable for prising items apart, levering and pulling nails.

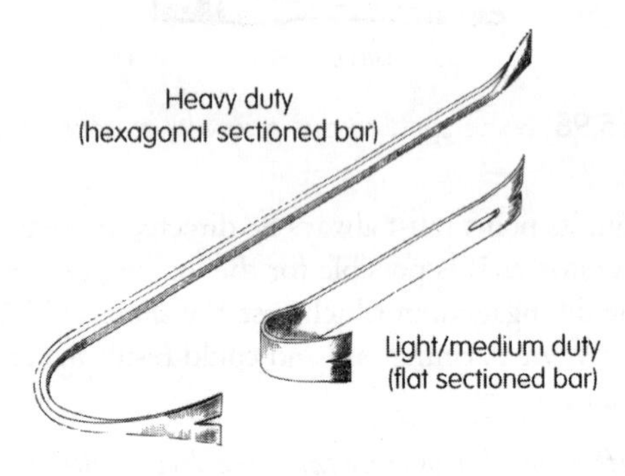

Figure 5.100 Wrecking bars

5.9.2 Pincers (Figure 5.101)

These are used to withdraw small nails and panel pins. As shown in Figure 5.101, by using a rolling action the necessary leverage is provided. The jaws of these pincers are very strong and therefore capable of cutting through soft metal and small sections of mild steel wire.

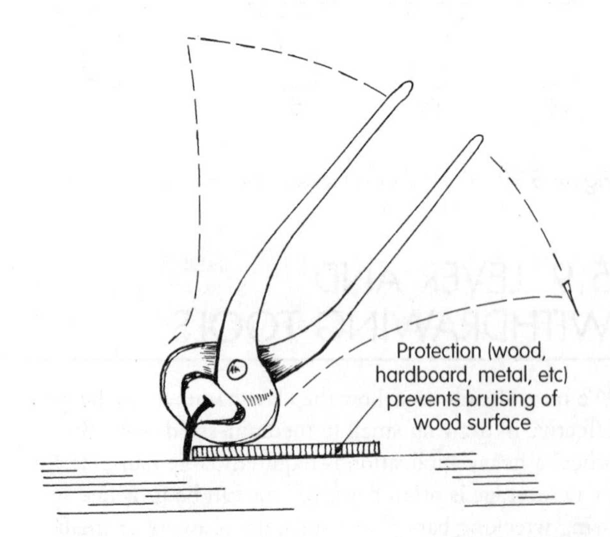

Figure 5.101 Pincers being used to withdraw a small nail

5.10 FINISHING TOOLS AND ABRASIVES

The final cleaning-up process will be determined by whether the grain of the wood is to receive a transparent protective coating or is to be obscured by paint. In the former case, treatment will depend on the type and species of wood whereas the treatment for painting is common to most woods.

Hardwood

Because its main use is decorative, hardwood is usually given a transparent protective coating. Unfortunately, the grain pattern of many hardwoods makes them difficult to work since planing often results in torn or ragged grain. A scraper can resolve this problem and should be used before finally rubbing down the surface with abrasive paper.

NB: Always follow the direction of the grain when using abrasive paper before applying a transparent finish (see Figure 5.105).

Softwood

Softwood surfaces that require protection are usually painted but there are some exceptions. In preparing a surface for paint, flat surfaces must be flat but not necessarily smooth as minor grain blemishes, etc. will not show. Surface ripples caused by planing machines do show, however and will require levelling with a smoothing plane, after which an abrasive paper should be used. The small scratch marks left by the abrasive help to form a key between the wood and its priming paint (first sealing coat).

5.10.1 Scraper

This is a piece of hardened steel sheet, the edges of which have been turned to form a burr (cutting edge) (see Figure 5.102a).

A scraper is either held and worked by hand or is set into a scraper plane which locks and is held like a large spokeshave. Flat scrapers must always be kept bent during use, to avoid digging their sharp corners into the wood (Figure 5.102b). Scraper planes, however, can be pre-set to the required cutting angle and then simply pushed.

5.10.2 Sanding

Sanding is the application of abrasive paper or cloth to the surface of wood.

There are several kinds of grain used in the manufacture of abrasive sheets, the most popular for hand sanding being glass and garnet (which is harder than glass). Both are available in sheet size 280 mm × 230 mm and are

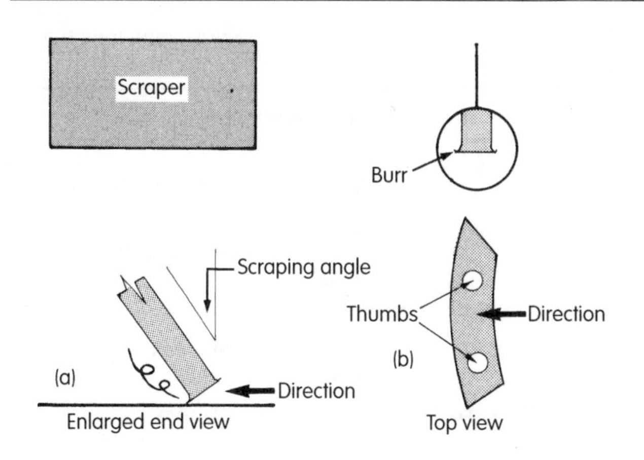

Figure 5.102 Handheld cabinet scraper

graded as shown in Table 5.6 according to their grit, which ultimately determines the smoothness of the wood. This table also shows aluminium oxide as an abrasive used for general machine use; it can also be used for hand sanding.

Most of the terms used in Table 5.6 are self-explanatory but Figure 5.103 illustrates the main difference in grain configuration between 'open coat' and 'closed coat'. Table 5.7 should help you to relate those with application.

If you look on the back of abrasive sheets you will notice various codes (Figure 5.104). Tables 5.6 and 5.7 have been devised to help you unravel this sometimes mystifying information.

When sanding by hand a sanding block should always be used to ensure a uniform surface. Typical examples are shown in Figure 5.105, together with a method of dividing a sanding sheet into the appropriate number of pieces to suit the blocks.

Table 5.6 Comparative grading of abrasive sheets

Product description					End forms					Grits available																						
Grain	Backing	Product code	Bond	Coat	Sheets	Coils	Rolls	Belts	Discs	1200	1000	800	600	500	400	360	320	280	240	220	180	150	120	100	80	60	50	40	36	30	24	16
Glass	Paper C	117	G	C	●														FL	0		1	1½	F2		M2*		S2	2½	3		
	Paper C	121	G	C	●														00	0		1	1½	F2		M2*		S2	2½	3		
Garnet	Paper A	129	G	O	●												9/0		7/0	6/0	5/0	4/0	3/0	2/0	0							
	Paper C+D	128	G	O	●																5/0	4/0	3/0	2/0	0	½	1	1½				
Aliminium oxide	Paper C	134	G	O	●																	●	●	●	●	●	●	●	●			
	Paper E	156	R/G	C	●		●	●	●						●		●	●	●	●	●	●	●	●	●	●	●	●	●	●	●	
	Paper E	157	R	C				●	●										●	●	●	●	●	●	●	●	●	●	●			
	Cloth X	141	R	O	●	●	●	●	●													●	●	●	●	●	●	●	●			
	Cloth X	147	R	C			●	●	●								●	●	●	●	●	●	●	●	●	●	●	●	●		●	
	Cloth FJ	144	R	C				●								●	●		●	●	●	●	●	●	●	●						
	Cloth J	145	R	C				●									●	●	●	●	●	●	●	●	●	●						

Note: **Bond**: Glue – G; Resin – R; Resin/Glue – R/G; **Grain configuration:** Open coat – O; Closed coat – C
Grain size: *IF = 350 grit; M2 = 70 grit

	Product code	Description
Abrasives for hand use		
Glass paper – C	117	Cabinet quality for joinery and general wood preparation
Glass paper – C	121	Industrial quality for joinery and general wood preparation
Garnet paper – A	129	For fine finishing and contour work on furniture
Garnet paper – C+D	128	Durable product for fine finish stages of wood preparation
Abrasives for general machine use		
Aluminium oxide paper – C	134	For rapid stock removal with reduced clogging. For hand and orbital sander use
Aluminium oxide paper – E	156	For sanding medium and softwoods. A durable product recommended for heavy hand sanding, orbital sanders, and pad and wide belt sanders
Aluminium oxide – Cloth X	141	General belt sanding of wood, particularly recommended for portable sanders, open coat configuration reduces clogging

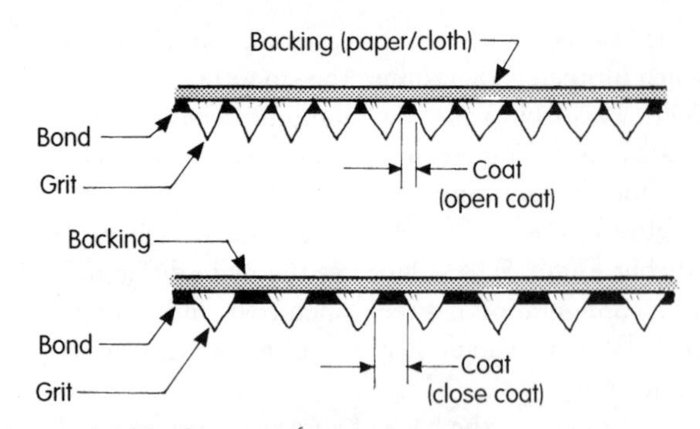

Figure 5.103 Grain configuration

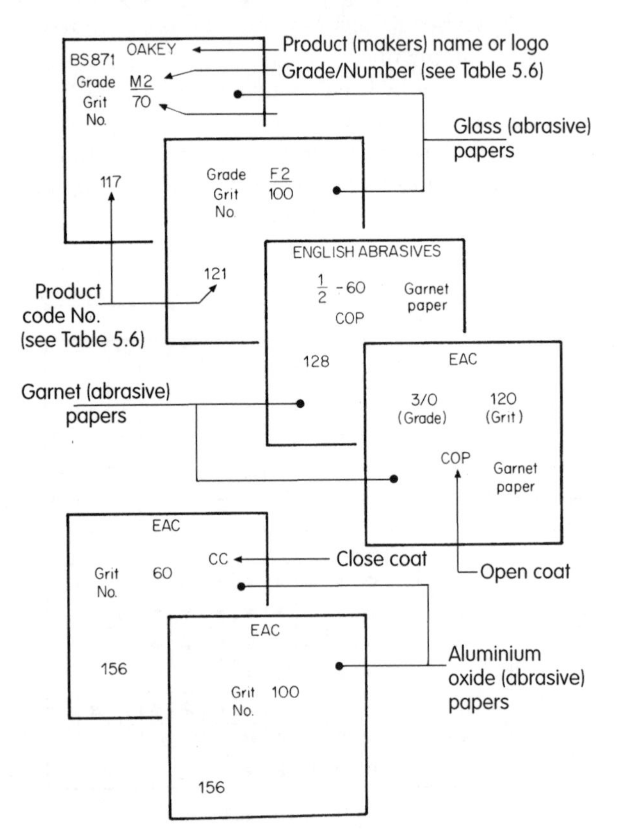

Figure 5.104 Understanding the coding used on the back of coated abrasives as identification

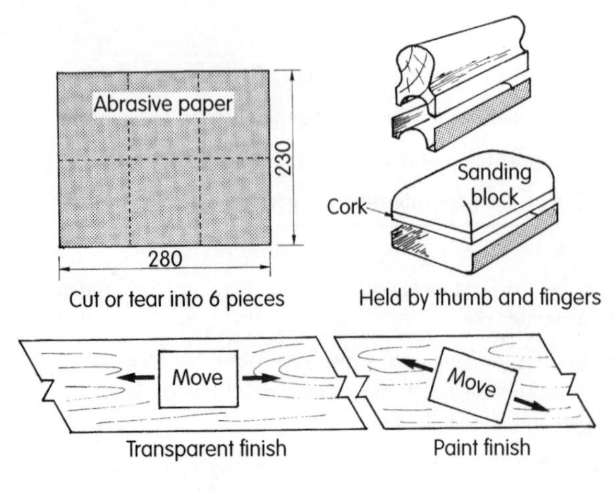

Figure 5.105 Sanding by hand

Table 5.7 Choosing the correct coated abrasive for the type of work

Key	Application
Backing	
Paper	A lower cost material for less demanding uses
Cloth	Higher tensile strength, strong and durable
Bond	
(G) All glue	Used for hand and light machine applications, a flexible bond
(R/G) Resin over Glue	Intermediate flexibility and durability
(R) All resin	Higher band strength and heat resistance, for more demanding applications
Grain Configuration	
(O) Open coat	Reduced clogging, coarser finish
(C) Closed coat	Finer finish, faster cutting

Grit size
The coarser the grit (larger particles), the smaller the grit size number, according to international FEPA 'P' standards. The non-technical abrasives are graded according to different standards and are equivalent as shown in the table.

Note: grit (particle) size – the smaller the grit number, the more coarse the grit. For example, grit number 24 is very coarse and grit numbers 50–70 medium, while grit number 400 is very fine

Figure 5.106 Bench holdfast (with kind permission from Record Tools Ltd)

5.11 HOLDING EQUIPMENT (TOOLS AND DEVICES)

Holding tools are general purpose mechanical aids used in the preparation and assembly of timber components. Holding devices have usually been contrived to meet the needs of a specific job or process and Table 5.8 gives examples.

Items (a) to (o) in Figure 5.113 are mentioned throughout the text and are almost self-explanatory. The mitre block and box, (j) and (k), are devices used to

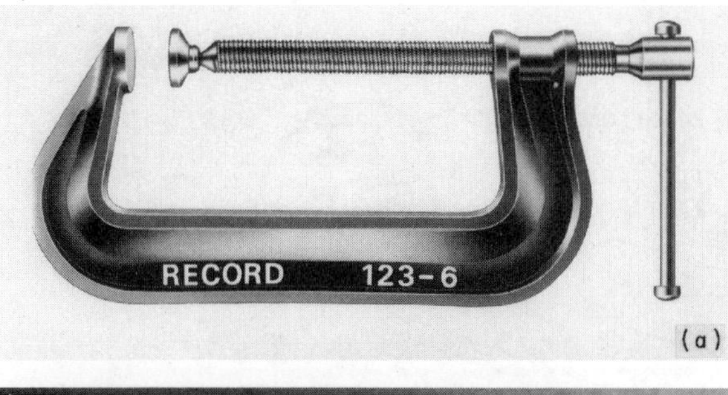

Figure 5.107 'G' cramp (with kind permission from Record Tools Ltd)

Figure 5.108 Bench holdfast and 'G' cramp in use

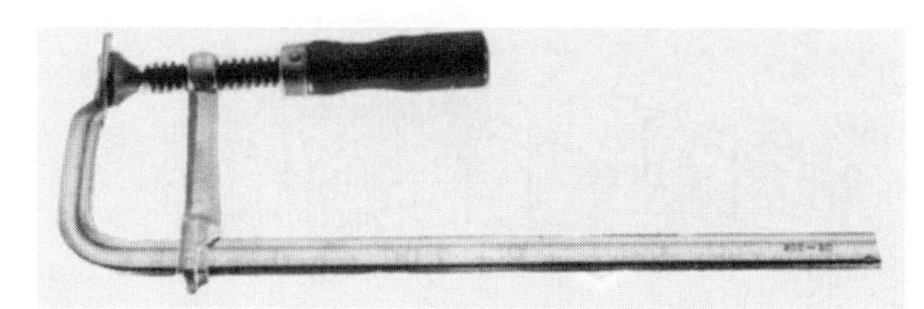

Figure 5.109 'L' cramp

Figure 5.110 Sash cramp and lengthening bar (with kind permission from Record Tools Ltd)

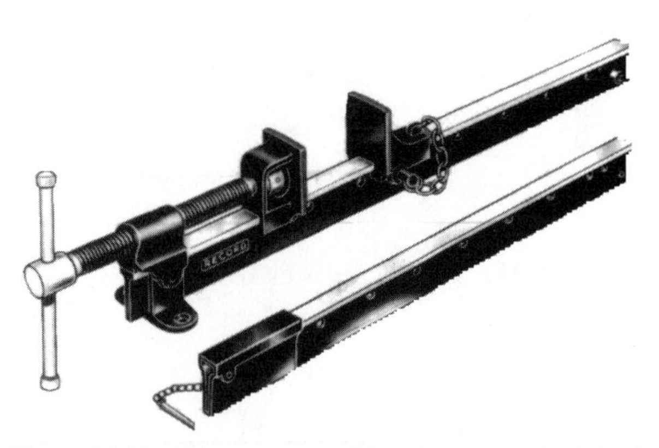

Figure 5.111 'T' bar cramp and lengthening bar (with kind permission from Record Tools Ltd)

Figure 5.112 Dowelling jig (with kind permission from Neill tools)

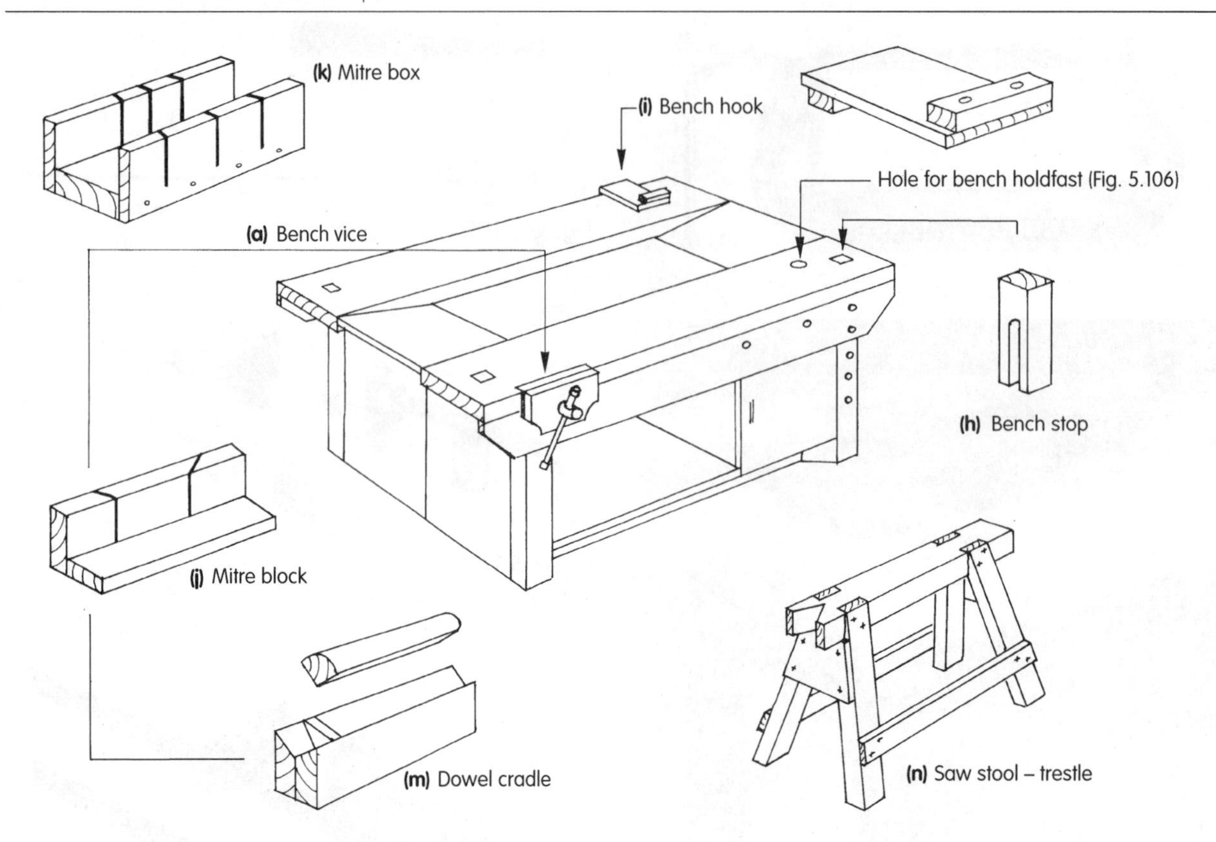

Figure 5.113 Workbench and holding equipment

Table 5.8 Holding equipment (see Figure 5.113)

Equipment (see Fig. 5.113)	Use: Preparing material	Use: Assembly aid
Holding tool		
a) Bench vice	Yes	Some models
b) Bench holdfast (Figs. 5.106 and 5.108)	Yes	No
c) 'G' cramp (Fig. 5.107)	Yes	Yes
d) 'L' cramp (Fig. 5.109)	Yes	Yes
e) Sash cramp (Fig. 5.110)	Yes	Yes
f) 'T' bar cramp (Fig. 5.111)	No	Yes
g) Dowelling jig (Fig.5.112)	Yes	
Holding devices		
h) Bench stop*	Yes	No
i) Bench hook*	Yes	No
j) Mitre block	Yes	No
k) Mitre box	Yes	No
l) Mitre box and saw (Fig. 5.114)	Yes	No
m) Dowel cradle	Yes	No
n) Saw stool (trestle)	Yes	Yes
o) Stanley workmate (Fig. 5.115)	Yes	Yes

Note: * Provision can be made in their construction for left- or right-handed users

support squared or moulded sections while they are sawn to an angle of 45°. The dowel cradle, (m), holds squared or rounded sections. Saw stools or trestles, (n), are usually used in pairs, either to form a low bench or for support while sawing.

Figure 5.113 features a double-sided work bench, together with items (a), (h), (i), (j), (k), (m) and (n) (as featured in Table 5.8), most of which can be made by the joiner. In recent years proprietary mitre boxes with attached saws and folding trestles with adjustable work-tops have become very popular. Figure 5.114 shows a

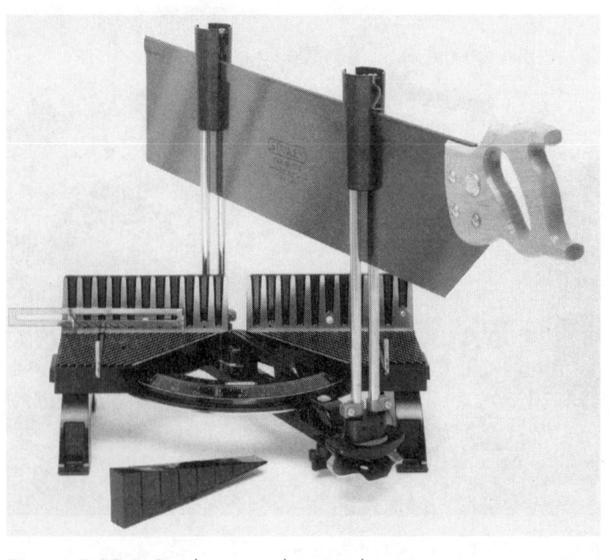

Figure 5.114 Stanley mitre box and saw

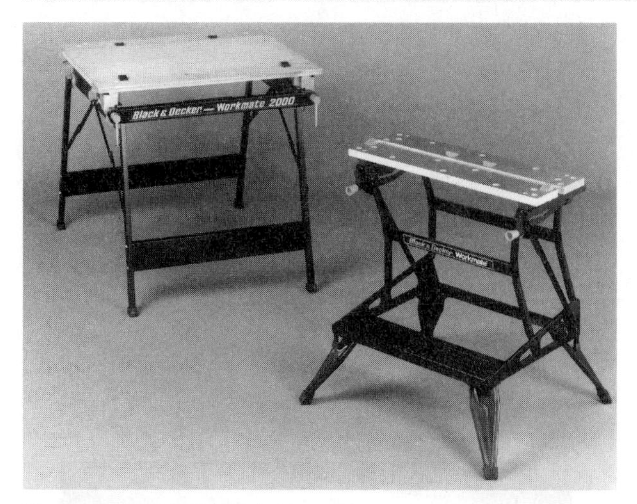

Figure 5.115 Stanley Workmate

Stanley mitre box and saw, capable of accurately cutting the most complex of moulded profiles to a mitre or bevel. Figure 5.115 shows the very popular Workmate which is a strong folding trestle. The top is in two parts which serves as a vice and/or cramp capable of holding the most awkward of workpieces.

5.12 TOOL AND ACCESSORY STORAGE, CONTAINERS AND PROTECTION

The variety and condition of the tools used by the craftsperson often reflect the quality and type of work they are capable of undertaking. A basic tool kit may consist of:

- four-fold metre rule
- flexible steel tape
- combination square
- claw hammer
- handsaw (panel saw) and possibly a tenon saw
- jack and smoothing planes
- assortment of chisels – firmer/bevel-edged
- carpenter's brace
- assortment of twistbits and countersinks
- hand drill (wheel-brace)
- assortment of twistdrills
- screwdrivers
- bradawl.

NB: Sharpening accessories (oilstones, etc.) should also be included in this kit.

However, irrespective of the kind of work you do, many different types of tool will be acquired, all of which will require some form of protection and safe storage provision, whether you are to be site or workshop based.

Extending the number and type of tools in your kit will depend on your job specification. Broadly speaking, tool kits can be broken down into three groups (see Tables 5.1–5.5):

1 everyday use – basic tool kit
2 occasional use – more common of the special use tools
3 specific use – special use tools.

In time you will find that you are duplicating some of your everyday tool kit. For example, having two panel saws should ensure that one is always kept sharp.

No matter what means of storage is chosen, the most important factors to consider about a container are that each item is housed separately or is individually protected from being knocked against another and that all cutting edges are withdrawn or sheathed in some way. Of course, the body of the tool must also be protected from damage. Unfortunately many joiners seem to regard this form of protection as secondary.

Always bear in mind that a well-used tool is not one sporting the battle scars of its container but one which, after many years of active service, appears virtually unscathed. There are a variety of ways in which you may want to house your tools and the method usually chosen will reflect the type of work your company undertakes. Different methods of containing tools include:

- traditional toolchest (Figure 5.116a)
- traditional toolbox (Figure 5.116b)
- Porterbox (Figure 5.116c)
- tool case (Figure 5.116d)
- tool tray (usually racked) (Figure 5.116e)

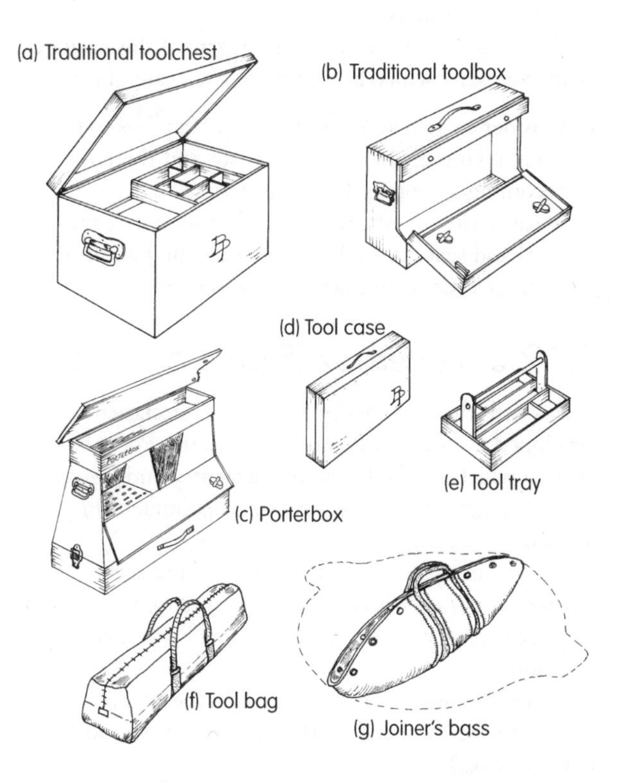

Figure 5.116 Methods of containing and carrying tools and small ancillary equipment

- tool bag (holdall) (Figure 5.116f)
- joiner's bass (Figure 5.116g).

A combination of the above is usually chosen.

5.12.1 The Traditional Toolchest (Figure 5.116a)

Traditionally a very strong, secure, top-opening rectangular box of solid construction. It is designed to accommodate all of the joiner's tools. Larger tools are housed within base compartments, whereas smaller tools are separated within a series of compartmentalised sliding lift-out trays. The two upper trays are half the box length, thereby allowing access to contents below without their removal.

Transporting such a chest is a two-man operation so heavy-duty drop-down chest handles are essential.

5.12.2 The Traditional Toolbox (Figure 5.116b)

Traditionally these boxes are strong, rectangular and dropfronted, long enough to house the longest handsaw and generally suitable for both workshop and site use. Top, bottom and sides of timber are dovetail jointed at the corners. Plywood is glued, screwed and/or nailed to this framework to form the front and back. Once set, a front portion is cut out to form a hinged flap which houses the handsaws. The interior can accommodate one or more drawers.

For a number of reasons I have regarded this arrangement as unsatisfactory, but there are two main objections.

1 In order to gain access to tools, the flap has to be left open, taking up a lot of floor space and thereby becoming a tripping hazard to passers-by.
2 Because the top of the box is often used as a saw stool and seat during break periods, the handle and closing/security mechanism is in the way.

5.12.3 Porterbox (Figure 5.116c)

It was with these factors in mind that I designed and developed a system of boxes known as the Porterbox System which can be found in Carpentry and Joinery Book 1, together with the necessary manufacturing details.

This is an easily constructed, strong and secure stable box which not only houses all your basic tool kit, with easy access, but can also serve as a small workbench and saw stool.

As can be seen from Figure 5.117, the working height can be increased by the inclusion of the add-on Portertray or Portercase.

Finally, easy portability can be provided by fitting a Portercaddy which is an add-on base unit.

Figure 5.117 The Porterbox

5.12.4 Tool Case (Figure 5.116d)

There are occasions when the intended job does not warrant taking all your basic tool kit and a smaller, lighter box may be the answer. In this case the length of the saw may not be the main criterion in its design, because a separate saw case (see Figure 5.118a) could be used along with the tool case. However, it should be possible to fit a tenon saw within the lid of the tool case.

5.12.5 Tool Tray (Figure 5.116e)

This is an open-top box with a box-length carrying handle, used for transporting your tools from job to job within the confines of the work area (site). It may also serve as a means of keeping tools tidy when working with other trades, and a rack for chisels, etc. is a useful feature. After each work period tools are transferred back to their secure main tool box or chest.

5.12.6 Tool Bag (Holdall) (Figure 5.116f)

This is an elongated bag of hardwearing material long enough to contain a hand saw and strong enough to hold a variety of tools. Its great advantage over a box or case is its lightness and portability. Its disadvantages are that each tool needs individual protection and it is not secure.

5.12.7 Traditional Joiner's Bass (Figure 5.116g)

This is a near circular shape of heavy-duty, canvas-type fabric with two rope handles attached across the fabric which, when brought together, form a half-moon bag. Leather straps or eyelets around the edge lashed with cord provide the means of closing the bass. The only advantage over the bag is that when the bass is fully open all its contents, with the exception of those contained in side pockets or pouches or wrapped in rolls, etc., are instantly revealed.

NB: The bass used by a plumber is usually smaller in diameter.

5.12.8 Individual Tool and Tool Set Protection

All tools should be prevented from moving when being stored, particularly if they are to be transported. (The cutting edges and sharp points are the most vulnerable to damage.) All planes should have the cutting face of their blades retracted before storage. Other tools requiring special treatment are:

a saws (saw teeth)
b chisels (cutting edge)
c twistbits (cutters, spurs (wings) and thread).

a) Saws

Where box protection is not available, a saw bag (Figure 5.118a) made from a strong material is a good alternative. In both cases the teeth will still need protection which can be provided by a plastic sheath (Figure 5.118b) available from the saw manufacturer and tool suppliers. Alternatively a timber lath with a groove down the length of one edge (Figure 5.118c) and held onto the saw by a band will do just as well or perhaps you could design your own.

b) Chisels

Most new chisels are now provided with a plastic end cover (sheath) (Figure 5.118d) by the manufacturer to protect the cutting edge of each blade. End covers are also available (usually in sets) from suppliers as a separate item.

Where chisels are not racked or safely compartmentalised, a strong purpose-made chisel roll is a useful storage alternative. As shown in Figure 5.118e, when the roll is unrolled, selection is quick and easy, as each chisel is easily identified within its own individual pocket. To pack the roll away, simply fold down the top flap over the handles, roll up (Figure 5.118f) and secure the roll by using the ties or straps provided.

NB: All chisels must have their cutting edges safely covered with an end protection before being put into a chisel roll.

c) Twistbits

End protection is essential and plastic sheaths (Figure 5.118g) are available. Unlike chisels, the cutting edges only have a limited resharpening capability. If the spurs (wings) or thread become damaged the cutting effectiveness is reduced.

Bit rolls (Figure 5.118h) are purpose-made bit holders. Similar in design and function to chisel rolls, the main difference is that the pockets are slimmer to accommodate individual bits.

5.12.9 Tool Holders and Pouches

The modern site worker and those involved in repetitive nailing work will be familiar with the belted hammer holder and nail pouch. Figure 5.119 illustrates a typical linked example of nail pouches and a hammer holder.

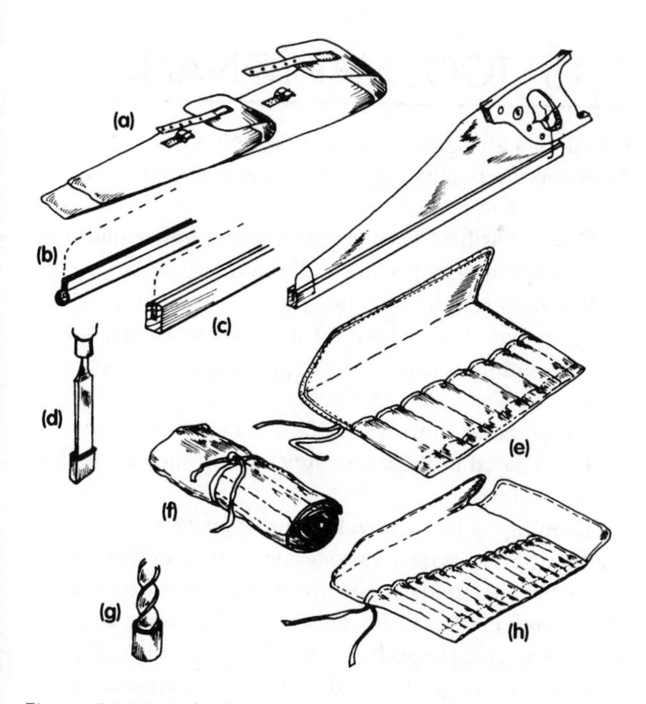

Figure 5.118 Individual tool protection and storage

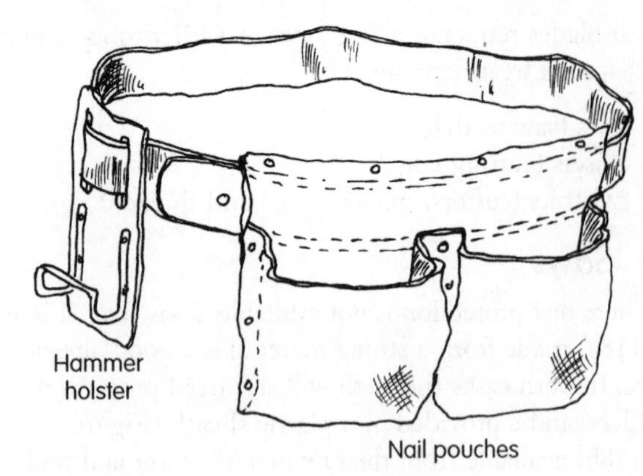

Figure 5.119 Tool holders and pouches

5.12.10 Fixings

Ideally the various sizes of nails and screws should be kept separately. Screws are generally boxed or packaged and labelled according to gauge, size and head style, so all that is needed is a tray or box to contain them. Nails, on the other hand, tend to be bought in bulk and therefore require storing in bins or trays, etc., according to size. However, on site nails tend to get mixed, which can be both frustrating and time consuming. At the outset it is well worth constructing a compartmentalised nail tray similar to the one shown in Figure 5.116e which can be incorporated within a chest and could also house items such as fixing plugs and small items of hardware. It may be worth considering making a small chest such as the Porterchest shown in Figure 5.120.

Figure 5.120 The Porterchest

Joiners frequently acquire odd screws and nails and these could be kept in a screw and nail bag like the one shown in Figure 5.121. This is nothing more than a circle of leather with holes around the circumference for a drawstring. This is used to draw the circle and the items laid within it into a bag – although mixed up, they are safely contained. To retrieve an item it is simply a matter of opening the drawstring, whereupon all the items will be fully exposed ready for easy selection. This method of containing odd loose nails and screws has been used for many years, particularly by the jobbing joiner, who regularly requires the often unexpected odd size of nail or screw.

Traditionally nail and screw bags were sited one at each end of the joiner's bass to help balance the load when it was carried over the shoulder.

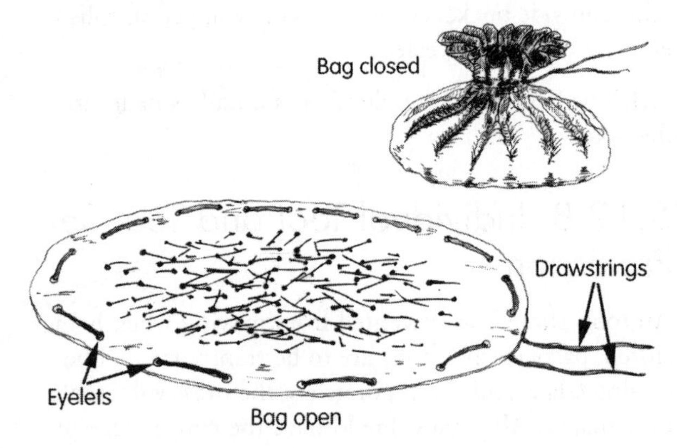

Figure 5.121 Screw and nail bags

5.13 TOOL MAINTENANCE

Of all the cutting tools mentioned, only saws and Surforms are purchased ready for use (the remainder will need sharpening).

Once tools have been sharpened, it is only a question of time before they become dull (blunt) again. Some makes of tools dull much more quickly than others due to the quality of steel used for the blade or cutter, but general dulling is caused either by the type of work or by the abrasive nature of the material being cut. However, other contributory factors include foreign bodies encased in the material being cut, such as hidden nails or screws.

Keeping tools sharp will require varying amounts of skill in the use of devices associated with tool maintenance. The techniques used for tool maintenance can differ from craftsman to craftsman and in some cases take many years to perfect. Table 5.9 gives a list of equipment thought necessary for the maintenance of the cutting tools discussed in this chapter.

Table 5.9 Tool maintenance equipment

Equipment	Tool					
	Drills	Saws	Planes	Bits	Chisels	Screwdrivers
Grindstone (machine)	Yes	–	Yes	–	Yes	Yes
Oilstone	–	–	Yes	–	Yes	Yes
Slipstone	–	Yes	Yes	–	Yes	–
Stropstick	–	–	Yes	–	Yes	–
Oil can	–	–	Yes	–	Yes	–
Mill saw file	–	Yes	–	–	–	–
Saw file	–	Yes	–	Yes	–	–
Needle file	–	–	–	Yes	–	–
Saw vice-clamp	–	Yes	–	–	–	–

5.13.1 Equipment

Grinding Machines

Every time a blade is sharpened, part of its grinding angle (see Figures 5.131 and 5.137) is worn away, making sharpening more difficult. The grinding angle must therefore be re-formed by machine. The grinding machine, together with the relevant regulations, is dealt with in Section 9.10.

Oilstones

These are used to produce a fine cutting edge (a process called honing) and may be manufactured from natural stone, such as Arkansas, or artificial which, being the least expensive, is the most common. Artificial stones can be made up from grit particles derived from carborundum (silicon carbide) or alundum (aluminium oxide). Grit sizes produce stones of either fine, medium or coarse texture. It is possible to purchase a stone with a coarse or medium surface on one side and a relatively fine texture on the other, called a combination stone.

Oilstones are very brittle and will break or chip unless they are protected. Figure 5.122 illustrates a method of constructing a protective box out of hardwood. Half of the oilstone is mortised into the base and half into the lid and allowance can be made for packings at the ends to enable the full length of the stone to be used and to reduce the risk of sliding off.

Rubber washers or blocks have been let into or pinned on to the underside to prevent the box sliding about during use.

Waterstones

These require soaking in water before use. Their keen surface can produce a sharp cutting edge quickly and with relative ease. There is a good range of grades from very fine to coarse.

Diamond-surfaced Stones

These stones are used with water and produce a very keen cutting surface. Stones are colour coded according to their harshness:

- green – extra fine
- red – fine
- blue – coarse
- black – extra coarse.

As you would expect, they are much more expensive than conventional stones, but they should keep their sharpening surfaces for many years without losing their efficiency.

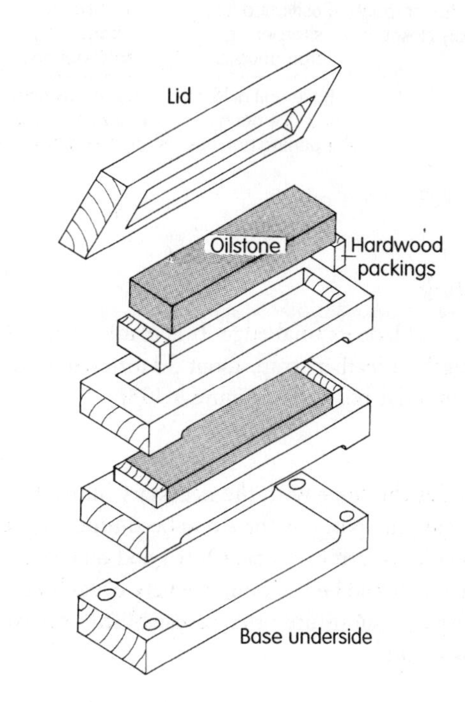

Figure 5.122 Oilstone box

Slipstones

These are composed in a similar manner to oilstones, but are designed to hone the curved cutting edges of gouges and moulding-plane cutters, etc. Figure 5.138 shows a slipstone in use – handheld (not recommended). If a holding device were used, it would leave both hands free to position the gouge or cutter against the stone. Such a device is illustrated in Figure 5.123 which can also serve as a protective box.

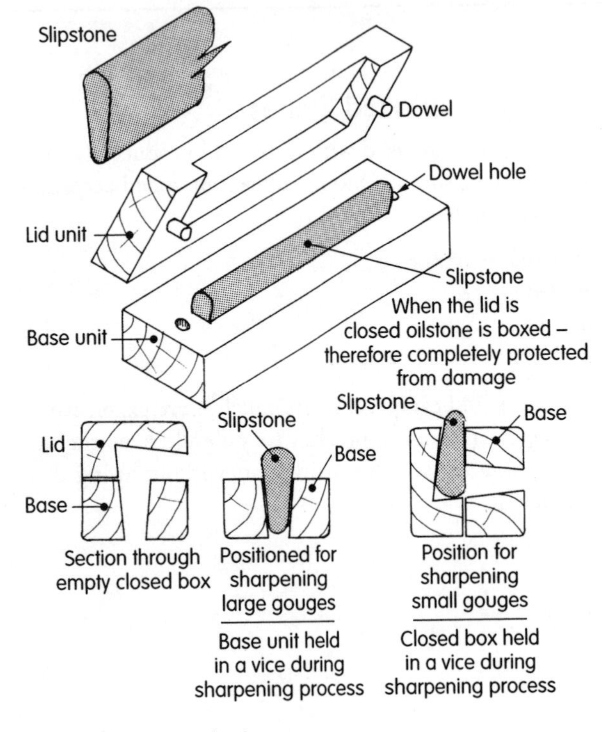

Figure 5.123 Porterslip box

Stropstick

This gives a blade its final edge. It is made from a thick, short length of leather strap, about 50 mm wide, stuck or fixed to a flat board (see Figure 5.133f).

Oil

Oil provides the stone with the necessary lubrication and prevents the pores of the stone becoming clogged with particles of grit or metal. Only good quality thin machine oil should be used but the very occasional use of paraffin as a substitute often re-establishes the stone's keenness to cut.

Files

These are used to sharpen saws and bits. At least two sizes of saw files will be required, together with a flat or mill saw file. Saw files can also be used to sharpen

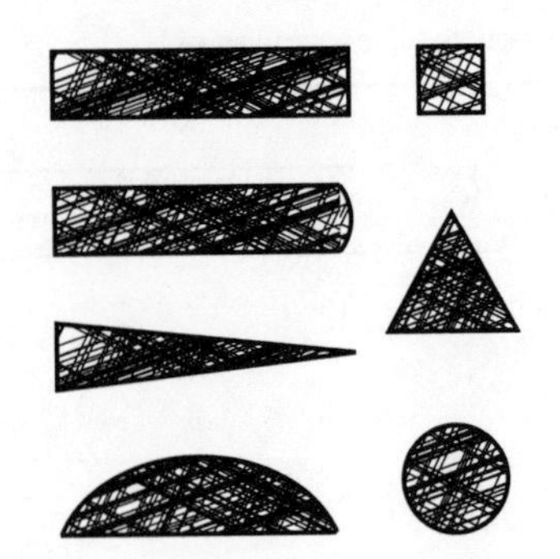

Figure 5.124 File sections

twistbits, but one or two small needle files would prove useful for the smaller sizes of bit. Figure 5.124 shows some of the different file sections available.

Holding Equipment

There are various aids for holding a tool or cutter while it is being sharpened. When sharpening a saw, it is important that it is held firm and positioned in such a way that the teeth can be seen, preferably without stooping. Figure 5.125 illustrates a device for this purpose which can be either fixed into a bench vice or cramped to a solid object.

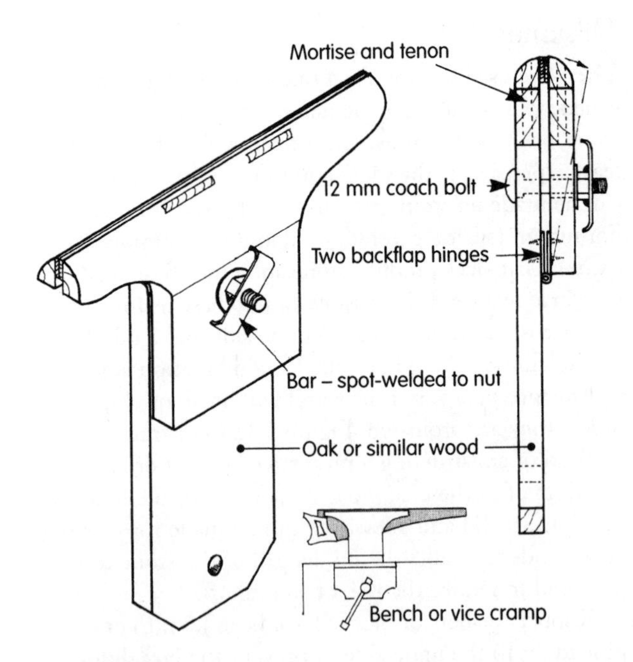

Figure 5.125 Bench saw vice

Sharpening a spokeshave blade can be difficult and dangerous, due to its shortness. A holder, as shown in Figure 5.126, provides the answer.

5.13.2 Saws

Of all tools, saws are the most difficult to sharpen. The problem can be lessened if sharpening is carried out at

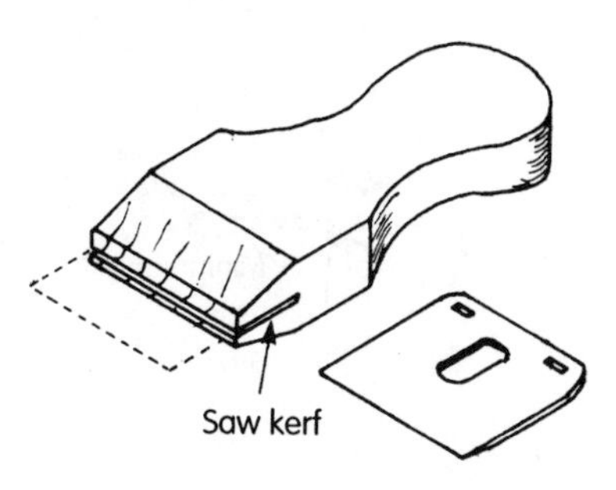

Figure 5.126 Spokeshave blade holder

frequent intervals, because as the condition of the saw worsens, so does the task of resharpening. Signs of dullness are the teeth tips becoming shiny or extra pressure being needed during sawing. The sharpening technique will vary with the type of saw.

If teeth become very distorted, due to inaccurate sharpening or accidentally sawing nails, etc., then the following processes should be undertaken:

- topping
- shaping
- setting
- sharpening.

Topping

This involves bringing all the teeth points in line (Figure 5.127a) by lightly drawing a long mill saw file over them. A suitable holding device is shown in Figure 5.127b.

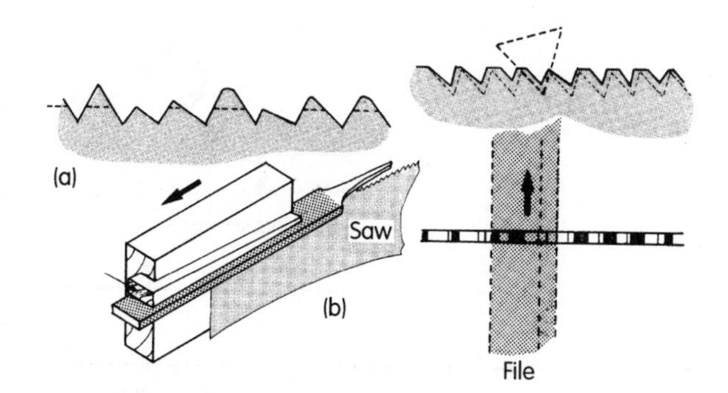

Figure 5.127 Topping a saw blade

Shaping

This means restoring the teeth to their original shape and size. The file face should be just more than twice the depth of the teeth and held level and square to the saw blade throughout the process (see Figure 5.128). Shaping is the most difficult process to perfect.

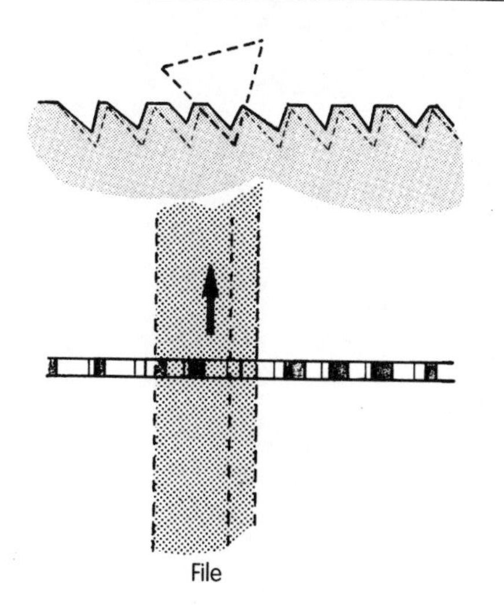

Figure 5.128 Shaping saw teeth

Setting

This means bending over the tips of the teeth to give the saw blade clearance in its kerf (see Figure 5.23). Manufacturers and saw doctors use a crosspein hammer and a saw anvil for this purpose. A suitable alternative is to use saw set pliers, which can be operated with one hand simply by pre-selecting the required points per 25 mm of blade, placing over each alternate tooth and squeezing (Figure 5.129).

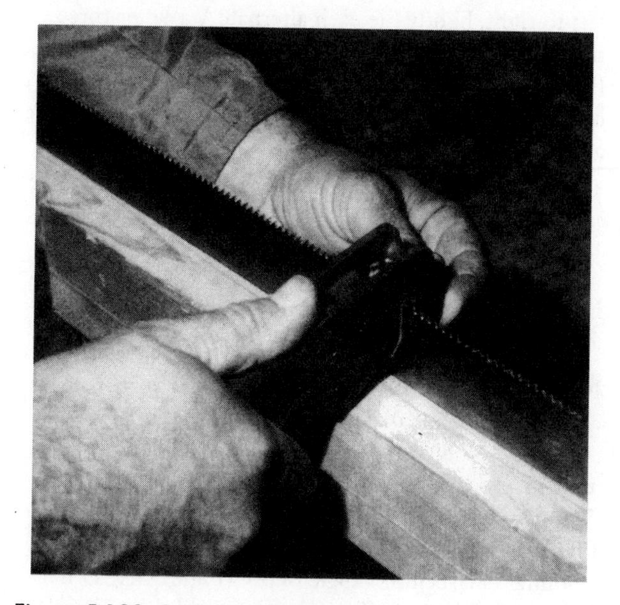

Figure 5.129 Setting saw teeth

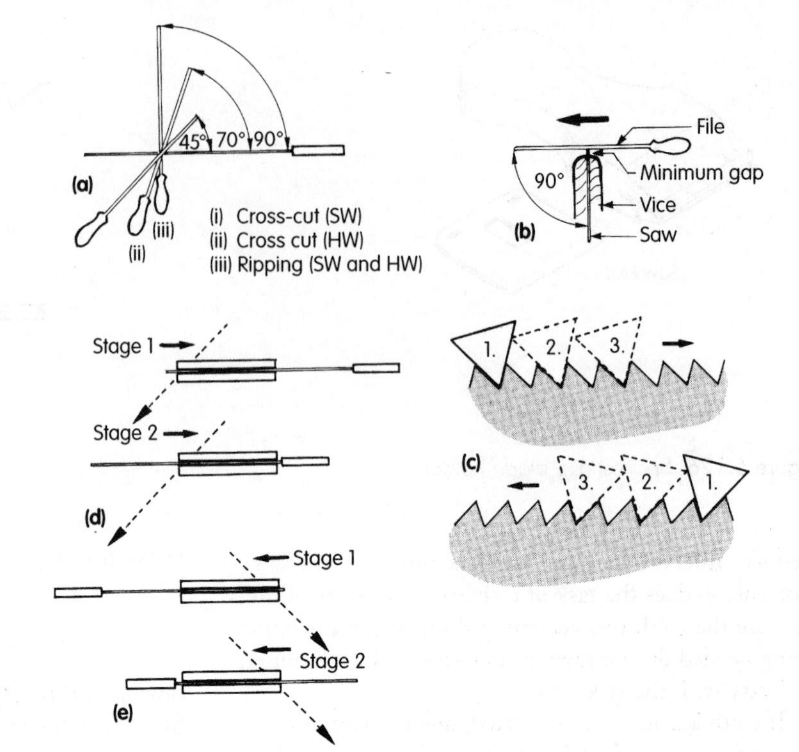

Figure 5.130 Sharpening saw teeth. Note: saw teeth of the overhanging portion of the saw should be covered with a saw blade sheath to protect the operator from the sharp teeth

Sharpening

This involves putting a cutting edge on the teeth. Clamp the saw, with its handle to your right, as low as practicable in its vice. Saw teeth on the overhanging part of the saw should be covered (use a saw blade sheath – Figure 5.118b or c). Place your file against the front face of the first 'toe' tooth set towards you and the back edge of the tooth facing away from you, then angle the file to suit the type of saw or work (Figure 5.130a) and, keeping it flat (Figure 5.130b) and working from left to right, lightly file each alternate V two or three times (Figure 5.130c and d). After reaching the handle (heel), turn the saw through 180° and repeat as before, only this time working from right to left (Figure 5.130c and e).

NB: Stages c,d and c,e may be reversed.

If the teeth are in a very bad condition, you would be well advised to send your saw to a saw doctor (a specialist in saw maintenance) until you have mastered all the above processes.

NB: Whenever a saw is not in use, ensure that its teeth are protected.

5.13.3 Planes

Plane irons (blades) require a grinding angle of 25° and a honing (sharpening) angle of 30° (Figure 5.131) if they are to function efficiently. However, there is one exception: plough planes use a common 35° grinding and honing angle.

The blade shape should correspond to one of those shown in Figure 5.132. A smoothing plane blade has its corners rounded to prevent digging in, whereas a jack plane blade is slightly rounded to encourage quick and easy removal of wood. A rebate plane blade must be square to the iron, for obvious reasons.

Honing

Honing (sharpening) using an oilstone requires much care and attention. The following stages should be carefully studied:

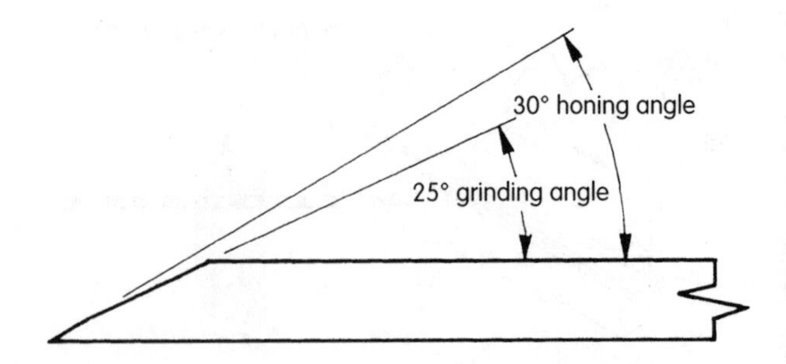

Figure 5.131 Plane blades – grinding and honing angles

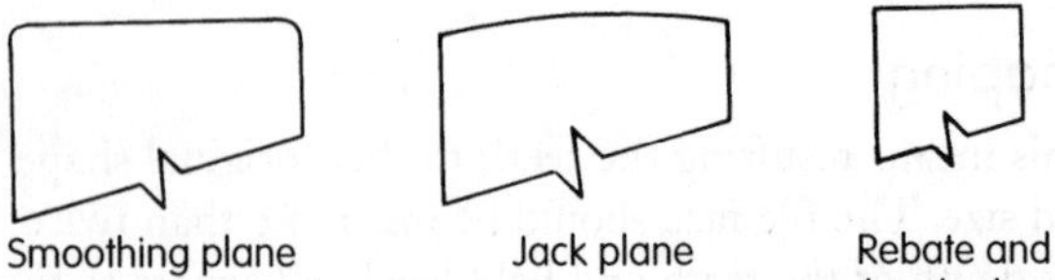

Figure 5.132 Plane blade shapes

1 Remove the cap iron from the cutting iron. Use a large screwdriver to remove the cap screw, not the lever cap, otherwise its chrome coating will soon start to peel.
Note: This operation should only be carried out whilst the cap iron and cutting iron are laid flat on the bench with some provision made to prevent the cutting edge rotating dangerously, e.g. held within and up against the side of the bench well.
2 Ensure that the oilstone is firmly held or supported.
3 Apply a small amount of oil to the surface of the oil-stone.
4 Holding the cutting iron firmly in both hands, position its grinding angle flat on the stone, then lift it about 5° (Figure 5.133a).

Figure 5.133 Plane blade sharpening process

5 Slowly move the blade forwards and backwards (Figure 5.133b) until a small burr has formed at the back. It is important that all the oilstone's surface is covered by this action, to avoid hollowing it (Figure 5.133c).
6 The burr is removed by holding the iron *perfectly flat*, then pushing its blade over the oilstone two or three times (Figure 5.133d).
7 The wire edge left by the burr can be removed by pushing the blade across the corner of a piece of waste wood (Figure 5.133e).
8 If a white line (dullness) is visible on the sharpened edge, repeat stages 5–7. If not, proceed to stage 9.
9 Holding the iron as if it were on the oilstone, draw it over the stropstick (Figure 5.133f).

The practice of showing off by using the palm of the hand as a strop is dangerous and silly. Not only can it result in a cut hand or wrist, but may also cause metal splinters to enter the skin, not to mention the associated problem of oil on the skin.

Blade Positions

Bench plane irons should be positioned as illustrated in Figure 5.134 to ensure effective cutting. The position and angles of cutting irons without cap irons can vary, as can be seen in Figure 5.135, which shows the arrangement for rebate, plough and block planes.

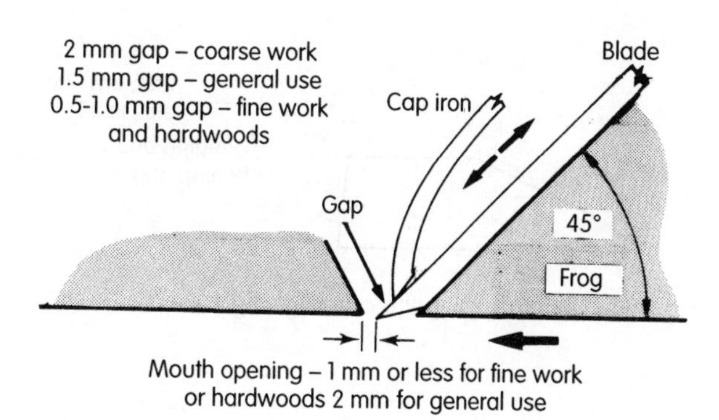

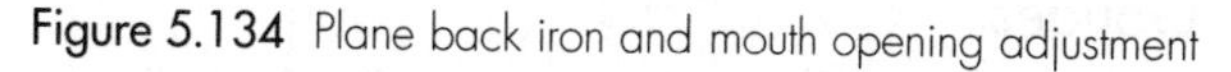
Figure 5.134 Plane back iron and mouth opening adjustment

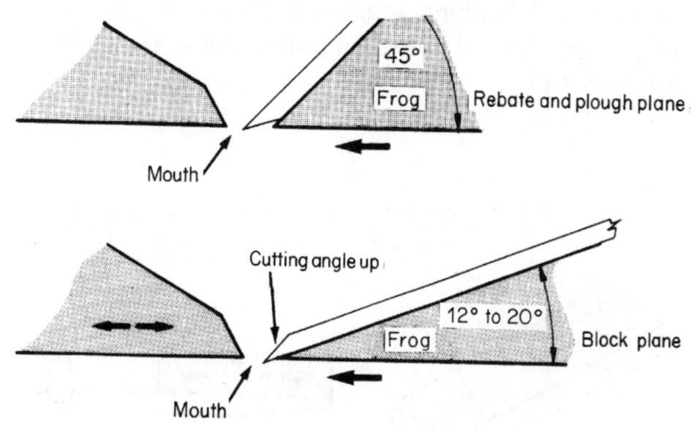

Figure 5.135 Rebate, plough and block plane arrangement

5.13.4 Chisels

Flat-faced Chisels

These should be ground and honed to suit the wood they are to cut, as shown in Figure 5.136. However, where access to a grinding machine is difficult, i.e. on-site work, it is often possible to extend the useful life of the grinding angle as shown in Figure 5.137.

The same principles of honing apply as for plane irons, although extra care must be taken not to hollow the oilstone.

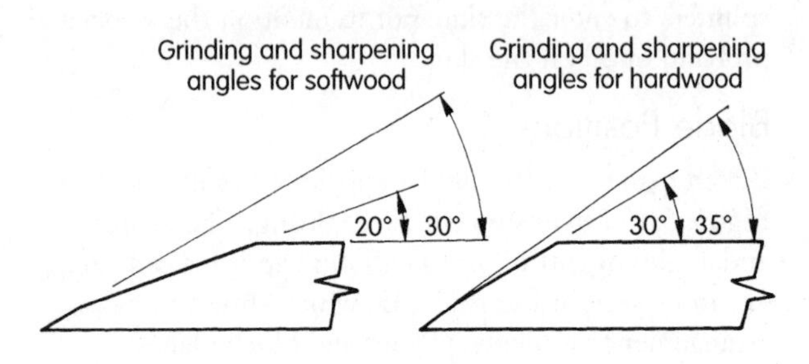

Figure 5.136 Chisel grinding and honing angles

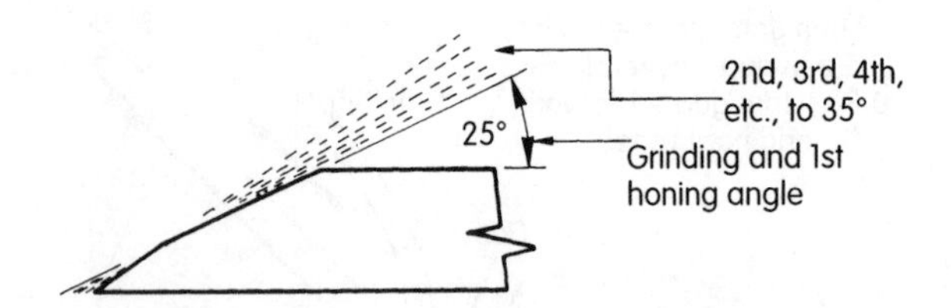

Figure 5.137 Extending the grinding angle life

Gouges

Firmer gouges are ground on a conventional grinding machine, slowly rocking the blade over its abrasive surface. Figure 5.138 shows the method of honing and burr removal. Scribing gouges are ground on a special shaped grinder and honed on a slipstone as shown in Figure 5.139.

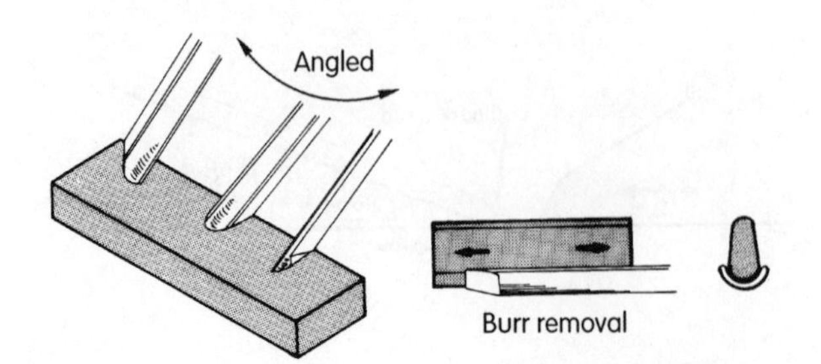

Figure 5.138 Sharpening an out-cannel firmer gouge

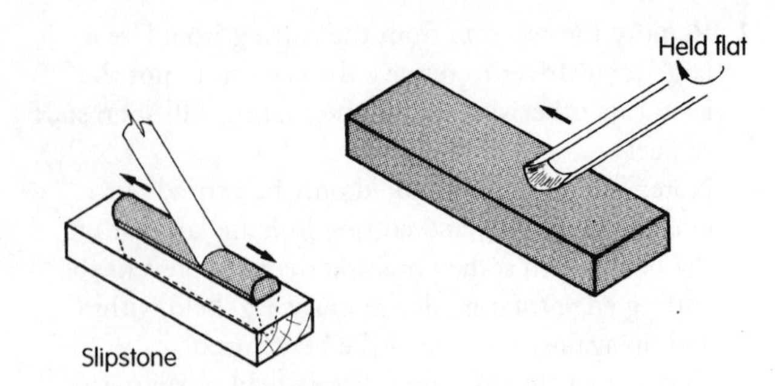

Figure 5.139 Sharpening an in-cannel scribing gouge

5.13.5 Bits

A twistbit's life is considerably shortened every time it is sharpened, so be sure that sharpening is necessary. The spurs (wings) are usually the first to become dull, followed by the cutters. If the screw point becomes damaged, replace the bit. Figure 5.140 shows the method of sharpening a twistbit. Failure to keep it sharp will cause the bit to tear the wood instead of cutting it as it progresses through the bore hole.

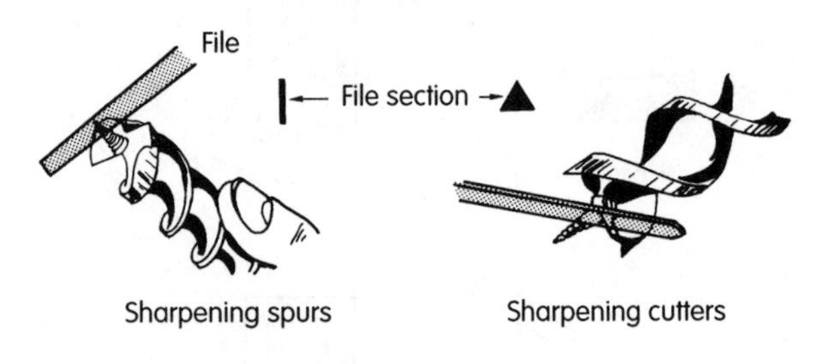

Figure 5.140 Sharpening a twistbit

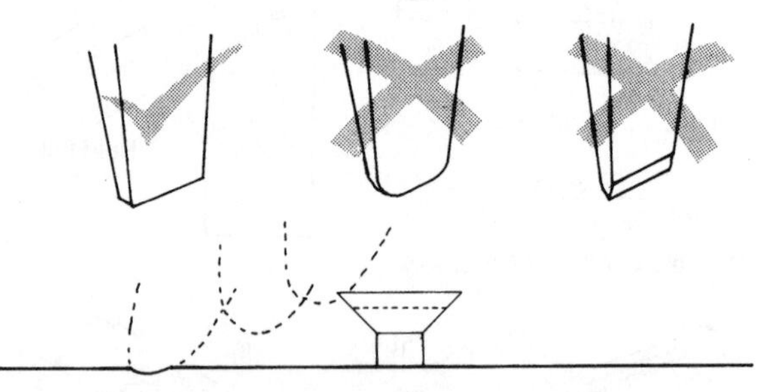

Figure 5.141 Screwdrivers for slotted screws

5.13.6 Screwdrivers

These are the most misused of all hand tools. They have been known to be used as levers, chisels and scrapers. This is not only bad practice, it is also dangerous. Points do, however, become misshapen after long service, even with correct use. Figure 5.141 illustrates how a point can become misshapen and the possible consequences. The point should be re-formed to suit the screw eye by filing (using a fine-cut file), rubbing over an oilstone or regrinding (see Section 9.10) on a grindstone.

6

Setting-out Tools and Procedures for Site Work

Setting-out involves different processes, one or more of which will be necessary before starting any constructional work. They include the following:

- linear measurement
- working to a straight line
- setting-out angles
- setting-out concentric curves
- establishing a datum
- levelling
- vertical setting-out.

The accuracy with which these are carried out will determine the final outcome of the work.

6.1 LINEAR MEASUREMENT

Measuring distances greater than 1 metre generally means using a tape.

6.1.1 Retractable Tapes

These tapes are available in a variety of lengths, made of steel or fibreglass coated with PVC (linen tapes are now obsolete because of their tendency to stretch). Tapes over 5 metres in length have a built-in rewind mechanism.

In use, the accuracy of a tape will depend on:

- the clarity of its gradations
- whether it is held in the correct plane (Figure 6.1)
- the amount of tension being applied to the tape. This is very important when measuring long distances
- readings being correctly taken from left to right.

Measurement errors frequently occur when measurements are transferred from drawing to site. An example of how errors can be reduced is shown in Figure 6.2. Figure 6.2a represents a plan view of individually dimensioned wall recesses. If, on transfer, one of these distances was wrongly measured it would not only affect the overall length of the wall but would also have a cumulative effect on all those measurements which followed. For example, an error between points B and C would result in points D, E, F and G being wrongly positioned.

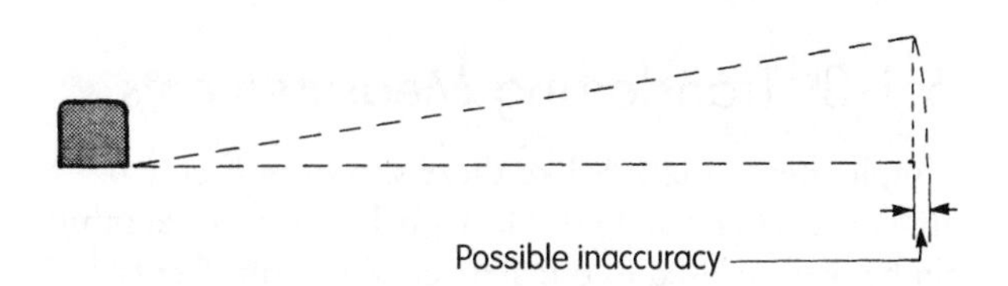

Figure 6.1 Using a tape measure

Figure 6.2b shows how measurements can be transferred by using 'running measurements'. The tape is run once from A to G (total length) and all intermediate measurements are referred back to A as shown in the build-up of the running total.

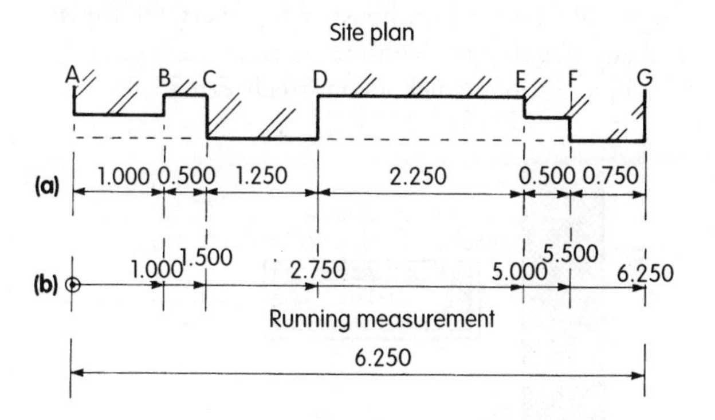

Figure 6.2 Transferring measurements from a drawing to site

6.1.2 Sloping Sites

Figure 6.3 shows how horizontal measurements over sloping or obstructed ground are carried out. Pegs or posts are positioned on, or driven into, the ground to act as intermediate measuring stations.

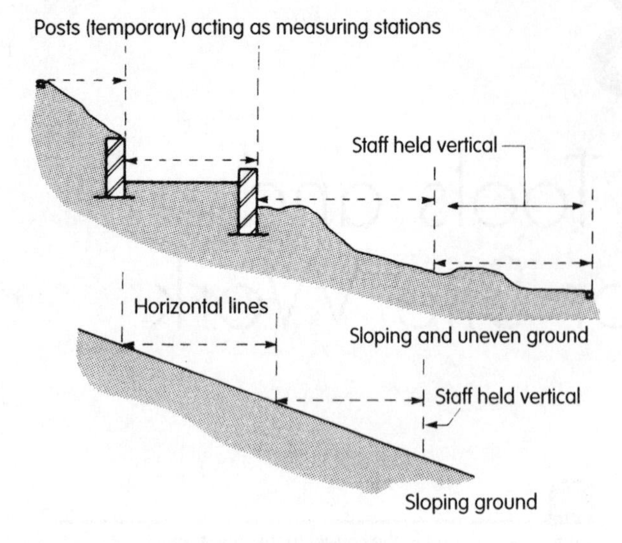

Figure 6.3 Measuring sloping sites

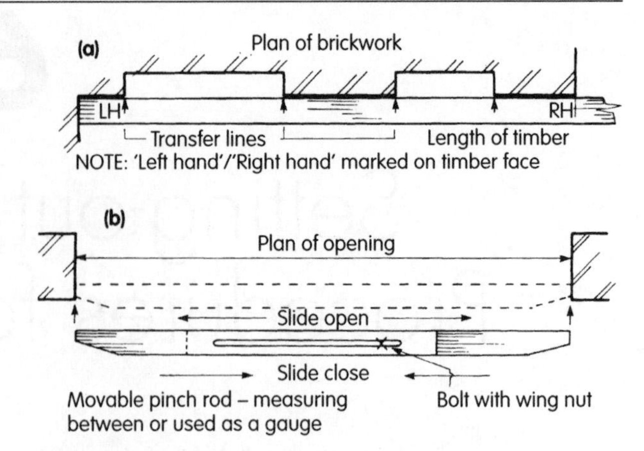

Figure 6.4 Aids for transferring measurements

6.1.3 Transferring Measurements

Simple measuring aids like those shown in Figure 6.4 are very accurate and useful. A small sectioned length of timber can be used to transfer actual measurements. Transfers of this nature are often made from site to workshop where an item of joinery has to fit a specific space or opening. A *pinch rod* (Figure 6.4b), on the other hand, consists of two short lengths of timber, one of which slides on the other to enable its length to be adjusted. This is ideal for measuring between openings; once set, the pieces are held in that position with nails or a cramp, etc.

A more sophisticated method of taking either vertical or horizontal internal measurements is by using a Rabone 'Digi-rod'. The rod is fabricated from steel and consists of a body which houses a liquid crystal display, level and plumb vials to ensure accurate readings and telescopic sections which are manually extended.

As the rod is extended the reading is simultaneously registered on the LCD display shown through the window, as can be seen in Figure 6.5.

6.2 STRAIGHT LINES

Most work carried out by the carpenter and joiner relates to or from a straight line. Straight lines are easily established by using the following methods:

- a length of string, cord or wire held taut
- a pre-determined flat surface such as a straight-edge
- visual judgement – sighting.

6.2.1 Building Line

Figure 6.6 demonstrates how a line is used and how small obstacles can interfere with the line, particularly over long distances. A builder's line can be used to mark a chalk line on a floor; the line is coated with chalk (blackboard chalk or powdered chalk in a proprietary chalking device), stretched tight and then lifted just off

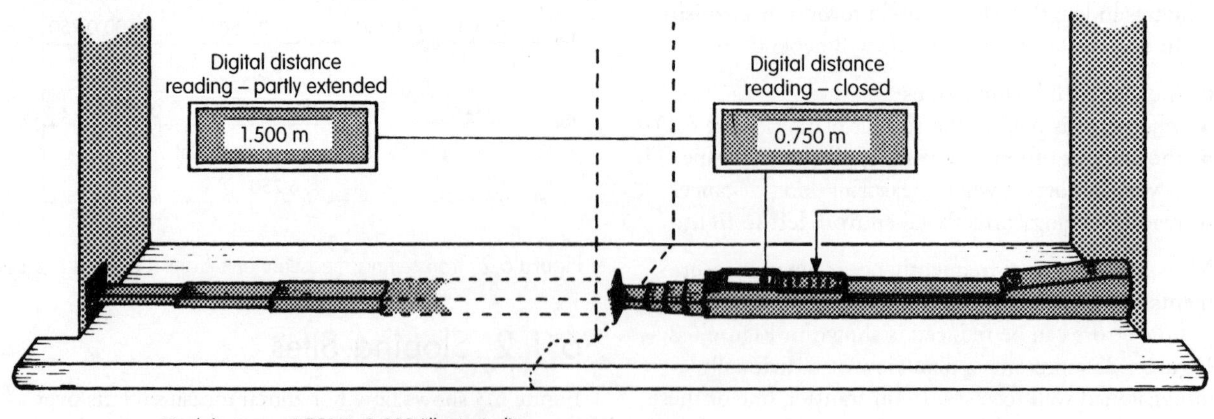

Figure 6.5 Taking internal measurements with an electrical digital measuring rod

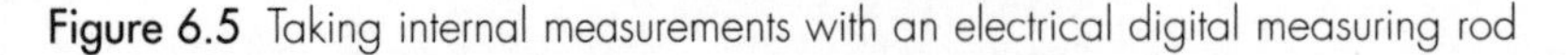

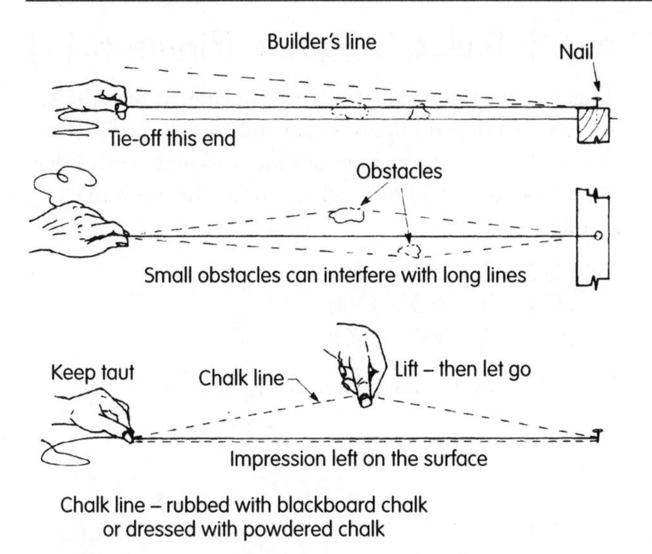

Figure 6.6 Using a builder's line

the floor midway along its span and let go. On removing the line, it will be found that the chalk has left a clear impression of the line on the surface of the floor.

6.2.2 Straight-edges

This is a length of timber or other material which can retain a straight edge.

If a straight-edge is to be used in conjunction with a spirit level both edges must be straight and parallel. Figure 6.7 gives an example of a straight-edge in use. In this case misaligned posts, etc. can quickly be recognised.

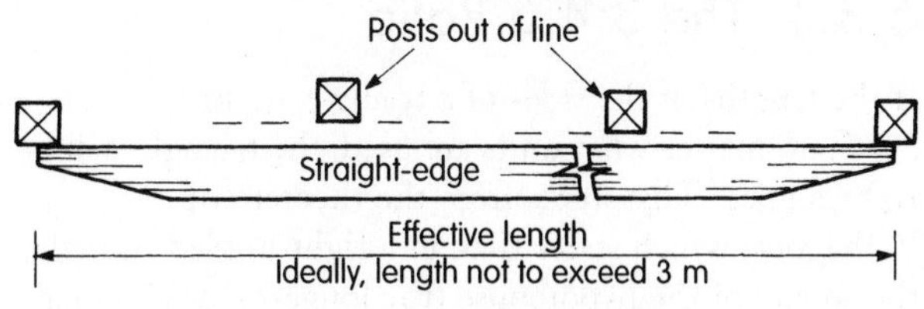

Figure 6.7 Testing for misalignment with a straight-edge

6.2.3 Sighting (Boning Rods)

Neither the line nor the straight-edge is suitable for covering long distances since the line would sag and the straight-edge would be impracticable. Sighting over long distances is possible by a method known as *boning*. Figure 6.8 shows how boning is carried out by using a minimum of three *boning rods*, usually T-shaped, which are held or stood above the required line. The middle one is positioned at a pre-determined point and adjusted until it is aligned with the line of sight, at which point the tops and bottoms of all the boning rods are in a straight line with each other.

Note: A straight line is not necessarily a level line.

6.3 SETTING-OUT RIGHT ANGLES

Let us first consider right angles (90° angles) and three simple methods of forming them:

1 by measurement, using the principle of a 3:4:5 ratio
2 by using a builder's square
3 by using an optical site square.

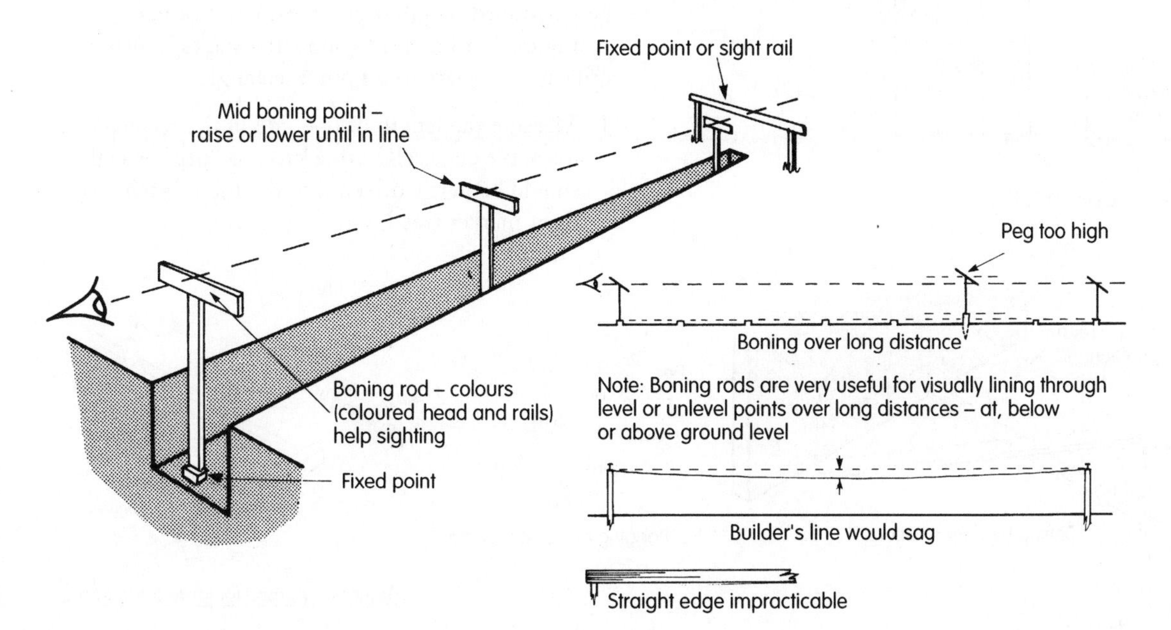

Figure 6.8 Sighting/lining through a straight line over a long distance

6.3.1 The 3:4:5 Ratio

If the lengths of the sides of a triangle are in the ratio 3:4:5, no matter what units are used, the triangle will be right-angled. This stems from the theorem of Pythagoras which states that, in a right-angled triangle, the square of the hypotenuse (the longest side) is equal to the sum of the squares of the other two sides. Consider the triangle in Figure 6.9, where side *a* is 3 units long, side *b* is 4 units long and side *c* (the longest side) is 5 units long:

$$a^2 = 3^2 = 9$$
$$b^2 = 4^2 = 16$$
$$c^2 = 5^2 = 25$$
i.e. $$a^2 + b^2 = c^2$$

In other words, in a triangle with sides in the ratio 3:4:5, the square of the longest side is equal to the sum of the squares of the other two sides; therefore, such a triangle must be right-angled. Figure 6.10 shows how this principle can be applied to a practical situation. Note: The same principle will also work with sides in the ratio 5:12:13.

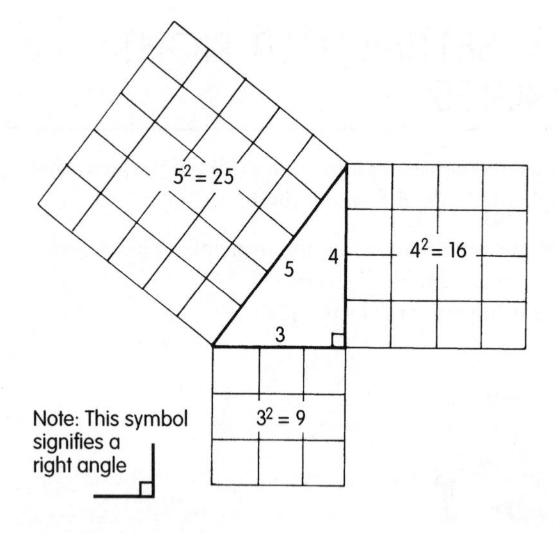

Figure 6.9 Ratio of 3:4:5

6.3.2 Builder's Square (Figure 6.11)

This is a purpose-made timber right-angled set square. Its size will depend upon where and for what it is used. A useful size could be built around a triangle with sides 900, 1200 and 1500 mm, which fulfils the 3:4:5 rule. For example:

side $a = 3 \times 300 = 900$
side $b = 4 \times 300 = 1200$
side $c = 5 \times 300 = 1500$

Its application is shown later in Figure 6.15.

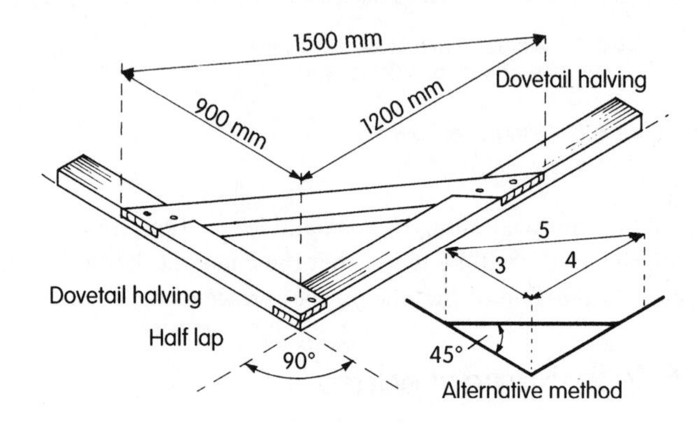

Figure 6.11 The builder's square

6.3.3 Optical Site Square (Figure 6.12)

This consists of two fixed-focus telescopes permanently set at right angles to one another, each being capable of independent movement within a limited vertical arc. A circular spirit level in its head enables the datum rod to be positioned plumb over a fixed point or peg.

Figures 6.13 and 6.14 show the stages in using an optical site square to set out a rectangle.

1 Measure the length of one side of the rectangle and mark the ends AB with a 'cross' or 'peg'. A nail should be partly driven into the top of each peg as a sight marker (see insert).

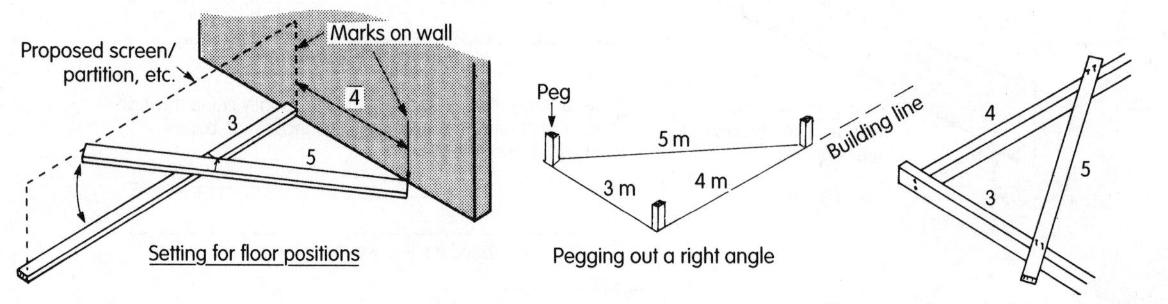

Figure 6.10 Application of the 3:4:5 principle

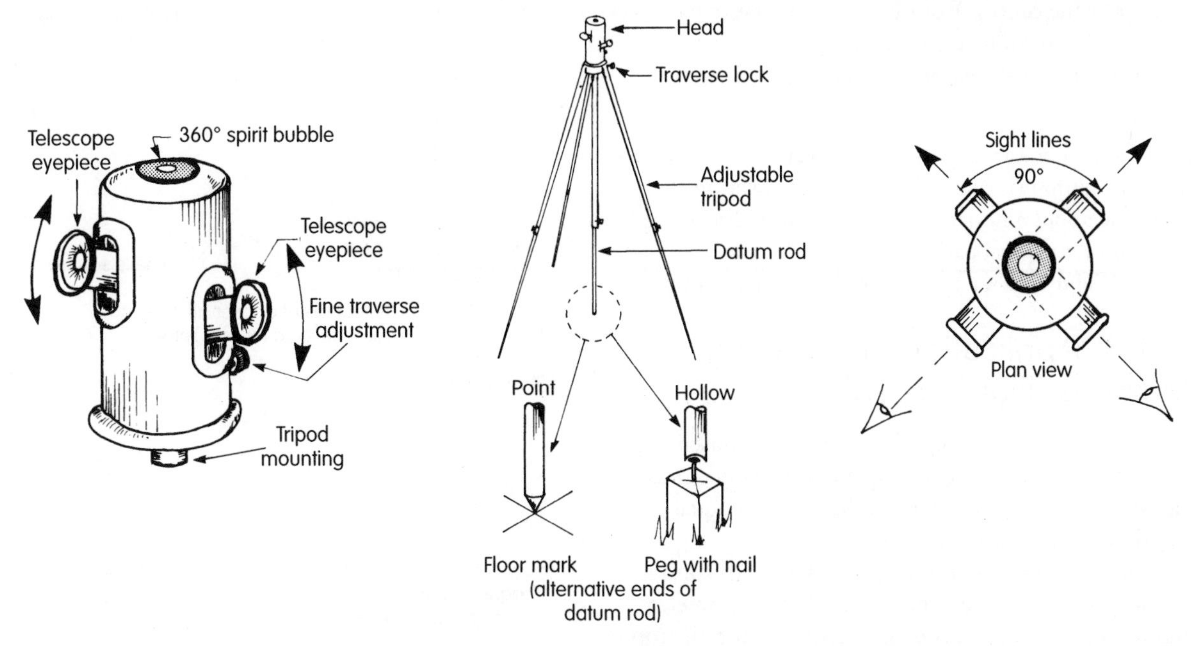

Figure 6.12 Optical site square

2 Set up the square at A, positioning the datum rod over the cross or nail. Plumb the datum rod using the spirit level and by adjusting the tripod legs.
3 From position A, rotate the head and lock it when the telescope is in line with B. Use the fine-adjustment knob to bring the crosshairs (viewed through the telescope) exactly over the cross, nail head (see insert) or floor mark (Figure 6.13).
4 Keeping the head in the same position, move round 90° to look through the other telescope. Place a mark or position a peg C, in line with the cross-hairs of the telescope and the correct distance from A (Figure 6.13). The two sides AB and AC are now at right angles.

a First sighting through lower telescope – using the fine traverse adjusting screen to bring the 'cross-hairs' in line with the front edge of the concrete raft (slab)

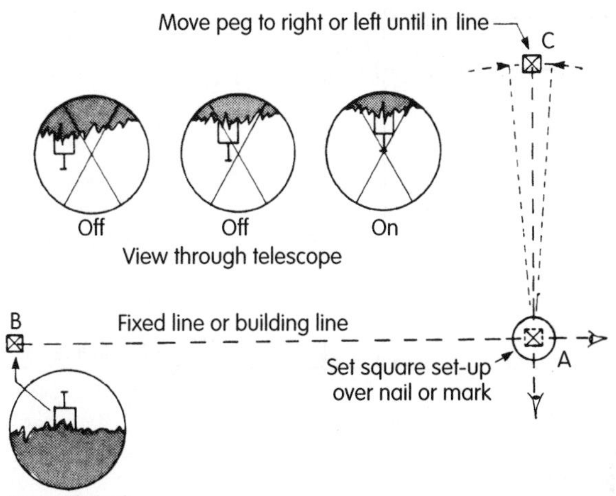

Figure 6.13 Site square application

b Sighting through the upper telescope to ensure that the side edge of the concrete raft is at a right angle (90°) to the front

Figure 6.14 Site square in use

5 Repeat the process from B or C or measure two lines parallel with those right-angled sides formed.
6 Diagonals must always be checked as shown in Figure 6.15.

As an added bonus, this instrument can also be used to set work plumb (vertical), by raising or lowering the telescope about a horizontal axis so that its line of sight will follow a vertical path.

NB: Figures 6.14a and b show a site square in use.

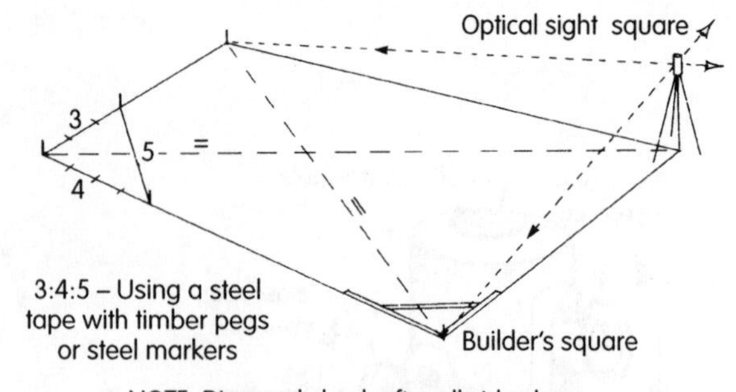

Figure 6.15 Squaring techniques

6.3.4 Application of Squaring Methods (Figure 6.15)

No matter which method of setting-out is chosen, the perimeter or framework of a square or rectangle must have opposite sides parallel and diagonals of equal length, as shown in Figure 6.16a.

Where corner squaring pegs interfere with constructional work, lines are extended outside the perimeter and repositioned on nails or saw kerfs cut into the top of pegs or boards as shown in Figure 6.16b. The corners are now located where the lines cross each other and the lines may be removed or replaced at will.

Figure 6.17 shows how a rectangular building plot can be set out from a known building line. A building line is a hypothetical line, usually set down by the local authority to determine the frontage of the proposed building. It may relate to existing buildings or be a fixed distance from a roadway.

Once the building line is established via pegs 1 and 2, the following procedures can be followed:

- the frontage distance can be determined with pegs 3 and 4

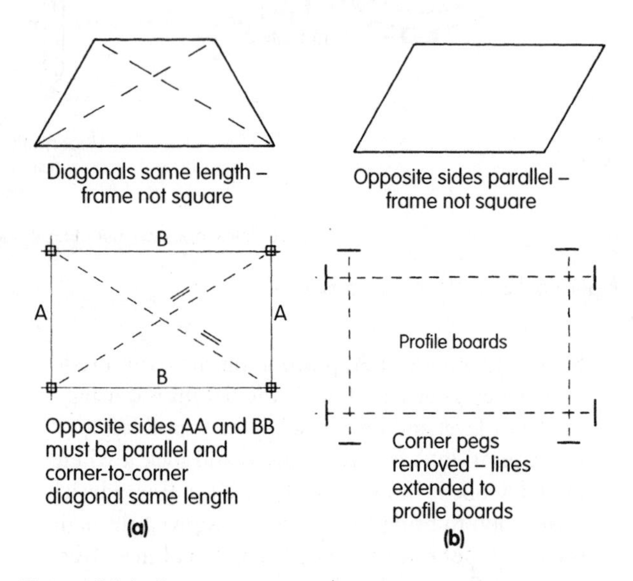

Figure 6.16 Setting a rectangular area

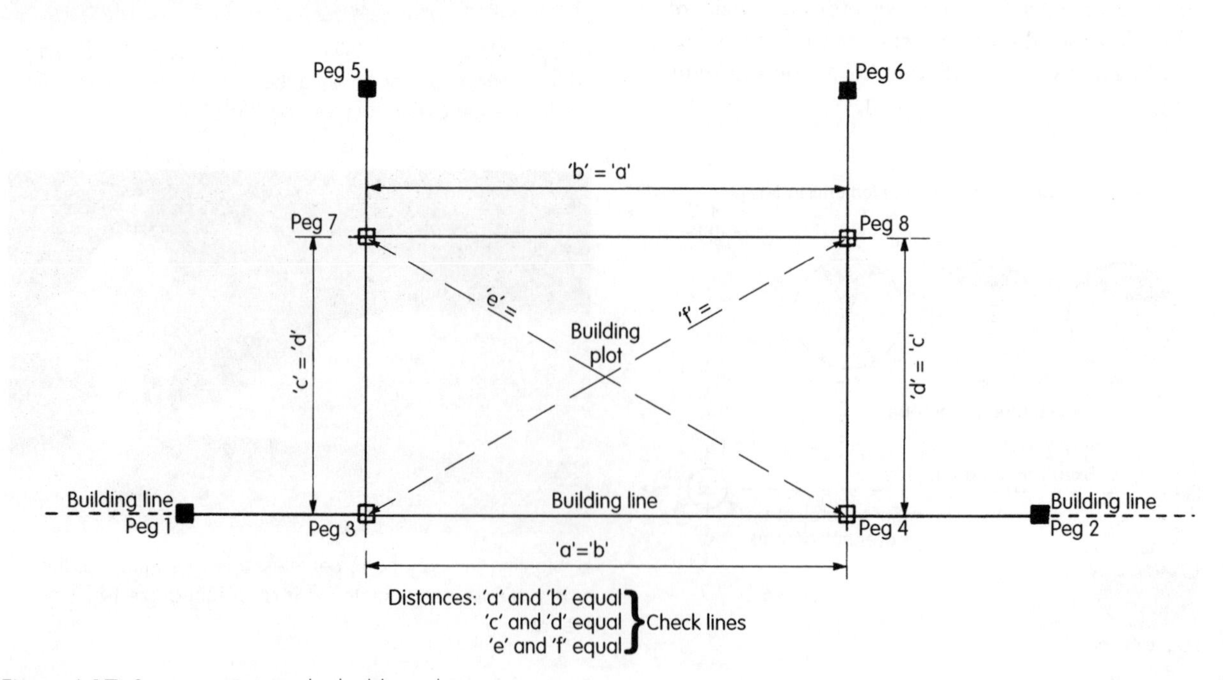

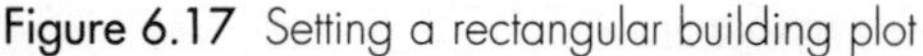
Figure 6.17 Setting a rectangular building plot

- the left-hand side line is set square from the building – line peg 3 to peg 5
- the right-hand side line is set square from the building – line peg 4 to peg 6
- the depth of building is marked with pegs 7 and 8
- the check for square as previously stated in Figure 6.16a.

6.3.5 Profile Boards (Figure 6.18)

These indicate, with the aid of lines, where corners, junctions or changes of direction occur. Figure 6.18 shows how these boards are made and positioned in line with walls or foundations, etc.

Depending on the type of project and site conditions, these boards may be made up on site from short timber posts (say of 50 × 50) with rails (say of 25 × 100) nailed to them; the faces of these rails should be positioned on opposing faces to resist the pull on the lines. Because posts are not normally left in the ground for long periods special treatment may not be necessary.

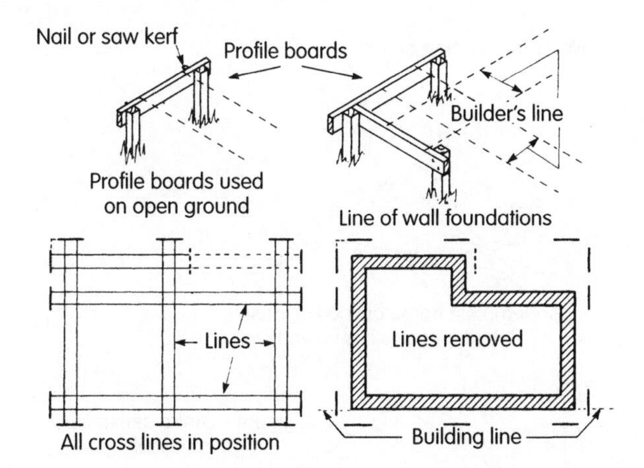

Figure 6.18 Using profile boards to line out the position of a wall or foundation

6.4 ANGLES OTHER THAN RIGHT ANGLES

These are set out by using either angle templates (a full-size pattern for forming shapes) or on site geometry. Figure 6.19 shows a builder's equilateral triangle and its application. Figure 6.20 shows examples of where and how angles are applied to simple regular and semi-regular polygon shapes.

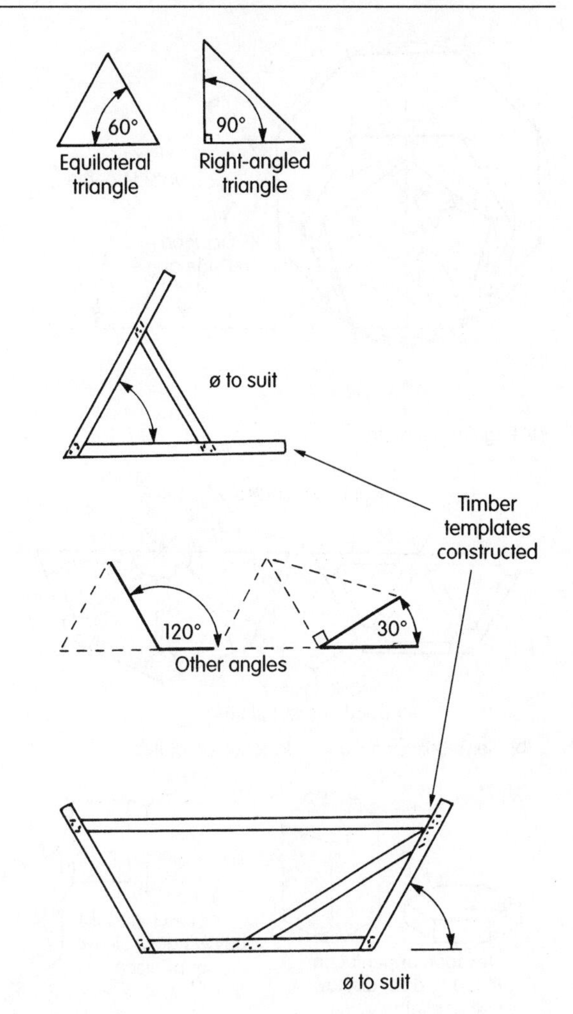

Figure 6.19 Angle template

6.4.1 Regular Polygons

These are figures of more than four sides all of which are equal in length. Each polygon has a specific name in relation to the number of sides:

Number of sides	Name
5	Pentagon
6	Hexagon
7	Heptagon
8	Octagon
9	Nonagon
10	Decagon

Figure 6.20a shows how five-, six- and eight-sided figures are set out. The procedure is as follows:

1 Draw side AB.
2 Bisect AB to produce the centre line ℄.
3 Erect the perpendicular BC.
4 Using radius AB, scribe arc AC across the centre line to produce point 6.

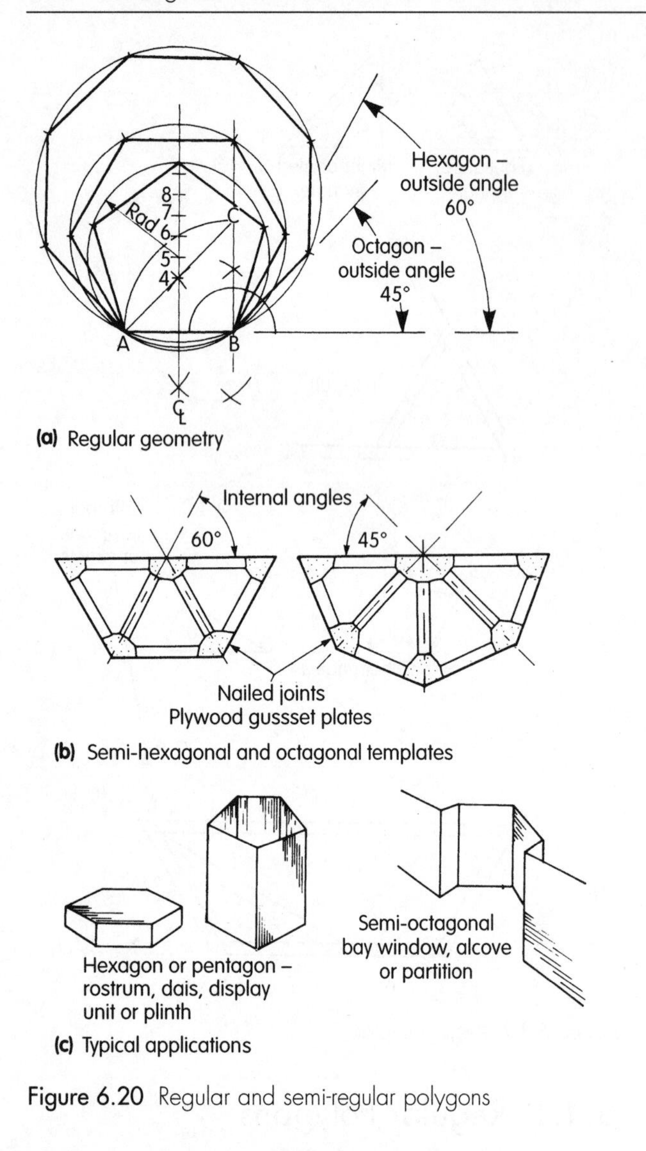

Figure 6.20 Regular and semi-regular polygons

5 Join AC to produce point 4.
6 Point 5 is midway between points 4 and 6.
7 Points 4, 5, 6, 7, 8, etc. are the same distance apart.

These numbered circle centre points are used to circumscribe the equivalent number of sides AB of the polygon.

Figure 6.20b shows how semi-regular hexagonal and octagonal templates can be simply made by using nailed plywood gusset plates over both faces where all the members intersect.

Figure 6.20c shows how these shapes can be applied practically.

6.5 CIRCLES, SEGMENTS AND ARCS

Figure 6.21a shows how large circles, semicircles and the arcs from segments can be scribed by using a wood trammel. We came across the use of trammel heads in Section 5.2.1 and this is a good opportunity to check back to see if trammel heads on a bar could be used here. Figure 6.21b shows how a triangular frame made of timber can be used as a template to mark out and scribe a large segmental arc.

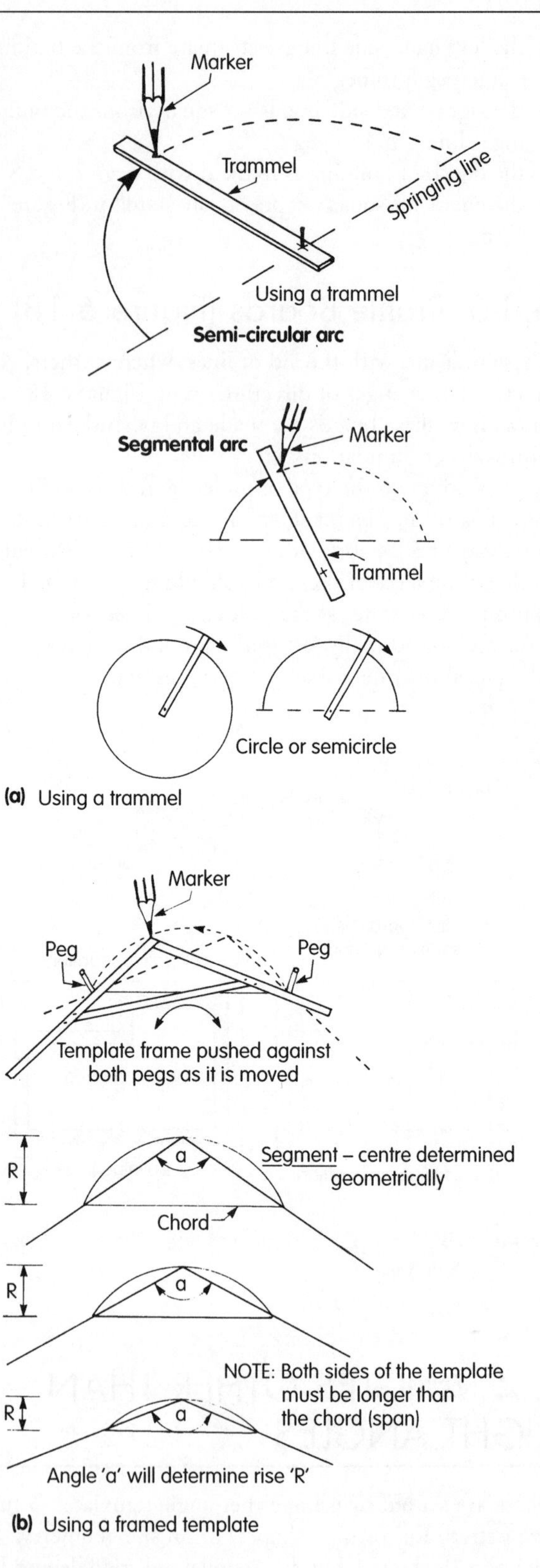

Figure 6.21 Curved work

6.6 DATUMS

A datum is a fixed point or horizontal line to which height or depth can be referred.

Before any levels can be taken on a building site, a fixed datum point must be established as shown in Figure 6.22. This point is known as a temporary benchmark (TBM) and it must relate to a true Ordnance Survey benchmark (BM) which can be identified by an arrow pointing up to a line. Benchmarks are recorded on Ordnance Survey maps and are found cut into rocks or the walls of buildings.

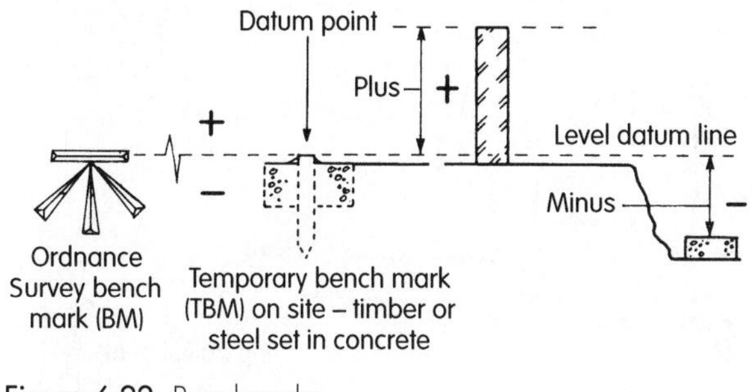

Figure 6.22 Benchmarks

6.6.1 Datum Lines

Datum lines shown in Figure 6.23 need only relate to the work in hand. These are temporary horizontal lines which have been struck at a convenient height so that floor and/or ceiling slopes can be measured and their dimensions recorded.

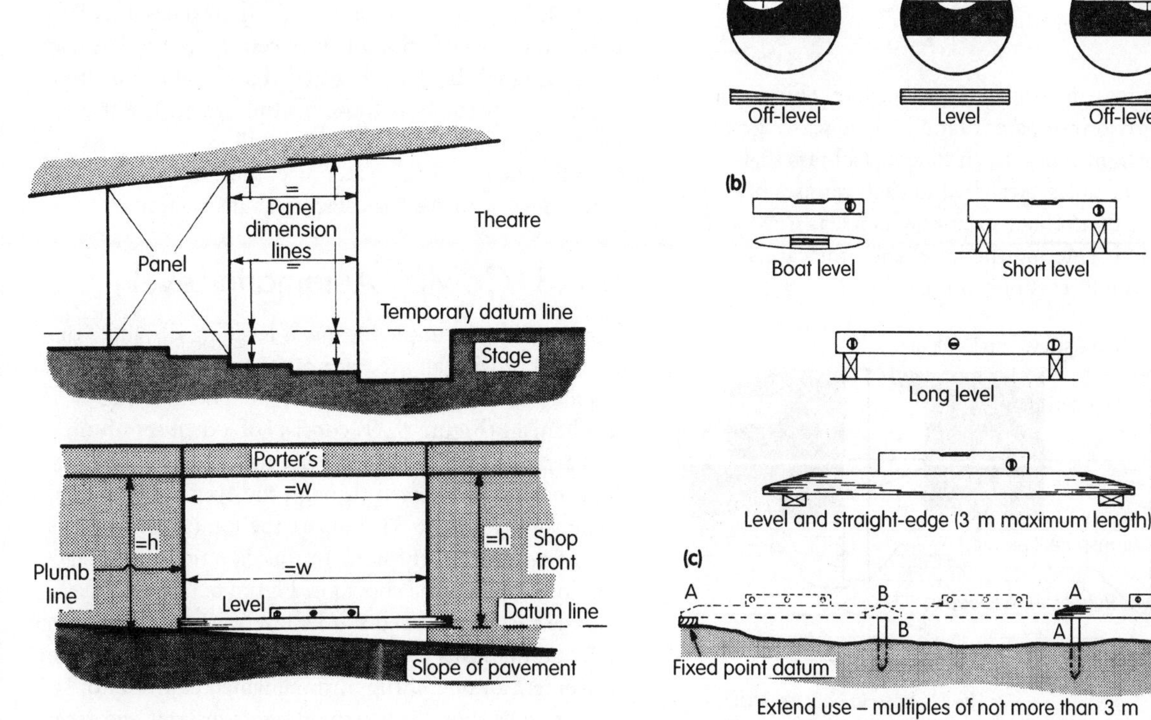

Figure 6.23 Datum lines

6.7 LEVELLING

Levelling is the act of producing a line or surface which is horizontal (level). Any one of the following aids can be used to establish a level line:

- a spirit level
- a water level
- the Cowley automatic level.

6.7.1 Spirit Levels (see also Section 6.8.3)

These consist of a wood or metal body (rule) with parallel edges into which are inset one or more curved glass or plastic tubes known as vials containing spirit and a bubble of air. The position of the air bubble indicates whether the spirit level is horizontal (Figure 6.24a).

The accuracy of the level will depend on the trueness of the bubble tube, the level's effective length and the skill of the operative. Figure 6.24b shows how, with the aid of a straight-edge, its efficiency can be increased for lengths up to 3 m. Greater distances can be covered by moving both level and straight-edge forward together. There is, however, a danger of minor errors accumulating. This can be avoided if, at each stage, both level and straight edge are turned through 180° as shown in Figure 6.24c, i.e. A–B, then B–A, then A–B and so on.

Figure 6.24 Levelling with a spirit level

6.7.2 Water Levels

This works on the principle that water always finds its own level, which is true when it is contained in an open system.

Consider Figure 6.25a where water is contained in a U-tube and, no matter which way the tube is tilted, the water level remains in a horizontal plane. In Figure 6.25b the tube has been divided and connected by a flexible hose. Once again, provided the hose is unobstructed, the water level will be the same at both ends. This is the basic principle on which the modern water level is founded. However, as can be seen from this model (Figure 6.25c), if one end is raised or lowered too far, water will spill out – a problem which in the main has been overcome with the modern water levels.

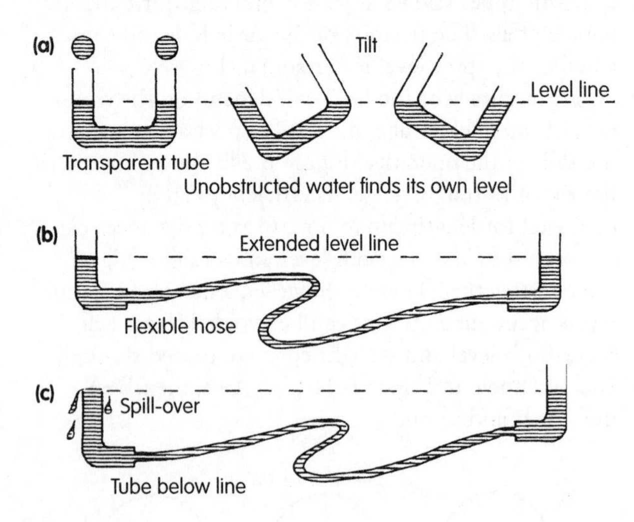

Figure 6.25 Basic principles of a water level

A modern two-man-operated water level (Figure 6.26) consists of two transparent (and in some cases graduated) plastic sight tubes. Each tube has a brass sealable cap with vent holes. Attached to each cap is a brass rod (plunger) with a rubber stopper which acts as a valve by preventing air entering the hose when the cap is shut down and the level is not in use.

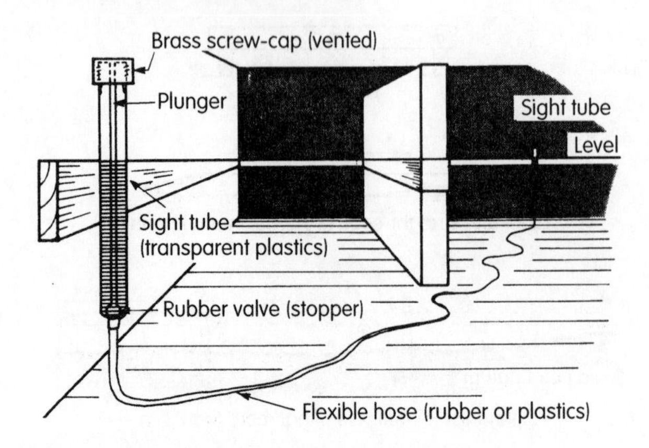

Figure 6.26 Two-man-operated water level

The long rubber or plastic hose (up to 18 m) makes this level particularly useful for levelling floors, ceilings, formwork, etc., especially around corners or obstacles. Figure 6.27 shows the use of a one-man-operated water level. In this case there is a main static water vessel, which is positioned locally, whilst the handheld sight tube and staff can be moved around at varying distances from it. The operational principle is the same as for the two-man level.

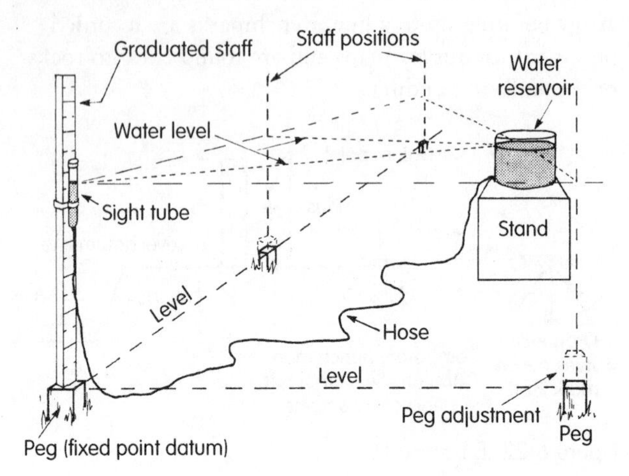

Figure 6.27 One-man-operated water level

Another type of one-man water level (electronic water level), which has a working distance of up to 18 metres, is illustrated in Figure 6.28.

Once the electronic water level is fixed to a known levelling point (datum) the water-filled hose is moved to where the level is to be transferred. When the water (seen through the transparent tubing) reaches a level that coincides with that of the electronic water level, an audible signal (bleep) is emitted. This level can be used in similar situations as those for the two-man water level, even round corners.

NB: The electronic water level must only be used in accordance with the manufacturer's instructions.

6.7.3 Cowley Automatic Level

This provides a simple means of levelling distances of up to 50 m with an accuracy of 6 mm in every 30 m. It requires no setting-up or alignment. The levelling mechanism (Figure 6.29) consists of a dual set of mirrors arranged inside a metal case in such a way that a reflected 'target' viewed through the eyepiece will appear to be split (Figure 6.31a) unless the top of the target is in line with the 'sight line'. In this case the split images will join together, as shown in Figure 6.31b.

The metal case need not be level, since any variations will automatically be compensated for by the pendulum movement of one of the mirrors, which comes into operation as soon as the tripod pin is inserted into the base of the metal case.

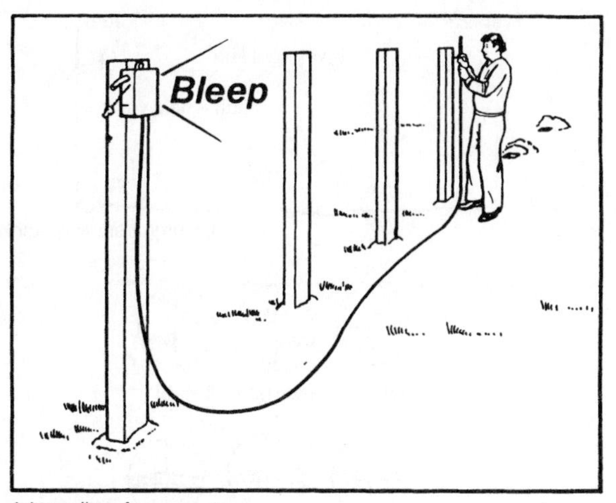

(a) Levelling fence posts

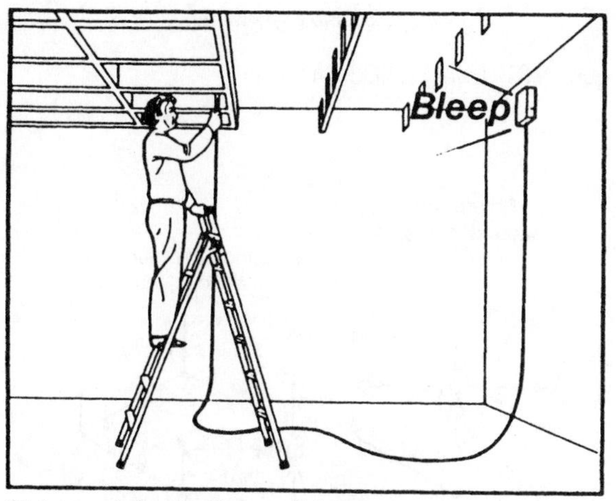

(b) Suspended ceiling from wall datum

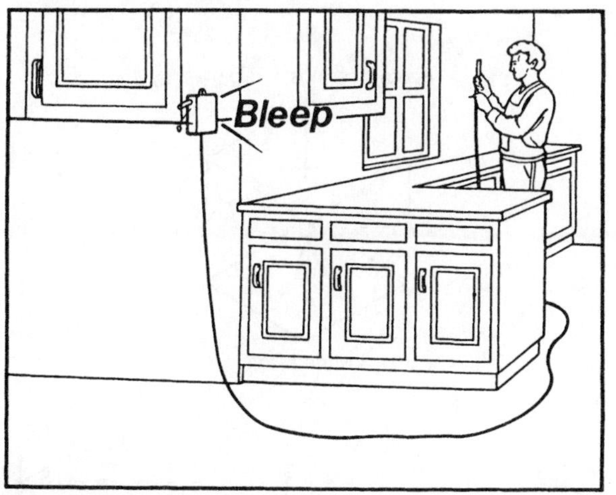

(c) Levelling joinery fitments

Figure 6.28 Rabone Chesterman's one-man-operated electronic water level – examples of use

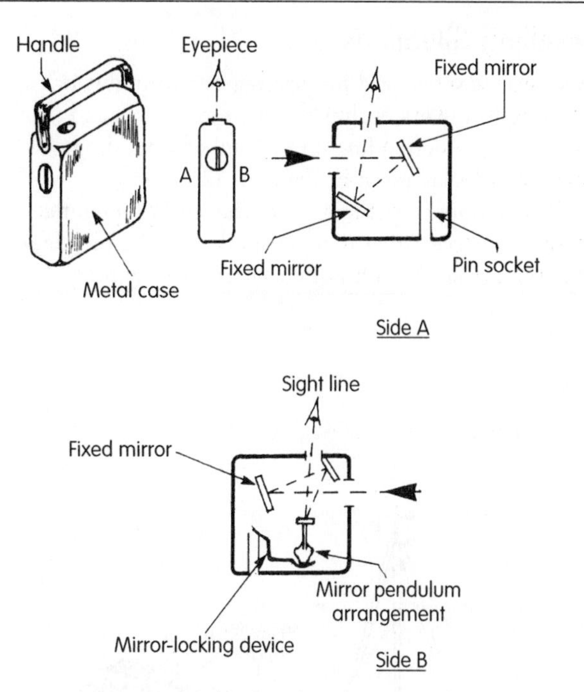

Figure 6.29 Cowley level and mirror arrangement

Note: Never carry the level while it is attached to a tripod or stand because the mirror mechanism could become damaged. Once the level is detached from the tripod pin, the mechanism becomes locked and is therefore safe to travel.

Using the Level

The procedure for using the Cowley level on a tripod is as follows:

1 Set the tripod (Figure 6.30a), pin uppermost, on a secure footing; a base board may be necessary on smooth, flat surfaces (Figure 6.30c).
2 Position the level on the tripod and see that it is free to rotate; the levelling mechanism is now unlocked.
3 Position the staff and target on the datum or starting point. The staff must be held as near to vertical as possible.
4 Direct the level towards the dayglow coloured face of the target.
5 Sight on to the target. An assistant moves the target up or down (Figure 6.30d) until the 'on-target' position appears through the eyepiece, as shown in Figure 6.31b.
6 As soon as the 'on-target' position is achieved, the target is clamped to the staff ready to be transferred to other points requiring the same level, as shown in Figure 6.32. Figure 6.33 shows that, by taking a reading of the graduated staff, different levels can be measured.

Levelling Situations

A Cowley level is ideal for pegging out groundwork, levelling formwork, wallplates, floor joists, etc. and establishing a datum line on the wall of a building. A few examples are given in Figure 6.32.

If it is more convenient to set the level up on a flat elevated surface rather than a tripod, then the bricklayer's stand (Figure 6.30b) offers alternative support.

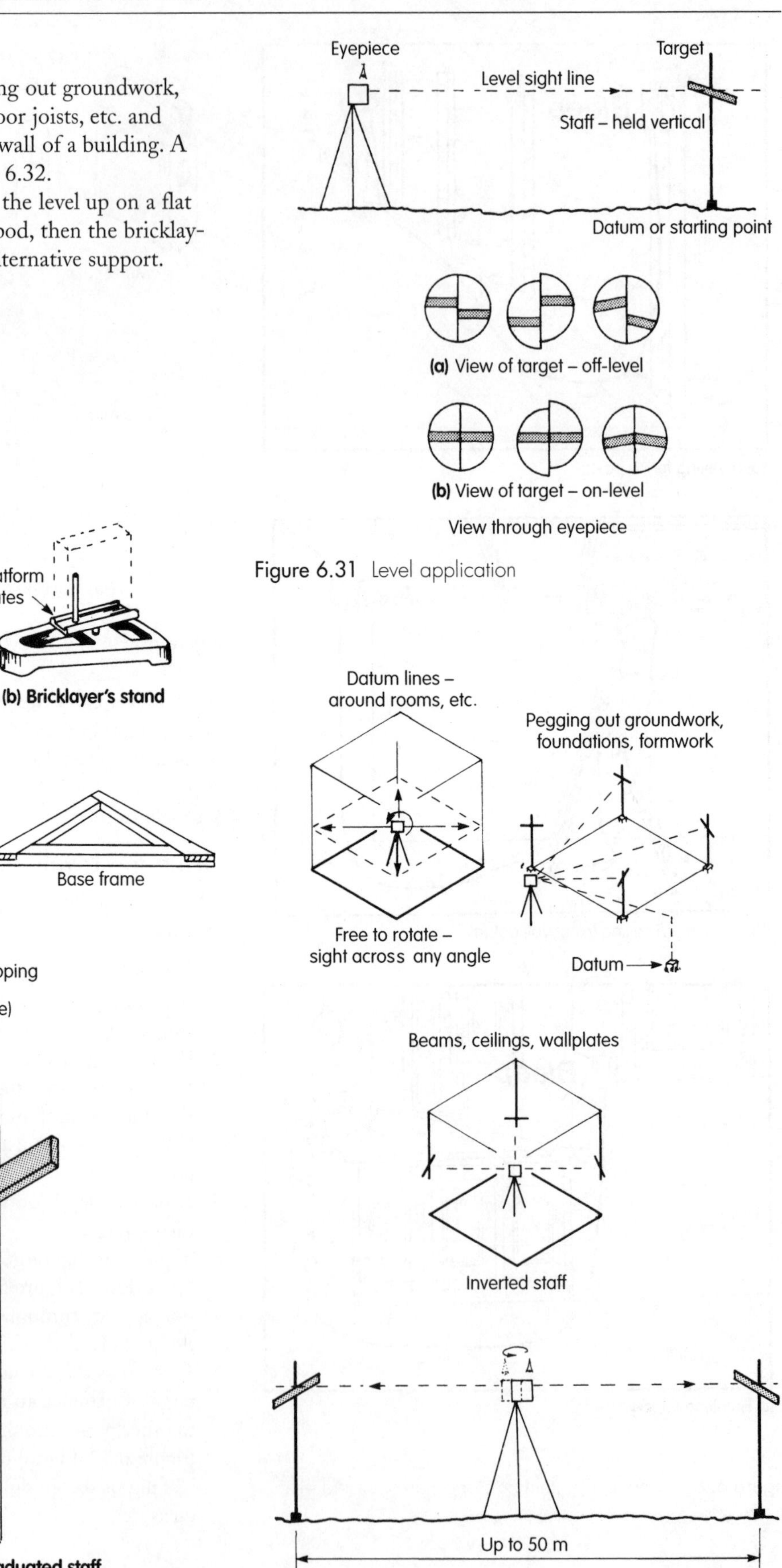

Figure 6.31 Level application

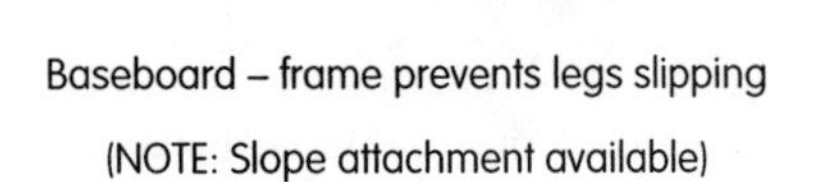

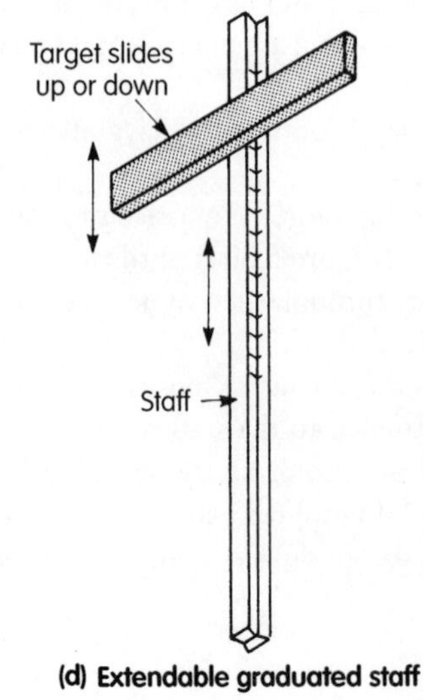

Figure 6.30 Cowley level accessories

Figure 6.32 Levelling situations

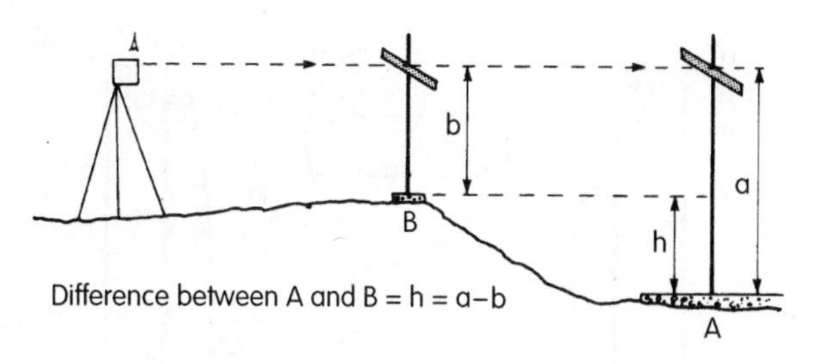

Figure 6.33 Measuring differences in height

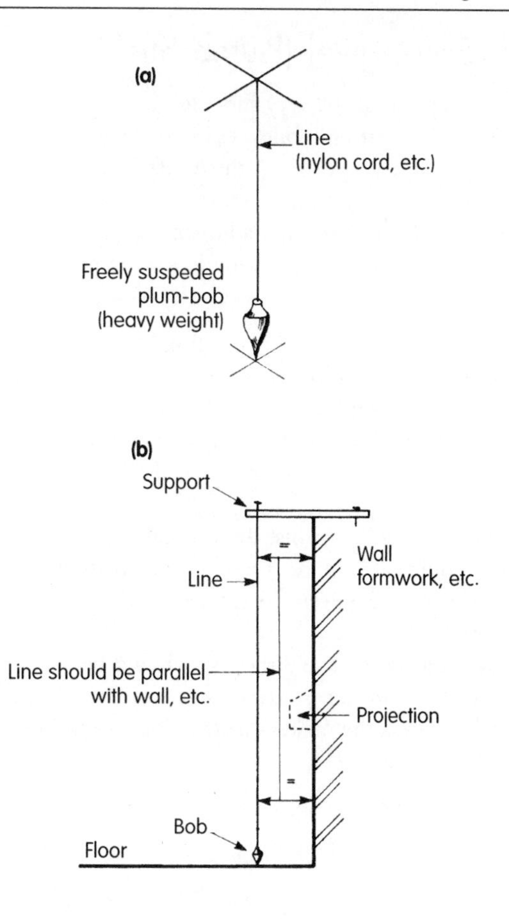

Figure 6.34 Plumb-bob and line

6.8 VERTICAL SETTING-OUT

It is essential that all constructional work is carried out with plumbness in mind, not just for the sake of appearance but as an assurance of structural balance and stability. Plumbness can be achieved with the aid of:

- a suspended plumb-bob
- a plumb-bob and rule
- a spirit level
- a site square (as already mentioned in Section 6.3.3).

The more specialised optical instruments in this field are known as theodolites.

6.8.1 Plumb-bob

A plumb-bob is a metal weight which, when freely suspended by a cord or wire, produces a true vertical line. It is ideal for indicating vertical drop positions (Figure 6.34a) or as a vertical margin line from which parallels can be measured or referred, such as a wall (Figure 6.34b) or a column box (an arrangement of formwork for casting concrete columns).

Metal bobs can weigh between 56 g and 5 kg or more and the choice will largely depend on the type of work and the conditions under which they will be used. A very long plumb line may have to have its bob submerged in a container of water to help reduce the amount of swing.

6.8.2 Plumb-bob and Rule

A plumb-bob and rule (Figure 6.35) allows the plumb line to make contact with the item being tested for plumbness. The rule is made from a straight parallel-sided board. A hole is cut through one end to accommodate the bob and allow the line to make contact with the face of the rule and saw kerfs are cut in the top to grip the line. A central vertical gauge line is used as the plumb indicator.

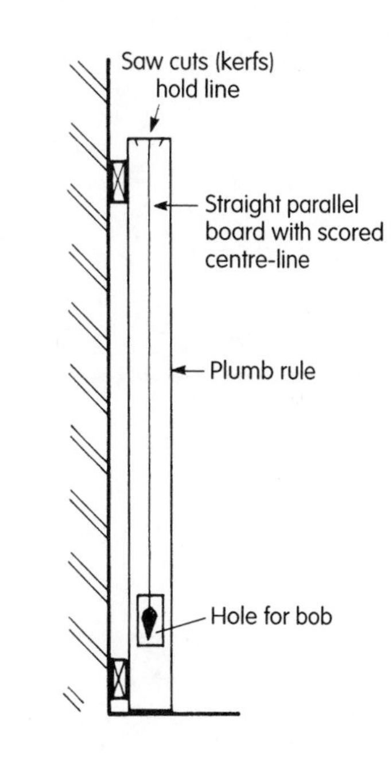

Figure 6.35 Plumb-bob and rule

6.8.3 Spirit Level (Plumb Stick)

With the exception of the very short level, all spirit levels can be used to test plumbness (Figure 6.36). Accuracy will depend greatly on the level's length and bubble setting; virtually all are factory sealed but some may be provided with means of adjustment.

To test a bubble for accuracy, the following procedure is used:

1 Position the level against a firm, straight vertical object. Note the position of the bubble in relation to the central mark on the tube (vial).
2 Keeping the level in a vertical plane, turn it through 180°. The bubble has now been turned about (end for end).
3 Reposition the level against the vertical object. The bubble should take up exactly the same position in the tube as before – even the smallest variation is unacceptable.
4 If adjustment is necessary, the level should be returned to the maker or agent; in the case of adjustable levels, the maker's instructions should be carefully followed.

Inaccuracies usually only occur as a result of accidental damage or misuse.

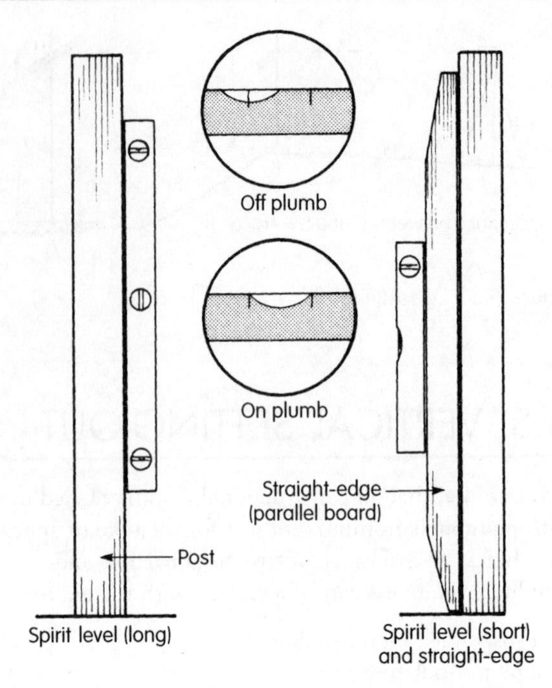

Figure 6.36 Testing plumbness with a spirit level

NB: By substituting the horizontal plane for the vertical plane, the same testing principles can be applied to levelling.

7

Joints

7.1 BASIC WOODWORKING JOINTS

This section deals with the types of joint formed in solid timber using hand tools. It should be noted that all these joints can be made with the aid of woodworking machines (see Chapter 9). To be successful, a carpenter and joiner must be able to select the type of joint that is most suitable for a given situation, understand the principles of how they are made and have the practical skill to form them.

When choosing and making an appropriate joint, it is best to remember five things.

- The smallest possible amount of timber should be cut away from the pieces being jointed.
- Simple joints accurately cut are the best.
- Each part of the joint has equal strength.
- The form of the joint allows for the varying strength and movement characteristics of the timber (see Section 1.10.4).
- The bearing surfaces of the joint must be able to transfer the loads across the joint without failure.

Woodwork joints can be divided into five basic groups:

1. *Lengthening* – end or lengthening joints are used when timber is required in longer lengths than those which occur naturally.
2. *Widening (glued)* – the edges of boards are jointed and glued together to increase the width of timber when wide panels are required.
3. *Widening (dry)* – joints are formed on the board edges which are fitted together dry and secured to a background frame to form large boarded areas such as floors. The function of these joints is to align the edges of the boards, while allowing some movement between the boards as they shrink or swell.
4. *Framing* – these joints are used to hold individual timbers together where they meet when set at an angle to each other and to transfer loads from one member to another without the joint failing.
5. *Hinged* – used when one timber member is required to pivot or turn against another, e.g. doors in door frames, casements in window frames. In the majority of cases ironmongery is used in forming these joints (see Section 7.6).

7.2 LENGTHENING JOINTS

7.2.1 Lap Joints

The two timbers being jointed overlap each other and are secured by nails, screws or bolts (Figure 7.1).

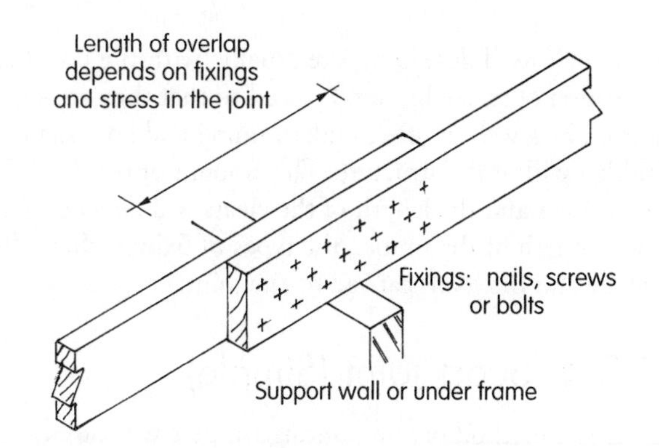

Figure 7.1 Lap joint and support

7.2.2 Butt Joint (Single Cleat)

The two timbers being endjointed are butted together end on and connected by a single cleat secured by nails, screws or bolts (Figure 7.2).

7.2.3 Butt Joint (Double Cleat)

The two timbers are butted together end on and a cleat is placed on either side to sandwich the joint, secured by nails, bolts or screws (Figure 7.3).

Note: In certain circumstances fish plates (flat metal plates), nail plate connectors, are used in place of the

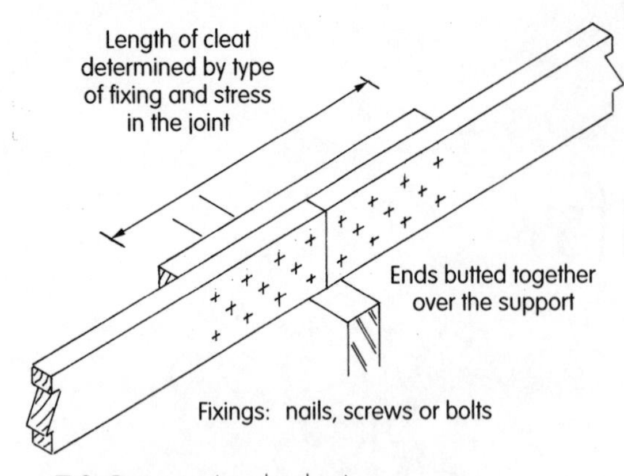

Figure 7.2 Butt joint (single cleat)

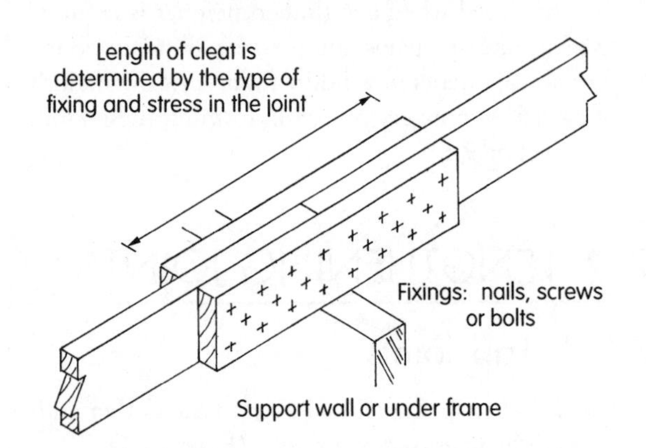

Figure 7.3 Butt joint (double cleat)

timber cleats. These joints are usually (with the possible exception of trussed rafters) placed so that they are supported by a wall or other underframing and are usually hidden within the structure. The amount of overlap of the timber and the length of the cleats will depend upon the strength of the timber, the types of fixings, their efficiency and the loads carried by the joint.

7.2.4 Scarf Joint (Simple)

This is a method of end jointing timbers without increasing the cross-sectional area of the timber at the joint. Figure 7.4 shows a simple scarf joint. The slope of

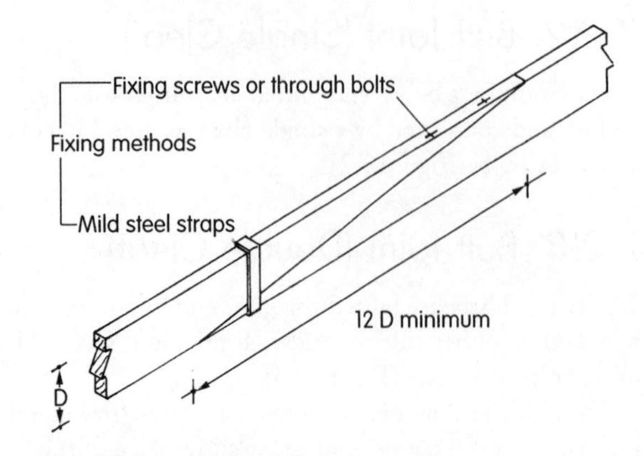

Figure 7.4 Simple scarf joint proportions

the bearing surfaces should not be steeper than 1 in 12. This joint can be secured by nails, screws or metal plates.

7.2.5 Scarf Joint (Hooked and Wedged)

Figure 7.5 shows a hooked and wedged scarf joint being tightened and secured by a pair of hardwood folding wedges.

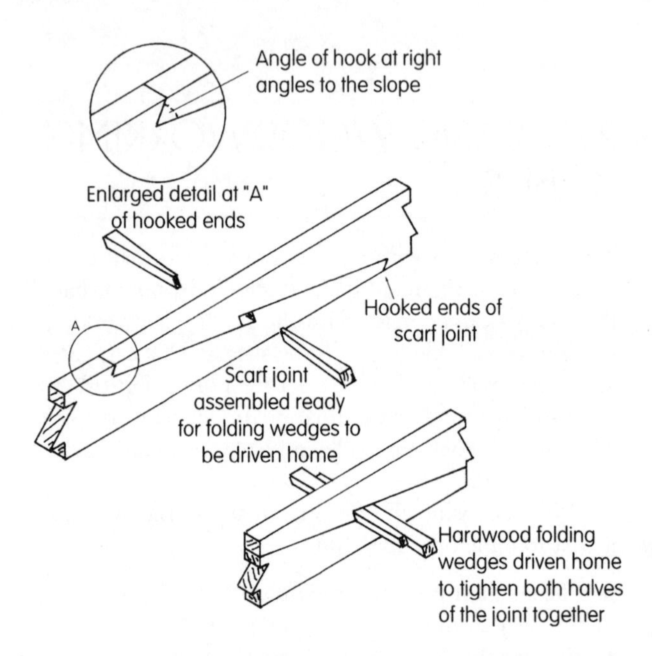

Figure 7.5 Scarf joint (hooked and wedged)

7.2.6 Laminated Joints

Laminated joints are formed by overlapping (laminating) varying lengths of timber of uniform cross-section, glued together in the form of a sandwich. Long structural members of large and varying cross-section can be made using this jointing technique (Figure 7.6a, b).

7.2.7 Finger Joints

The finger joint (Figure 7.7) is machine made. The fingers are machined before being glued and assembled by controlled end pressure in an automatic cutting and assembly process. This jointing method is used for making long lengths of timber from short ends of uniform cross-section or for rejointing after cutting out defects in the mid-section of timbers.

7.2.8 Counter Cramp Joint

This is a simple and effective way of securing and tightening butt joints when lengthening wide boards for counter tops and stair strings. The butt joint may be

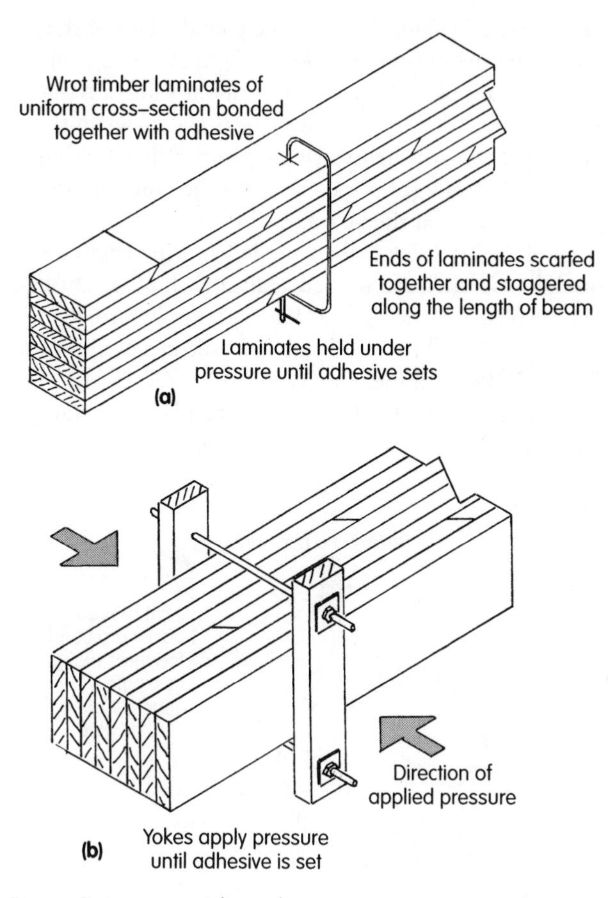

Figure 7.6 Laminated member

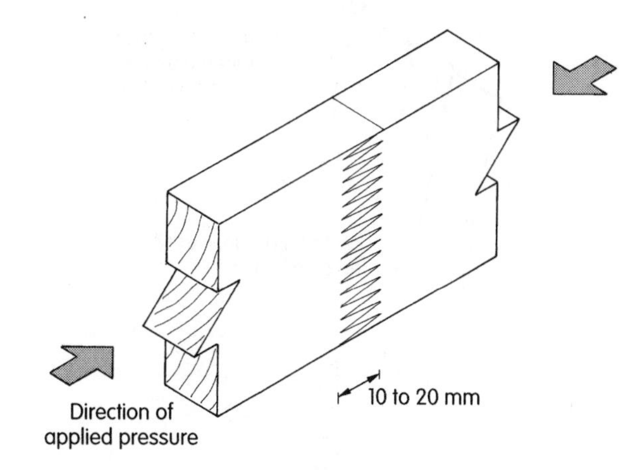

Figure 7.7 Finger joint

tongued, dowelled or biscuit (an elliptical disc of compressed, laminated Beech) jointed across the endgrain to keep the board surfaces in line (Figure 7.8a). After forming the butt joint on the board ends, three cleats each have a trench cut in them to about half their thickness (Figure 7.8b). Cleats 1 and 3 are screwed to the back of the left hand board to form a mortise with the cut trenches in line. The middle cleat, 2, is screwed to the back of the right hand board in the same manner so that the trench is slightly out of line with the other two. Folding wedges are driven home to align the three

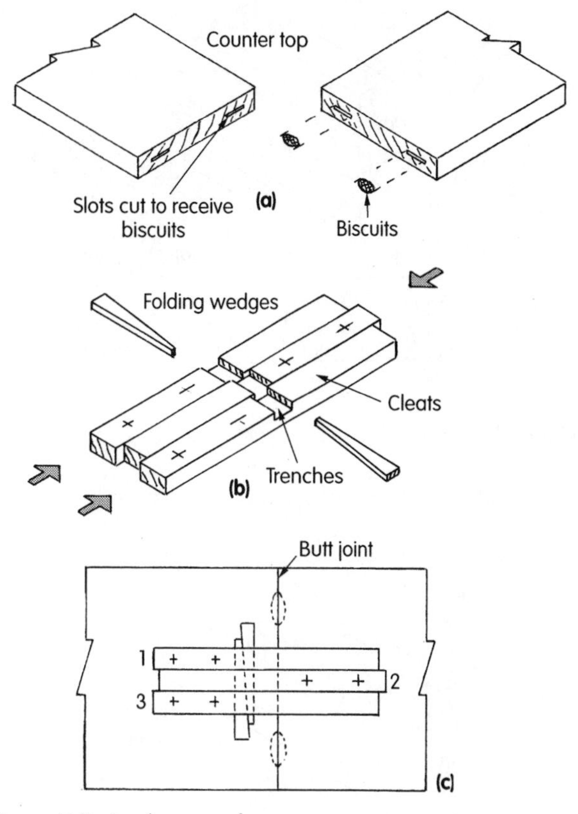

Figure 7.8 Application of counter cramp

trenches in the cleats, which at the same time has the effect of drawing the edges of the butt joint together (Figure 7.8c). For very wide boards, two or more sets of counter cramps can be used.

7.2.9 Half-lap Joints

Figure 7.24a shows a half-lap joint formed end to end. It can be used as a locating and lengthening joint for timber wallplates.

7.3 WIDENING JOINTS (EDGE GLUED)

Examples of these are shown in Figure 7.9. They allow the width of the timber to be increased by glueing together the jointed edges of the boards to form wide panels.

7.3.1 Butt Joint (Figure 7.9a)

This is the simplest of all the edge joints, as the glued square edges are brought together in a perfect fit. Figure 7.10 shows how the joint is formed. The boards are marked to identify their faces, then the edges are matched in pairs and identified by drawing one or more pencil lines on the face of the board across the edges

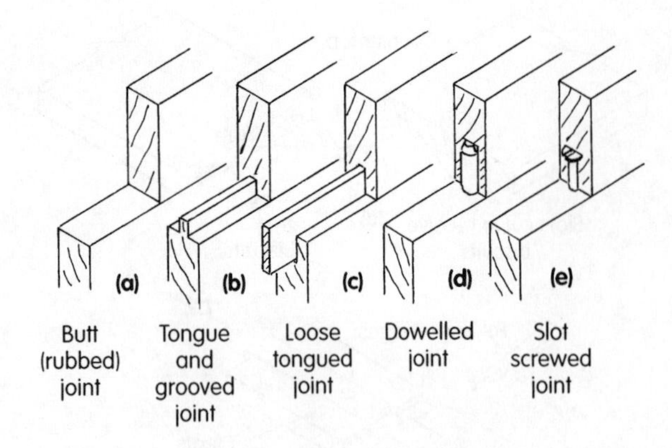

Figure 7.9 Types of glued widening joints

being jointed (Figure 7.10a). The paired edges of the boards are placed side by side and held in a vice while the edges are planed square and straight using a try-plane (Figure 7.10b). The edges fit perfectly when the faces of the boards lie in a single plane. When held up to the light, no light is seen through the joint and the joint line between the boards is almost invisible (Figure 7.10c). Glue is then applied and the edges are brought together and rubbed against each other until air and surplus glue are excluded and a bond is formed (Figure 7.10d). The boards are then stacked as shown in Figure 7.10e until the glue has set. If a non-animal type of adhesive is used the boards must be placed in cramps until the adhesive has set (Figure 7.10f).

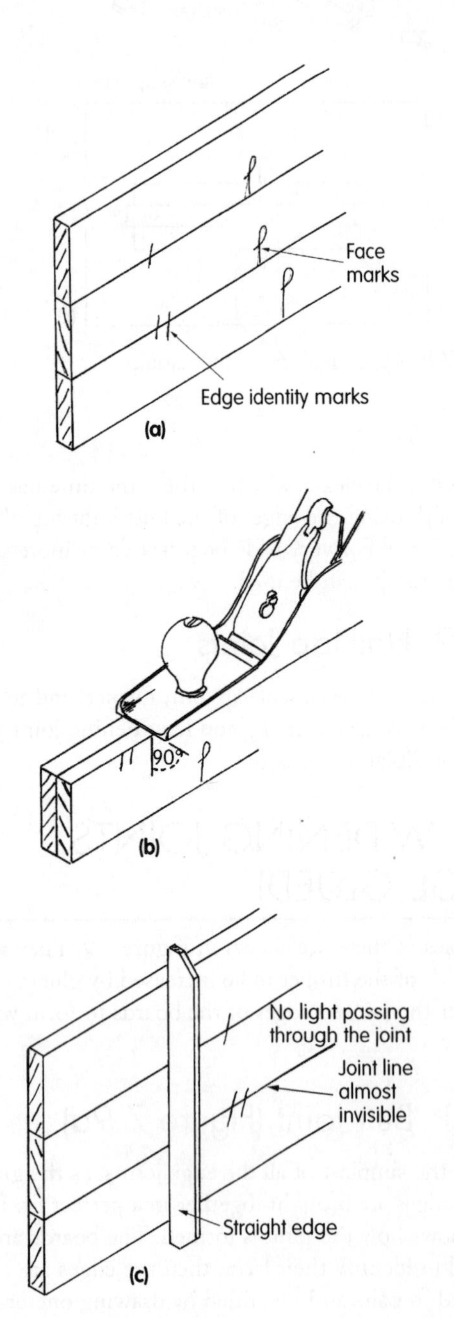

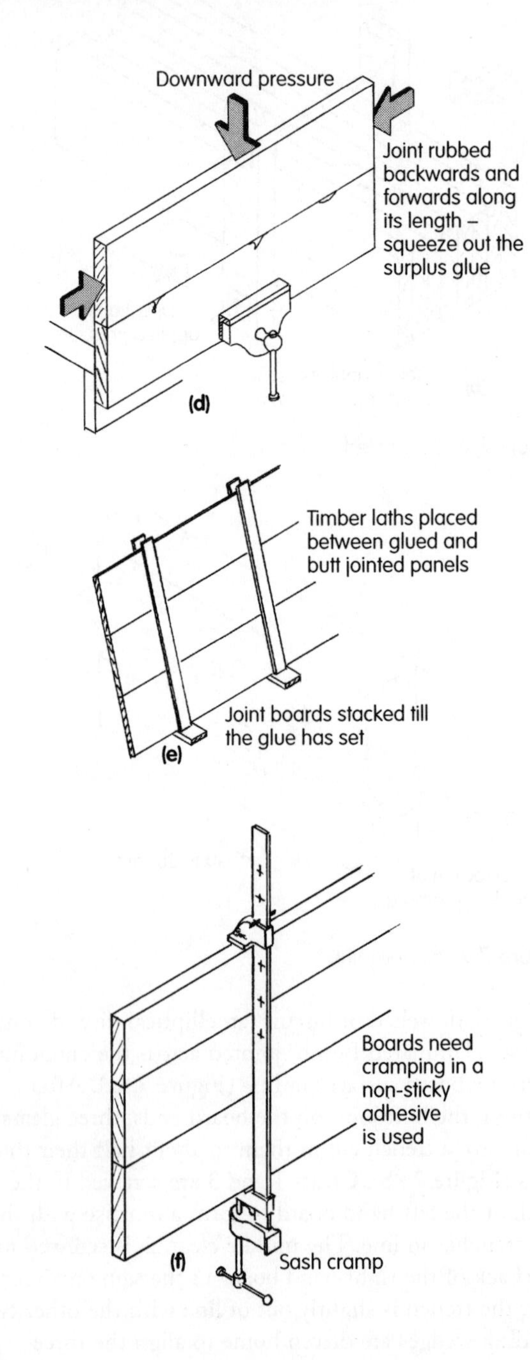

Figure 7.10 Forming a glued butt joint

7.3.2 Dowelled Joint (Figure 7.9d)

Dowels with a diameter of approximately one-third of the thickness of the board are inserted into the contact edges of the boards at reasonable intervals along the length of the joint. The edges of the joint are prepared as in Figure 7.10a, b, c. The positions of the dowels are then marked out as in Figure 7.11a. A hole equal to the diameter (D) of the dowel is drilled into each edge of the joint to a depth slightly greater than half the length of the dowel (Figure 7.11b).

The dowels are cut to length, chamfered off at each end and a groove formed down their length as shown in Figure 7.11c. The dowels are glued into the holes formed to receive them, surplus air escapes along the groove cut in the dowel and the joint is clamped together. Alternatively purpose-made dowels with chamfered ends and fluted sides can be used (Figure 7.11d).

7.3.3 Tongued and Grooved Joint (Figure 7.9b)

This joint is formed by cutting a groove in one edge and a tongue to fit into the groove on the other edge (Figure 7.9b). The groove is formed by using a plough plane fitted with a blade equal in width to the groove being cut (Figure 7.12a), while the tongue is formed using the same plane set with a tongue cutter (Figure 7.12b). As an alternative the tongue can be formed on the edge being jointed by cutting a rebate on the back and front of the board (Figure 7.12d). The joint is then glued and assembled.

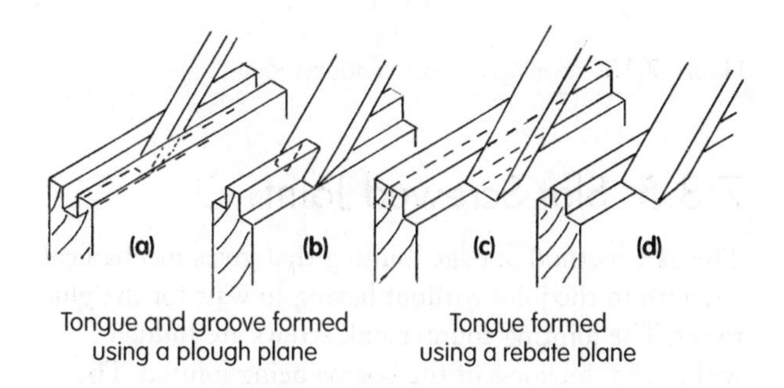

Figure 7.12 Forming a tongued, grooved and glued joint

7.3.4 Loose Tongued Joint (Figure 7.9c)

Both edges of the board have a groove worked into them and a loose tongue cut from plywood is inserted into the slot formed by bringing the two grooved edges together. To make a permanent joint glue or adhesive must be used (Figure 7.13).

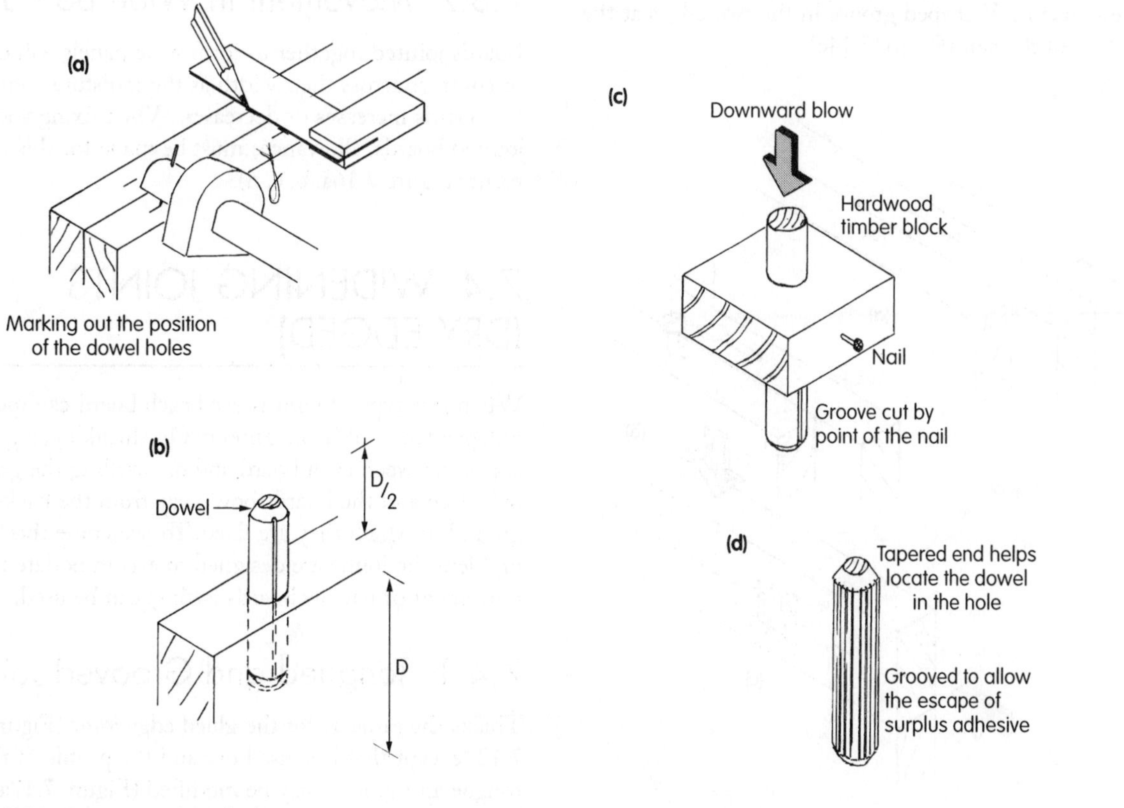

Figure 7.11 Forming a dowelled joint

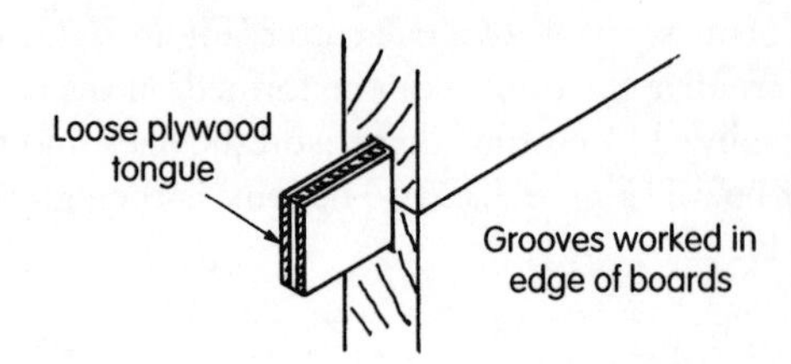

Figure 7.13 Forming a loose tongued joint

7.3.5 Slot Screwed Joint

This is a method of edge jointing that gives mechanical strength to the joint without having to wait for the glue to set. The jointing countersunk screws are hidden within the thickness of the boards being jointed. The edges of the joint are planed square and straight as for a butt joint (Figure 7.10a–c).

The boards are placed side by side and offset in their length by the distance the screw will be driven along the slot, between 10 and 20 mm (Figure 7.14a). The key hole slots and screw positions are set out and cut at between 300 and 450 mm intervals along the joint length. The two edges of the joint are brought together and lightly cramped. The head of the screw, which projects some 12 mm, being located in the circular eye of the key hole slot (Figure 7.14b). One board is driven along the other until the ends align and the shank and countersunk head of the screw is forced along the slotted portion of the key hole, causing the head of the screw to cut a V-shaped groove in the two edges at the bottom of the slot (Figure 7.14c).

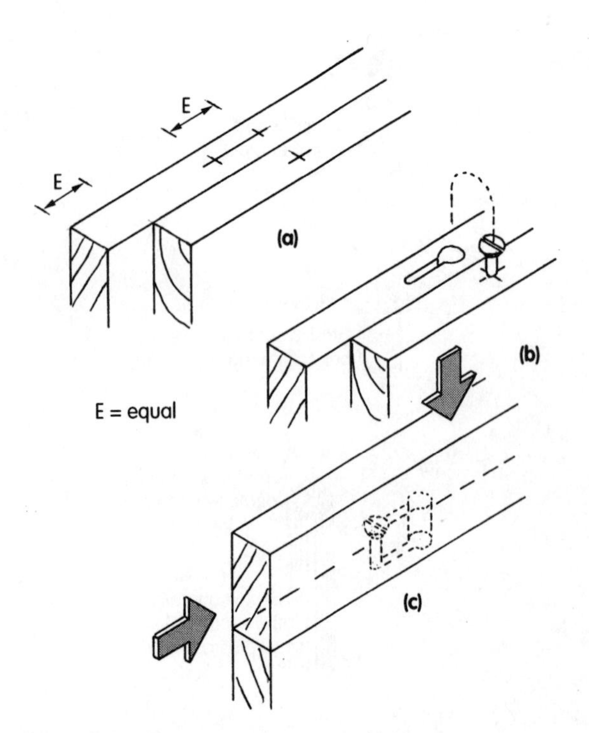

Figure 7.14 Forming a slot screwed joint

7.3.6 Distortion in Wide Boards

Whatever method is used to increase the width of boards by edge jointing, distortion of the timber will occur with variation in moisture content. As stated in Section 1.7.5, radial sawn boards are more stable than those tangentially sawn, which are liable to cup as the moisture content is reduced. Figure 7.15 shows ways to minimise this distortion by using narrow boards and alternating the heart side of the boards between the back and front face of the panel.

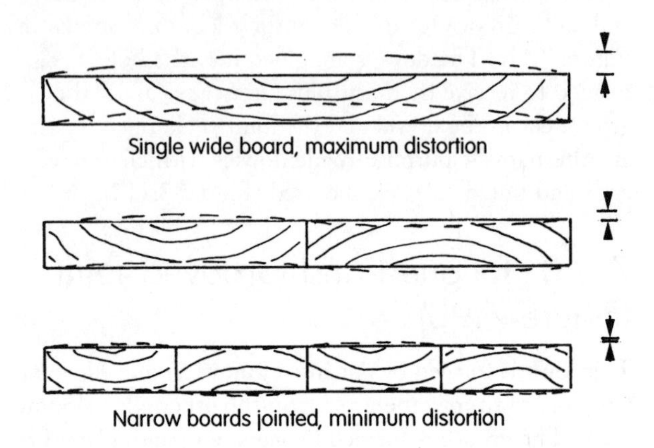

Figure 7.15 A method of minimising distortion in jointed panels

7.3.7 Movement in Wide Boards

Boards jointed together to form wide panels will expand or contract across their width as the moisture content of the boards increases or decreases. When fixing wide jointed boards, allowances must be made for this movement (Figure 7.16a, b, c, d).

7.4 WIDENING JOINTS (DRY EDGED)

When this type of joint is used each board can move independently of its neighbour. On shrinking, a gap will appear between each board and on swelling the gaps will close and the boards bow away from the background to which they are fixed. To overcome this problem the joints are designed to accommodate this movement or narrow boards (strips) can be used.

7.4.1 Tongued and Grooved Joint

This is the same as for the glued edge joint (Figure 7.12) except that it is used dry and the profile of the tongue and groove may be modified (Figure 7.17a, b). To keep shrinkage gaps to a minimum, narrow radially

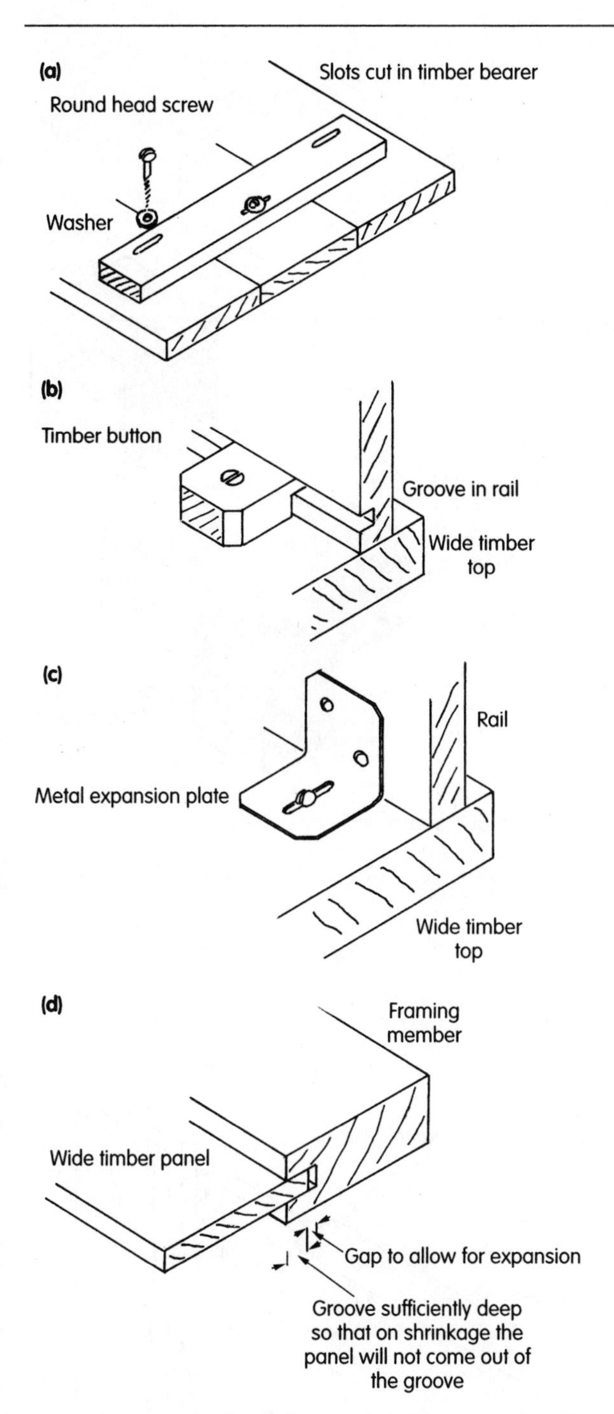

Figure 7.16 Methods of allowing for movement when fixing wide boards

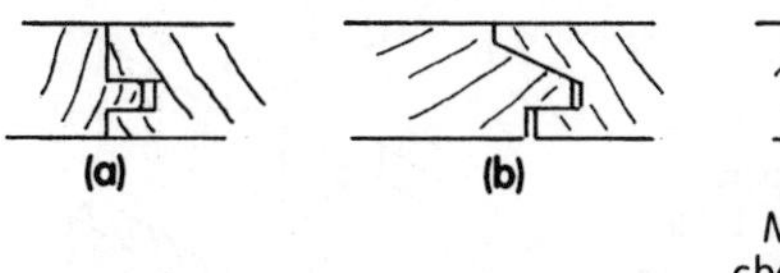

Figure 7.17 Sections of tongued and grooved boards and matchboard

cut boards are used and the moisture content of the timber kept at a constant level. Each board is cramped hard against its neighbour before it is fixed. Working a small chamfer on the edges of the board makes a feature of the joint and masks the gap (Figure 7.17c).

7.4.2 Vertical Boarding (Cladding)

Vertical boards of various cross-sectional shape are secured to a background frame to form large areas of boarding on walls, etc. (Figure 7.18a, b, c, d, e).

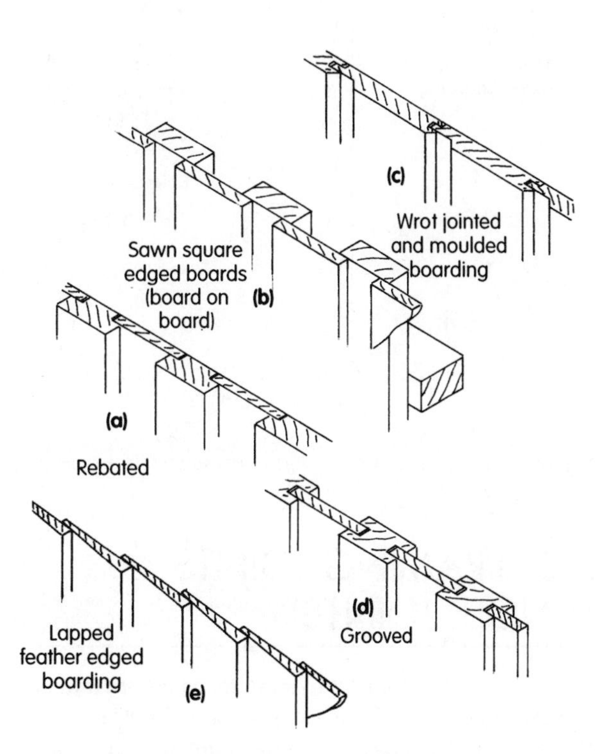

Figure 7.18 Types of dry jointed vertical timber cladding

7.4.3 Horizontal Boarding (Cladding)

Boards of various cross-sectional shape are secured horizontally to a background frame to form large areas of boarding on walls, etc. (Figure 7.19a, b, c).

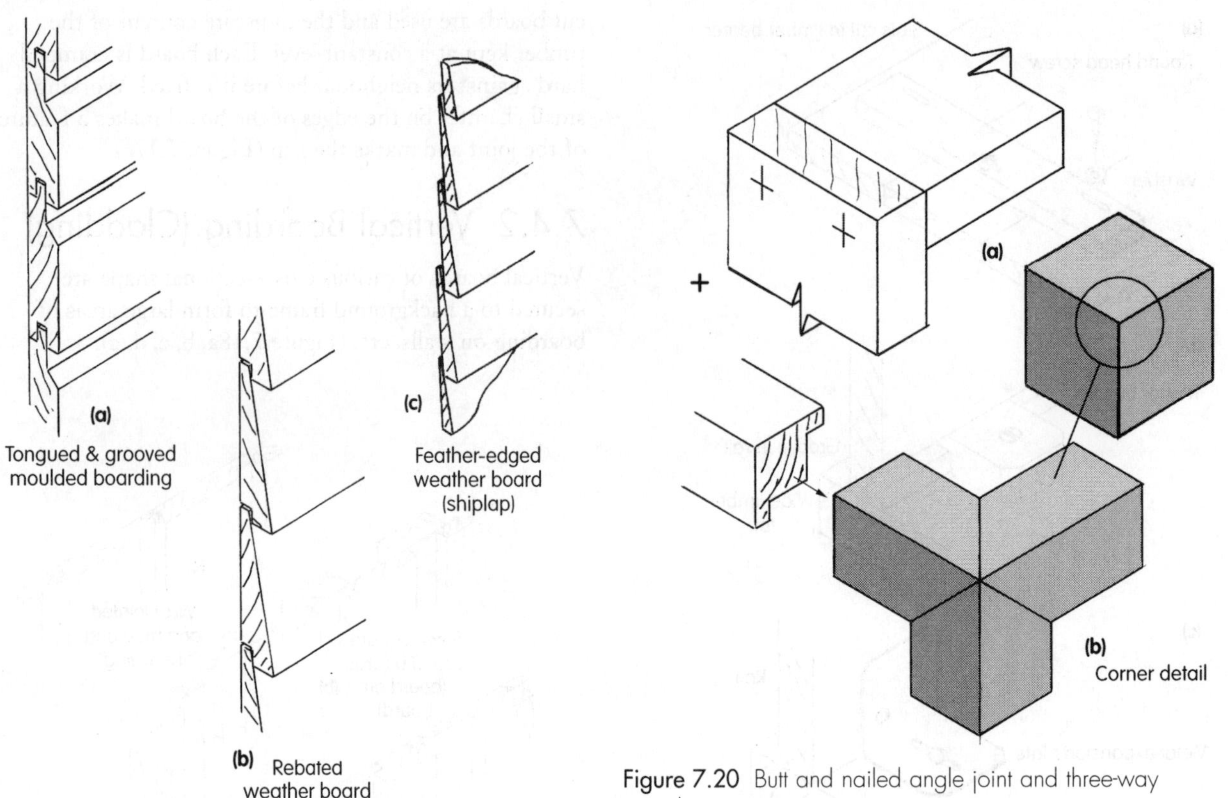

Figure 7.19 Types of dry jointed horizontal timber cladding

Figure 7.20 Butt and nailed angle joint and three-way mitred corner joint

7.5 FRAMING JOINTS (ANGLE JOINTS)

These are used to connect framing members at an intersection and can vary from a simple nailed butt joint (Figure 7.20a) to a complex mitred joint at the junction of a show case (Figure 7.20b). Framing joints fall into groups:

a bridle
b dovetail
c dowelled
d halving
e housing
f mitred
g mortise and tenon
h notched and cogged
i scribed.

a) 7.5.1 Bridle Joints

In bridle joints a central open slot is cut approximately one-third the width of the member, into which a tenon fits. Traditionally they were used for jointing large timber members in wooden roof trusses. This joint is now used mainly in the construction of light timber frames. Figure 7.21 shows various types of bridle joints.

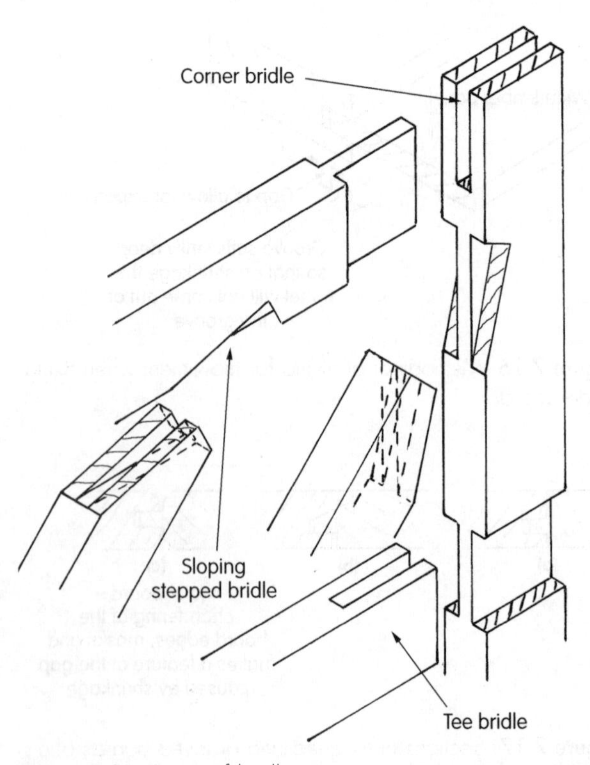

Figure 7.21 Types of bridle joint

b) 7.5.2 Dovetail Joints

Most woodworking joints are designed for and perform most efficiently when subject to compressive (crushing) forces. The dovetail is an exception, being designed to resist tensile (pulling) forces (Figure 7.22a, b, c). The strength of the dovetail joint is in the self-tightening

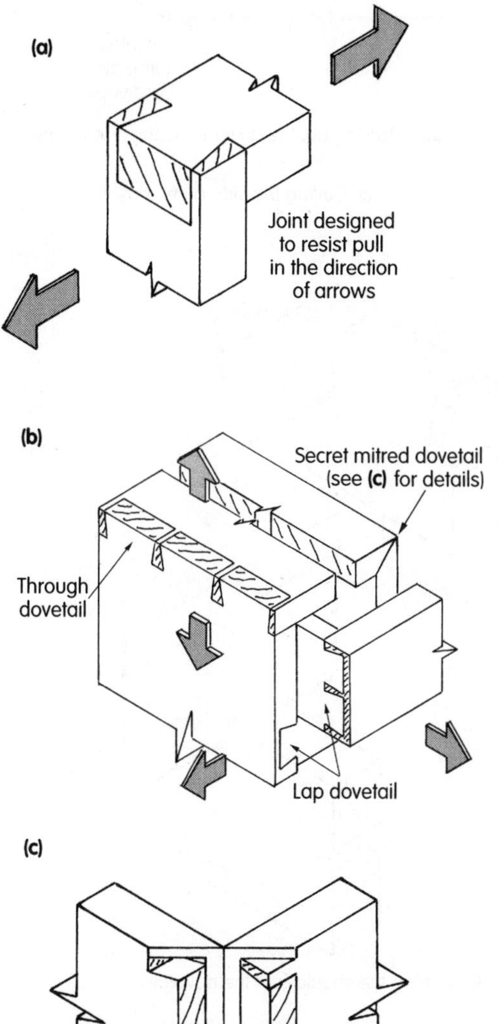

Figure 7.22 Types of dovetail joint and their construction

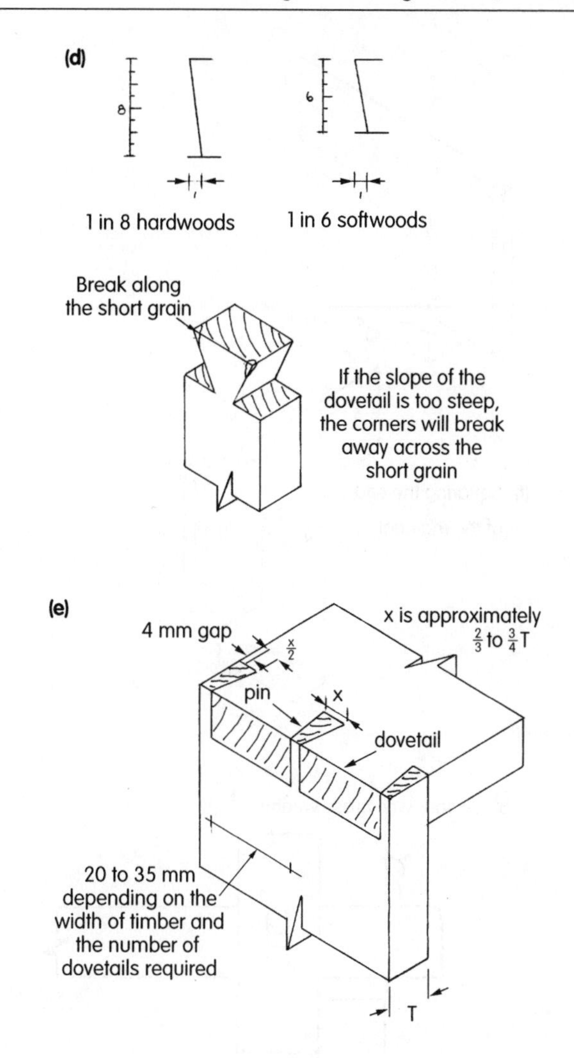

Figure 7.22 (continued) d, e Construction of dovetailed joints

effect of the dovetail against the pins (Figure 7.22a). The slope of a dovetail varies between 1 in 6 and 1 in 8, depending on the hardness of the timber (Figure 7.22d). If the slope is greater, the outer corners of the tails tend to break off across the short grain (Figure 7.22d).

The number of tails and their size will depend on the width of the boards being jointed (Figure 7.22e). Tails are rather larger than the pins, except for those produced by machinery where tails and pins are of equal size. Dovetails are divided into through, lapped and secret mitred (Figure 7.22b, c).

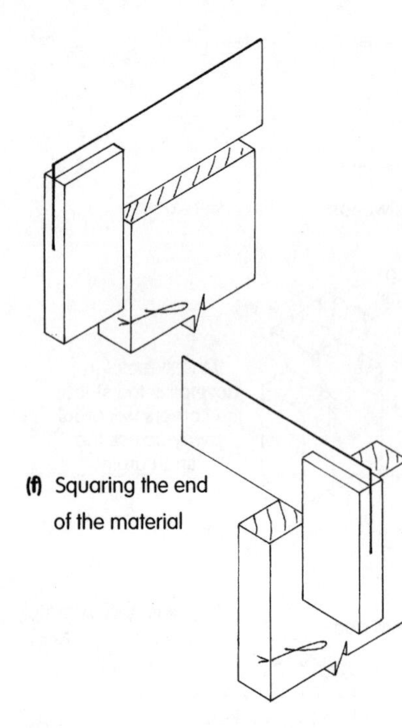
(f) Squaring the end of the material

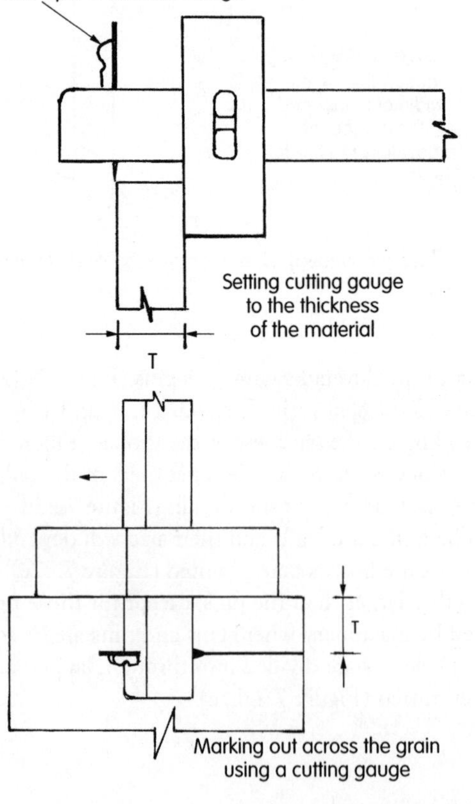

(g) Scribing the dovetail depth using a cutting gauge

Figure 7.22 (continued) f–g Construction of dovetailed joints

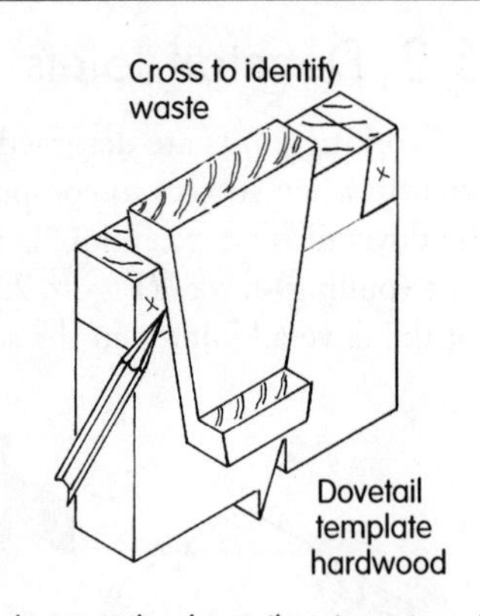

(h) Marking out the dovetails using a template

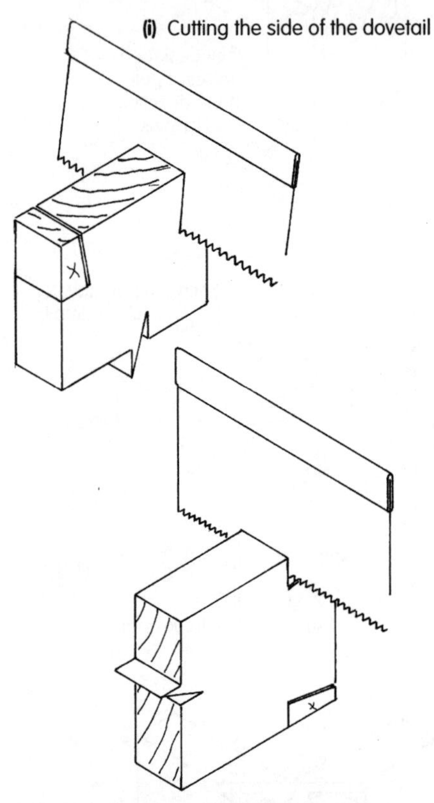
(i) Cutting the side of the dovetail

(j) Cutting the shoulder of the dovetail

Figure 7.22 (continued) h–j Construction of dovetailed joints

There are two methods in use for cutting dovetails:

1 cutting the pins and then the tails
2 cutting the tails and then the pins.

There is some controversy over which is the best method as there are advantages and disadvantages with both procedures. Figure 7.22f–p shows a method of marking out and cutting a single through dovetail by cutting the tail first.

- Square off the ends of the timber being jointed (Figure 7.22f).

(k) Marking out the pins from a cut dovetail

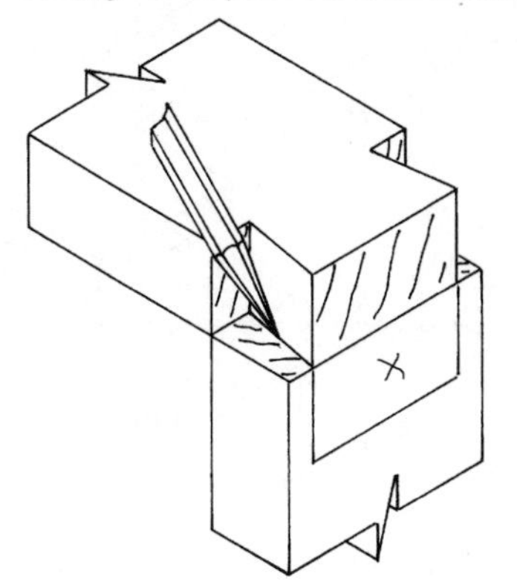

(l) Sawing down the waste side of the pins

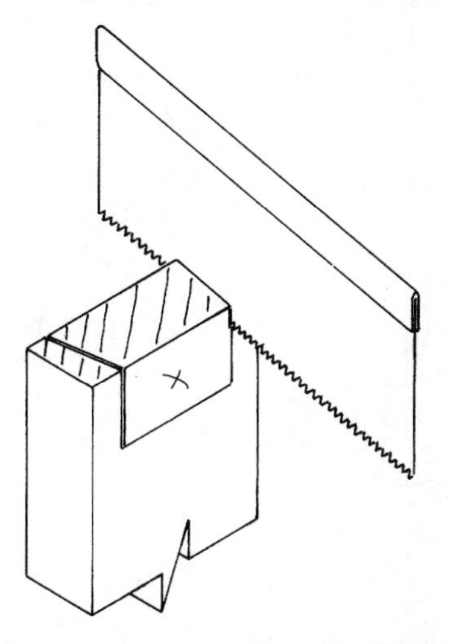

Figure 7.22 (continued) k–l Construction of dovetailed joints

- Set a cutting gauge to the thickness of the material and gauge round with the stock of the gauge against the endgrain (Figure 7.22g).
- Using a dovetail template or a bevel, mark out the dovetail (Figure 7.22h).
- With a dovetail saw cut down the line of the dovetail to the shoulder line keeping the saw cut on the waste side of the line (Figure 7.22i).
- Cut along the shoulder line to remove the waste material (Figure 7.22j).
- Mark the position of the pins from the cut dovetail (Figure 7.22k).
- Saw down the side of the pin to the shoulder line, the saw cut being made in the waste material (Figure 7.22l).
- Remove the waste by sawing across the grain with a coping saw, then pare square with a chisel, or using only a chisel, remove the waste by chopping out from both sides and back to the shoulder line (Figure 7.22m).

(m) Methods of removing the waste between the pins

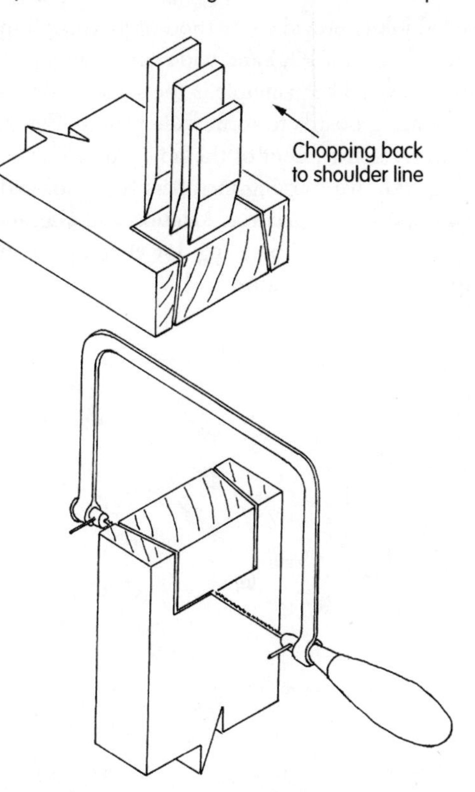

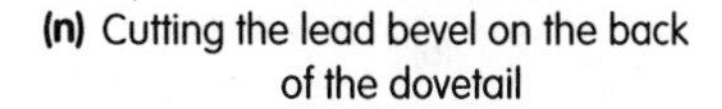

(n) Cutting the lead bevel on the back of the dovetail

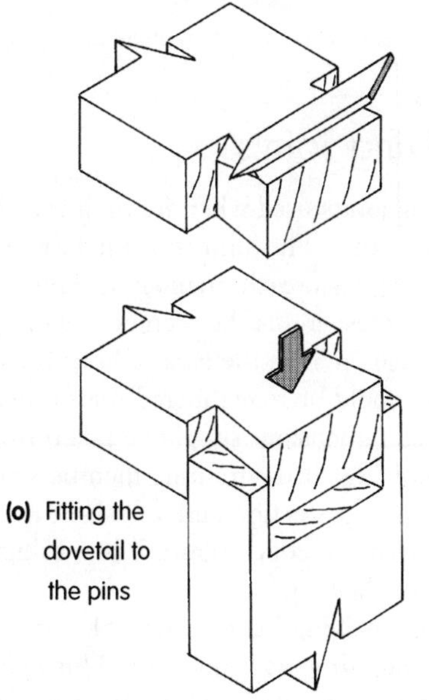

(o) Fitting the dovetail to the pins

Figure 7.22 (continued) m–o Construction of dovetailed joints

- Cut a lead bevel with a chisel on the back corners of the dovetail (Figure 7.22n).
- Without any further adjustment the joint should fit together with a slight tap from a hammer (Figure 7.22o).

7.5.3 Dowelled Joints

Dowelled joints provide a method of jointing framed members (Figure 7.23a) and wide carcasing panels (Figure 7.23b). The principle is the same as that for dowel jointing boards to form wide panels (Figure 7.11). Correct alignment of the holes to receive the dowels in the two framing members being jointed is of critical importance. Accurate location and alignment of the holes are achieved by using doweling jigs – one example is shown in Figure 5.112.

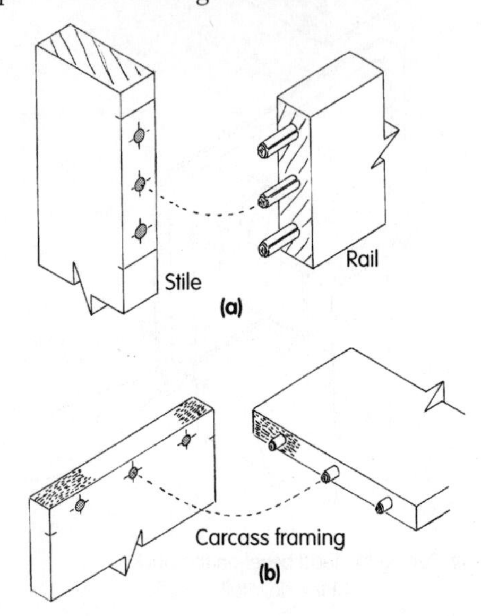

Figure 7.23 Dowelled joints in wide panels and framing

7.5.4 Halving Joints

These are locating joints used when framing members cross or abut each other. The joint is formed by removing half the material from each member at their point of intersection. In these joints the members abut face on, rather than edge on as in the case of housing joints (Figure 7.25). It should be remembered that in marking out these joints all dimensions should be taken from an identified face and edge if the framing members are not accurately sized in cross-section and a flush fit is required on one face and edge. Figure 7.24a–f illustrates the various types of halving joint.

When cutting a halving joint, make a face and edge mark on the framing timbers being used. This helps in the handing of the frame and identifies the face and edge you wish to finish flush. Mark out the shoulders of the joint using a try-square and pencil, keeping the stock of the square against the face side or edge (Figure 7.24g). Set the marking gauge to half the depth of the timber being jointed and gauge, keeping the stock of the gauge always against the face side. Mark the waste material to be removed (Figure 7.24h).

Secure the material in a vice, to a joiner's stool or Workmate (Table 5.8) and make a guide cut across the

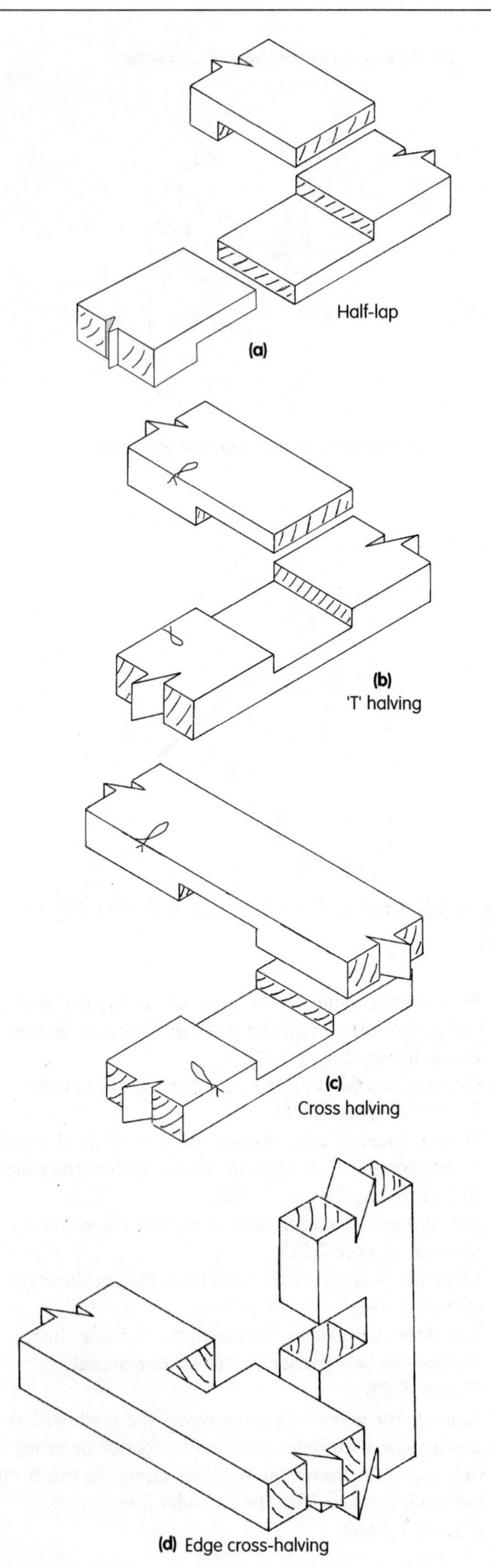

Figure 7.24 Types of halving joints and their construction

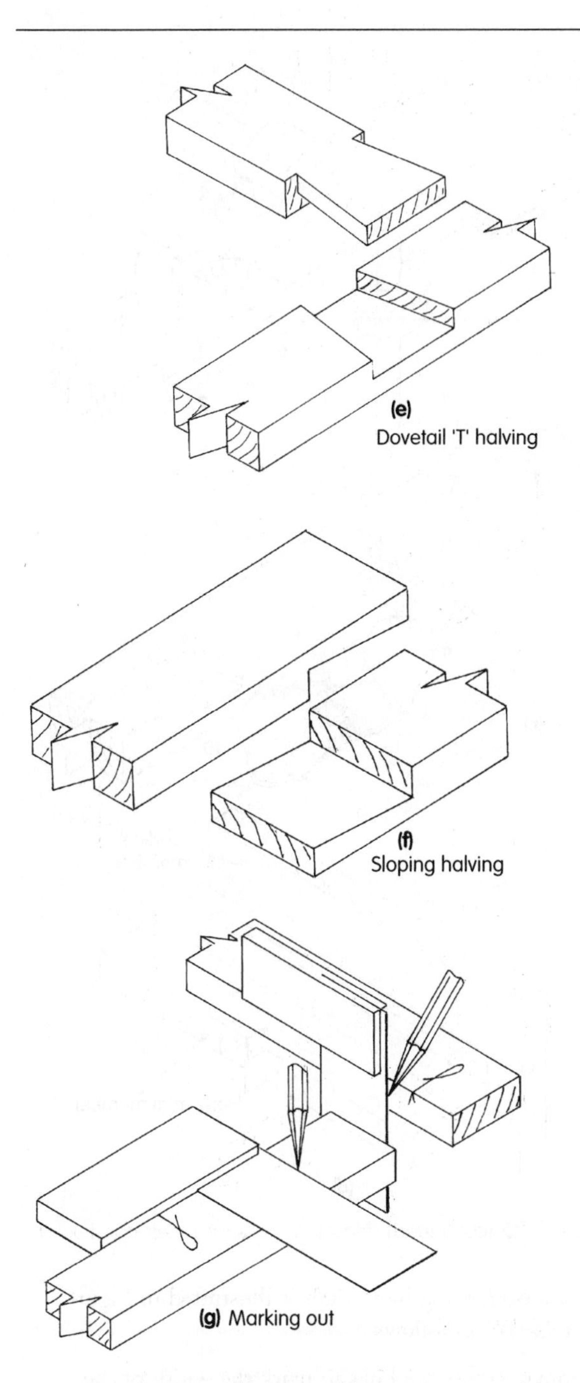

Figure 7.24 (continued) Types of halving joints and their construction

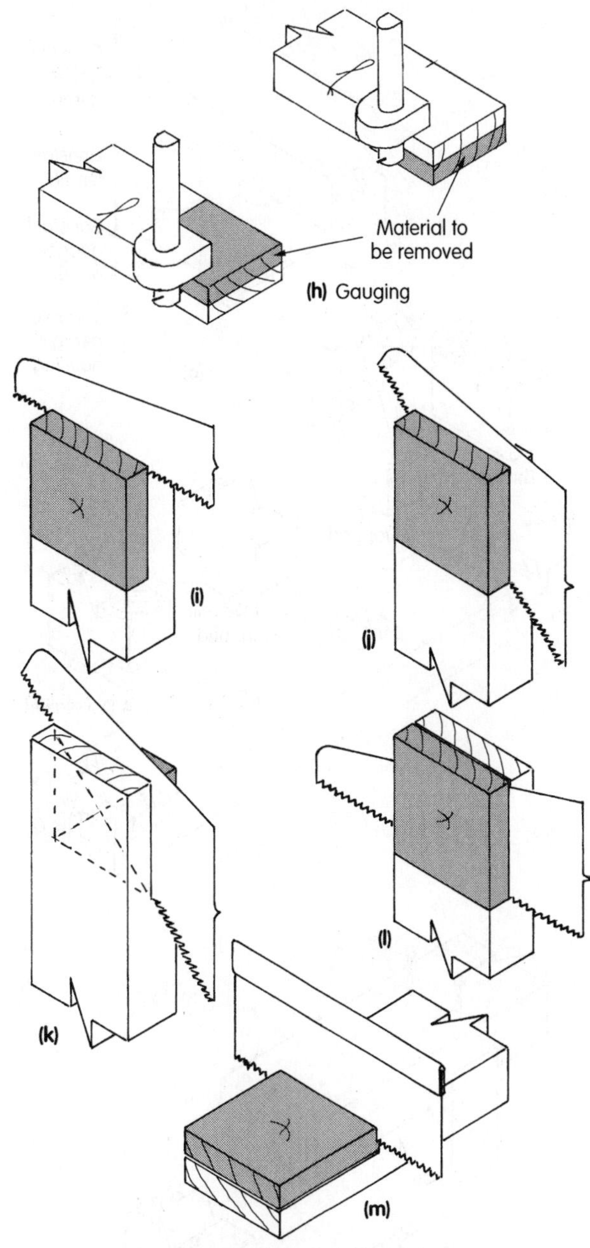

Figure 7.24 (continued) Types of halving joints and their construction

end of the joint on the waste side of the line (Figure 7.24i). Keeping the front of the saw in the guide cut, saw down forming an angled cut using the line on the edge you can see as a guide line for the saw (Figure 7.24j). Rotate the timber until you see the other side of the joint with the line that requires cutting and follow the same procedure (Figure 7.24k). Holding the saw square to the timber and using the saw cuts previously made as a guide, saw down to the shoulder line (Figure 7.24l). Finally, saw across the grain along the shoulder line to remove the waste piece (Figure 7.24m).

7.5.5 Housing Joints

These are used as locating joints in framing. Joints of this type are useful when nominal sized (unwrot) timber framing is used, where the thickness of the timber varies. On the horizontal members the depth of the housing can vary with the varying thickness of the timber, while the wood under the housing remains at a uniform thickness. This allows all the vertical framing members to be cut to the same length (Figure 7.25a). Figures 7.25(1–4) show the various types of housing joint. These joints require fixing with nails or anchor framing plates (Figure 7.25b). The method of cutting a

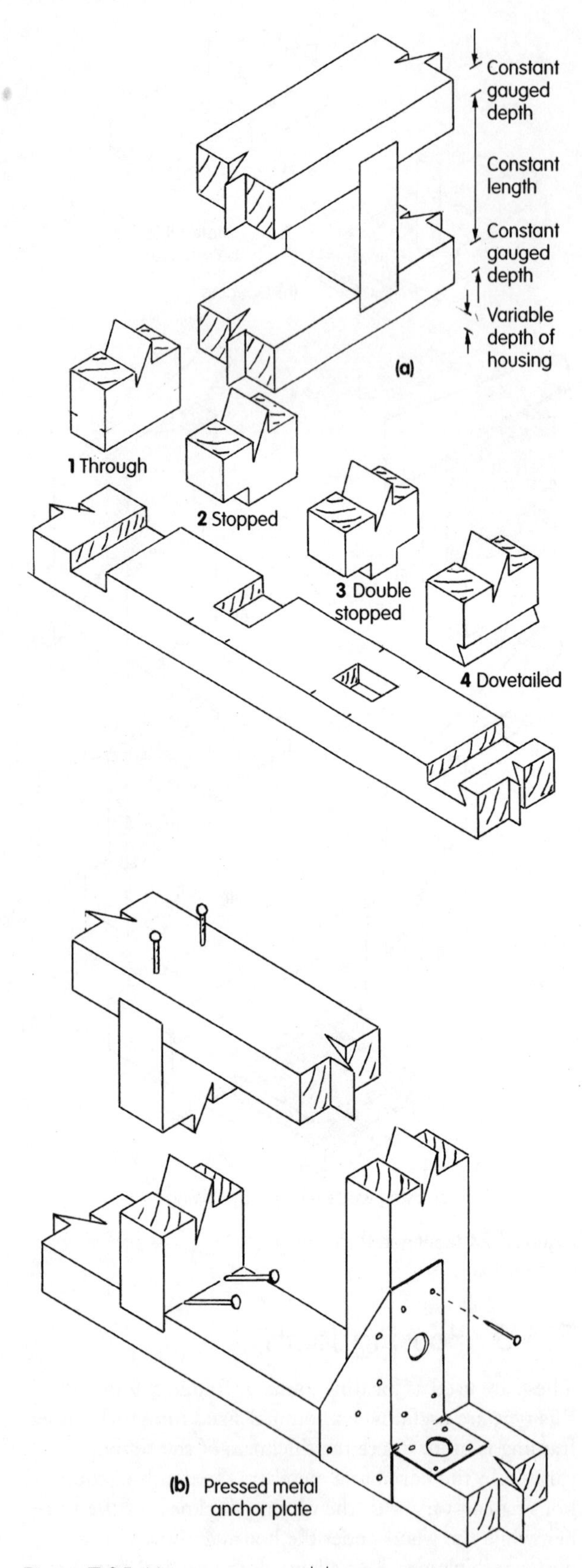

Figure 7.25 Housing joints and their construction

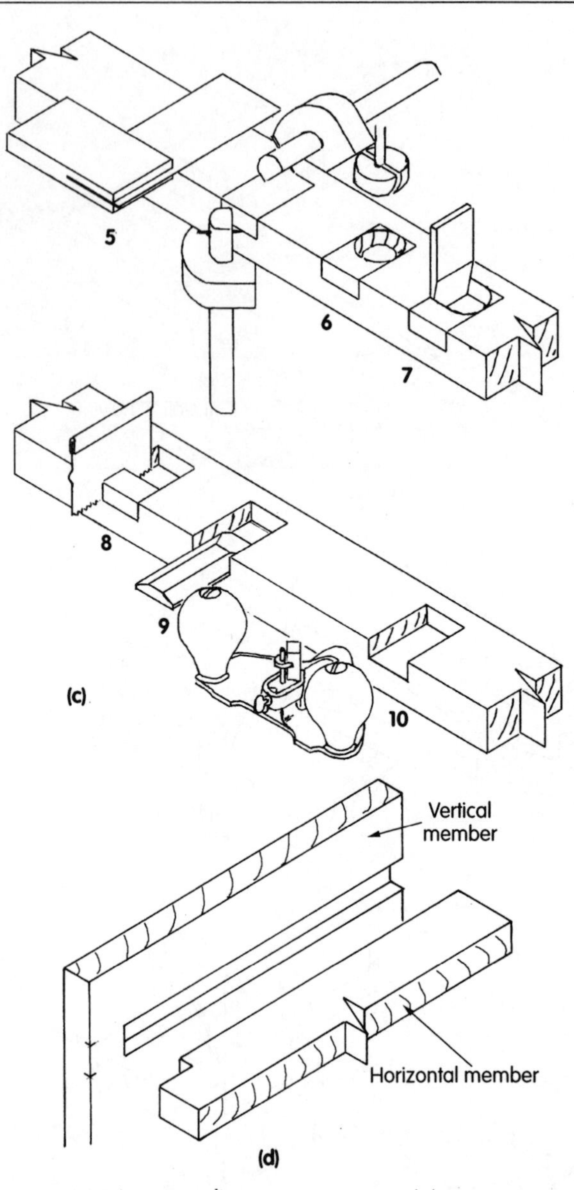

Figure 7.25 (continued) Housing joints and their construction

stop housing using hand tools is illustrated in Figure 7.25c (5–10) as follows.

- Square across the lines to mark the width of the housing and gauge the width and the depth. Depth does not usually exceed one-third the thickness (5).
- Bore one or two holes to the depth of the trench. Alternatively the material can be removed by using a chisel as it might be when cutting a stub mortise (6).
- If the holes have been bored, chisel the edges of the hole square and back to the marking-out lines (7).
- Using the toe of a tenon saw, cut along the marking-out lines to the depth of the housing (8).
- Remove the waste wood with a mallet and chisel (9).
- Level off the bottom of the housing using a hand router (10).

Housing joints are used in framing shelves, cabinets and partitions (Figure 7.25d).

7.5.6 Mitred Joints

These joints are used when moulded timbers such as skirting boards and architraves meet at an external angle, usually a right angle. The angle for the mitre cut is found by bisecting the angle made by the intersecting members. For members meeting at right angles (90°) the angle for the mitre cut is 45° (Figure 7.26 a–d).

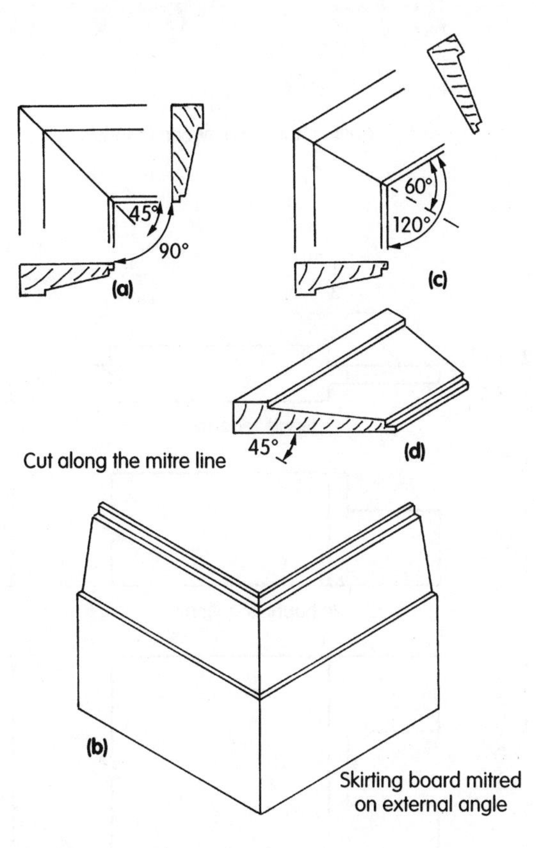

Figure 7.26 Mitred joints and setting out

7.5.7 Mortise and Tenon Joints

These are the most commonly used framing joints, being versatile, easily concealed within the framing and the most efficient. Figure 7.27a–f shows examples of the various types of mortise and tenon. The names that identify them are as follows.

- **a** *Through mortise and tenon* – in which the mortise hole goes completely through the timber member.
- **b** *Stub mortise and tenon* – in which the mortise hole is only cut partially through the timber member.
- **c** *Double tenon* – in which two tenons are cut side by side in the thickness of the framing member.
- **d** *Single haunched tenon* – see 7.5.7a.
- **e** *Barefaced tenon* – includes a tenon with only one shoulder.
- **f** *Twin tenon* – in which two tenons are cut one above the other in the depth of the framing member.

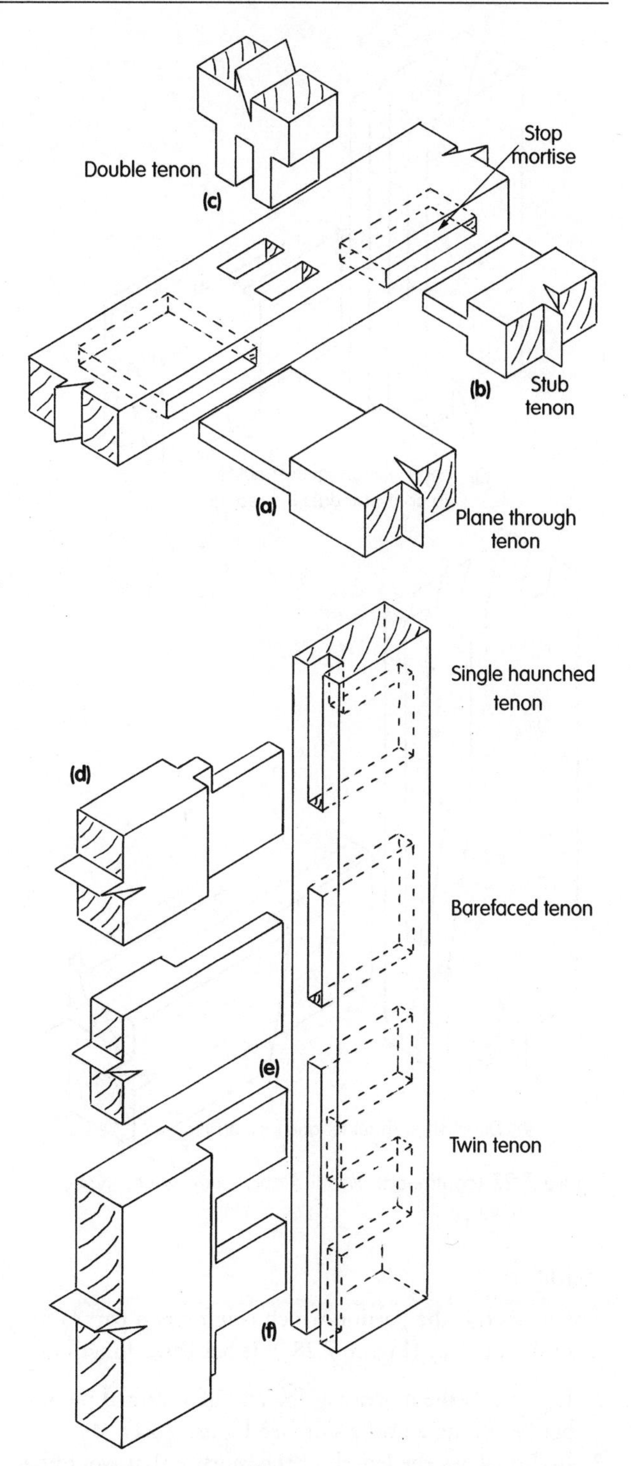

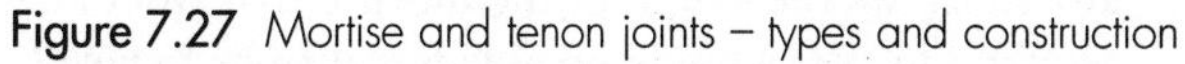
Figure 7.27 Mortise and tenon joints – types and construction

- **g** *Unequal shouldered tenon* – in which one shoulder is set out of line with the other.
- **h** *Diminished shoulder tenon* – depending on whether the stile is rebated diminished, or stop rebated, one or both shoulders are cut at a sloping angle.

NB: There is some confusion in the use of the terms double and twin tenons which are often interchanged. BS 565 identifies a double tenon as shown in Figure 7.24c.

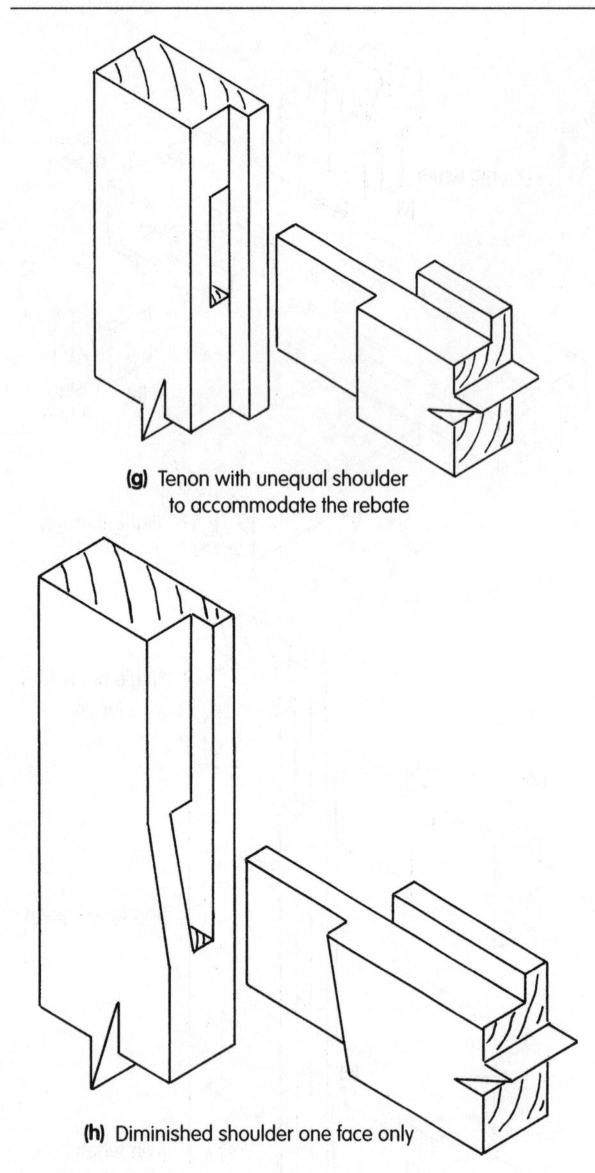

(g) Tenon with unequal shoulder to accommodate the rebate

(h) Diminished shoulder one face only

Figure 7.27 (continued) Mortise and tenon joint – types and construction

Haunch

The haunch is the portion which is left when part of a tenon is cut away (Figure 7.28a). It has three functions.

1 It prevents the mortise at the end of a framed member becoming a bridle slot (see Figure 7.21).
2 It also allows the length of the mortise that would be required to take a tenon on a wide rail to be reduced, by splitting a single tenon into a twin tenon (Figure 7.27f). If the depth of a tenon and length of a mortise are excessive, shrinkage of the framing members would cause the tenon to become loose in the mortise, while the long mortise would weaken the framing member in which it is cut.
3 The haunch also stops light showing through the joint between the vertical and horizontal framing members, if the member containing the mortise shrinks.

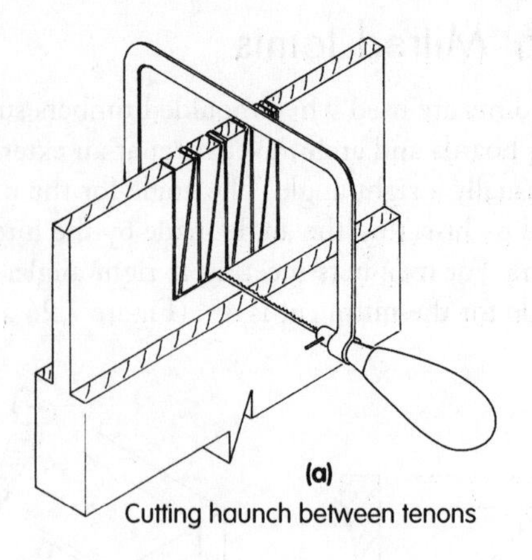

(a) Cutting haunch between tenons

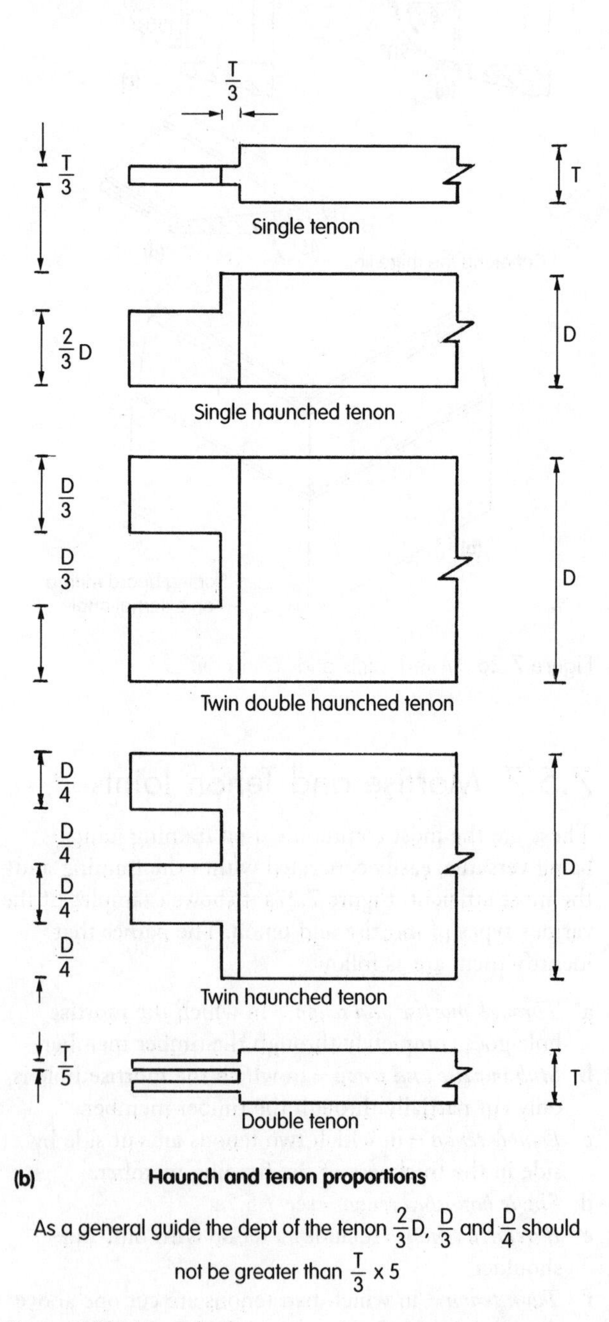

(b) **Haunch and tenon proportions**

As a general guide the dept of the tenon $\frac{2}{3}$D, $\frac{D}{3}$ and $\frac{D}{4}$ should not be greater than $\frac{T}{3} \times 5$

Figure 7.28 Haunched tenons

As a general guide, the depth of a tenon should not be greater than five times its thickness, although in some circumstances this ratio may be varied (Figure 7.28b).

Franking (Figure 7.29)

This is the reverse of haunching and is used when the groove worked in the member to receive the haunch weakens it by removing excessive material in the area of the mortise i.e. narrow sash stile.

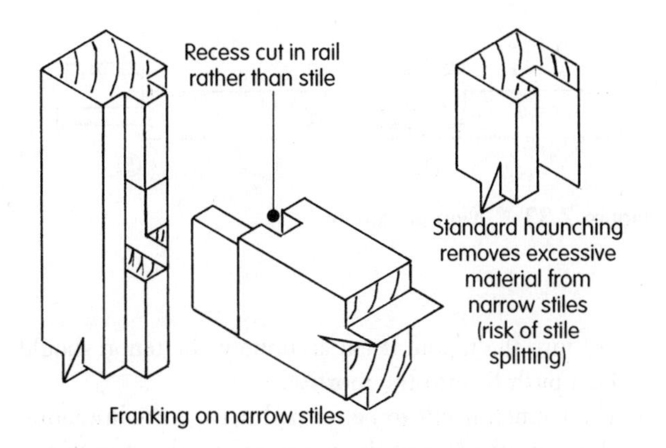

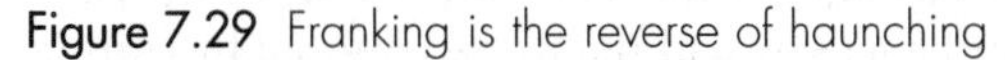

Figure 7.29 Franking is the reverse of haunching

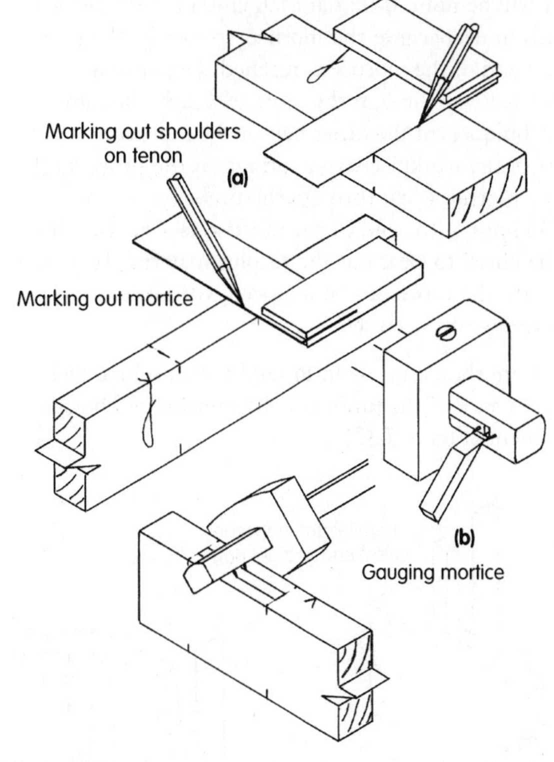

Figure 7.30 A mortise and tenon joint using hand tools

Making a Mortise and Tenon Joint Using Hand Tools (Figure 7.30)

- Mark the face side and edge of the material and square a line round the timber to mark the position of the shoulder line for the tenon. On the other piece square across the edges two lines to mark the position of the mortise. Keep the stock of the try-square against the face side or edge when marking-out (Figure 7.30a).
- Set the distance between the two points of the mortise gauge equal to that of the chisel you are going to use to cut the mortise. Scribe on the timber with the points of the gauge, keeping the gauge stock against the face side, to mark both the width and position of the mortise and tenon (Figure 7.30b).
- To cut the mortise (Figure 7.31) secure the material in a vice or on a bench. Starting at the centre of the length of the mortise and holding the chisel vertically with its cutting edge across the grain between the gauged lines, strike the chisel with the mallet, driving it into the timber 3 or 4 mm, using a slight pendulum motion in line with the mortise to remove the chisel.
- Repeat the operation 3–4 mm nearer the end of the mortise. Continue this until the end of the mortise is reached.

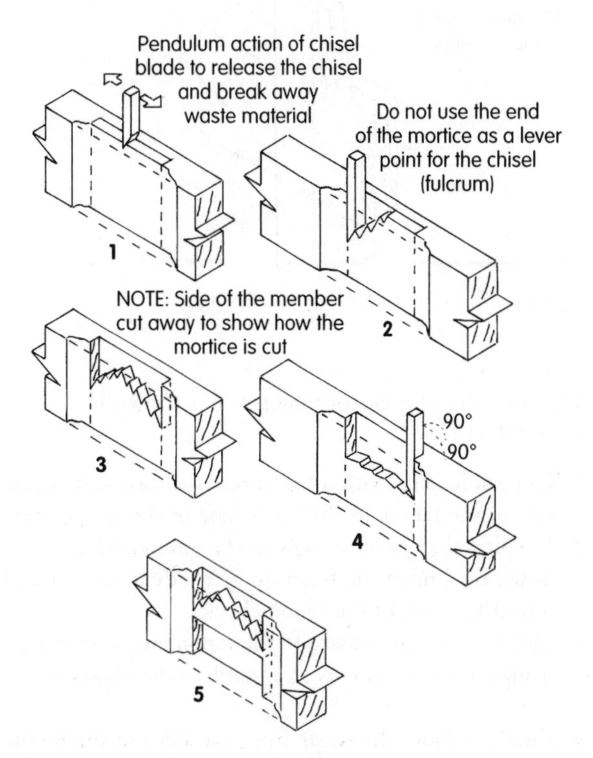

Figure 7.31 Sequence of cutting a mortise

- It will be noticed that at each chisel cut the blade is driven deeper into the mortise, so that by the time the end of the mortise is reached the chisel will have cut halfway through the mortise. Using the same technique, cut the other half of the mortise.
- Turn the workpiece over and repeat the process till the mortise is cut through the timber.
- Do not use the end of the mortise as a fulcrum for the chisel to clear out the surplus material. If cut correctly, the waste can be removed with little or no levering of the chisel.

NB: Care should be taken to hold the chisel parallel with the face of the timber or the mortise will be out of alignment (Figure 7.32).

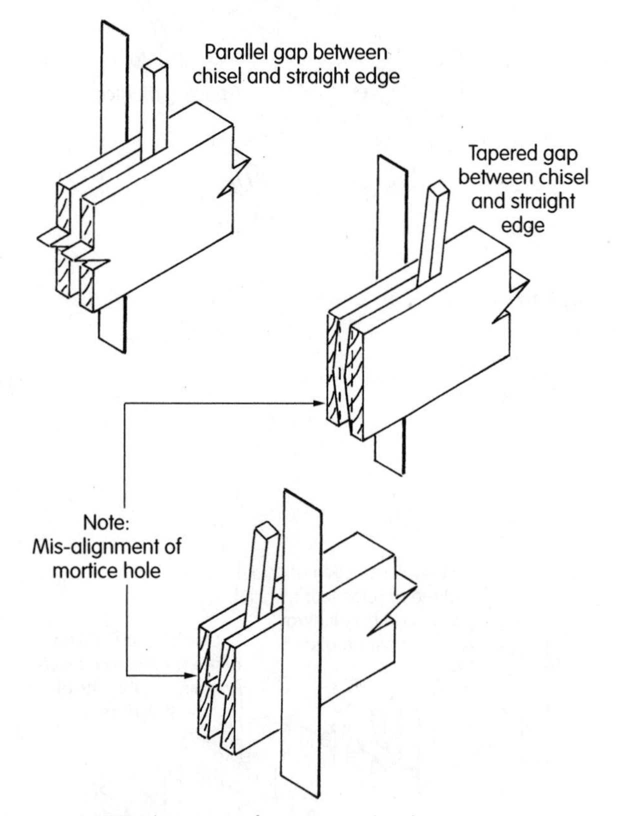

Figure 7.32 Alignment of a mortise chisel

The procedure for cutting the tenon is shown in Figure 7.33.

1 Cut guide kerfs with a tenon or panel saw across the end of the tenon on the waste side of the gauge lines.
2 Keeping the toe of the saw in the guide kerf, saw down the line of the tenon to give a diagonal cut and repeat for each of the tenon cheeks.
3 Holding the saw square to the tenon end, saw down using the diagonal cuts as a guide to the shoulder lines.
4 Finally remove the waste from the sides of the tenon by sawing along the shoulder lines, taking care not to cut into the tenon. If cut accurately, the tenon should be a push fit into the mortise.
5 If a haunch needs to be formed between twin tenons, the vertical edges of the tenon are cut down with a tenon or panel saw and the waste removed by cutting along the line of the haunch with a coping saw (Figure 7.28a). That part of the tenon cut away to form the haunch may be cut into wedges for wedging up the mortise and tenon joint.

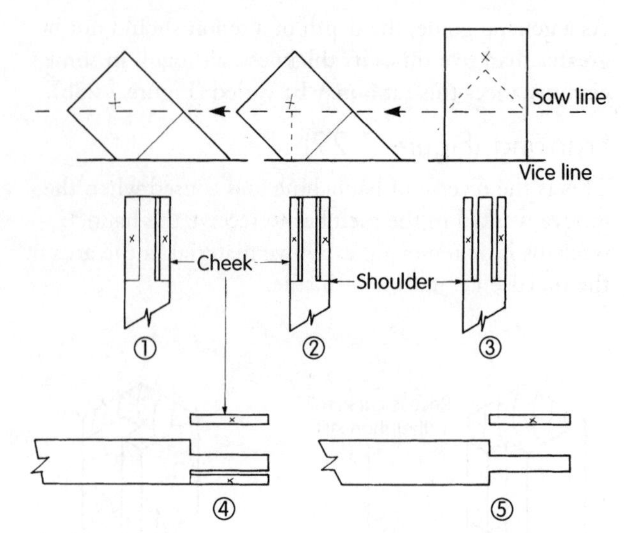

Figure 7.33 Cutting the tenon

7.5.8 Notched and Cogged Joints

These joints are used in heavy structures framed in large sectioned timbers, to locate the members and ensure uniformity of depth and line when nominally sized timbers are used. An example of this would be the notching of joists over a wall plate to keep the top of the joists in line to provide a flat frame to receive the decking (Figure 7.34a, b). Cogged joints have the same function, but are a little more complex to cut, although less timber is cut away when forming the joint (Figure 7.34c, d). The method of cutting these joints is the same as for housing and halving joints, except that the timbers are much larger and heavier.

7.5.9 Scribed Joints

These joints are used when moulded timber trims meet at internal corners or at the intersection of mouldings worked on framed members such as doors and windows. To form this joint a reverse profile of the mould is cut and butted over the profile of the moulded section to which it is being jointed. The profile or outline shape of the mould is obtained by mitring the moulded member, then cutting the timber square to the face while following the outline formed by the mitre cut (Figure 7.35).

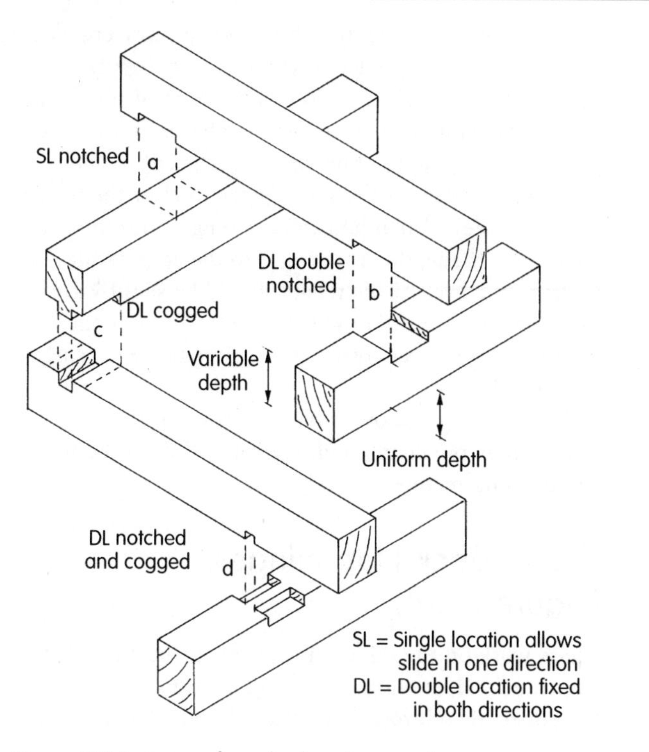

Figure 7.34 Types of notched and cogged joints

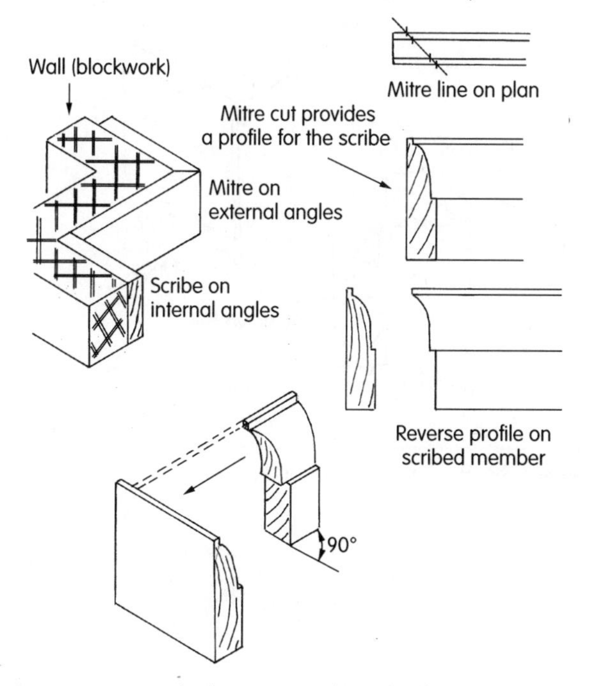

Figure 7.35 Scribed joints – methods for forming

7.5.10 Scribing

Where timber trims such as skirting boards and cover moulds have to be fitted against uneven surfaces, such as walls, floors and ceilings, scribing is used to ensure a close fit between the cover mould and its abutment. Figure 7.36 shows that by marking a pencil line on the

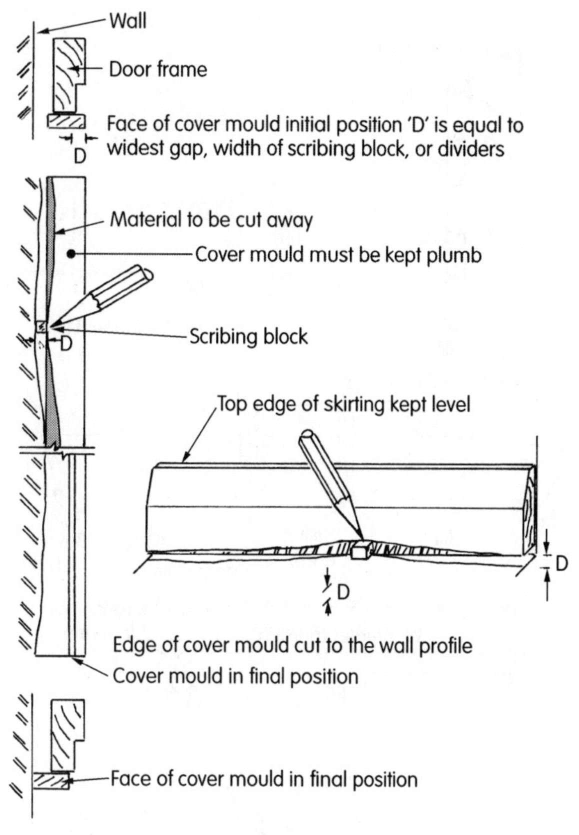

Figure 7.36 Scribing cover trim to uneven surfaces

abutting face of the timber trim running parallel to the uneven surface, an identical profile of this surface can be produced, and when cut, will fit without showing gaps.

7.6 HINGED JOINTS

This type of joint is used where opening and shutting actions are required in items of joinery such as windows and doors, drop-down leaves and counter flaps. In one or two specialist cases associated with furniture making the hinges are formed in timber but in general, hinges made from ferrous metal (iron content) e.g. steel and non-ferrous metal, e.g. brass are used. The best but most expensive types of hinge are made from cast or drawn metal and machined with great accuracy to ensure uniformity of size. Less expensive hinges are made from pressed metal.

7.6.1 Butt Hinges (Figure 7.37)

These are the most common form of hinge used in joinery work and consist of a central pin that passes through the knuckle formed on one edge of the hinge wings (leaves) (Figure 7.37a).

When fixing hinges, it should be remembered that the opening edge furthest from the hinge traces an arc as it swings (Figure 7.37b), causing the trailing edge to

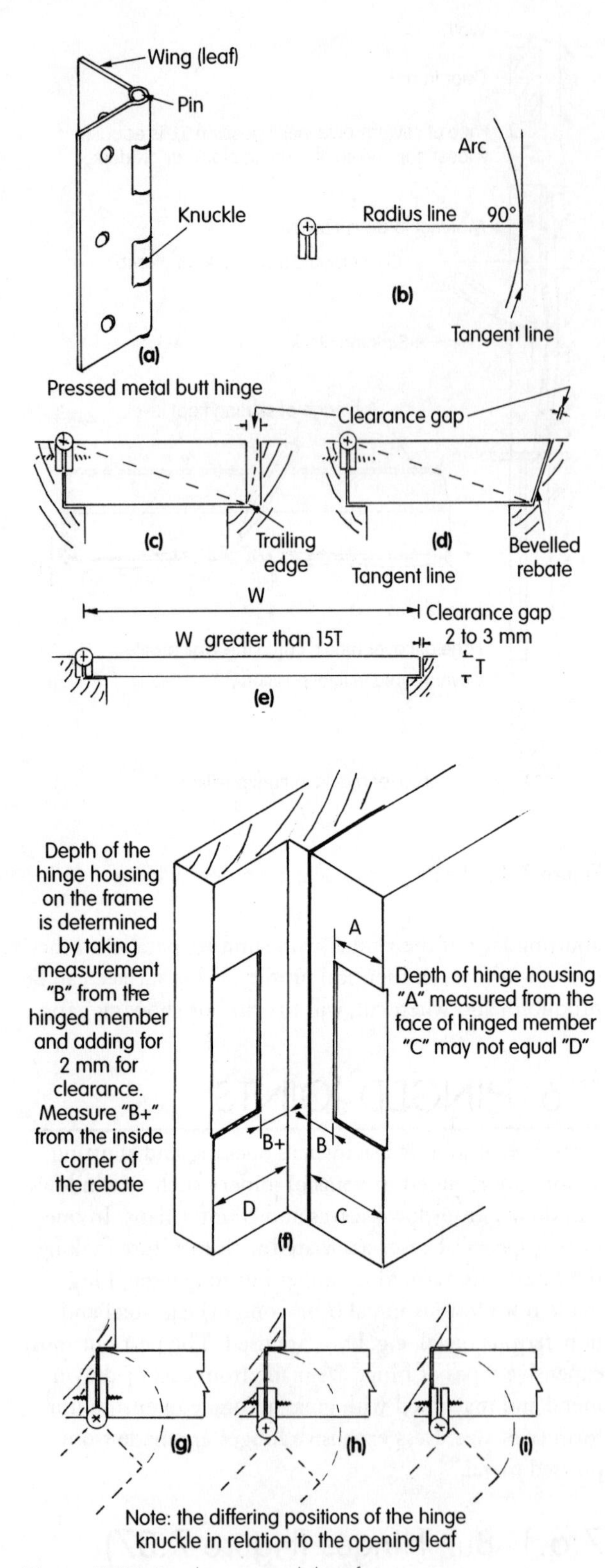

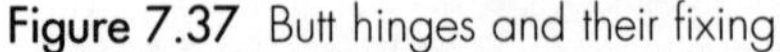

Figure 7.37 Butt hinges and their fixing

catch in the receiving rebate if there is insufficient clearance, or the opening edge or rebate is not slightly bevelled (Figure 7.37c, d). When the ratio of the thickness to the width of the hinged member is at least 1:15 a clearance gap of 2–3 mm will be sufficient to allow the trailing corner of the square edge to clear the rebate (Figure 7.37e). When housing the hinges into the rebate of a frame, the distance from the edge of the hinge to the back of the rebate should be slightly greater than the distance from the inside edge of the hinge housing to the inner corner of the edge of the closing member (Figure 7.37f).

Figure 7.37g–i shows what occurs when the position of the hinge pin changes in its relationship to the frame and opening member.

7.6.2 Back Flap Hinges (Figure 7.38)

These work on the same principle as the butt hinge, but have much wider wings (leaf) and are usually fixed to the underside of drop-down flaps (Figure 7.38a,b).

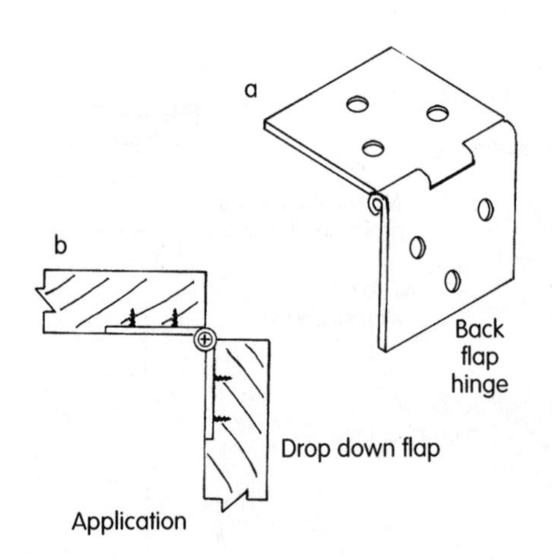

Figure 7.38 Back flap hinges and their use

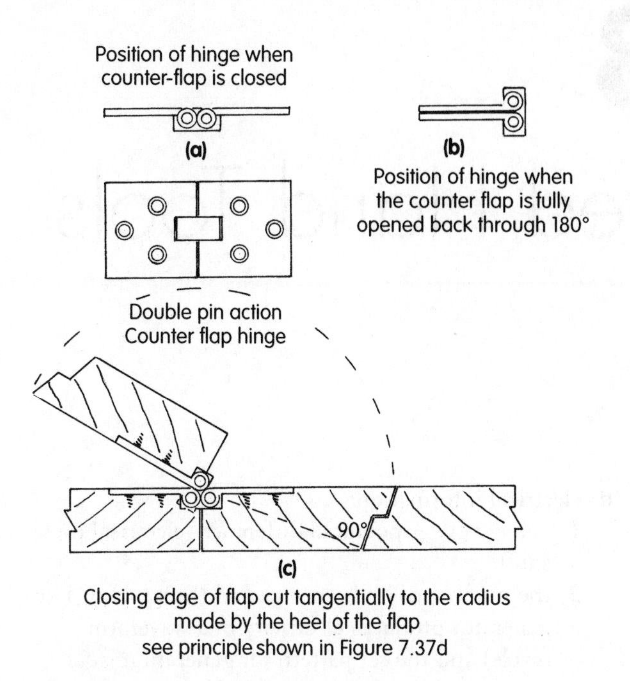

Figure 7.39 Counter flap hinges and their use

7.6.3 Counter Flap Hinges (Figure 7.39)

These are similar to back flap hinges, but have a double pin action so that they can be fitted flush without the knuckle protruding on the top surface of a flap which can then be rotated through an angle of 180° (Figure 7.39a, b, c).

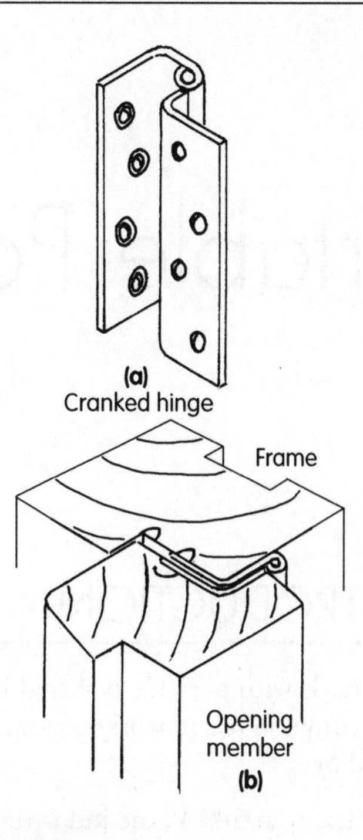

Figure 7.40 Cranked hinges as applied to an opening casement

7.6.4 Cranked Hinges (Figure 7.40)

In these hinges the wings have a right-angled crank in them and they are used when the opening member projects and overlaps the frame (Figure 7.40a, b). The example shown illustrates how this type of hinge is applied to an opening casement of a window. Notice the anti-capillary grooving cut into the rebates of the sash stile and casement frame.

8

Portable Powered Hand Tools

8.1 INTRODUCTION

This section deals with portable powered hand tools which are in common use in workshops and on building sites, powered by:

- mains electricity at 240 V; the industrial use of tools supplied with power at this voltage is not recommended as an electric shock received from the tool or cable can be fatal
- mains electricity at 240 V stepped down via a transformer to 110 V for use with power tools; this system must be used on all building sites
- electricity supplied from portable petrol or diesel generators at 240 or 110 V
- battery, 9–15 V (cordless power tools)
- compressed air; the air is supplied from a fixed or portable air compressor
- cartridge-operated fixing tools (ballistic tools).

Power tools are inherently more dangerous to use than hand tools as you do not have complete and immediate control of the power source. Power tools speed up manual work and make it easier, but training and practise in their safe use are necessary to obtain the advantages these tools offer to the carpenter and joiner.

8.2 SPECIFICATION PLATES

These are permanent labels of soft metal (aluminium) securely attached to each power tool and on which are stamped or engraved (see Figure 8.1a–f):

a the manufacturer's name
b identification and reference numbers: model, production and serial numbers to be used when ordering replacement parts
c capacity: chuck size, spindle speed in revolutions per minute (abbreviated to rev/min or RPM)
d electrical information:
  1 voltage (V) – potential difference (electrical pressure)
  2 the number of cycles per second (hertz, Hz). The generator produces electricity in a waveform (cycle) and the set pattern for generation is 50 waves or cycles per second (50 Hz)
  3 amperes (A) – the rate at which the electrical current flows
  4 watts (W) – the amount of electrical energy being used.
  As watts = volts × amperes it will be seen that if the potential difference (voltage) is reduced from 240 to 110 V the rate of flow of electricity (amperes) must be increased if the power output is to remain the same. This explains why heavier cables are used with lower voltage (i.e. 110 V): to allow for increased current flow.

  Watts = volts × amperes
  400 = 240 × 1.76
  400 = 110 × 3.6

e insulation symbol if the equipment is double insulated
f manufacturing standard – kitemark and standards number, if manufactured to a British Standards specification.

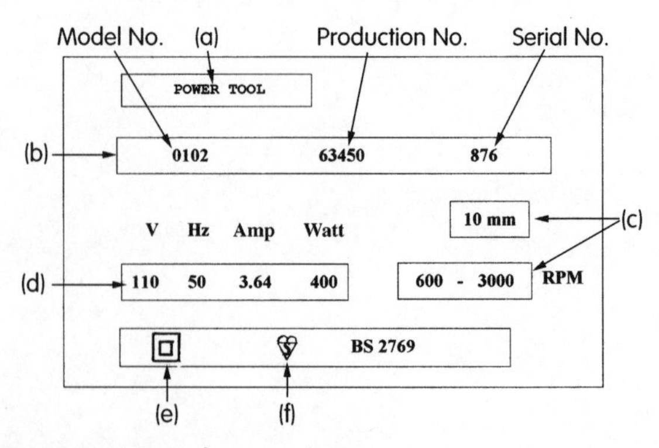

Figure 8.1 Specification plate

8.3 EARTHING AND INSULATION

8.3.1 Mains Electricity Supply and Earthing

Any apparatus which is not double insulated must be earthed via a low resistance path (earth wire) which allows the electricity to flow to earth without passing through the operator if anything should go wrong. The apparatus must also be protected by a fuse, usually contained in the line of the live wire (brown in colour) in the plug attached to the tool. The rating of this fuse must be similar to the rating of the machine. A tool rated at 4.5 A should have a fuse no greater than 5 A (Figure 8.2). If there is a fault and the power tool casing becomes live the yellow and green sheathed low-resistance conductor provides a path to carry the current safely to earth. The increased flow of electrical current across the high-resistance wire in the fuse causes it to burn out (blow) (Figure 8.2a). For the earthing system to work the earth conductor must have continuity from the tool to the earthing point and fuses of the same rating as the tool must be used. The system will fail if the earth wire running in a striped yellow and green sheath has:

- not been connected to, or become disconnected from, the plug
- become damaged in the lead to the machine or the trailing cable
- become disconnected from the casing of the machine or come loose.

For more detailed information about electricity for powered tools see the Health and Safety at Work Act 1974 and the Electricity at Work Regulations 1989.

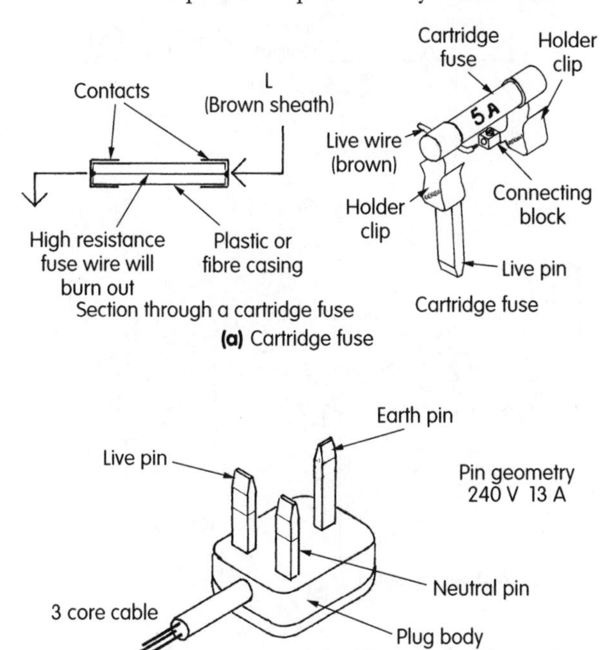

(a) Cartridge fuse

(b) Three pin 13 amp plug

(c) Three pin connection to a power tool

(d) Earth wire symbol

Figure 8.2 Three pin plug system

8.4 DOUBLE INSULATED POWER TOOLS

Electrically driven power tools are now manufactured with double insulation. Figure 8.3. shows how a double insulating barrier is formed round all the components in the tool that carry an electrical current. The double insulation is achieved by constructing the body of the tool from non-conductive materials and further isolating the metal inner parts carrying an electrical current by an inner lining of a similar non-conductive material. Portable power tools which are double insulated have their specification plates stamped with the kitemark and BS numbers 2745 and 2769 (Figure 8.3). Double insulated power tools can be used without earth leads under the Electricity at Work Regulations 1989, provided they are manufactured to the above British Standards. Although double insulated power tools are safer than single-insulated and earthed tools, there is still danger to the operator if the cables carrying the electricity supply at 240 V are faulty.

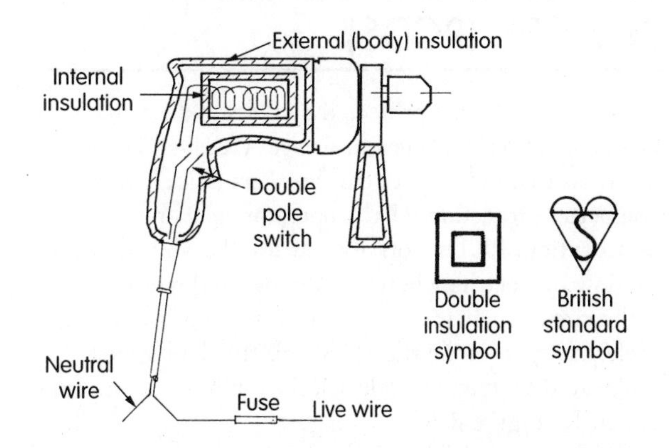

Figure 8.3 Double insulation and symbols (BSS and double insulation)

8.5 110 V POWER TOOLS

To reduce further the danger from electric shock from tools or faulty cables, the electrical pressure (volts) is reduced via a transformer from 240 to 110 volts, the specified voltage for the use of power tools on building sites. Figure 8.4 shows a single-insulated power tool supplied by a 110 V transformer with a centrally tapped earth, so that if the body of the tool becomes live the operator will only receive an electric shock of 55 V.

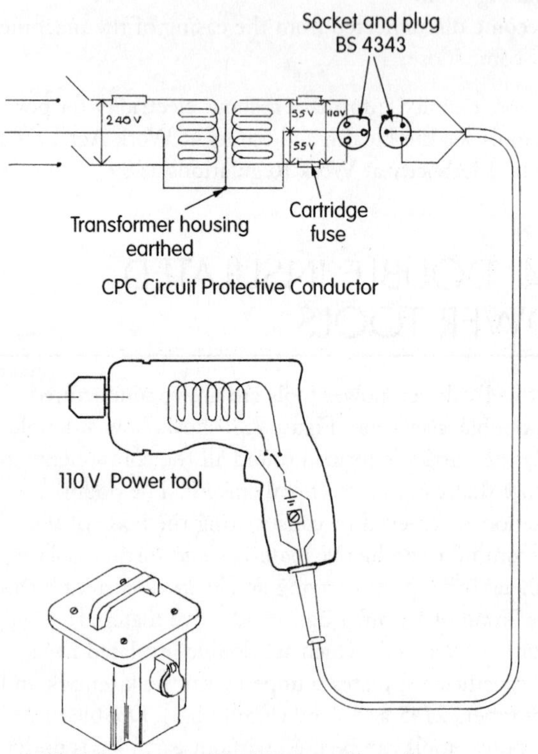

Figure 8.4 110 V transformer

8.6 RESIDUAL CURRENT DEVICES (RCDS)

Electrical circuits are protected by fuses and circuit breakers, but residual current devices (RCDs), formerly known as earth leakage circuit breakers (ELCBs), provide better protection for the operator against electrocution. RCDs work by detecting an imbalance in the flow of electricity between the live and neutral wires in the circuit such as would occur if there was a leakage of electricity to earth; they then trip the double pole switch to disconnect the electrical supply almost instantaneously (Figure 8.5a).

RCDs are available:

- as socket outlets in a ring main system
- as adapters fitted between the plug on the tool lead and the socket outlet
- combined with the plug, being permanently attached to the tool lead.

Adapters are not recommended as they may inadvertently be left out of the system. As RCDs depend on the functioning of their electrical and mechanical components, it is necessary to test them at frequent intervals and before each use. A test and reset button, together with an indicator light, are incorporated in the unit so that the test can be easily carried out (Figure 8.5b).

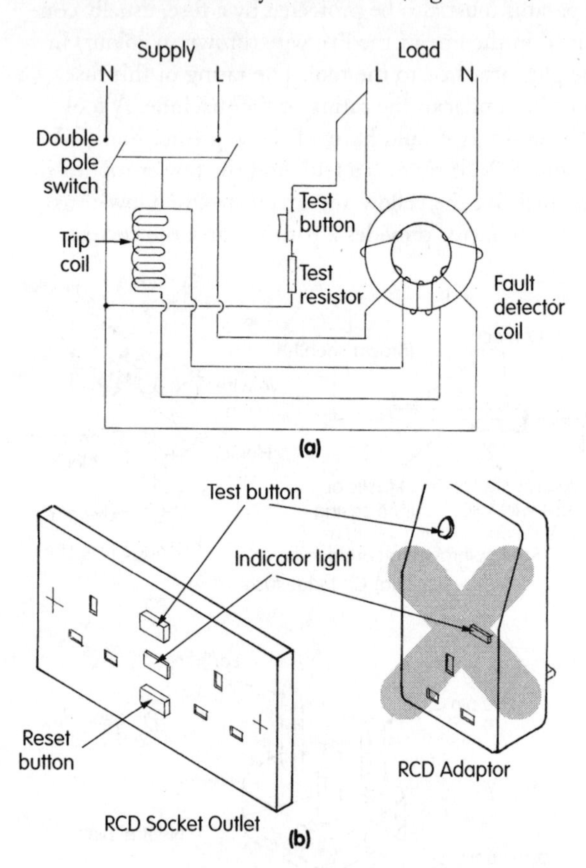

Figure 8.5 Residual current device

8.7 THE SAFE USE OF ELECTRICALLY OPERATED POWER TOOLS

Before using portable power tools the operator must be certain that all necessary precautions have been taken to ensure their own safety and that of others working nearby. The following safety checklist must be applied.

1 You must be familiar with the working characteristics of the tool and be competent in its use and have received supervised training from a designated person.

2 Check that the power tool is visually in good repair and authorised for use.
3 Check that it is fitted with an undamaged plug of the correct type (Figure 8.6).
4 Is the electrical cable of the correct rating and in good condition, with the cable clamps and sleeves correctly and securely fixed?
5 Are all the necessary guards correctly fitted and working easily?
6 Are the cutters and drills being used in good order and sharp?
7 Are trailing cables as short as possible, clear of the floor, protected against traffic and clear of the cutters?
8 Is the necessary protective equipment for head, ear, eye, nose and mouth being worn?
9 Loose clothing should not be worn; ties and shirt tails should be tucked in.
10 Isolate (disconnect) the tool before making adjustments, and changing cutters and drills.
11 If the tool does not function correctly or appears damaged, say so and remove it from use till it can be repaired and tested.
12 If an injury occurs, no matter how slight, seek first aid and report it to the person responsible for safety. Prompt action may help to avoid later complications.

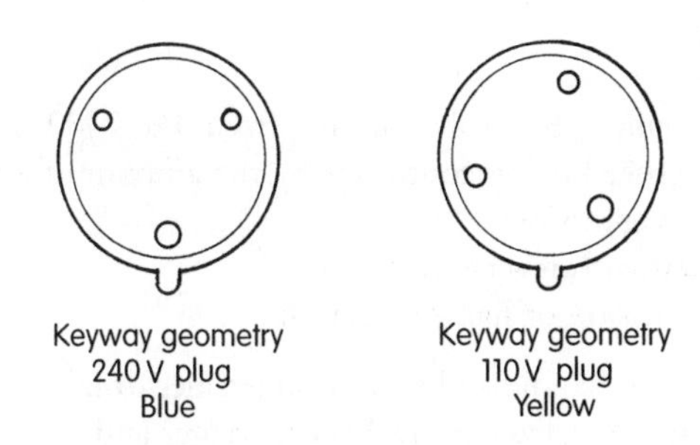

Figure 8.6 Electric plug geometry for transformers

8.8 PORTABLE ELECTRICITY GENERATORS

If for some reason mains electricity is not available, electricity can be provided by the use of portable generators, powered by petrol or diesel engines. Figure 8.7 shows a portable generator giving an output of 4400 watts, 240 V/18 A or 120 V/36 A with a generating frequency of 50 Hz, driven by a 388 cc petrol engine. The generator must be sited to minimise pollution caused by exhaust fumes and noise. It should be sited and installed in accordance with BS 7375 by a competent person (electrician) and must be effectively earthed.

Figure 8.7 Portable electricity generator (with kind permission from Makita UK Ltd)

8.9 USE OF POWER TOOLS

Provided there is a suitable source of power available and the amount of work warrants their purchase, these tools can speed up carpentry and joinery operations. Portable power tools are manufactured for two distinct markets: 'do it yourself', where light and occasional use of the tool is expected, and heavy industrial use. Although the tools carry out the same operations, industrial models have more powerful motors and are of more robust construction to stand up to heavy and constant use.

Power tools are divided by their function into the following categories:

- drills – rotary, rotary and impact (percussion), hammer drills and drill attachments
- screwdrivers and socket wrenches
- sanding machines, belt and pad
- circular saws
- jigsaws
- planers
- routers
- staple and nailing guns.

8.9.1 Drills

The choice of a drill will depend on:

- the amount of use
- diameter and depth of the holes required
- the type of material being drilled and its characteristics.

Figure 8.8a Drill chucks – key-operated and keyless (courtesy Elu)

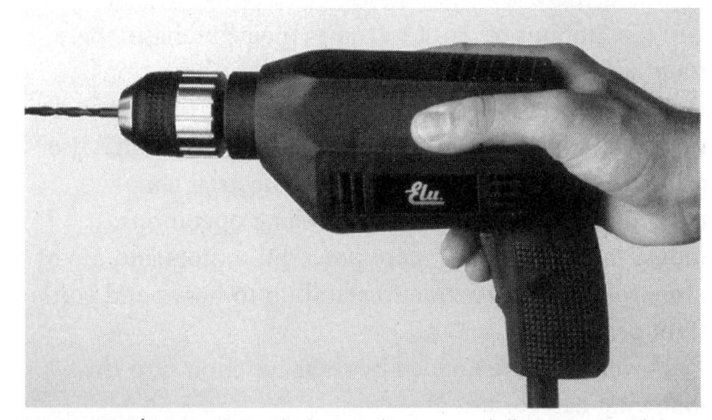

Figure 8.8b Compact lightweight rotary drill (courtesy Elu)

The portable power drill consists of:

- a chuck which is key operated or, in the more recent machines, operated without the use of a key (a keyless chuck, see Figure 8.8a); chucks are of two sizes, the smaller taking drill shanks from 1 to 10 mm in diameter and the larger from 1.5 to 13 mm
- a body containing the drive motor and any incorporated gearing. The power rating of the motors vary from 400 watts for the lighter to 1000 watts for the heavier drills. Drills may include one or more of the following functions:

    1 single-speed drills having a non-load rotational speed of 2400 rpm
    2 dual-speed drills which can be run at 900 and 2400 rpm via gearing engaged by a turnover lever

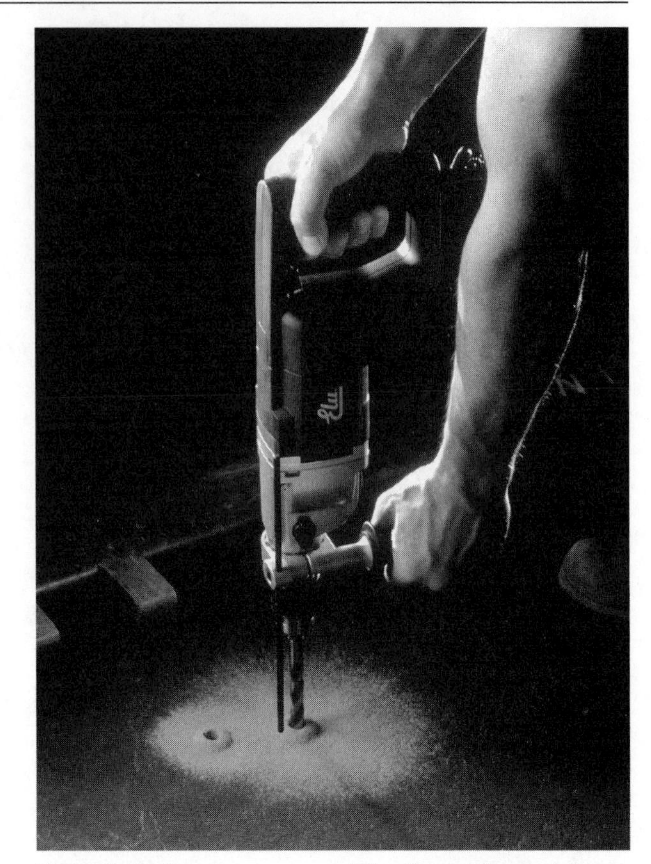

Figure 8.8c Rotary percussion drill with an 'D' grip drill handle (courtesy Elu)

    3 variable speed drills running from 0 to 2400 rpm, the speed being controlled by the amount of trigger depression
    4 drills with reverse rotation
    5 percussion or hammer action

- pistol grip handle and trigger on smaller drills (Figure 8.8b), back or 'D' handle trigger and adjustable side handle on the larger drills (Figure 8.8c).

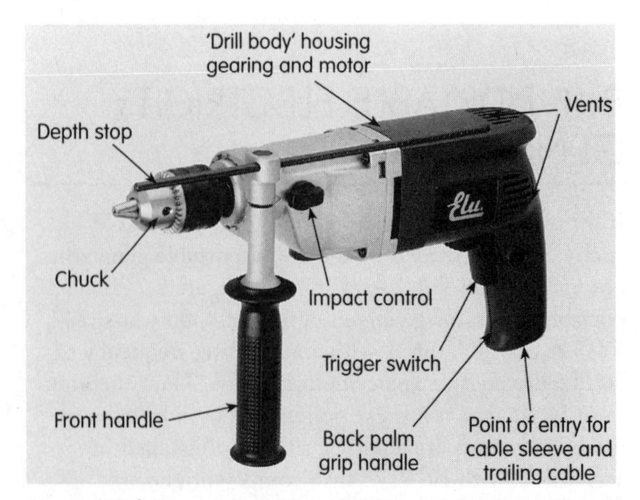

Figure 8.8d Rotary percussion drill (courtesy Elu)

8.9.2 Drill Cutting Speeds

Although drill speeds are quoted in revolutions per minute, it is the cutting speed of the drill bit at its outer edge that is important for effective drilling and this is stated in metres per second (m/s) To determine the cutting speed of a drill we need to know:

- the distance a point on the outer edge of the drill bit travels in one revolution (Figure 8.9)
- the number of times the drill rotates in one second.

The following example shows how the cutting speed for a 10 mm diameter drill rotating at 2000 RPM is determined.

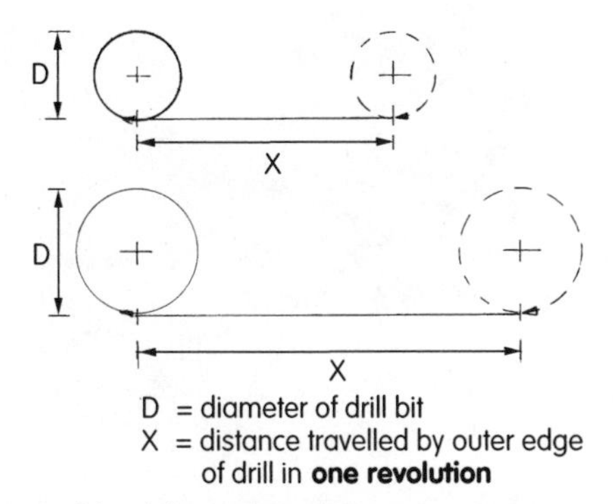

Figure 8.9 Drill speeds

1 The distance travelled by the outer edge of the drill in one revolution = πD

$$3.142 \times \frac{10}{1000} = 0.03142 \text{ m}$$

2 The number of revolutions per second =

$$\frac{\text{the chuck speed in RPM}}{60}$$

3 From the given information, the cutting speed in metres per second (m/s) will be

$$\pi \times 0.01 \times \frac{2000}{60} = 1.05 \text{ m/s}$$

Table 8.1 gives a list of recommended cutting speeds for the drilling of metals of various hardness. Drilling holes in timber usually requires a cutting speed about twice that used in drilling mild steel. Drill bit manufacturers issue tables giving rotational speeds recommended for various sized drills used for drilling holes in a variety of metals and other materials. The general rule is the larger the hole, the slower the drill rotates to achieve the same effective cutting speed.

Table 8.1 Recommended cutting speeds

Material	Cutting speeds (m/s)
Aluminium	1.00 to 1.25
Mild steel	0.40 to 0.50
Cast iron	0.20 to 0.40
Stainless steel	0.15 to 0.20

8.9.3 Using the Drill

If a portable power drill requires both hands to hold the drill, the workpiece will need to be securely held in position by some form of clamping device, so that it does not rotate under the force of the circular cutting action of the drill. Sufficient pressure must be applied to allow the drill bit to cut into the material; this pressure is released when the drill point is about to break out or the drill has to be withdrawn from the hole to clear waste particles (swarf – the metal shavings cut by a drill) created by the cutting action of the drill. If the applied pressure is insufficient to cause the cutting edge of the drill to bite into the material, the drill point will rub rather than cut the material and this frictional contact between the two surfaces will generate heat.

8.9.4 Drill Attachments

Although many attachments are made to allow the drill to be used for other functions, only those that aid the drilling process should be considered for industrial use. Attachments to turn the drill into a lathe or circular saw are intended only for light occasional use by the hobbyist.

a) Angle Drilling

This can be done using an attachment for the drill or by using a purpose-made drill (Figure 8.10). The angle drill allows for the drilling of sides of joists or in corners where space is limited along the axis of the drill bit.

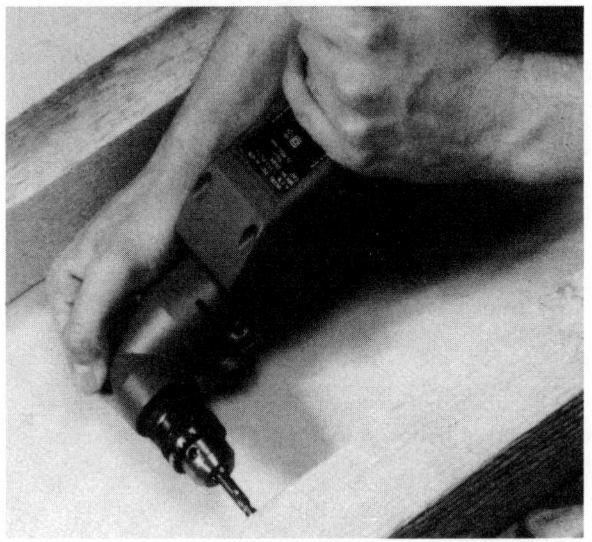

Figure 8.10 Angle drill attachment

b) Drill Stands

These hold the body of the drill in a vertical position and allow for greater accuracy when drilling (Figure 8.11).

NB: A chuck and drill bit guard must be used to conform to the safety regulations under the Health and Safety at Work Act 1974 and other current legislation.

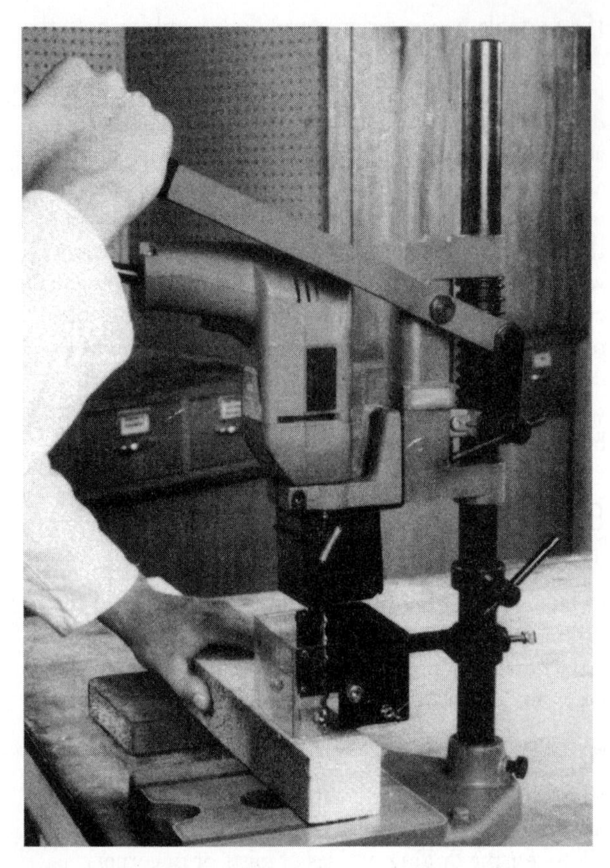

Figure 8.11 Drill and drill stand

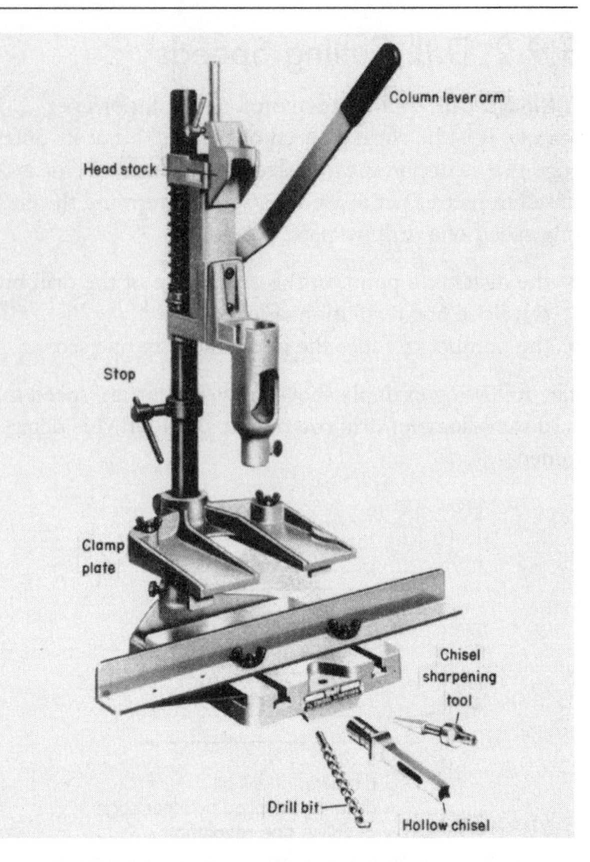

Figure 8.12a Mortising attachment (parts)

c) Drill Stand and Mortising Attachment

This attachment works in a similar way to the drill stand, except that there is a further attachment to hold a hollow square chisel. Inside this an auger bit cuts away the majority of the timber before the square hollow chisel is forced downwards into the material to form a square hole. By moving the material being mortised along a horizontal axis between the downward drill strokes a rectangular slot or mortise can be cut (Figure 8.12a, b).

Operational procedures are the same as those when using a fixed hollow chisel mortising machine (see Section 9.8.1).

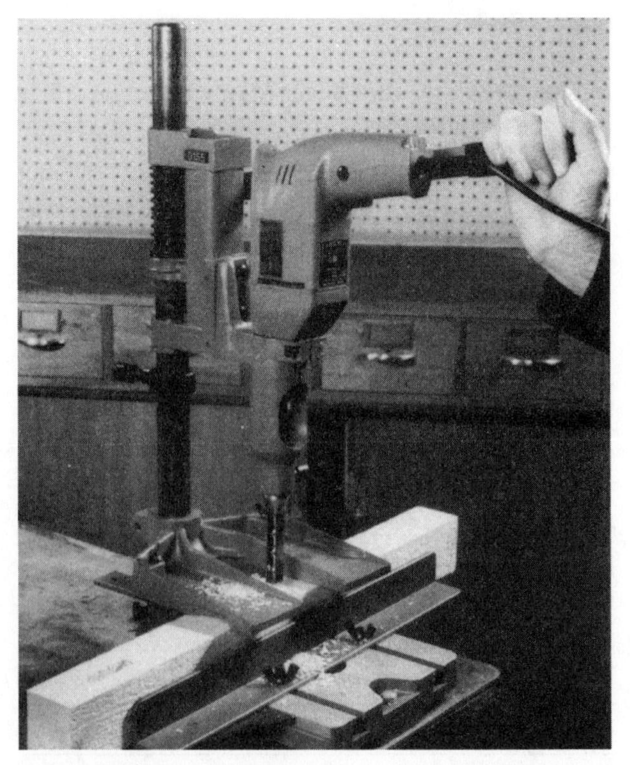

Figure 8.12b Mortising attachment and stand for drill

8.10 POWER OPERATED SCREWDRIVERS

This power tool (Figures 8.13 – 8.15) provides the operator with a fast and virtually effortless way of driving screws. Screwdriver bits are available in different sizes and shapes to engage all types of screw head driving slots and indents (see Table 11.2).

The tool incorporates a bit cover which, when in contact with the surface into which the screw is being driven, disengages a positive clutch with a depth-setting device which allows the screw head to be flush with, or driven below, the surface before disengaging the drive bit. It should be remembered that pilot and clearance holes still have to be drilled before the screw is inserted unless fixing to plaster board, soft panel products or using special types of screws (see Figure 11.12).

There are three types of power screwdriver:

1 drywall screwdrivers (Figure 8.13) with a single clutch setting which allows the screw head to finish flush with the surface when fixing drywall panels (such as plasterboard or softboard) to background studding of timber or light metal sections (self-tapping screws)
2 power screwdrivers with adjustable clutch settings to allow screw heads to be driven to predetermined depths below the surface before disengaging
3 power screwdrivers with automatic screw feeds. The screws are attached to a plastic cartridge belt which automatically feeds a screw into the driving position, where, with the exception of 1 above, it is located in the pre-drilled hole and driven home, all in one operation (Figures 8.15).

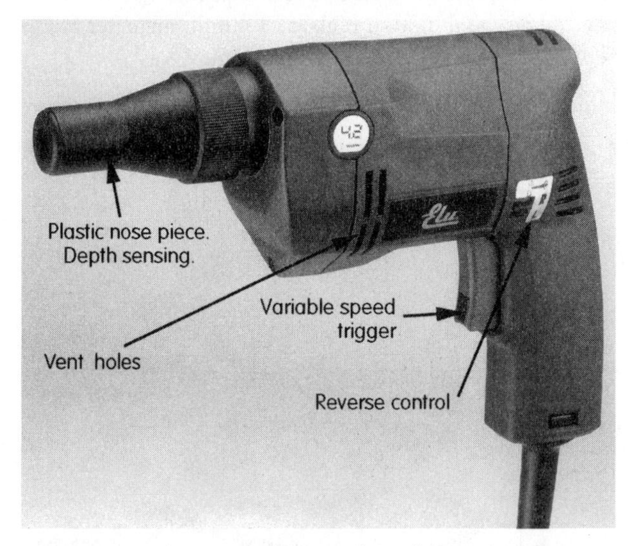

Figure 8.13 Electric screwdriver (drywall) (courtesy Elu)

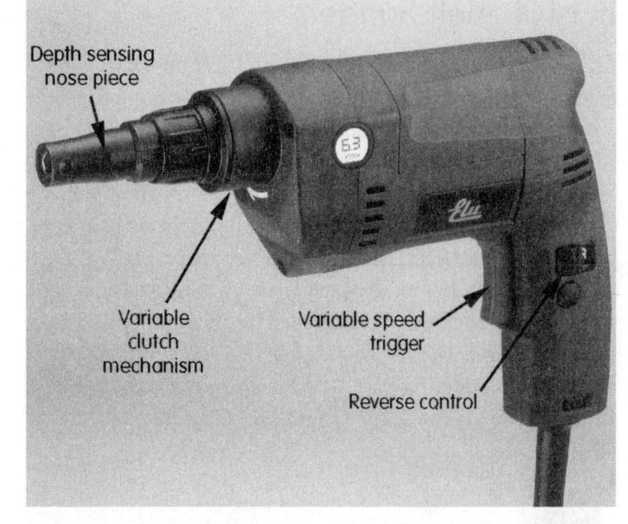

Figure 8.14 Electric screwdriver (variable clutch settings) (courtesy Elu)

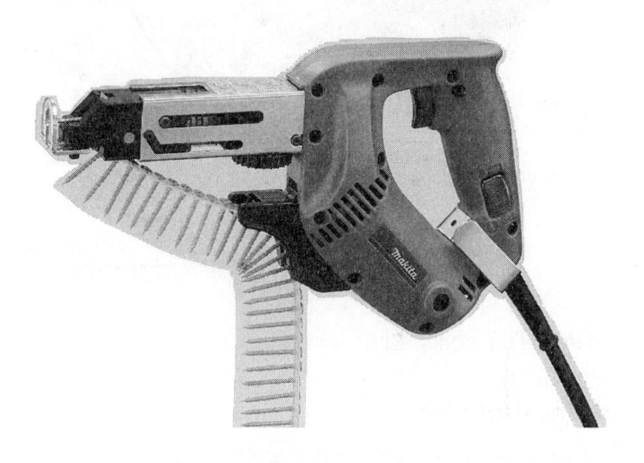

Figure 8.15 Electric screwdriver with automatic screw feed (courtesy Makita)

8.11 POWERED SANDING MACHINES

There are four basic types of sanding machine for bench work:

1 belt sanders
2 orbital sanders
3 random orbit sanders
4 heavy-duty belt and disc floor sanders (these are specialist machines outside the scope of this book).

8.11.1 Belt Sanders

These are designed for heavy bench sanding work. The machine consists of a main body which contains the motor, drive mechanism and dust extractor. An endless abrasive belt is driven by the drive or heel roller and tensioned by the front (toe) or belt tension roller, and runs over a steel base plate faced with a cork or rubber pad situated between the heel and toe rollers. It is this area of the belt below the base plate that makes contact with and abrades the workpiece. The dust generated by this abrading action is discharged via a suction exhaust into a dust collecting bag, which retains the dust while allowing the air to escape (Figure 8.16a).

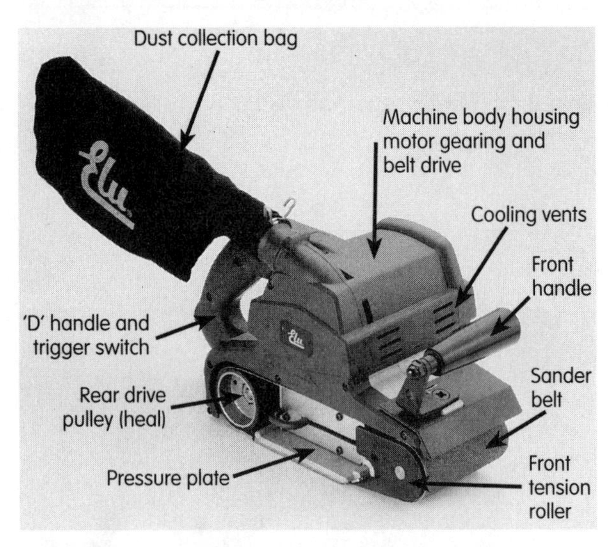

Figure 8.16a Portable belt sander (courtesy Elu)

Figure 8.16b Portable belt sander in use (courtesy Atlas Copco)

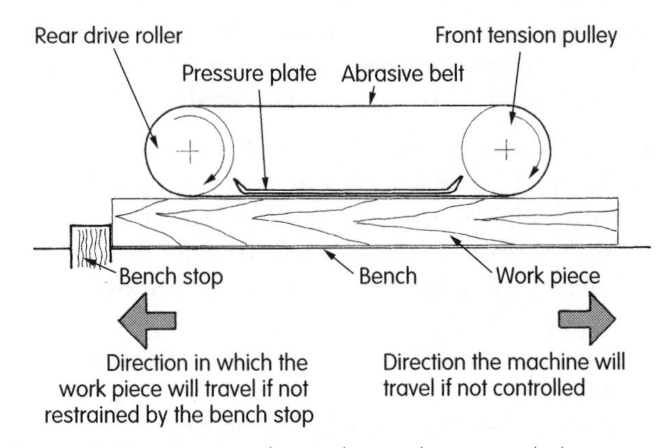

Figure 8.16c Securing the workpiece being sanded

Method of Use

Place the material to be sanded against a bench back-stop. Lift the sander clear of the workpiece and allow the belt to reach its full working speed before gently lowering it down on the workpiece heel first. This will avoid kick-back of the workpiece. As full contact is made there will be a tendency for the sander to move forward and force the workpiece back against the stop due to the frictional grip of the abrasive belt. Sanders of this type need to be held with a firm grip to control forward movement (Figure 8.16b). The surface finish of the workpiece will depend upon what grade of abrasive has been used (see Chapter 5) and the skill of the operator in traversing the machine over the workpiece.

8.11.2 Orbital or Finishing Sanders

The orbital sander takes its name from the orbital path traced by the abrasive pad when working. The machine consists of a body containing the drive motor, a pistol grip handle and trigger at the rear. Attached to the body is the base which houses the mechanism that provides the orbital motion, the fixing clips for the abrasive sheets, the sanding pad faced with felt or rubber, a front handle and a dust-collecting mechanism. In this type of machine, sheets of abrasive paper are stretched over the base pad and held in place by front and rear clips attached to the base. The orbital motion is about 5 mm in diameter at between 12 000 and 14 000 RPM (Figure 8.17a).

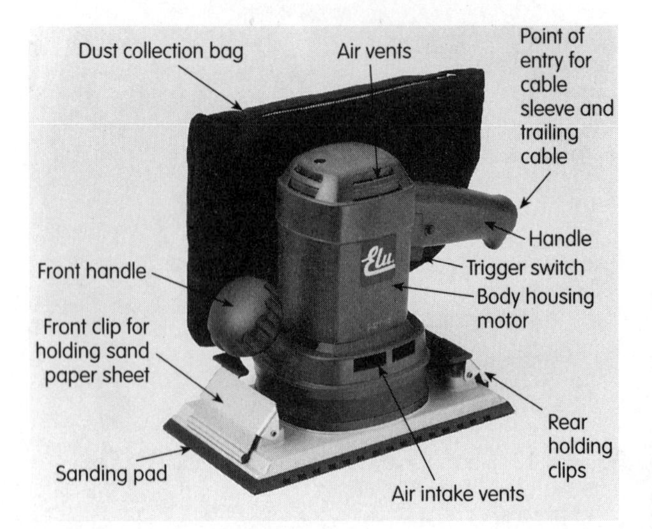

Figure 8.17a Orbital sander (courtesy Elu)

Method of Use

Start the sander and lightly apply to the surface being sanded. The machine's own weight should provide sufficient pressure for sanding to take place. Excessive pressure may cause scratching of the surface and clogging of the abrasive grit and the body of the sander may rotate while the pad remains stationary. Select the correct grades of abrasive paper for the finish required (see Table 5.6). This will involve changes of paper during the sanding operation. Although orbital sanders are fitted with dust extraction, dust is a hazard, and an approved mouth and face mask should always be worn.

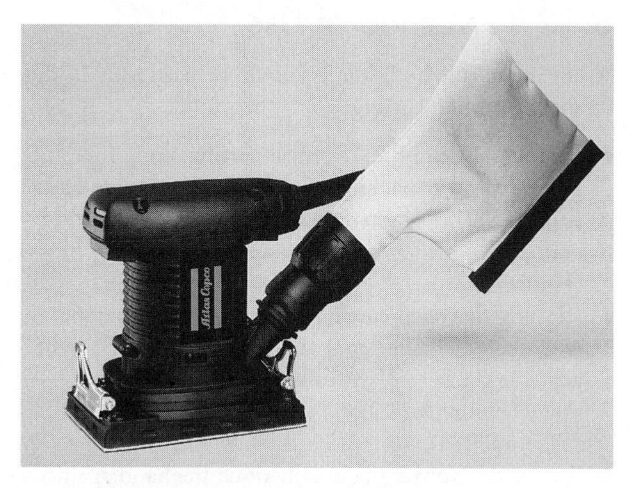

Figure 8.17b Random orbital palm grip sander (courtesy Atlas Copco)

8.11.3 Random Orbit and Palm Sanders

The random orbit sander (Figure 8.18) works with an eccentric (offcentre) action plus rotation. The base of the sander is circular with a flexible sanding pad which is perforated to allow the dust to be drawn into the collecting bag or extraction system. The self-adhesive perforated sanding discs have a peel-off backing for attaching to the base pad. As an alternative, some sanding discs have a 'Velcro' backing for easy fixing and removal.

Palm grip sanders are similar to random orbit sanders, but the sanding base rectangular and instead of a pistol grip handle, there is a palm grip on top of the motor casing (Figure 8.17b).

8.12 POWERED CIRCULAR SAWS

The handheld portable circular saw with blade diameters of 175–250 mm gives vertical cuts to a depth of 65–90 mm and angle cuts to reduced depths. Figure 8.19a gives

Figure 8.18 Random orbital sander (with kind permission of Makita)

the names and functions of the various parts of the saw. These saws are used for cutting:

- solid soft and hardwood timbers
- a range of wood-based manufactured boards
- plastic laminates.

Table 8.2 is a general guide to the types of circular saw blade available for these machines and the materials the various teeth profiles will cut best. The quality of the cut, the ease of cutting and safe operation of the saw will depend on the saw blade teeth being of the correct profile, sharp and adequately set, if the teeth are not tungsten carbide tipped.

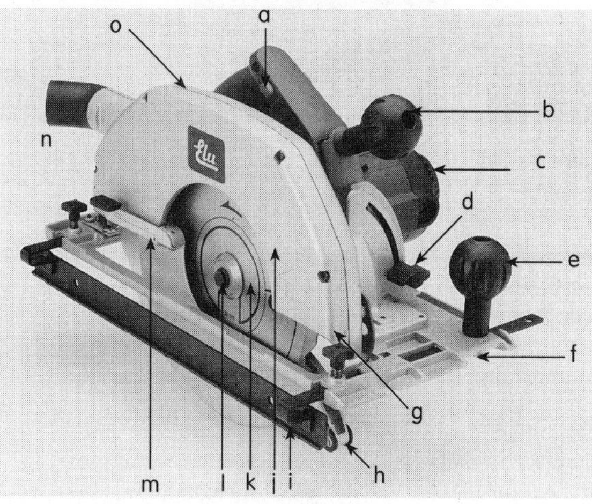

Figure 8.19a Portable powered circular saw (parts): **a** – back handle and trigger switch; **b** – front body handle; **c** – motor and motor housing; **d** – scale and adjustment for bevelled cutting; **e** – front handle on base plate; **f** – base plate; **g** – top guard; **h** – antifriction rollers on bottom guard; **i** – adjustable fence; **j** – direction of rotation of saw blade indicator; **k** – saw blade; **l** – spindle bolt and collar; **m** – spring-loaded retractable bottom saw blade guard; **n** – waste extraction nozzle; **o** – depth of cut adjustment behind top guard (courtesy Elu)

NB: The blade in a portable saw has an anti-clockwise rotation and cuts in an upward direction (Figure 8.20). The direction of the saw rotation is shown by an arrow on the top guard and on the saw blade itself.

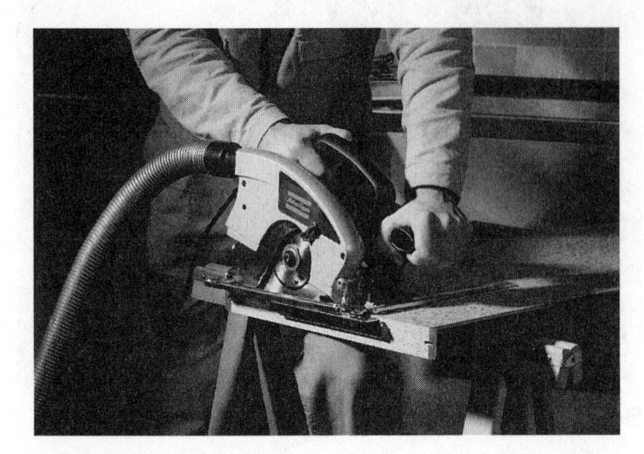

Figure 8.19b Circular saw in use (crosscutting) (courtesy Atlas Copco)

Figure 8.19c Circular saw in use (bevel crosscutting) (courtesy Atlas Copco)

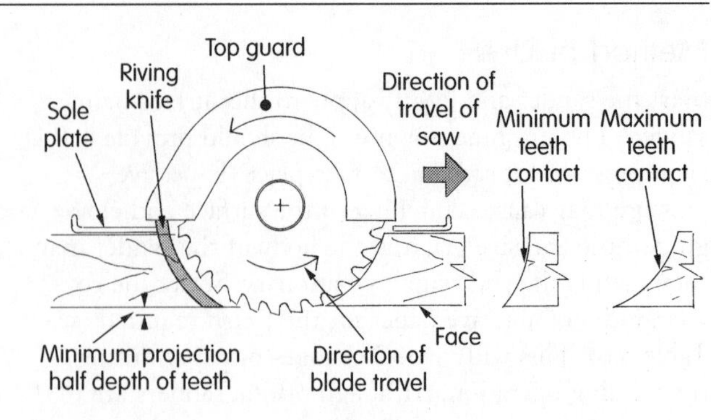

Figure 8.20 Circular saw (blade rotation and setting)

8.12.1 Method of Use

As a general rule both hands should be used to hold this type of saw when cutting.

- Ripping (Figure 8.21) vertical cutting along the grain) uses the machine's own adjustable guide fence (Figure 8.21a) or a temporary guide fence tacked or cramped to a wide board or panel that is being cut (Figure 8.21b).
- Ripping (bevelled cutting along the grain) is done with the saw blade set at an angle other than a right angle to the base plate, using the machine's own guide (Figure 8.21c) or a temporary guide (Figure 8.21d).
- Square crosscutting is usually done freehand against a guide line with the saw blade in the vertical position or using the machine's own fence as shown in Figure 8.19b.
- Bevelled crosscutting (Figure 8.19c) can be single (bevelled on face, square on edge) or compound (bevelled on both face and edge) (Figure 8.21e). For single bevel cutting the blade is set in the vertical position and for compound cutting it is set at an angle to the base plate.

Table 8.2 General guide to circular saw blades

Blade type	Suitable for cutting	Operation	Remarks
Combination blade	Softwood and hardwood	Ripping and crosscutting	General purpose
Crosscut blade	Softwood and hardwood	Crosscutting	Fine finish
Composition and wall board blade	Composite boards of most types and aluminium sheet		Fine-toothed blade
Flooring blade	Reclaimed timber	Capable of cutting the occasional nail	Specially tempered blade
Planer and mitre blade	Softwood and hardwood	Ripping and crosscutting	Produces an almost planed-like finish to the sawn edge
Tungsten-carbide-tipped blades (TCT)	Soft and hardwoods manufactured boards, laminates, etc.	Ripping and crosscutting	Sawn finish depends on the tooth profile and pitch*

Note: *See Section 9.4

- Mitre cuts are similar to single bevels, with the guide template set at 45°.
- Rebating is done with two saw cuts along the grain, to a pre-determined depth and at right angles to each other (Figure 8.21f).
- Plough grooves consist of a series of parallel saw cuts to a pre-determined depth and as close together as possible along the grain on the edge of the timber (Figure 8.21g).
- Trenching consists of a series of parallel saw cuts across the grain of the timber to a pre-determined depth and as close together as possible, using a temporary fence or jig to guide the saw (Figure 8.21h). Saw kerfing can be used as an effective way of bending timbers to a radius.

When using the saw the depth of blade below the sole plate of the machine should be such that the teeth should project about half their depth below the underside of the material being cut (Figure 8.20). This allows the maximum number of teeth to be in contact with the material and reduces the angle of cut and the tendency of the teeth to splinter the material at the edges of the cut as they cut out through the upper surface. If possible, it is better to cut the material with the face side down.

8.12.2 Safe Operation

1. Check the machine and adjust the fences and guards to their correct positions for the work being executed, before connecting the machine to the power supply.
2. Adequately secure the workpiece and any necessary jigs or templates.
3. Allow the full rotational speed of the blade to build up before starting to cut.
4. Always follow through the cut rather than draw the saw back while cutting.
5. Keep both hands on the machine handles while the blade is in motion.

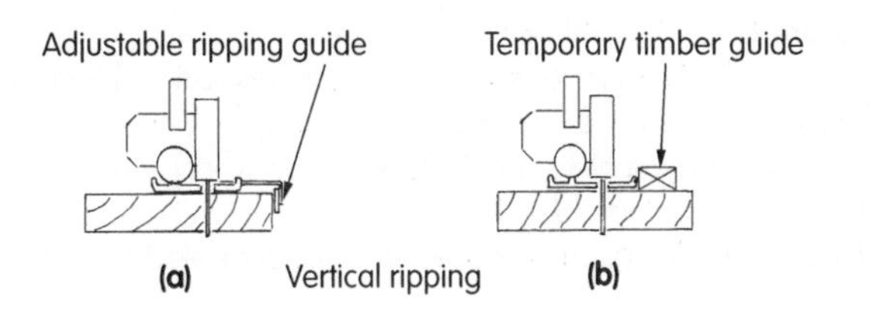

Adjustable ripping guide
Temporary timber guide
(c) Bevelled ripping (d)

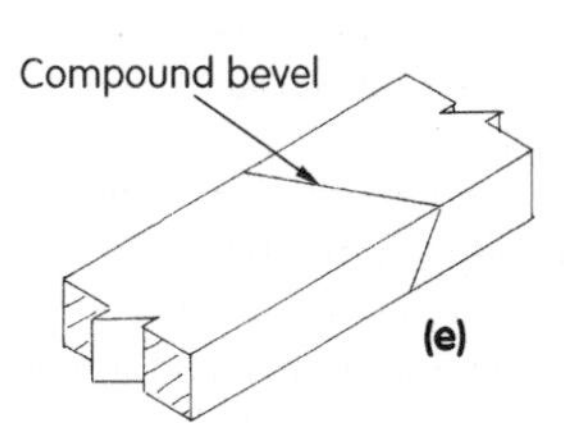

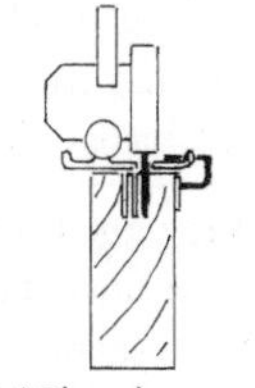

Temporary timber guides
Trench formed by saw cuts
(h) Multi-kerfing to form trench

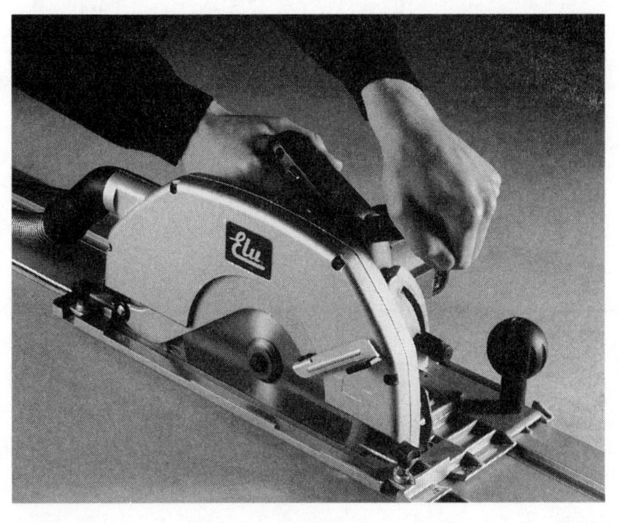

Figure 8.21 Use of portable powered circular saw for vertical ripping, with associated guide fences. Notice provision for dust extraction

6 Make sure the springloaded bottom guard moves freely and operates at all times.
7 Disconnect the machine from its power source before making adjustments.
8 Use only those saw blades recommended by the machine manufacturer for the particular material being cut.

8.13 POWERED JIGSAWS

A powered jigsaw (Figure 8.22) drives a bayonet-shaped blade with teeth cut along its leading edge, using a reciprocating action (backwards and forwards, or up and down movement). In addition, the blade can have a pendulum motion, so that it moves away from the cutting face on the downward non-cutting stroke (Figure 8.23). This action helps clear the waste from the saw cut and prolongs the life of the blade by reducing friction on the downward stroke. Cutting occurs only on the upward stroke of the blade.

Industrial standard jigsaws with a power rating of 500–600 watts will cut timber up to 60 mm in thickness and metals of various hardness from 2 to 10 mm thick, provided the correct type of blade is used. Table 8.3 gives a general guide to the types of jigsaw blade available. There has been some rationalisation in the design of the end fixing of the blade into the machine and most makes of machine use the Euro pattern fitting, but one or two machine manufacturers use an alternative pattern (Figure 8.24).

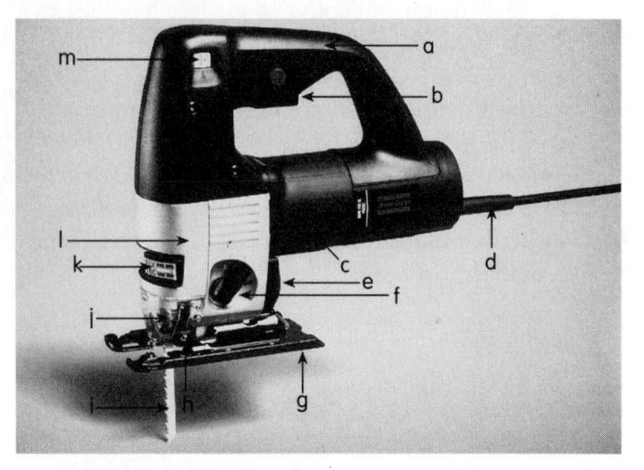

Figure 8.22 Portable powered jigsaw
a – handle; **b** – trigger switch; **c** – motor housing; **d** – cable sleeve and cable; **e** – dust extraction port; **f** – blade pivoting lever; **g** – adjustable sole plate; **h** – thrust wheel; **i** – saw blade; **j** – blade holder; **k** – sawblade quick release locking mechanism; **l** – gearbox cover; **m** – variable blade speed

Table 8.3 General guide to the various types of jigsaw blade

Blade type	Description of material and thickness	Teeth per 25 mm	Tooth pitch (mm)	Blade length (mm)
Bi-metal high performance blades HSS teeth welded to a tough flexible blade body	Metal up to 2.3 mm thick	24	1.0	75
	Metal 1.2 to 5.0 mm thick	18	1.4	75
	Metal over 3 mm thick and thin wood	14	1.8	75
	Fine cutting wood and plastic	10	2.5	100
	Fast cutting wood fibre board and plywood	6	4.1	100
	Down cutting blade for laminated plastics surfaces	10	2.5	100
Standard blades for cutting wood	Very fine finish	10	2.5	100
	Fine finish fast cut	6	4.1	100
	Coarse cut	8	3.0	100
	Fine finish thin material	12	2.0	76
	Fast cut thin material	12	2.0	76
	Curved cutting thin material	12	2.0	76
	Fast cut fine finish thick material	6	4.1	100
	Curved cutting thick material	6	4.1	76
Standard blades for cutting metal	Non-ferrous metal up to 10 mm thick	13	1.9	100
	Non-ferrous metal 3.6 mm thick plastic	8	3.0	100
	Curved cutting. Metal 3.6 mm thick	8	3.0	100
	Stainless steel 1.4 to 4 mm thick	21	1.2	76
	Very thin metal	36	0.7	76
Standard blades special purpose	Tungsten coated for glass fibre			76
	Double cutting blade for superior wood finish			76
	Coarse tungsten coated blade for slates			76
	Tungsten tipped teeth	6	4.0	96

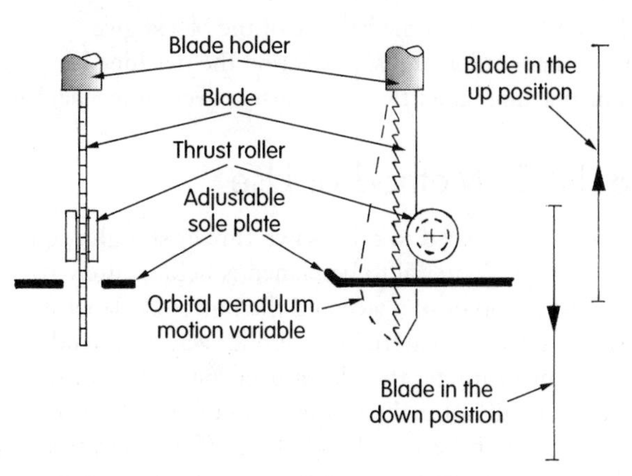

Figure 8.23 Orbital motion of the jigsaw blade

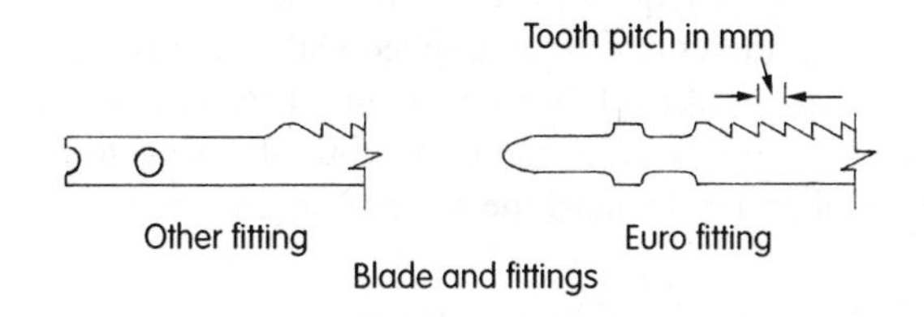

Figure 8.24 Types of jigsaw blade fitting

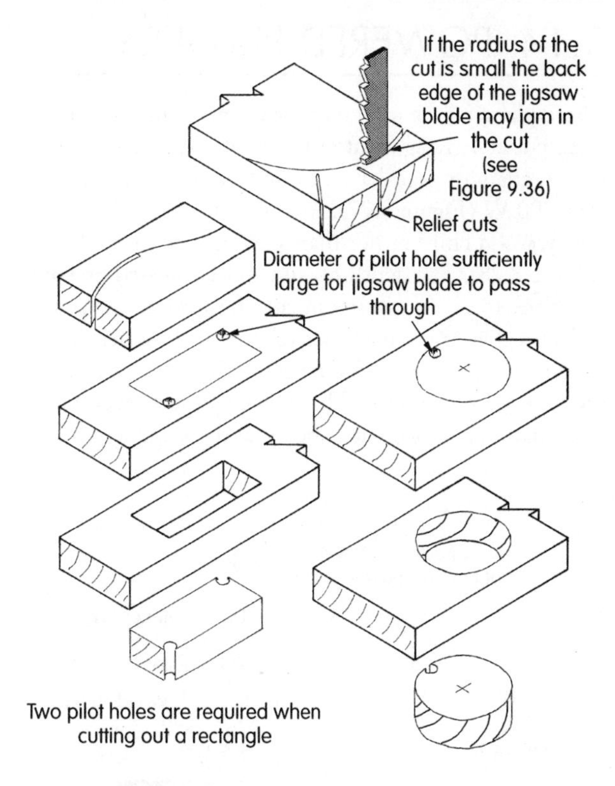

Figure 8.25b Cutting curves with a jigsaw

8.13.1 Method of Use

The saw can be used for straight cutting along and across the grain of timber and other wood-based materials, but it is mainly used for freehand curved cutting or curved cutting using a jig (Figure 8.25a, b). When cutting a curve with a small radius which may trap the blade, make a series of relief cuts before cutting to the outline of the curve. The cut can be vertical or angled by adjustment of the base plate.

8.13.2 Safe Operation

1 Hold the workpiece securely and make sure there is no obstruction of the saw blade as it projects below the underside of the cut. Remember that the part of the blade projecting below the workpiece is unguarded.
2 Inspect, set up and adjust the jigsaw before connecting to the power supply.
3 Allow the saw blade to reach its full cutting speed before starting the cut.
4 Use pre-bored holes to start internal cuts. Do not attempt to make plunge cuts (Figure 8.26).
5 Keep your hands on the machine handle and away from the underside of the workpiece while the blade is in motion.
6 Do not put the saw down until the blade is at rest.
7 Keep the trailing lead cable to the rear of the saw and clear of the saw blade.

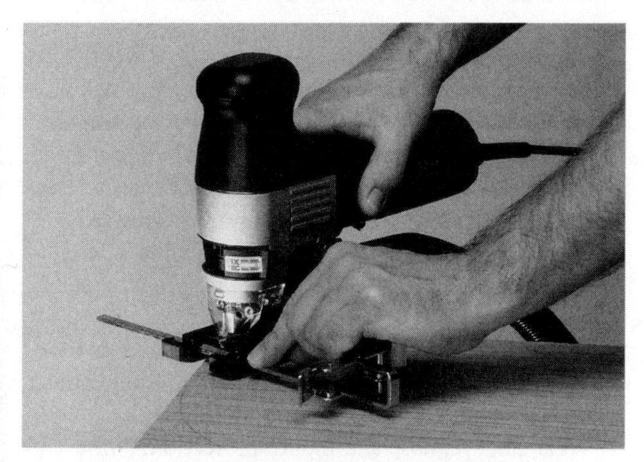

Figure 8.25a Use of the jigsaw (courtesy Atlas Copco)

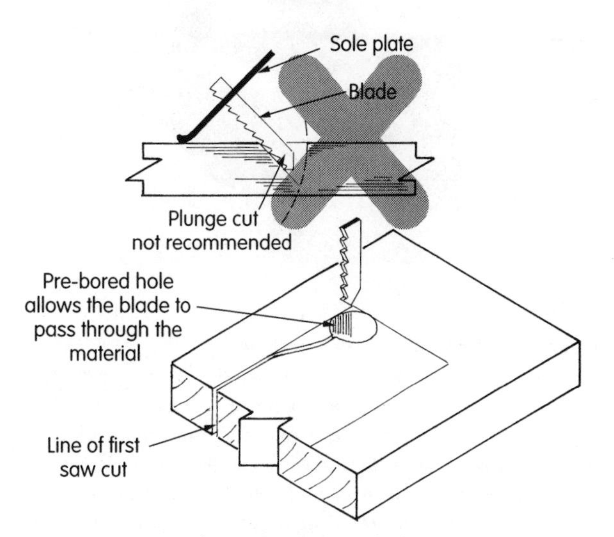

Figure 8.26 Use of pre-bored starter holes

8.14 POWERED PLANERS

These planers have a rotary block containing two or three cutters, giving a cutter width of 80 mm. The block is belt, chain or gear driven by an electric motor of 500–800 W (Figure 8.27). As the cutting action is rotary, wood chips rather than shavings are formed (Figure 8.28). The depth of cut can be adjusted on some machines to give a maximum cut of 3 mm in one pass.

With the larger machines rebates up to a depth of 24 mm can be formed. As the cutters are set in a circular rotating block they should have equal projection and be kept sharp, with each cutter being of the same weight.

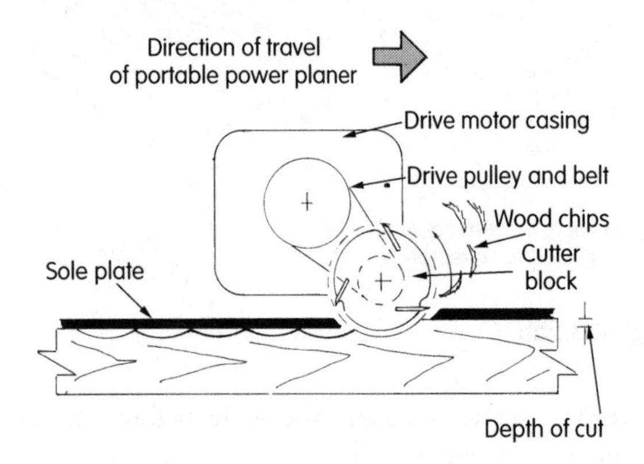

Figure 8.27 Portable powered planers (rotary cutting)

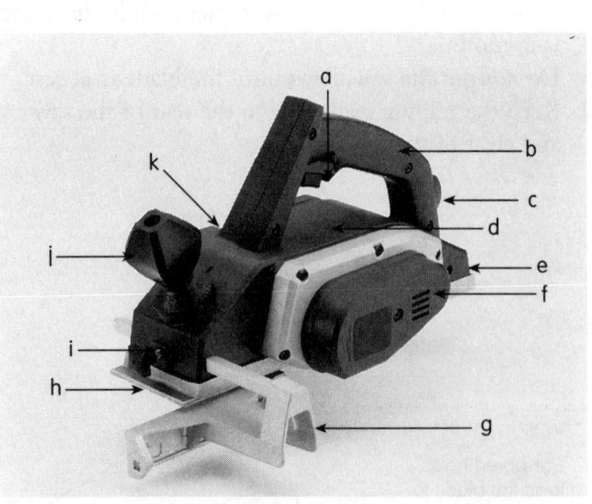

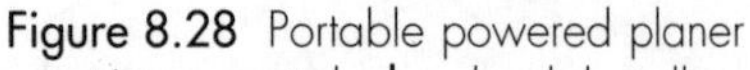

Figure 8.28 Portable powered planer
a – trigger switch; b – back handle; c – housing for cable sleeve; d – motor casing and motor; e – back sole plate; f – housing for drive belt; g – guide fence (adjustable); h – front sole plate (adjustable); i – locking handle for guide fence; j – front knob and adjustment for front sole plate; k – adaptor for shaving collection (hidden by body of plane)

This will maintain the balance of the block, give smoother cutting and save wear on the machine bearings. The machines are fitted with dust collection mechanisms.

8.14.1 Method of Use

The planer is used to dress sawn timber straight, square and to size. Material to be planed is held securely against a stop or in a vice and the powered planer is moved over the material in a similar way to a hand plane. Just running the planer over the surface of the material will not make it square, straight or true. You need to eye the material and plane off the high spots till the surface is satisfactory. One edge is then planed square and true to the dressed face, before the material is planed to the required width and thickness. Using temporary guides or fences supplied with the machine, rebates can be formed. Some machines have one or two V-shaped grooves cut in the front adjustable shoe to help in aligning the machine for cutting chamfers.

8.14.2 Safe Operation

1 Inspect and make all the necessary adjustments to the planer and ensure that the cutters are sharp before connecting to the power supply.
2 The workpiece must be secured in a vice or held against a bench stop using a bench knife, made from a piece of broken and sharpened bandsaw blade (Figure 8.29).

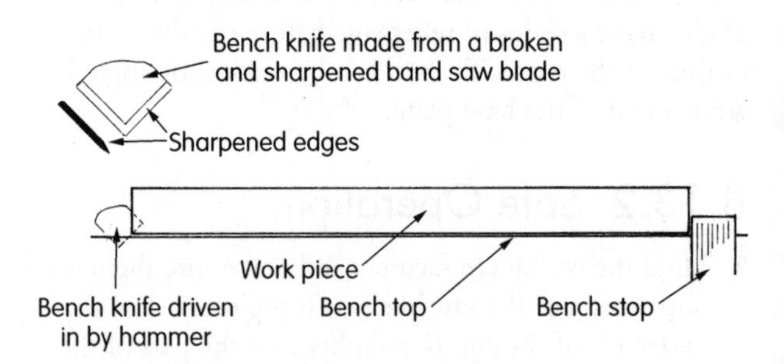

Figure 8.29 Securing the workpiece for planing

3 The machine is held in both hands, as for a jack plane. The hand on the rear handle controls the forward movement and the power switch, while the hand on the front handle (on some machines releases the bridge guard) provides downward or upward pressure, depending on where the machine is in relation to the start and finish of the cut (see Chapter 5 and Figure 5.42a).
4 Do not put the machine down until the cutters have stopped rotating and protect the cutters from contact with abrasive materials when not in use.
5 Wear ear, eye, nose and mouth protection. Noise pressure on the ear can be 90–100 dB.

Figure 8.30 Portable powered planer in use (extraction attachment not shown)

8.15 POWERED ROUTERS

These machines consist of a direct drive motor held vertically in a frame supported on a base (Figure 8.31a). The projecting end of the motor shaft is machined to take 6 or 12 mm collets into which cutters of various types can be fitted (Figure 8.32, Table 8.4). By use of an electrical frequency changer the motor will rotate at speeds of up to 24 000 RPM. Most routers are of the plunge type. This allows the cutter to be brought vertically downwards to contact the material being cut and then to rise out of the material into the body of the machine when the downward pressure is released. (Figure 8.31b). The cutter does not project below the base of the router except when it is working, unlike the fixed router where the cutter is permanently exposed below the base plate. The versatility of the router will depend on the range of cutters, jigs and templates available and the skill and experience of the operator.

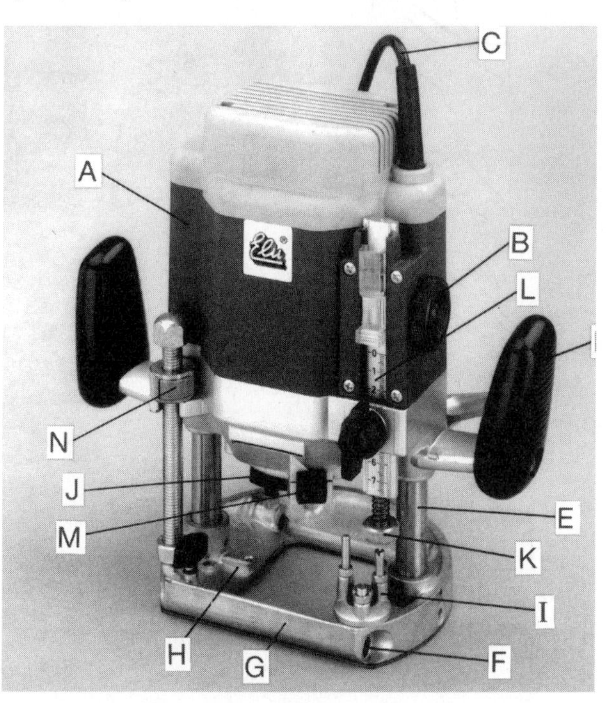

Figure 8.31a Portable powered router (parts)
A – motor and housing; B – depth stop; C – cable and cable sleeve; D – side handles; E – plunge guide bars; F – location holes for fence rods; G – router base; H – template guide housing; I – swivel stops (three depth settings); J – collet chuck (bit holder); K – stop buffer (controls depth of plunge); L – depth scale; M – spindle lock (used when changing cutters); N – adjustable return stop; the hose adaptor for dust extraction is not shown (see Figure 8.31b)

Figure 8.31b A router in use

8.15.1 Types of Operation

Operations that can be carried out with a router (Table 8.4) include:

- straight, curved and moulded grooves
- slots and recesses
- rebates on straight and curved edges
- beads and mouldings on straight and curved edges
- dovetailing, dovetail slots and grooves
- edge trimming of laminates and veneers.

8.15.2 Router Bits and Cutters (Figure 8.32)

For general woodworking use, including manufactured boards, three grades of metal are used in the manufacture of router cutters.

1 Solid tungsten carbide (STC) cutters are formed from a solid piece of tungsten carbide. These are very durable and can be used on wood, wood-based

Table 8.4 Guide to router bits and cutters

Bit/Cutter type	Operation/cutting	Profile	see Figure 8.32
Grooving			
	Ploughing (housing, trenching)	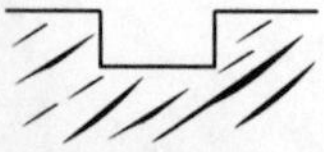	(a)
	Round core-box-bit groove		(b)
	Dovetail groove		(c)
	"V" groove		(d)
Edge forming bits			
	Round over (one quirk ovalo)		(e)
	Ovalo		(f)
	Coving		(g)
	Chamfering		(h)
	Ogee		(i)
	Rebating		(j)
Trimming			
	Square edge (flush)		(k)
	Chamfer (bevelled)		(l)
Edge cutters			
	Sawing		(m)
	Slotting	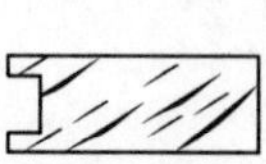	(n)

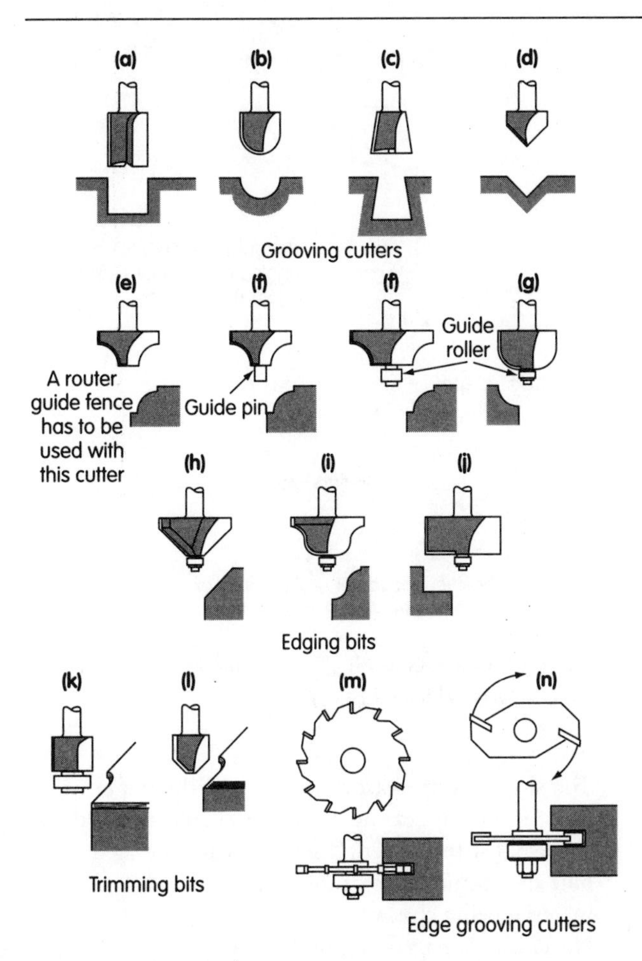

Figure 8.32 Types of router cutters

materials and plastics without being damaged by hard glue lines and abrasive materials.

2 Tungsten carbide tipped (TCT) cutters have tungsten carbide tips brazed onto the steel body of the cutter to form the cutting edge; they can be used with the same materials as STC cutters.

3 High speed steel (HSS) cutters are suitable for use on non-abrasive materials, e.g. softwoods, some hardwoods and certain plastics. They are the least expensive but require more frequent sharpening than the others.

8.15.3 Cutter Pins and Roller Guides

Edge trimming and moulding cutters have a pin or roller guide attached to the end of the cutter to follow the edge profile. The roller guide is easier to use than the fixed pin and is less likely to mark the edge with friction burns as it moves along (see Figure 8.32).

8.15.4 Accessories for the Router

These may include the following:

- a straight fence (Figure 8.33a)

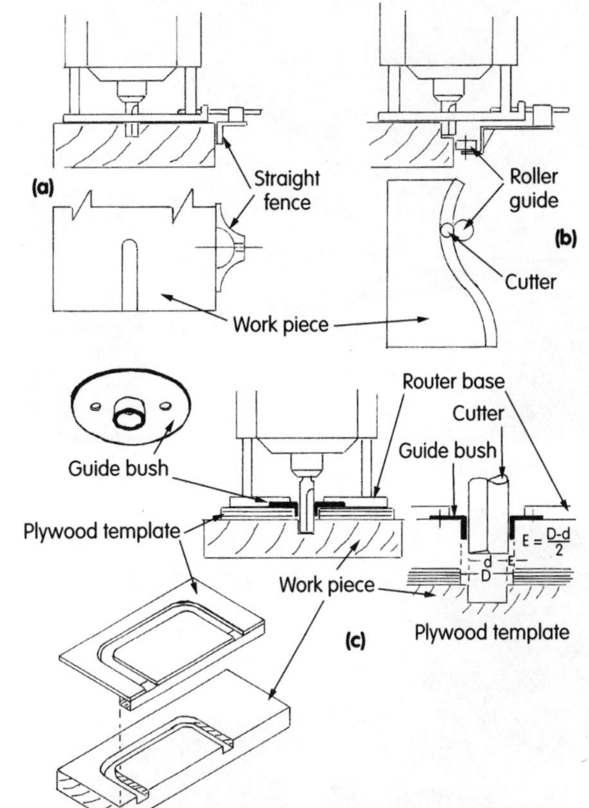

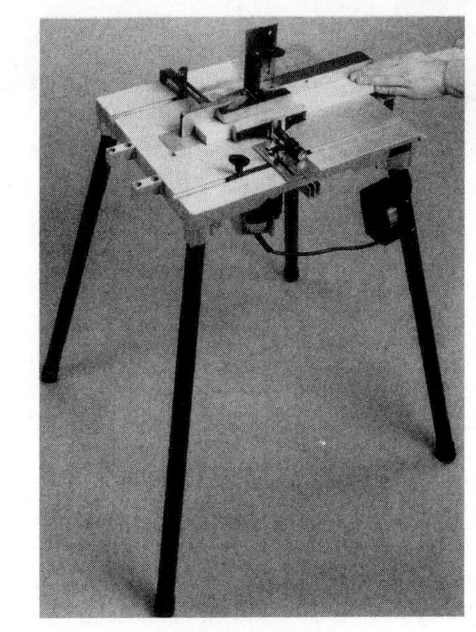

(d) Router table

Figure 8.33 Router accessories

- a roller guide (Figure 8.33b) for curved work
- guide bushes used in conjunction with jigs and templates (Figure 8.33c)
- dovetail templates (Figure 8.34)
- ellipse and circle cutting jigs (Figure 8.35)
- router table (Figure 8.33d).

Figure 8.34 Dovetail attachment (courtesy Elu)

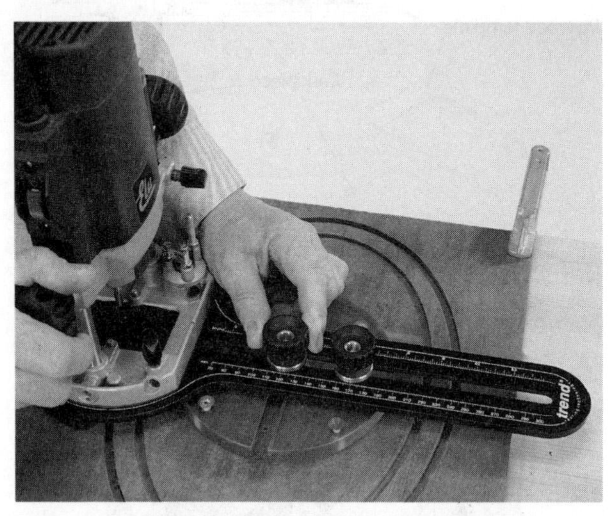

Figure 8.35 Attachment for cutting ellipses (courtesy Trend)

8.15.5 Safe Operation

1 Inspect and adjust the cutters and fences, including the adjustment of the plunge depth of the cutter. Check the cutter is revolving freely before connecting to the power supply.
2 Securely fix the workpiece to the bench or in the vice.
3 Start the motor and allow it to reach its working speed before engaging the cutter with the workpiece.
4 Deep cuts should not be made in a single pass, but as a series of shallow cuts. In cutting slots and grooves the depth of cut should be no greater than half the width of the cutter (Figure 8.36).
5 Move the router from left to right so that the bit cuts the edge of the material against its rotation (Figure 8.37).
6 Move the router along quickly enough to make a continuous cut without overloading the motor (causing it to slow down). Listen to the noise the motor makes. A reduction in speed of rotation of more than 20% will shorten the life of the motor.

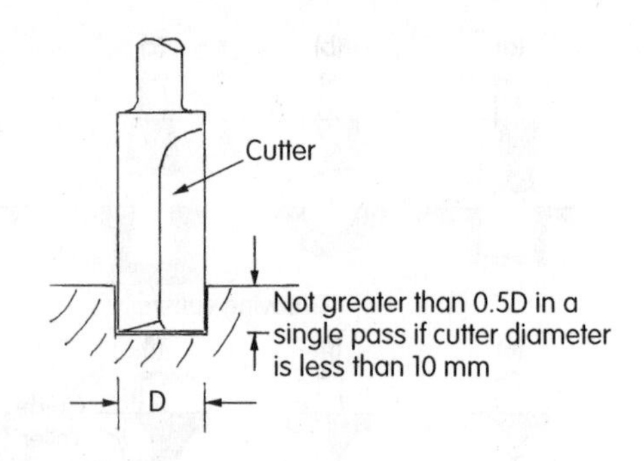

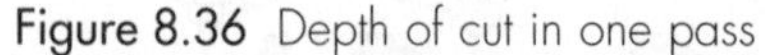

Figure 8.36 Depth of cut in one pass

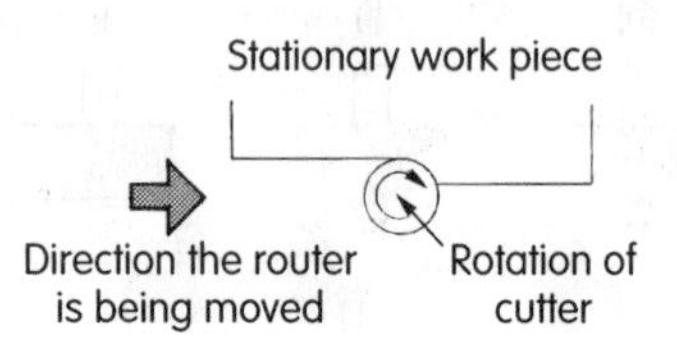

Figure 8.37 Direction of rotation of cutter in relation to the workpiece **(viewed from above)**

7 Too slow or stop–start movements of the router will cause frictional heating as the cutter rubs rather than cuts against the material and the heat generated will blue the cutting edge of the cutter and scorch the material being cut.
8 Never use damaged or blunt cutters.
9 Make sure the cutter is stationary before putting down or leaving the router.
10 Make sure the router and air intakes are clean and that the plunge mechanism moves freely.
11 Wear goggles, ear defenders and dust masks. Use the router's own dust collection bag and mechanism.

For further guidance consult The Supply of Machinery Safety Regulations 1992 and The Provision and Use of Work Equipment Regulations 1992.

8.16 STAPLER AND NAILING (TACKING) GUNS

These portable, electrically powered tools are used to tack fabrics and flexible plastic sheeting to background timber framing. An example would be the fixing of moisture and vapour barriers to timber partitioning. The larger machines will drive large staples and nails up to 25 mm long into timber and are much quicker than using a hammer.

They are used mainly for fabricating timber components in factories. An example of this would be domestic fencing panels. The staples and nails are loaded automatically into the gun. Figure 8.38 shows a typical electrically powered hand-operated tacker gun.

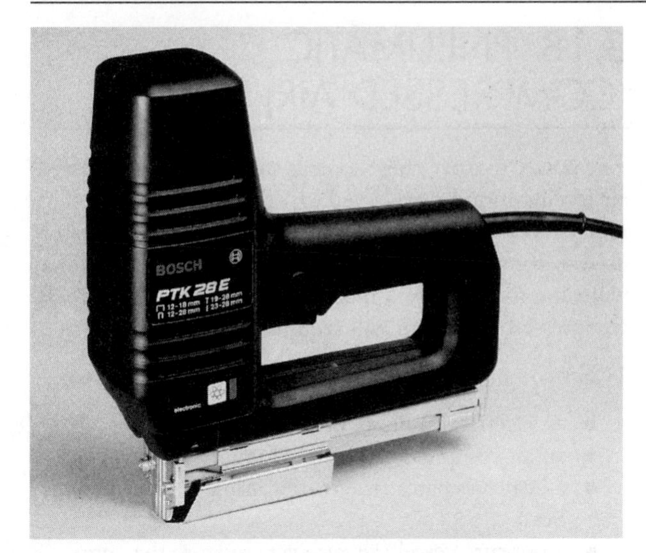

Figure 8.38 A Bosch PTK 28-E tacker (courtesy Robert Bosch Ltd)

8.16.1 Safe Operation

1 Inspect, adjust and load the staples or nails into the gun before connecting to the power supply.
2 These tools are dangerous if not used correctly and the nose mechanism must not be tampered with as it ensures that the tool cannot be operated until the lock-off mechanism is released by pressing the nose against the material being fixed.
3 Wear goggles and ear protectors.

Figure 8.39 Battery charger and battery for a cordless tool (courtesy Elu)

8.17 BATTERY-OPERATED (CORDLESS) HAND TOOLS

These tools are powered by rechargeable batteries which are normally located in the handle of the tool. When the battery has discharged it is removed to be recharged in a battery charger, while a spare fully charged battery is fitted into the tool. The length of time it takes a battery to discharge depends on the amount of use and the electrical power required to drive the tool.
Each tool is usually supplied with two batteries and a dual battery charger for normal one hour or fast 20 minute recharge. The charger is run off the mains 240 V supply, but some manufacturers will supply 110 V chargers by special order (Figure 8.39a). The batteries must be compatible with the charger being used. These tools are double insulated and driven by 7–15 V batteries, depending on the make and the power output required. Figure 8.40a shows the parts of a cordless drill, while Figure 8.40b shows the drill in use.
Cordless tools include:

- drills
- screwdrivers
- laminate trimmers
- sanders
- small grinders and circular saws with disc or blade diameters up to 100 mm
- jigsaws
- staplers.

Figure 8.40 Cordless drill, showing (a) parts and (b) in use (courtesy Elu)

8.17 Method of Use

The use of these tools is as for any hand-powered tools, except that there is freedom from trailing electrical cables and their attendant dangers.

8.17.2 Safe Operation

The same safety precautions should be taken as for other hand-powered tools and ear, eye, nose and mouth protectors should be worn as required. Inspect the tools and associated cutters to make sure they are in good working order, sharp and correctly fitted. Check the correct batteries are being used and you have the correct charger for the batteries.

8.18 PNEUMATIC (COMPRESSED AIR) TOOLS

The power to drive these tools is supplied by compressed air forcing round vanes fixed to a drive shaft to give rotary motion to the tool (Figure 8.41). The compressor used to compress the air can be a permanent fixture in a workshop complex or a mobile unit on a building site. A compressed air system will require two components.

1 For compressing the air, the following are required:

- an electrically, diesel or petrol-driven motor
- an air compressor (pump) (Figure 8.42)
- a heat dissipator (the compression of air generates heat)
- a pressure vessel (air receiver) to store the compressed air
- a condensation separator and draw-off tap

2 For delivering the compressed air to the power tools the following are required (Figure 8.43):

- rigid air lines and fittings and valve connectors for flexible hoses
- a filter to remove any remaining moisture from the compressed air as close to the delivery point as possible
- a pressure regulator which controls the pressure of the air reaching the individual tools
- a lubricator which allows a controlled amount of lubricant to reach the tool.

8.18.1 Method of Use

These tools are used in a similar way to other powered hand tools. The range of tools matches those driven by other power sources. The most common types of tools in use are (Figure 8.44):

- drills
- hammers
- sanders and grinders
- screwdrivers
- saws – circular and jig
- nail and staple guns (Figure 8.45).

8.18.2 Safe Operation

Tools are operated at an air pressure of 4–7 bar (60–100 lb/in^2). As there is no electricity involved you may think that there is no danger but a jet of compressed air at the working pressures required is sufficiently powerful to cause injury to the body and blood vessels. Air entering the blood stream can be fatal. Following a number of basic safety rules will ensure a safe working environment for you and others working in the vicinity.

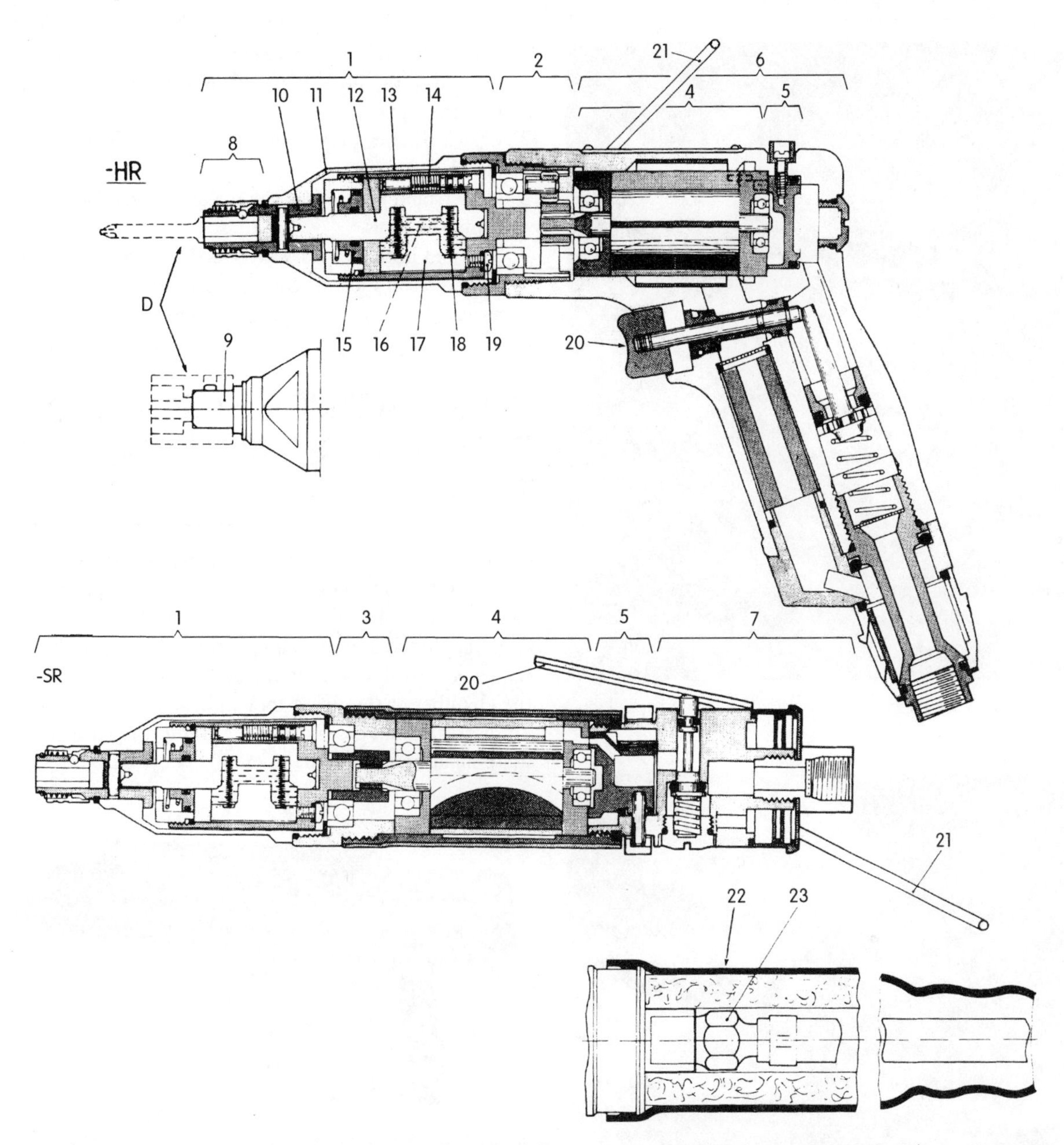

Figure 8.41 Exploded view of an ErgoPulse 5 pneumatic drill. **1** – pulse unit; **2** – planetary gear; **3** – driver; **4** – vane motor; **5** – reversing valve; **6** – pistol-grip handle/motor casing; **7** – back head; **8** – quick-change chuck; **9** – square drive (optional); **10** – driving spindle; **11** – housing; **12** – pulse-unit spindle; **13** – pulse-unit housing; **14** – adjustment screw; **15** – piston; **16** – oil chamber; **17** – vane; **18** – spring; **19** – oil filling screw; **20** – throttle valve; **21** – suspension yoke; **22** – exhaust hose; **23** – hose nipple

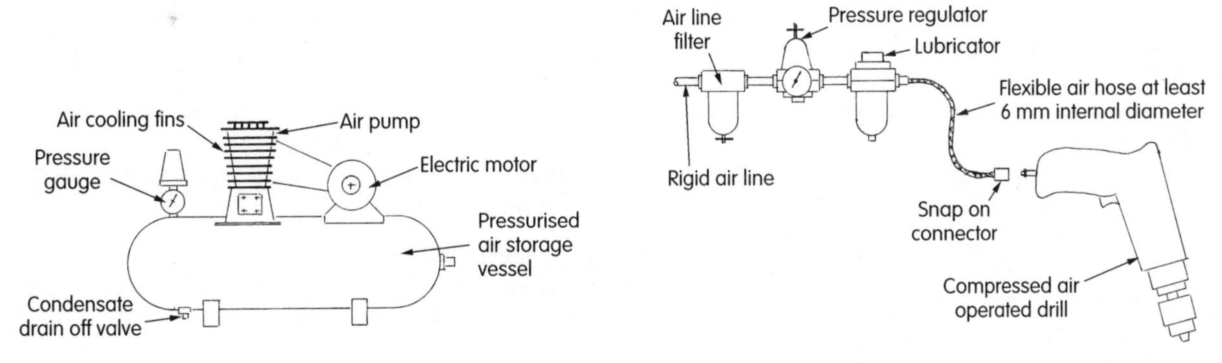

Figure 8.42 Compressed air pump

Figure 8.43 Air line filter, pressure regulator and lubricator

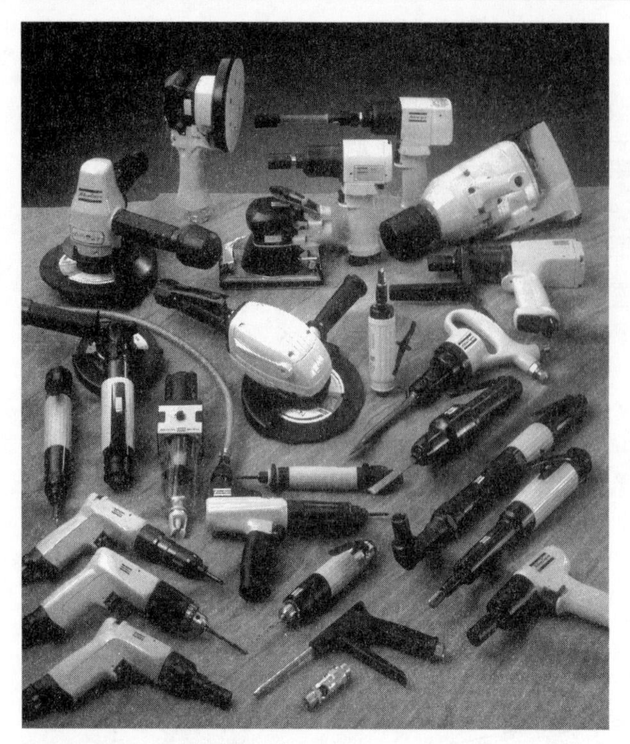

Figure 8.44 A selection of pneumatic tools (courtesy Atlas Copco)

Figure 8.45 Pneumatic nailing gun

1 Compressed air is for use with compressed air tools and is not the best way of blowing out dust from woodworking machines. Airborne dust can be a hazard to health.
2 Be sure the compressed air supply is at the correct pressure for the tool being used.
3 Check that the tools are in good working order, before and after use.
4 Make any adjustments to the tool or its cutters before connecting to the air supply.
5 Use the correct size of hose for the volume and pressure of air that is required to drive the tool.
6 Check that the hose connections are clean, correctly secured to the hose lines and securely pushed home into the valve connectors.
7 Check the flexible air lines are not corroded or abraded.
8 Understand and follow the maker's instructions when using the tools.
9 Training in the use of compressed air tools is necessary and a person must be competent in the setting up of air lines, ancillary equipment and use of the tools.

8.19 CARTRIDGE-OPERATED FIXING TOOLS (BALLISTIC TOOLS)

These tools are used to drive hardened metal fixing pins directly into hard materials such as brick, concrete or steel (Figure 8.46). This would normally require drilling, plugging or threading before a fixing could be made. A cartridge containing an explosive is fired into the chamber of the fixing tool and the resultant build-up of pressurised gases drives a pin into the hard base material at bullet speed.

Figure 8.46 Hilti cartridge tool with power regulator

There are two types of ballistic tool:

1 direct-acting (high velocity) tools in which the detonated explosive charge acts directly on the fixing pin, driving it out at high velocity from the barrel of the tool (Figure 8.47a)
2 indirect-acting tools in which the explosive charge drives a captive piston down the barrel to strike the head of the fixing pin, driving it into the base material (Figure 8.47b).

Depending on the manufacturer, there are three ways in which the piston and fixing pin can be arranged in the barrel of the ballistic tool (Figure 8.47c):

1 the piston at the firing end of the tool is driven down the barrel to strike the pin at the exit end.

2 the piston and the fixing pin are in contact, both being driven down the barrel.
3 the piston and the fixing pin are in contact at the exit end of the barrel with the point of the pin touching the material being fixed.

The tools have two distinct firing mechanisms:

1 those fired with a trigger mechanism which have the common characteristics of a pistol and may include a power control system allowing the force generated by the explosion to be modified; they are supplied as indirect-acting tools, of low velocity
2 those fired by striking with a club hammer; this type resembles a heavy metal cylinder with a plunger at one end which is struck with the hammer supplied by the manufacturers of the tool; it is indirect acting and is a low-powered tool (Figure 8.47d). This type is little used today and is no longer manufactured.

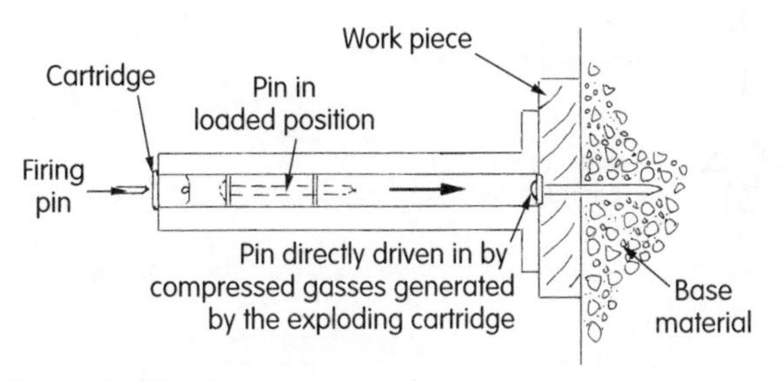

Figure 8.47a Direct-acting tools

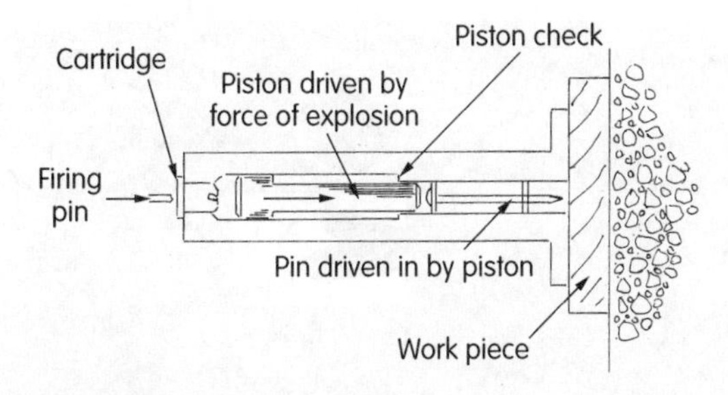

Figure 8.47b Indirect-acting tools

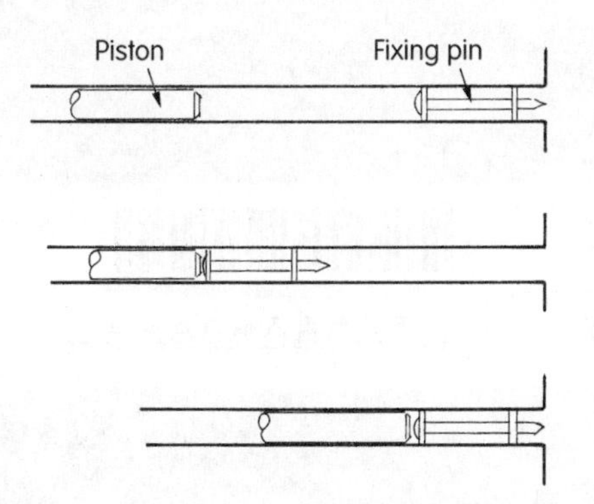

Figure 8.47c Positional relationship of piston to pins

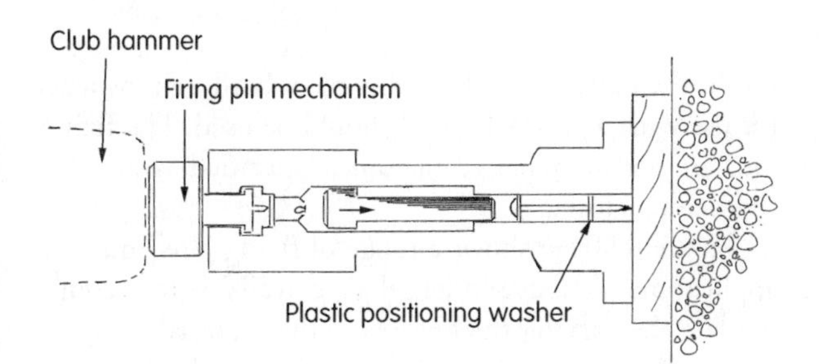

Figure 8.47d Hammer activated ballistic tool

A ballistic tool is classed as low powered when the exit speed of the fixing pin is not greater than 98.5m/s and high powered when the exit velocity exceeds 98.5m/s.

8.19.1 Cartridge Types

The cartridges have a range of explosive strengths to suit the firing of various types of fixing pin into a variety of backgrounds of variable hardness. Cartridge strengths are number and colour coded to BS 4078 Part 2: 1989 (Table 8.5).

Table 8.5 Colour coding of cartridges to BS 4078 Part 2: 1989

Cartridge strength	Code numbers	Identification colour
Weakest	1	Grey
	2	Brown
	3	Green
	4	Yellow
	5	Blue
	6	Red
Strongest	7	Black

Remember that the American and European coding systems for ballistic tool cartridges are different and it is important that the identification codes supplied by the manufacturer are used to identify their own cartridges. Cartridges are manufactured to fit the makers own ballistic tools and the cartridges are not interchangeable. Cartridges are supplied singly or assembled into plastic cartridge clips for automatic firing; this saves reloading after each pin is fixed (Figure 8.48). Cartridges must always be carried and stored in the maker's packaging until required. The outside of the packet must indicate its contents by:

- a colour-coded label indicating the cartridge strength
- the manufacturer's name
- the cartridge size
- the range of ballistic tools in which the cartridges may be used
- the number of cartridges contained in the box
- the manufacturer's batch identification serial number
- displaying the words 'SAFETY CARTRIDGE'.

Figure 8.49 illustrates a cartridge container label.

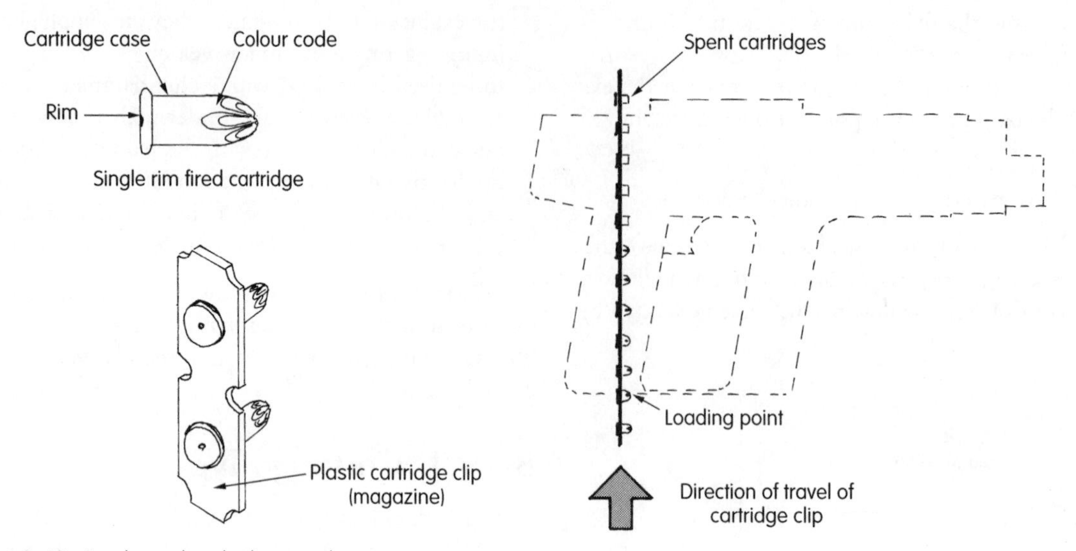

Figure 8.48 Single and multiple cartridge magazines

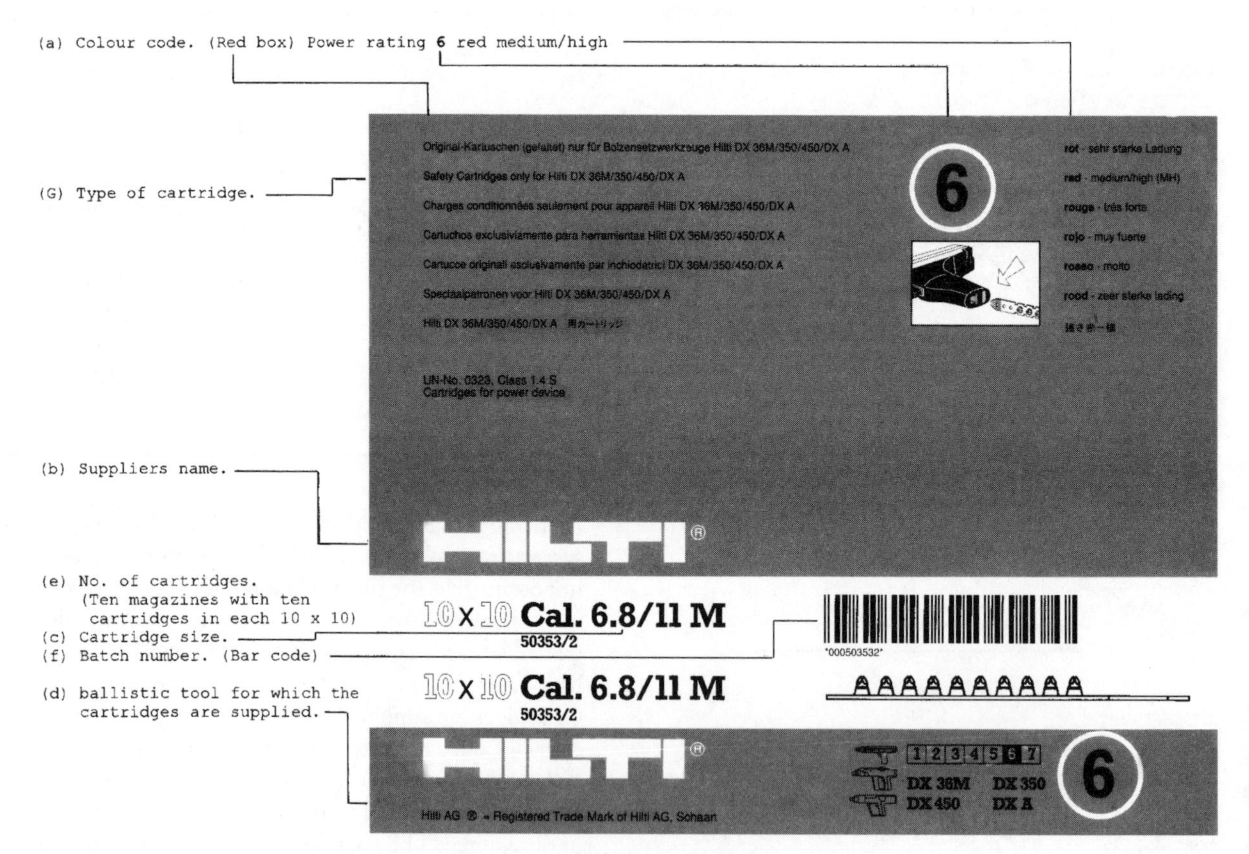

Figure 8.49 Cartridge container label

8.19.2 Fixing Pins Used with Ballistic Tools

As the fixing pins are required to penetrate brick, concrete and mild steel, the pin design and its physical properties must be such that a secure fixing to the hard background occurs and that, prior to firing, the fixing pin fits correctly into the barrel of the tool. The head of the pin, or sacrificial plastic washers, correctly align the pin in the barrel ready for firing so only those pins listed for use with a particular tool should be used. The fixing pins are not interchangeable unless specified as such by the manufacturer.

Figure 8.50a–e shows a range of fixing pins and their application. Various pin lengths are available to accommodate the varying thicknesses of the material being fixed and the depth of penetration into the background material.

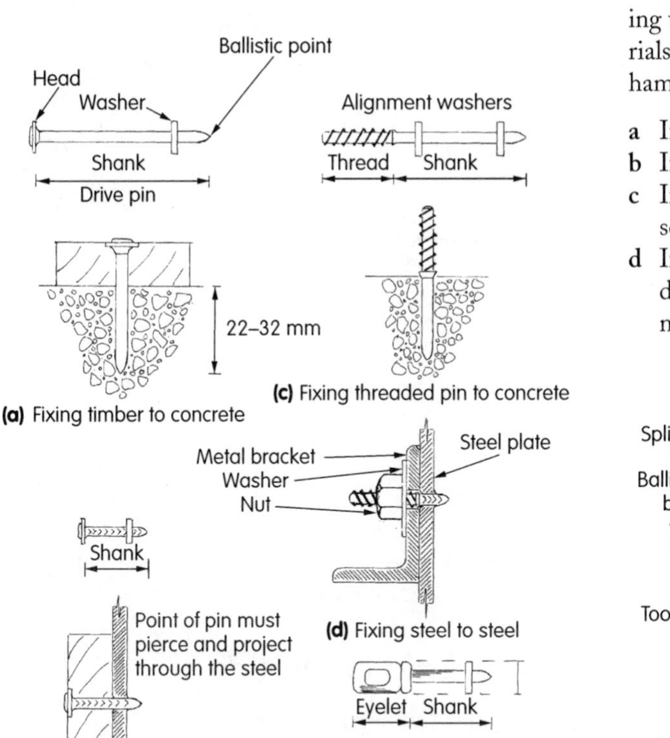

Figure 8.50a–e Types of fixing pins

8.19.3 Suitability of Base Materials

Suitable base materials are those which allow the pin to enter in a straight line to the correct depth and offer maximum resistance to withdrawal (good holding characteristics). Some types of base material are not suitable to take cartridge fired fixing pins:

- those materials which are too hard, such as hard bricks and rock (marble), hardened steel and cast iron; fixing into these materials may cause deflection, ricochet or shattering of the pin (Figure 8.51a)
- those materials which are brittle such as glass blocks, slates, glazed and hardened tiles, which will shatter as the pin is fired into them (Figure 8.51b)
- materials which are too soft, such as lightweight and hollow building blocks, hollow partitions and wood-based material that would allow the pin to pass completely through the material into free flight on the side away from the operative (Figure 8.51c).

Before using cartridge tools the background into which the pin is to be fired must be investigated. A solid looking wall may be a hollow plastered partition. Base materials can be tested for their suitability by using a hammer to drive in a fixing pin (Figure 8.52).

a If the pinpoint blunts, the material is too hard.
b If the surface begins to splinter, it is too brittle.
c If the surface is easily penetrated, the surface is too soft.
d If the pin fails to penetrate 2 mm or the above conditions are experienced seek advice and consult the manufacturer's literature.

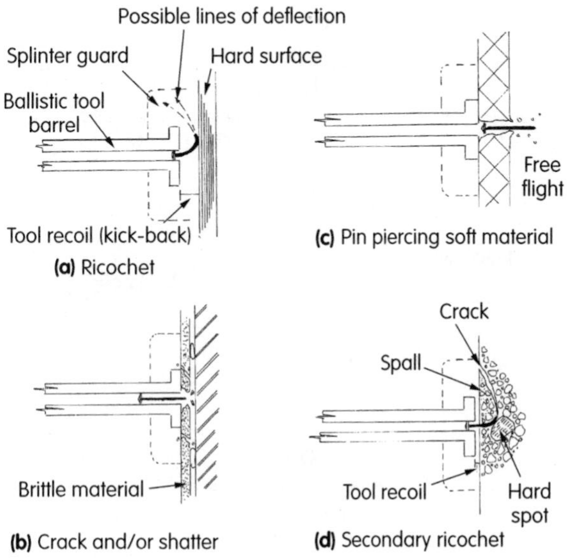

Figure 8.51a–d Fixing to base materials

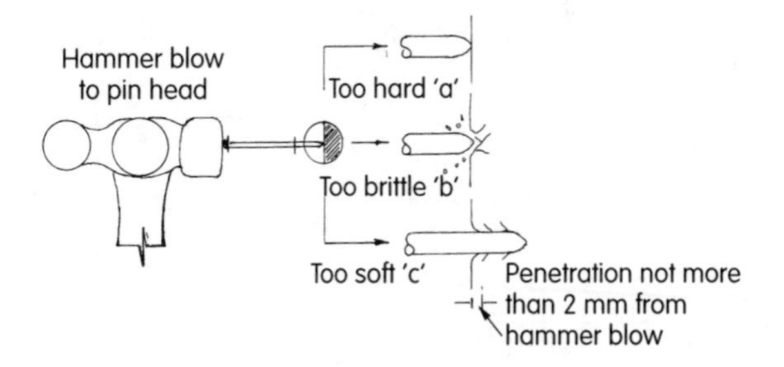

Figure 8.52 Testing base materials for suitability

8.19.4 Ballistic Fixing in Concrete

Pins fired into concrete and steel generate temperatures up to 900°C at the point of the pin. As a result of these high temperatures concrete and steel sinter (fuse together with the pin) which produces a holding force between pin and background of up to 15 kilonewtons.

To achieve maximum holding power:

a determine the thickness and characteristics of the concrete (Figure 8.53a)

b achieve the correct depth of penetration by using cartridges of the correct explosive strength (Figure 8.53b)
c fixings must always be 80 mm or more apart (Figure 8.53c)
d never fix a pin nearer than 80 mm to a spalled surface or edge (Figures 8.53a and 8.53d).

NB: Never attempt to fix a ballistic fired pin into prestressed concrete.

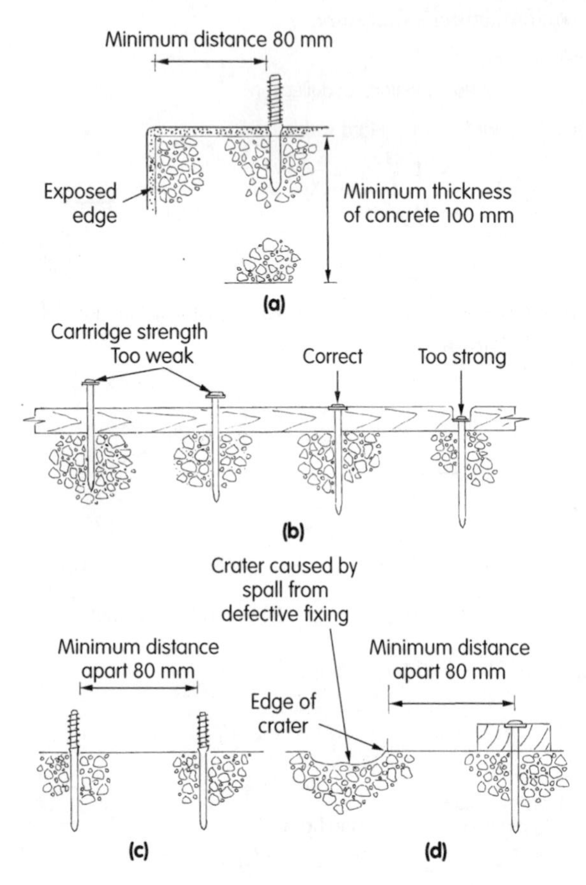

Figure 8.53a–d Fixing to concrete

8.19.5 Ballistic Fixings in Steel (Structural Mild Steel)

The points of pins fired into mild steel should penetrate and project just beyond the steel for maximum holding power (Figure 8.50b). If a pin does not penetrate through the steel the reactive compression forces at the point of the pin will tend to push it out of the steel.

To achieve maximum holding power:

a seek advice if the steel is anything but structural mild steel
b never place a pin closer than 15 mm to an edge or a drilled hole (Figure 8.54)
c the distance between fixing pins must not be less than 20 mm (Figure 8.54)

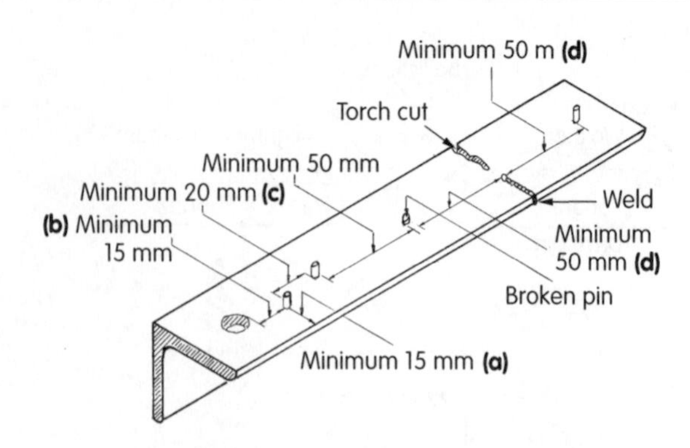

Figure 8.54 Fixing ballistic pins to structural mild steel

d never fire a pin closer than 50 mm to a broken-off pin, a weld or where the steel has been cut with a blowtorch (Figure 8.54).

8.19.6 Safe Operation

The operative (see Figure 8.55):

- must be at least 18 years of age
- must not be colourblind, as the cartridges are colour coded depending on explosive power
- must be trained in the use of the tool by the manufacturer's training instructor and obtain a certificate of competence. Note: Because of differences in makes and models of ballistic tools, each often requires a separate certificate of competence
- must use the tool in a competent manner, recognise potential hazards and take any necessary actions
- must implement the correct procedures for dealing with a misfire
- must wear the correct safety equipment, helmet, goggles and ear protectors

Figure 8.55 Hilti low velocity indirect acting ballistic tool in use

- must understand the dangers of working near flammable vapours and explosive gases and the need to remove these hazards and ventilate the work area
- must position the body to resist overbalancing if the tool recoils on firing. Never work from a ladder as both hands are required to operate the tool and only work from scaffolds that are securely anchored
- must never load a cartridge into the gun unless at the work area and ready to use it immediately and never carry the tool about when loaded with a cartridge
- must keep others away from the work area while the tool is in operation as a precaution against being injured by a possible ricochet
- if there is a possibility of the fixing pin passing through the background material on the side away from the operator, this area must be screened off to protect others from injury.

8.19.7 Misfire

A misfire can be caused by a fault in the firing mechanism, or a faulty cartridge. If a misfire does occur, take the following action:

- without drawing the tool from the face of the work, reactivate the trigger mechanism and fire the tool again
- if the tool still fails to fire, wait 15 seconds and remove the tool from the work face; keeping the muzzle pointing down and away from you, remove the cartridge, following the maker's instructions
- correctly replace the tool in its metal carrying case and return to the maker.

8.19.8 Servicing and Repair

To maintain the tool in good working order it will require regular cleaning and checking that any moving parts and locking mechanisms move freely and are lubricated as instructed by the manufacturer. Before and after use check the tool for defects; if damaged, take it out of service and return to the maker or the approved agent for repair.

8.19.9 Security

Ballistic tools and their accessories must be kept with the maker's case and under the control of an authorised operative. After use the tool must be unloaded and stored in its locked carrying case and placed in a secure store. Spare cartridges must be put back in the correct coded cartridge container. Be very careful not to mix up different cartridges. Never carry loose cartridges about; they should be correctly packed in their appropriate metal container.

8.19.10 References Associated with Ballistic Tools

- British Standard BS 4078 1966/1987, cartridge-operated fixing tools.
- Health and Safety Executive Guidance Note PM14, 'Safety in the use of cartridge-operated fixing tools'.
- manufacturer's handbooks giving instruction on the general characteristics and use of their cartridge-operated fixing tools.

8.20 IMPULSE CORDLESS NAILER

This is a gas operated nailing tool which uses a fuel cell rather than the ignition of a solid explosive from an individual cartridge to drive nails or staples into timber and timber-based products. It is not used for fixing to hard materials, such as brick, concrete or steel. The tool can take a variety of nail and staple fixings depending on the model used. Figure 8.56 shows an ITW Paslode impulse cordless nailer IM350 which takes smooth or ring shanked nails from 45 to 90 mm in length with a

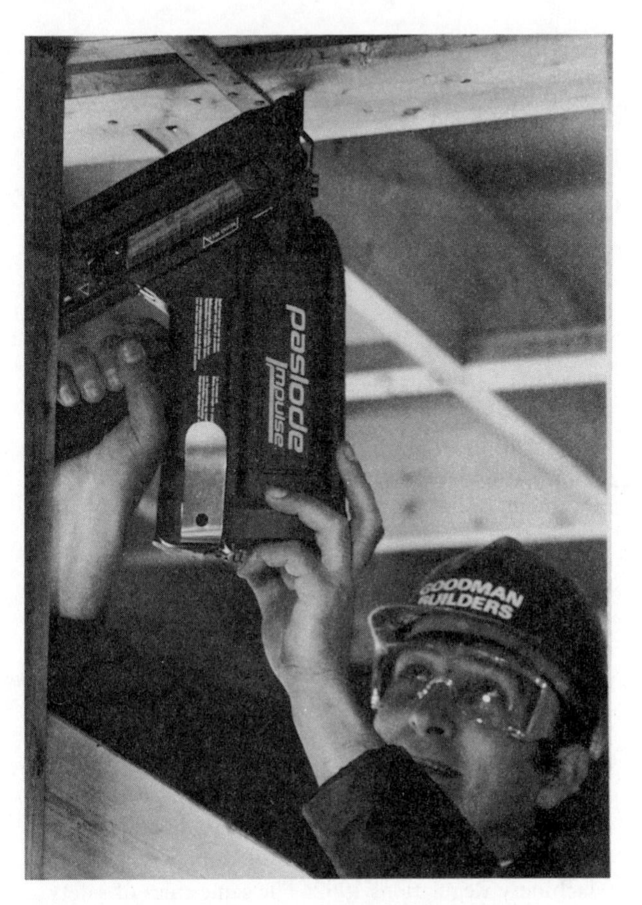

Figure 8.56 ITW Paslode impulse cordless nailer

magazine capacity of up to 48 nails, depending on their size.

The tool is powered by the ignition of a gas air mixture from an electrically generated spark. A gas cartridge (cylinder) loaded into the tool supplies the fuel gas. When the nose of the tool is pressed against the surface of the timber, a measured amount of fuel gas and air is injected into the combustion chamber, where it is ignited by a spark from the battery, as in a car engine, each time the trigger is operated. On ignition the expanding gases force a captive piston to drive the nail into the work piece.

A gas cartridge contains sufficient fuel to drive upto 1200 nails. The tool is capable of fixing up to three nails per second, but the rate of output will depend on the skill developed in the use of the tool and the positioning and spacing of the nails.

When you are using these tools, eye protection must be worn and training in the safe handling and correct use of the tool must be given. The tool has the usual built in safety devices, but it must be used according to the instructions provided in the manufacturer's literature. Fuel gas containers are colour coded and again must only be used in the tool for which they are designed.

8.21 MITRE SAWS

Mitre saws are portable electrically powered circular saws attached above the base by a pillar which, in turn, is fixed to a slotted receiver arm to take the projecting saw teeth as they cut through the material with a downward movement of the saw blade. This whole assembly can be rotated through a left- and right-hand arc of 45° in relation to the base and fence. After the base housing is locked at the required angle, the saw blade is brought into contact with the material to be cut by a downward pull on the saw handle. The latter also contains the on/off switch (Figure 8.57). The operator starts the saw and, when it is running at full speed, by a downwards movement, brings the blade into contact with the material being cut. The saw can be adjusted to give:

- square crosscutting
- bevelled crosscutting (mitring)
- compound bevelled crosscutting.

The saw cut gives a sufficiently fine finish to the cut surfaces for them to be abutted as a mitre without the need for planing.

There is provision for connection to a dust extraction unit and the machine conforms to the Unified European Standard (CE) under the Supply of Machinery Regulations 1992. The same rules of safety apply when using this portable power tool as to all others. Figure 8.57 shows an Elu PS274 crosscut mitre saw.

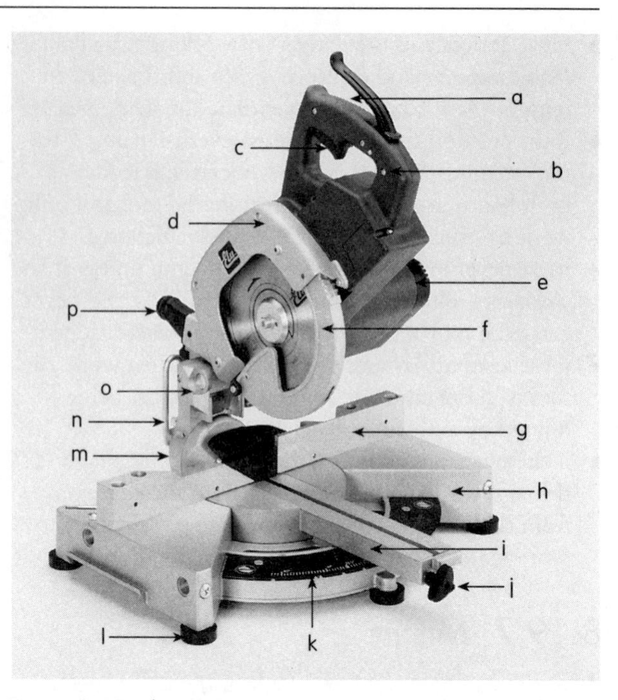

Figure 8.57 Elu PS274 crosscut mitre saw (courtesy Elu)
a – bottom saw guard release lever; **b** – handle; **c** – trigger switch; **d** – top saw guard; **e** – motor housing and ventilation slots; **f** – bottom saw guard; **g** – fixed fence; **h** – base; **i** – swivel base; **j** – swivel base locking handle; **k** – scale of degrees (for mitre angles); **l** – base foot pads; **m** – eccentric cam for setting bevels in the vertical plane; **n** – pillar locking lever; **o** – pillar attached to swivel base via eccentric cam; **p** – dust extraction port

8.22 GENERAL OPERATIONAL SAFETY RULES FOR ALL PORTABLE POWERED HAND TOOLS

- Use only those tools which you are competent to use. For others, seek training before use.
- Understand fully the function and limitations of the tool. Select the correct tool for the work.
- If electricity is the power source make sure the supply voltage is compatible to that of the machine. (Check the specification plate attached to the machine.)
- Always check the tool, its attachments, blades or cutters, cable, plug, socket and extension lead for compatibility and for any visible and obvious defects at the start of each work period. Defective tools and equipment must be taken out of service for repair or replacement.
- Effectively secure the workpiece before starting the work.
- Adopt a good stance and firm footing.
- Keep safety guards correctly adjusted.

- Keep work areas free from obstructions.
- Never wear items of clothing and jewellery, including rings, that may become caught in the moving parts of the tool.
- Do not let long hair obscure your vision or come in contact with the tool's moving parts.
- Cutters and blades must reach their working speeds before starting to cut.
- Moving parts of the tool must be stationary before it is put down.
- Never use tools which produce sparks or flame in unventilated spaces that may contain explosive gases or volatile vapours.
- Keep cables and hoses clear of cutting edges, abrasive materials or corrosive substances.
- Keep cutting tools sharp.
- Ensure that cutters, blades and fences are correctly mounted and securely fixed.
- Remove chuck keys, spanners and other adjusting tools and replace them in their holders before starting the machine.
- Never overload the machine.
- Never use electrically powered tools in damp conditions
- Use head, eye, ear, nose and mouth protectors.
- Keep your hands clear of the cutters and blades while they are in operation.
- The tools must be isolated from the power supply when making adjustments and when not in use.
- Keep your fingers clear of the switch of a charged or plugged-in tool until you are ready to use it.
- Make sure the tool is switched off before connecting it to the power supply. Take special care with tools that have a switch that can be locked on.
- Tools must be maintained at regular intervals by those appointed to do so and electrically powered tools must be tested for continuity of earthing and insulation.

9

Basic Woodworking Machines

Edited and updated by Eric Cannell

The aim of this chapter is to help the student to become aware of the more common types of woodworking machinery that the carpenter and joiner may encounter, to be able to recognise these machines by sight and to understand their basic function.

Undoubtedly the most important aspect of any woodworking machine is its safe use. To this effect, set rules and regulations are laid down by law and must be carried out to the letter and enforced at all times. The need for such strict measures will become apparent especially when one considers that, unlike most other industries, the majority of our machines are still fed by hand, thus relying on the expertise of the skilled operator who must concentrate and exercise extreme caution at all times.

Legislation concerning machinery is contained within the Provision and Use of Work Equipment Regulations 1992 (PUWER). Prior to this date the Woodworking Machine Regulations 1974 dealt specifically with machinery concerned with timber. Guarding and safe practices mentioned in the 1974 specific regulations are considered a minimum requirement when utilising the more 'traditional' machinery mentioned therein. From 1st January 1993, under the Supply of Machinery (Safety) Regulations 1992, the manufacturer or the importer into the European Community has to comply with this unified European standard (signified by the CE mark on the machine) or satisfy the health and safety legislation in force on 31st December 1992 (i.e. PUWER). Both sets of Regulations expect new machinery to have improved mechanical safeguards (e.g. interlocking guards, built-in automatic braking), thus reducing the risk of accidents to operators.

Particular attention should be given to PUWER Regulation 6 – Maintenance. It is a legal requirement in these regulations that all machinery is maintained in an efficient state and that a maintenance log is kept up to date.

Extracts from the Woodworking Machine Regulations 1974 are used in this chapter as a minimum requirement for guidance purposes only. Operators of woodworking machinery should, whenever reasonable and practicable, reduce the risk of accidents by assessing the situation in hand and they should at least adhere to the minimum guidelines.

Note: It is likely that the Health and Safety Executive will publish an Approved Code of Practice (ACOP) for various woodworking machines. The reader is advised to obtain and read these when available.

9.1 CROSSCUTTING MACHINES

These machines are designed to cut timber across its grain into predetermined lengths, with a straight, angled or compound-angular (angled both ways) cut. However, that primary function is to cut long lengths of timber into more manageable lengths. With a standard blade and/or special cutters, they can also be used to cut a variety of wood joints, for example housing, halving, mitred, dovetailed and birdsmouth joints (as shown later in Figure 9.5).

There are two main types of cross cut saws: the travelling-head (Figure 9.1) and radial-arm (Figure 9.2).

9.1.1 Positioning Timber to be Cut

The full length of the timber must always be fully supported at both ends, to avoid tipping once the cut has been made. Sawn ends must never be allowed to interfere with the saw blade. Figure 9.3 shows how this can be avoided.

'Bowed' boards (Figure 9.3a) should be placed round side down, with their crown in contact with the table over the saw-cut line, and packing should be used to prevent the raised portion creating a see-saw effect with the possibility of it rocking on to the blade. If the board is placed round side uppermost, cut ends may drop and trap the saw blade, creating a very dangerous condition.

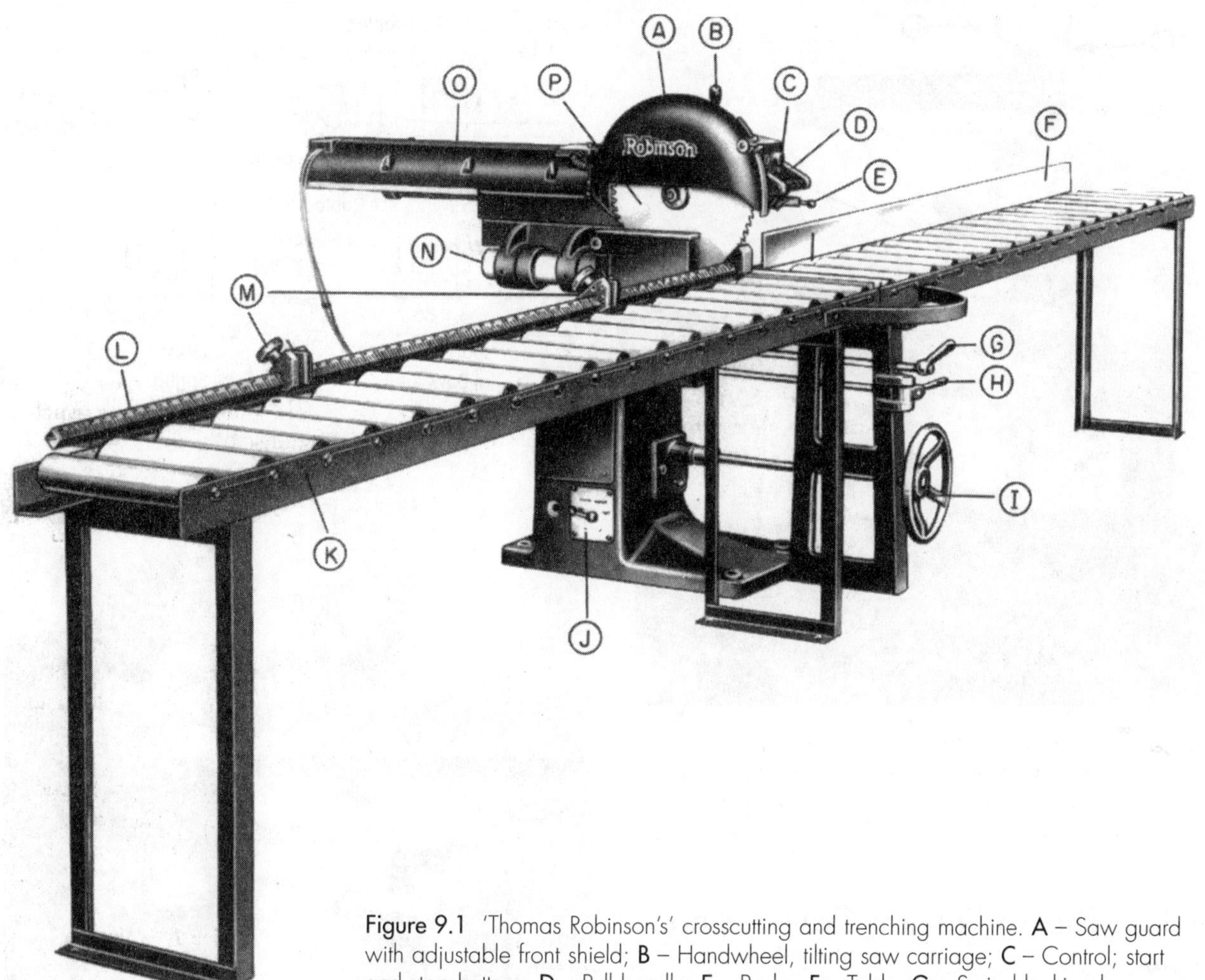

Figure 9.1 'Thomas Robinson's' crosscutting and trenching machine. **A** – Saw guard with adjustable front shield; **B** – Handwheel, tilting saw carriage; **C** – Control; start and stop buttons; **D** – Pull handle; **E** – Brake; **F** – Table; **G** – Swivel locking lever; **H** – Swivel adjustment; angle location lever; **I** – Handwheel; to adjust saw vertically (rise and fall); **J** – Machine isolator; **K** – Roller table;**L** – Cutting off gauge bar; **M** – Adjustable stops; **N** – Sawdust exhaust; **O** – Travelling carriage; **P** – Saw blade

Similarly, when dealing with 'sprung' boards (Figure 9.3b), contact with the fence at the saw-cut line is very important, otherwise the direction of saw blade rotation could drive the cut ends of the board back towards the fence, trapping the saw blade.

Note: Because the operative may use one hand to hold timber against the fence and the other to pull the saw, he must be constantly aware of the danger of a hand (particularly a thumb) coming in line with the saw cut.

9.1.2 The Travelling-head Crosscut Saw (Pull-over Saw)

Figure 9.1 shows a typical saw of this type. The saw, which is driven direct from the motor, is attached to a carriage mounted on a track, which enables the whole unit to be drawn forward (using the pull handle) over the table to make its cut. The return movement is spring assisted. The length of timber to be cut is supported by a wood or steel roller table and is held against the fence. For cutting repetitive lengths, a graduated rule with adjustable foldaway stops can be provided. Angle and height adjustments are made by operating the various handwheels and levers illustrated.

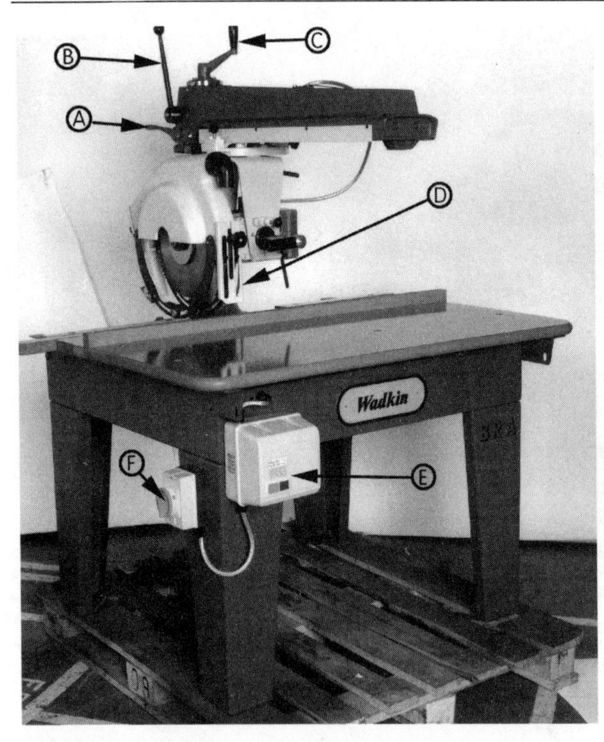

Figure 9.2 'Wadkin' universal radial crosscut saw.
A – Arm locating latch; B – Arm locking lever; C – Pillar rise and fall handle; D – Saw guard visor; E – Start-stop; F – Emergency shut-off

9.1.3 The Radial-arm Crosscutting Saw

This carries out similar functions to the travelling-head crosscut, but differs in its construction by being lighter, having its saw unit and its carriage, which are hung under an arm which radiates over the table, drawn over the workpiece.

Some of these machines (like the one shown in Figure 9.2) are generally more versatile than the travelling-head types; not only does the carriage arm swivel 45° either way but also the saw carriage tilts from vertical to horizontal and revolves through 360°, enabling ripping, grooving and moulding operations to be carried out.

Figure 9.4 shows a safer approach to utilising the machine, with both hands controlling the pulling of the saw unit across the timber. Note also the position of the front extension nosepiece, i.e. it should be adjusted as close as possible to the timber surface.

9.1.4 Crosscutting and Trenching Machine

Figure 9.5 shows just a few of the many different processes that can be carried out on this type of machine when the blade is either accompanied or replaced by different machining heads.

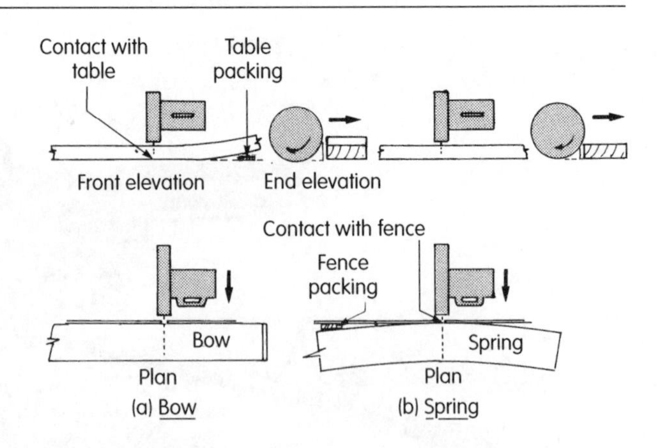

Figure 9.3 Crosscutting distorted timber

Figure 9.4 Trainee under instruction in the use of the radial arm crosscutting saw

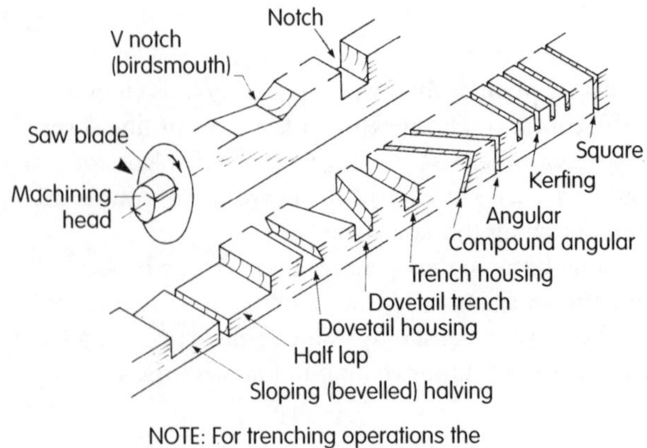

Figure 9.5 The versatility of the crosscutting and trenching machine

9.2 SAW BLADES

Saw blade size (diameter) in relation to the working speed of the saw spindle is very important, so much so that every machine must display a warning notice stating the minimum saw blade size which can be fitted for efficient operation (under PUWER Regulation 5 – Suitability of Work Equipment, and Regulation 24 – Warning).

Saw blades are designed to suit a particular type of work and in the case of a saw bench a rim speed (peripheral speed) of about 50 m/s is considered suitable. Lower or higher speeds could cause the blade to become overstressed and result in a dangerous situation. Crosscut saw blades usually require a higher rim speed than ripsaw blades.

To calculate the rim speed (peripheral speed), we must know the diameter of the saw blade and its spindle speed. Let us assume a saw blade diameter of 550 mm and a spindle speed of 1750 rev/min.

1 Find the distance around the rim (i.e. the circumference), using the formula for the circumference of a circle:

circumference = $\pi \times$ diameter

π may be taken to be 3.142 or 22/7

Diameter (D) = 550 mm = 0.55 m
Distance around rim = $\pi \times D$
= 3.142×0.55 m ~ 1.728 m

This is the distance travelled by a tooth on the rim in one revolution.

2 Find the distance travelled every minute by multiplying the distance around the rim by the number of revolutions per minute:

$1.728 \text{ m} \times 1750 \text{ rev/min} = 3024 \text{ m/min}$

3 To find the answer in metres per second (m/s), we must divide by 60:

$$\therefore \text{rim speed} = \frac{3024}{60} \text{ m/s} \sim 50 \text{ m/s}$$

The formula may be summarised as

rim speed (m/s) =

$$\frac{\pi \times \text{Dia of blade}}{1000} \times \frac{\text{Spindle speed (rev/min)}}{60}$$

9.2.1 Choosing the Correct Saw Blade

The relationship between a saw blade's size and the formation of its cutting edge will depend on one or more of the following:

- the type of sawing – crosscutting or ripping
- the type of material being cut, i.e. solid timber or manufactured boards, etc.
- the condition of the material being cut
- the finish required
- the direction of cut, i.e. across or with the grain.

Tooth shape and pitch (the distance around the circumference between teeth) greatly influence both the sawing operation and the sawn finish. Saw blades are generally divided into two groups: those most suited to cross cutting and those used for ripping and deeping (both cutting with the grain).

Figure 9.6 shows three different forms of tooth profile used for crosscutting. Notice that the front face of each tooth is almost in line with, or sloping forward of, the radius line. This is known as 'negative hook' and produces a clean cut.

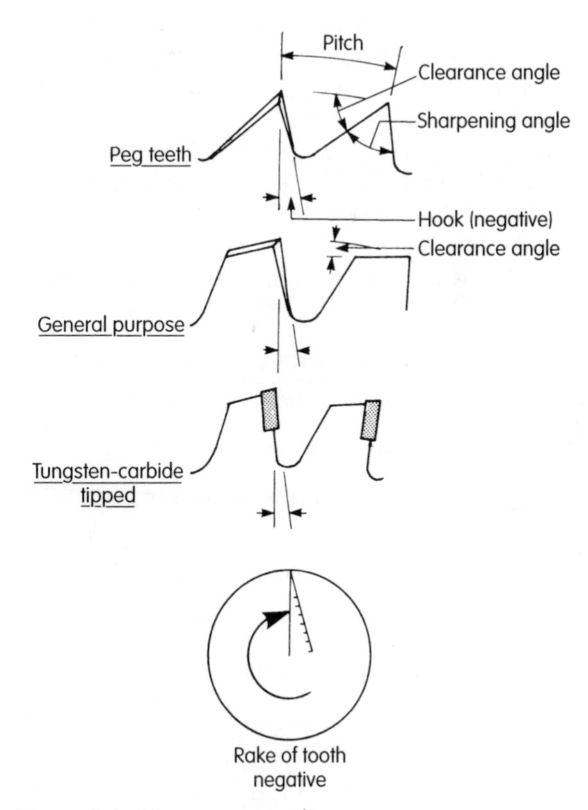

Figure 9.6 Crosscut saw teeth

Figure 9.7 shows how each tooth of a ripsaw blade slopes back from the radius line to produce a true hook shape, termed 'positive hook'. It is this shape which enables the tooth to cut with a riving or chopping action. The large gullet helps keep the kerf free of sawdust.

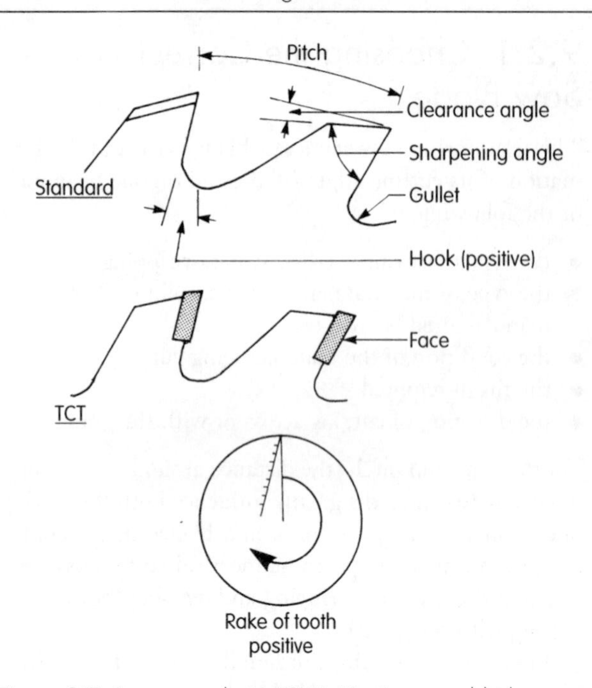

Figure 9.7 Ripsaw teeth. WARNING: A ripsaw blade must never be fitted to a crosscutting machine because this would lead to a self-feed effect of the blade towards the operator

A tungsten carbide tipped (TCT) tooth is ideal for cutting hard or abrasive material because its hardness means it stays sharp much longer than the conventional saw tooth. However, its eventual resharpening does involve the use of special equipment and, because of this, the maker of the saw blade or an appointed agent usually undertakes the task.

Knowing the number of teeth and the diameter of the saw blade will enable the tooth pitch to be determined:

$$\text{tooth pitch} = \frac{\text{rim distance (circumference of saw blade)}}{\text{number of teeth around the rim}}$$

where rim distance = $\pi \times$ diameter

For example, if a saw blade has a diameter of 500 mm and 80 teeth:

$$\text{Tooth pitch} = \pi \times \frac{\text{diameter}}{\text{number of teeth}} = \frac{3.142 \times 500\text{ mm}}{80} = 19.6\text{ mm}$$

Note: In general, the shorter the pitch, the finer the cut.

With the exception of certain saw blades which are designed to divide timber with the least possible waste, the kerf left by the saw cut must be wide enough not to trap the saw blade. Figure 9.8 shows how this is achieved. The hollow-ground saw blade (Figure 9.8a), which produces a fine finish and is used for crosscutting, uses its reduced blade thickness, whereas the parallel-plate saw blade (Figure 9.8b) uses its spring set (teeth bent alternately left and right). TCT saw blades (Figure 9.8c) rely upon the tip being slightly wider than the blade plate to provide adequate plate clearance.

Figure 9.8 also shows a saw blade mounting. The holes in the saw blade must match the size and location of both the spindle and the driving peg, which is incorporated in the rear flange. The front flange is positioned over the peg before the spindle nut is tightened.

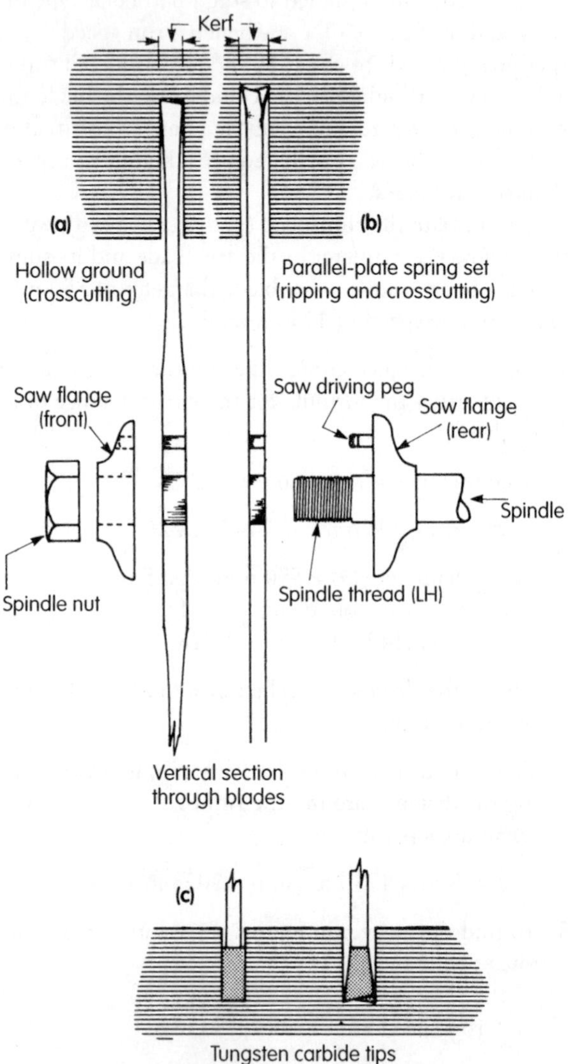

Figure 9.8 Circular-saw blades and their mounting

9.2.2 Guarding

Figure 9.9 shows how the riving knife should be positioned in relation to the size of the saw blade. Its thickness must exceed (by approximately 10%) that of the blade if a parallel-plate saw blade is used. However, its thickness should be less than the kerf. The purpose of the riving knife is to act as a guard at the exposed rear of the sawblade and to keep the kerf open, thus attempting to prevent the timber closing (binding) on

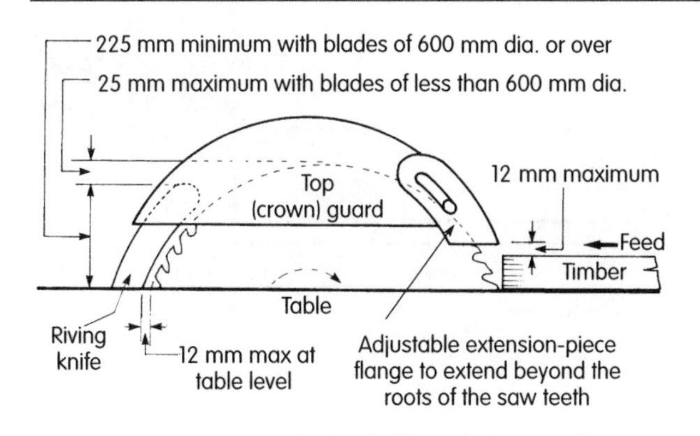

Figure 9.9 Arranging of riving knife, top (crown) guard, and extension piece

the saw blade. If the saw cut was allowed to close (with case-hardened timber, etc.), the upward motion of the back of the saw blade could lift the sawn material and possibly project it backwards towards the operator.

During the sawing process, the crown guard must provide adequate cover to the saw teeth. It should extend from the riving knife to just above the surface of the material being cut (it may be necessary to use a crown guard extension piece). The gap should be as narrow as practicable, but never more than 12 mm (Figure 9.9).

Figure 9.10 'Wadkin' circular saw bench. A – Saw blade; B – Riving knife; C – Crown guard; D – Crown guard adjustment; E – Extension guard; F – Adjustable fence (will tilt up to 45°); G – Table; H – Handwheel; saw spindle rise and fall; I – Isolator; J – Control; start and stop buttons; K – Mouthpiece (hardwood); L – Machine groove for cross-cutting gauge; M – Saw blade packing; N – Finger plate (access to saw spindle); O – Extension table (provision to comply with Regulation 11A)

9.3 HAND FEED CIRCULAR SAW BENCHES

These are primarily used for resawing timber lengthwise in its width (ripping or flatting) or its depth (deep-cutting or deeping). Some models can vary the depth at which the blade projects above the saw bench table.

Figure 9.10 shows a typical general purpose saw bench with provisions for a crosscutting fence. The dotted area at the back of the saw bench indicates where a backing-off table must be positioned when anyone is employed to remove cut material from the delivery end (see PUWER Regulation 11A – Dangerous Parts of Machinery).

Saw benches of this type use relatively large saw blades which require packings to help prevent the saw blade deviating from its straight path. Packings are pieces of oil-soaked felt, leather or similar materials specially made by the woodcutting machinist to suit the various type of blade and their relevant position above the saw bench table. Packings, together with a hardwood mouthpiece and backfilling, can be seen in Figure 9.11.

The mouthpiece acts as a packing stop and helps prevent the underside of sawn stock from splintering away. Like packings, it will be made to suit each size of saw blade. Backfilling protects the edges of the table from the saw's teeth; one side is fixed to the table, the other to the fingerplate. The fingerplate lifts out the table to facilitate changing a saw blade.

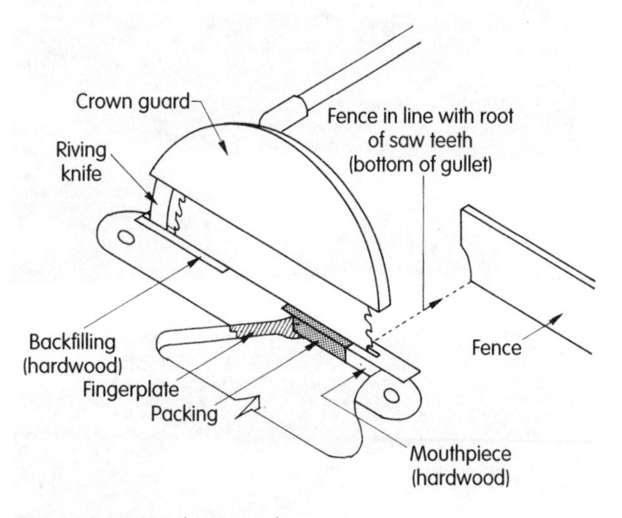

Figure 9.11 Packing circular saws

9.3.1 Method of Use

Figure 9.12 shows a ripping operation being carried out. In Figure 9.12a the cut is about to be started. As the cut progresses (Figure 9.12b and c), a pushstick (Figure 9.13) is used to exert pressure on the timber. During the last 300 mm of the cut only the pushstick is used (Figure 9.12d) and the left hand is moved out of harm's way. Hands are thus kept a safe distance away from the saw blade at all times.

(a)

(b)

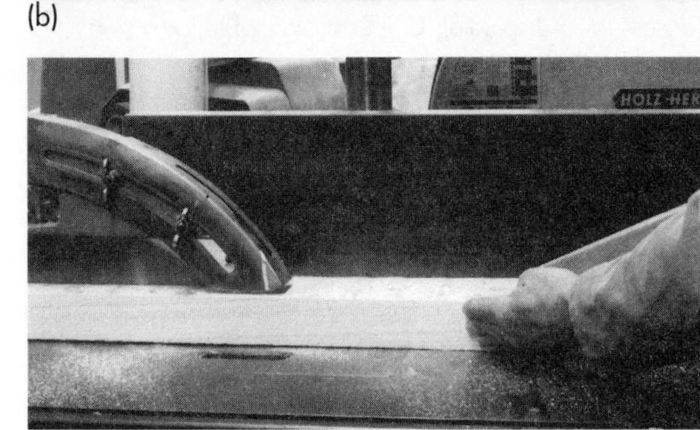

(c)

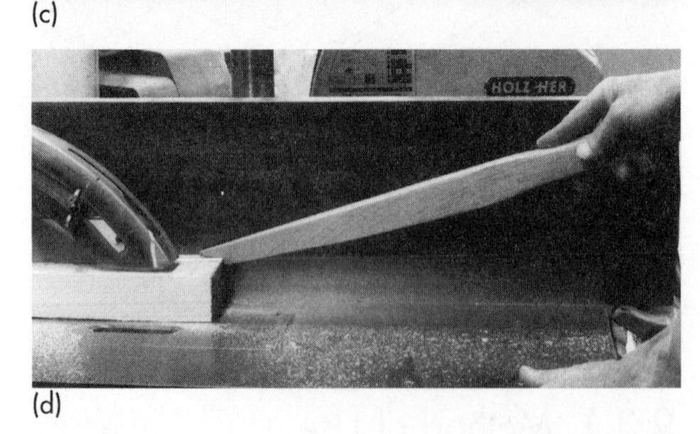

(d)

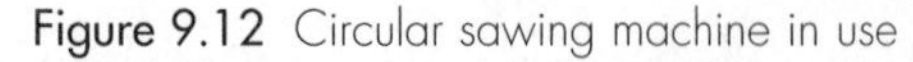

Figure 9.12 Circular sawing machine in use

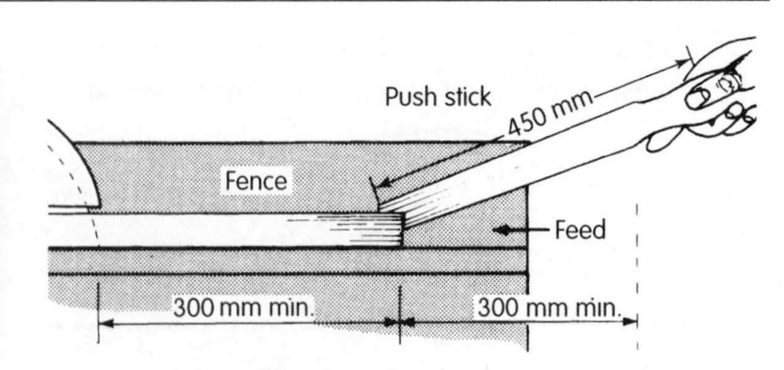

Figure 9.13 Push stick

Figure 9.14 Using a jig to cut wedges. Note: A push stick must be used to hold stock being used for wedges against the jig

By using a jig, jobs which otherwise could not be done safely are often made possible. Figure 9.14 shows a simple jig about to be used; while the cut is being made, a pushstick must be used to hold the short length of timber against the jig, as previously mentioned.

Unless the machine is of the movable type which cannot take a blade larger than 450 mm diameter or is used in conjunction with a travelling (rolling) table, the delivery end of a machine table must be not less than 1200 mm in length from the back of the saw blade (Figure 9.10).

9.4 DIMENSION SAW

This uses a smaller saw blade, thus limiting its maximum depth of cut to about 140 mm, depending on the size of blade and the saw-bench capacity. Dimension saw benches like the one illustrated in Figure 9.15 are capable of carrying out a variety of sawing operations with extreme accuracy and produce a sawn surface which almost gives the appearance of having been planed.

The main features which enable such a variety of operations to be done are:

- an adjustable double fence (tilt and length) – this adapts to suit both ripping and panel sawing
- a cutting-off gauge – this allows straight lengths, angles or single and double mitres to be cut
- a tilting saw frame and arbor (main spindle) – this facilitates bevelled cutting, etc.

Figure 9.15 'Thomas Robinson's' dimension and variety saw. **A** – Saw blade; **B** – Riving knife; **C** – Crown guard; **D** – Crown guard adjustment; **E** – Main table; **F** – Adjustable fence (will tilt up to 45°); **G** – Handwheel; saw tilt rise and fall; **H** – Handwheel; saw tilt adjustment; **I** – Control; start button; **J** – Combined brake and stop lever; **K** – Isolator; **L** – Tilting saw frame; **M** – Sliding table stop (adjustable); **N** – Mitre and crosscut fence and gauge; **O** – Sliding table (rolling)

- a draw-out table – for access to the saw arbor, for saw or cutter changing
- a rolling table – for panel cutting and squaring.

A more modern dimension saw is shown in Figure 9.16.

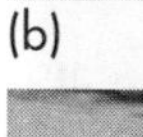

Figure 9.16 'Wadkin' AGS 250 tilting arbor dimensioning saw

9.4.1 Crosscutting on a Dimension Saw Bench

Figure 9.17 shows a dimension saw with a travelling (rolling) table being used in conjunction with a cross-cutting fence to make square (Figure 9.17a), angular (Figure 9.17b) and (tilting the saw blade) compound angle (Figure 9.17c) cuts.

(a)

(b)

(c)

Figure 9.17 Cross-cutting with a dimension saw, rolling table and fence. **Note:** The crown guard must be lowered before cutting begins – in (c) it is shown raised only for the sake of clarity

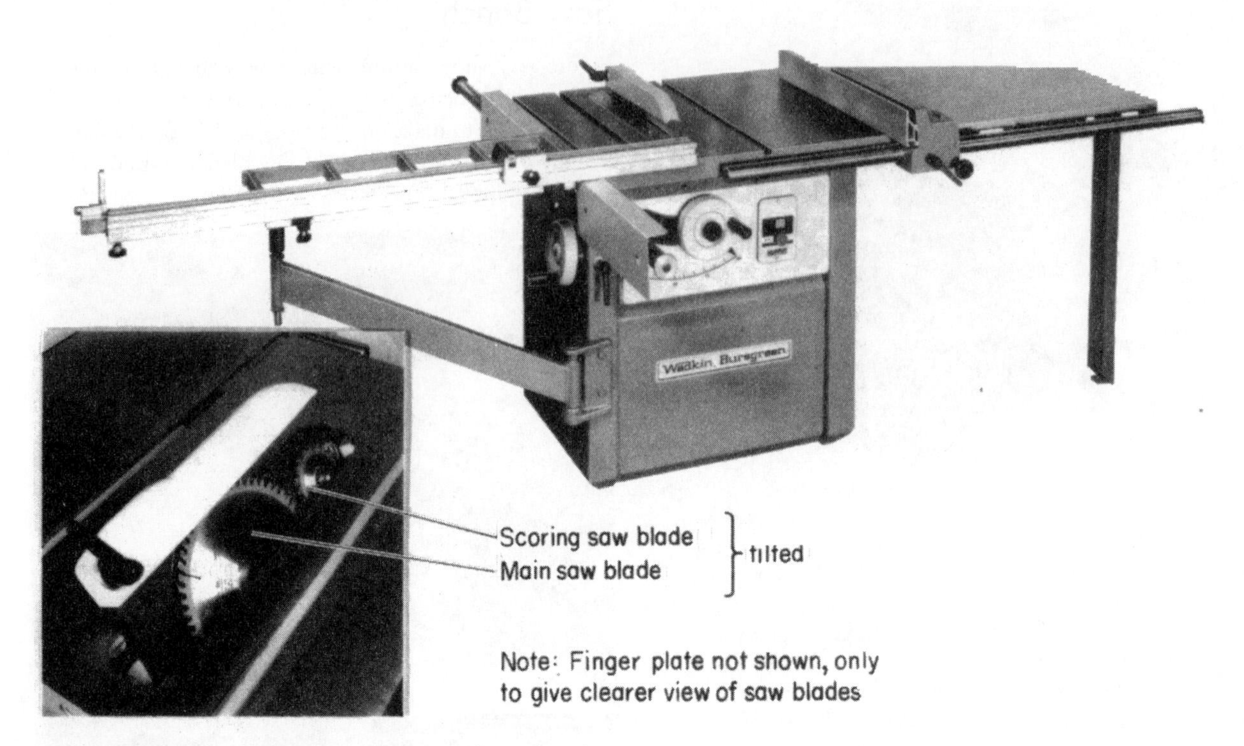

Figure 9.18 'Wadkin Bursgreen' AGSP panel saw bench

9.5 PANEL SAWS

Figure 9.18 shows a panel saw which is similar in design to the modern dimension saw, but has features dedicated to the cutting of sheet materials only. These features include:

- a smaller diameter blade – due to the limited thickness of the material to be cut
- a crown guard attached to the riving knife – removing the need for a support pillar (as on the dimension saw), restricting the sheet size that can be cut
- an extended sliding/rolling table – to cope with full sheet size
- a scoring saw (optional, see insert Figure 9.18) – a small diameter blade, pre-cutting the underside of veneered panels, thereby preventing breakout.

9.6 PLANING MACHINES

Planing machines generally fall into three groups:

1. hand-feed planers and surfacers
2. thicknessers or panel planers
3. combined hand- and power-feed planers.

The quality of finish produced by these machines will to a large extent depend upon the rate at which the timber is passed over the cutter or, in the case of the thicknesser, under the cutter, the speed at which the cutters revolve around their cutting circle (cutting periphery) – on average about 1800 m/min (30 m/s) – and the number of blades on the cutter block.

Close inspection of a planed surface will reveal a series of ripples left by the rotary cutting action of the blade or blades. These marks may have a pitch of 1–3 mm. As shown in Figure 9.19, the shorter the pitch, the smoother the surface finish; therefore speeding up the rate of planing the timber would mean degrading its surface finish.

These machines are used to smooth the surface of the wood and reduce sawn timber to a finished size. The first operation, known as flatting, must produce a face side which is straight, flat and twist free. This is followed by straightening a face edge which must be square to the face side, known as edging. The timber is then reduced in thickness by planing the opposite faces parallel throughout their length.

9.6.1 Hand-feed Planer and Surfacer

This is used for planing the face side and face edge. Figure 9.21a illustrates a typical traditional surfacing machine; Figure 9.21b shows a more modern machine. Its long surfacing table supports the timber as it is passed from the infeed table over the circular cutter block to the outfeed table. In accordance with the

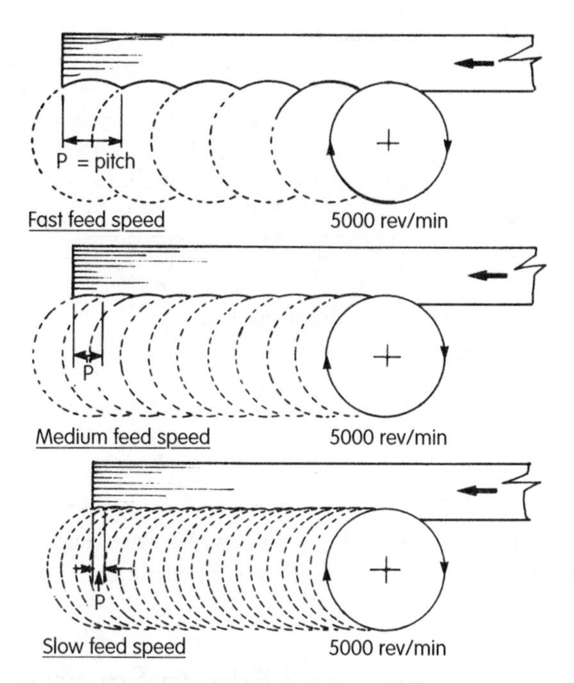

Figure 9.19 Examples of how a constant cutter speed can produce different surface finishes with different feed speeds

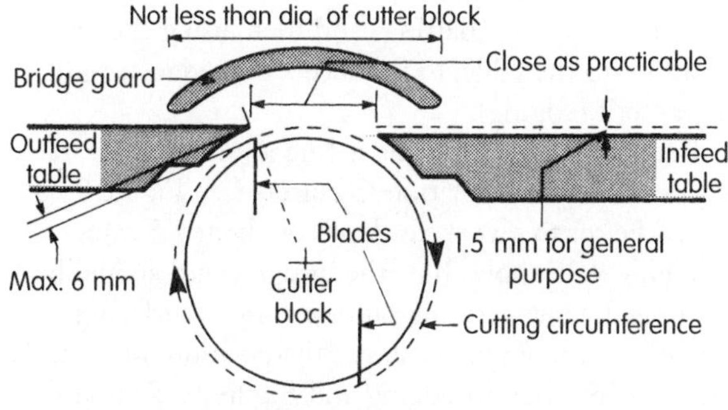

Figure 9.20 Cutters, table and bridge guard arrangement for surfacing

regulations, cutter guards (bridge and back guards) must always be in place during all planing operations. The fence can be moved to any position across the table and be tilted for bevelled work.

Some surfacing machines include in their design facilities to carry out such functions as rebating and moulding. Operations such as these require special methods of guarding the cutter and must not be carried out unless these are provided.

9.6.2 Thicknesser or Panel Planer

A thicknesser or panel planer such as the one illustrated in Figure 9.22 is used for the final part of the planing process, i.e. reducing timber to its finished size.

After the table has been set to the required thickness, timber is placed into the infeed end where it is engaged by a serrated roller which drives it below a cutter block.

(a) Thomas Robinson's hand feed planer and surfacer
A – Pressure bars (holding down springs); **B** – Adjustable fence (will tilt up to 45°); **C** – Infeed table; **D** – Handwheel, infeed table height adjustment; **E** – Handwheel, infeed table lock; **F** – Main frame; **G** – Control; start and stop buttons; **H** – Handwheel, outfeed table lock; **I** – Side-support table (extra support while rebating); **J** – Handwheel, outfeed table height adjustment; **K** – Outfeed table (delivery table); **L** – Telescopic bridge guard

(b) 'Wadkin' S400 surface planer

Figure 9.21 Hand-feed planer and surfacer machines

The machined piece is then delivered from the other end by a smooth roller. Two anti-friction rollers set in the table prevent any drag. The arrangement is shown in Figure 9.24.

9.6.3 Combined Hand- and Power-feed Planer

This is one machine, capable of both surfacing and thicknessing. Figure 9.23 shows a modern combined machine.

Unlike the purpose-made thicknesser, when these machines are used for thicknessing, that part of the cutter block which is exposed on the surfacing table must be effectively guarded throughout its length.

Figure 9.22 'Wadkin Bursgreen' roller feed planer and thicknesser. A – Thicknessing table (infeed); B – One of two anti-friction rollers; C – Cutter block guard and chip chute; D – Thickness scale; E – Control: start and stop buttons; F – Feed speed selector switch (4.5 m/min and 9m/min); G – Handwheel; raise and lower table; H – Lever, table lock; I – Outboard roller

Figure 9.23 'Wadkin' BAO/S Surface planer and thicknesser

9.6.4 Guarding the Machines

The thicknesser has its cutters enclosed but the surfacers and combined machines must be provided with a bridge guard (front and back) of a width not less than the diameter of the cutter block (Figure 9.20) and long

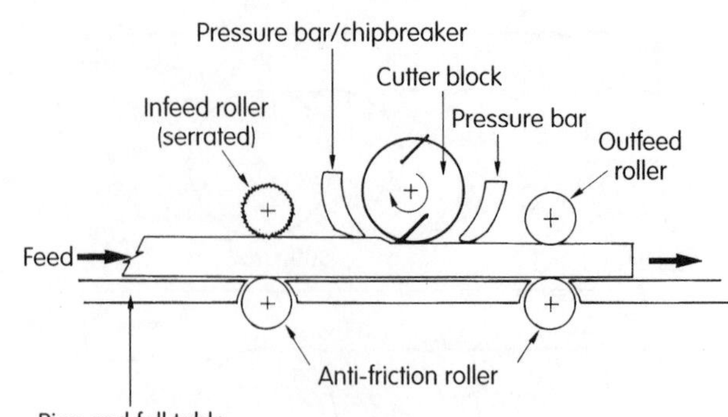

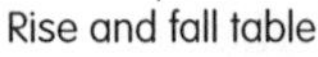

Figure 9.24 Thicknesser table and cutter block arrangements

enough to cover the whole of the table gap no matter what position the fence is in or what operation is being carried out.

9.6.5 Flatting and Edging Squared Timber

The object of this process is first to produce one perfectly flat face side (flatting) and then, using the fence set at 90° to the table, to produce a face edge at right angles to it (edging).

Before operations of this nature are carried out, table positions and their relation to cutters (see Figure 9.20) should be set to give a cut depth of about 1.5 mm.

Figure 9.25 shows how the bridge guard should be positioned when carrying out the various surfacing operations. Figure 9.25a shows the position for flatting, Figure 9.25b that for edging and Figure 9.25c that for when flatting and edging are carried out one after the other.

Hand positions for flatting are shown in Figure 9.26 – hands must never be positioned over the cutter. Figure 9.26a shows the approach position. Once the timber has passed under the bridge guard, the left hand is repositioned on the delivery side of the table (Figure 9.26b). As the process continues, the right hand follows (Figure 9.26c), the timber being pressed down on to the table during the whole operation. The whole process is repeated until the desired flatness is obtained.

In Figure 9.27 edging is being undertaken with the bridge guard set as shown in Figure 9.25c; again, the process is repeated until the edge is both straight and square to its face side.

When dealing with slightly bowed or sprung timber (badly distorted timber should be shortened or sawn straight, using a jig if necessary), it should be positioned fully on the infeed table (round side/edge up or hollow side/edge down) and then passed over the cutters by making a series of through passes until straightened.

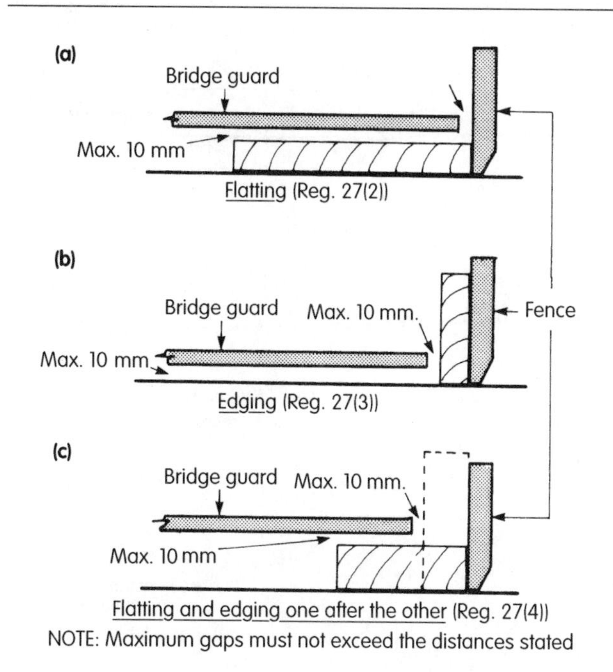

Figure 9.25 Positioning the bridge guard for flatting and edging operations

Where, for reasons of safety, short pieces of timber cannot be planed in accordance with Figure 9.25a, a pushblock (Figure 9.28) offering a firm and safe handhold should be used. Figure 9.29 shows a pushblock in use.

Wherever possible, the direction of wood grain should run with the cutters (Figure 9.30). In this way, tearing of the grain is avoided and a better surface finish is obtained.

9.6.6 Thicknessing

This process involves pushing a piece of timber, face down, into the infeed end of the machine, where it will be met and gripped by the fluted (serrated) infeed roller and mechanically driven under the cutters. Depending on the size and make of machines and the thickness scale setting, the cutters could remove up to 3 mm from the thickness of the timber, after which it is steadied by the outfeed roller (a smooth-surfaced roller, so as not to bruise the surface of the wood) and delivered from the machine with its upper surface planed smooth and parallel with its underside. Any friction between the underside of the timber and the table bed can be reduced by adjusting the two antifriction rollers to suit the condition of the timber. Provided the timber is dry, best results are obtained with the antifriction rollers set as low as practicable.

For effective and safe production techniques, the operative should be totally aware of the machine's capabilities (see the manufacturer's operating manual) and how it may be used safely.

(a)

(b)

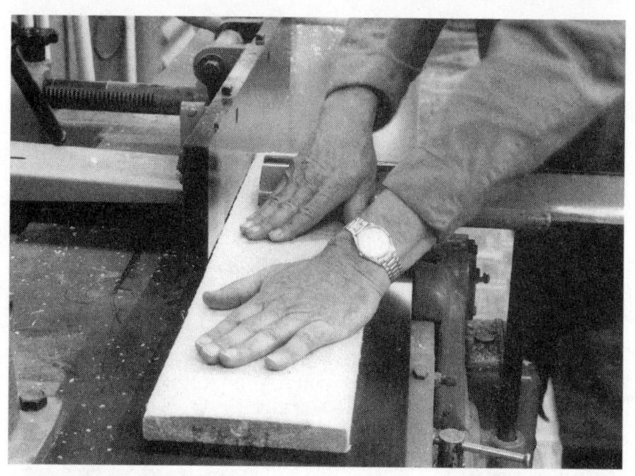

(c)

Figure 9.26 The process of flatting timber before edging

- For a smooth flat finish, the underface of the timber should be straight, flat and smooth, feed speed should be slow and the anti-friction roller should be set as low as practicable without causing the timber to stick.
- Feed timber so that it is cut with the grain.
- Timber must be allowed clear exit from the machine. Lengths must be limited to well within the distance between the outfeed outboard roller and any obstacle.

Figure 9.27 Edging after flatting

- Suitable means should be found for supporting long lengths of timber at the delivery end, otherwise the end of the cut will be stepped. Either a purpose-made stand or a proprietary roller stand could be used.
- Hands must be kept clear of the infeed end of the machine or they may become trapped between the timber and the table or even be drawn into the machine. Figure 9.31 shows two of the danger areas.

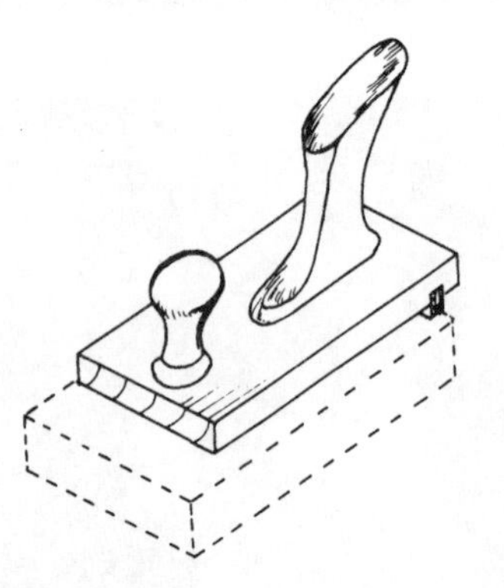

Figure 9.28 Push block with suitable handhold

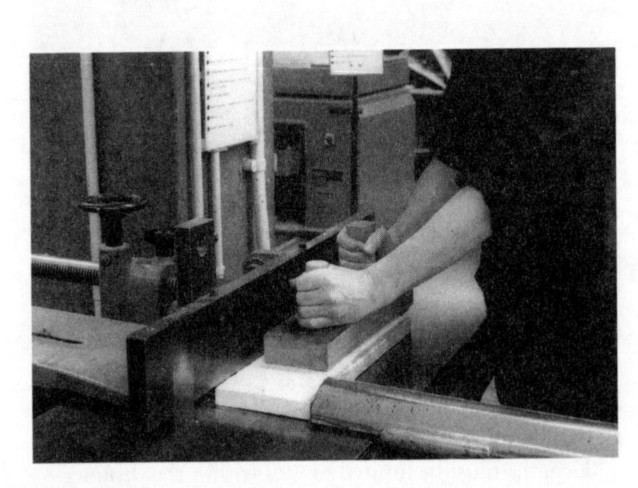

Figure 9.29 A push block in use

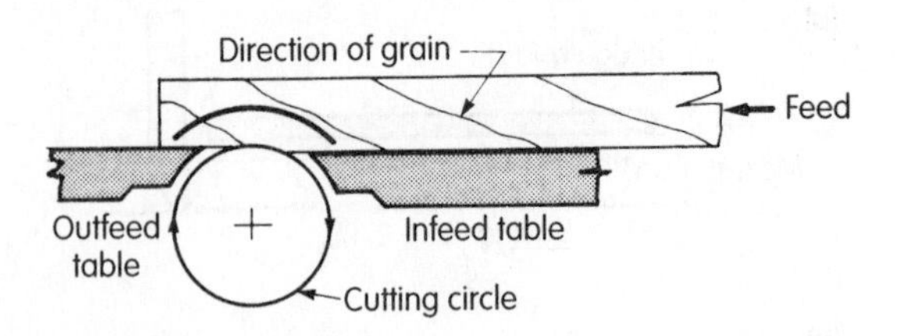

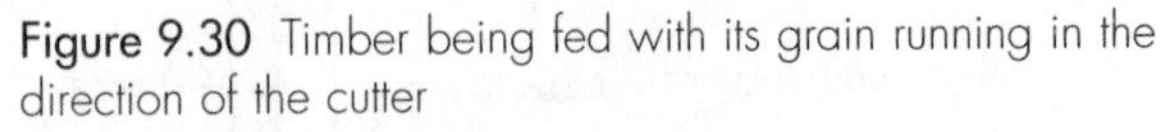

Figure 9.30 Timber being fed with its grain running in the direction of the cutter

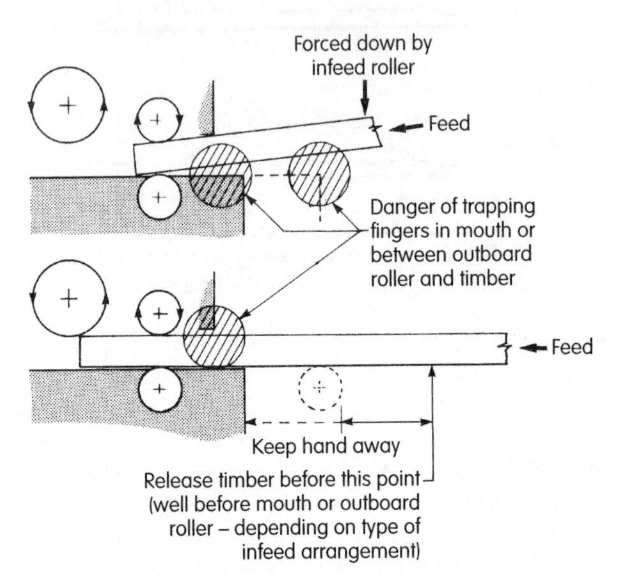

Figure 9.31 Possible danger areas arising from incorrectly feeding timber into a thicknessing machine

- The possibility of a machine ejecting timber back towards the operator, due to the direction of cutter rotation, is counteracted by using either a sectional infeed roller or an anti-kickback device which allows timber to travel towards the cutters but locks on to any backward movement.

Note: Unless suitably modified with anti-kickback devices, sectional feedrollers or other mechanical devices, pre-1974 thicknessing machinery must only be used to plane one piece of timber at a time (PUWER Regulation 5 – Suitability of Work Equipment). An easily understood notice must be prominently displayed to this effect (PUWER Regulation 24 – Warning) (Figure 9.32).

If machines are used to thickness more than one piece at a time, care must be taken to ensure that the position of the pieces on the table is such that they are restrained from being ejected, as shown in Figure 9.33. Figure 9.34 shows the motor drive and feed arrangement of a Wadkin BAO 300 thicknesser. The infeed roller is spirally serrated and its anti-kickback device is in the form of fingers.

Figure 9.32 Thicknesser displaying a notice stating that 'Not more than one piece of material should be fed into this machine at any one time'

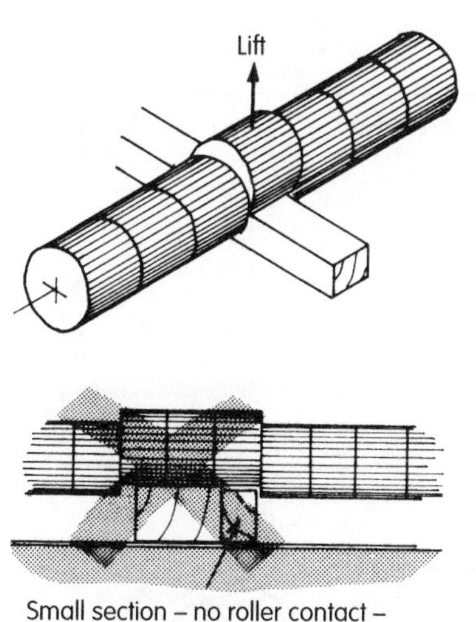

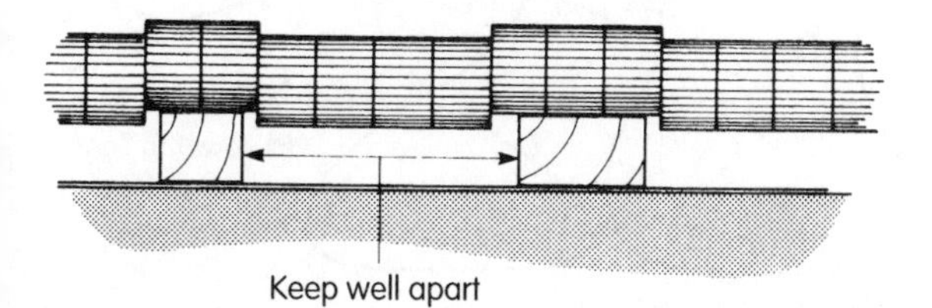

Figure 9.33 Thicknessing more than one piece of timber at a time with a recessional feed roller

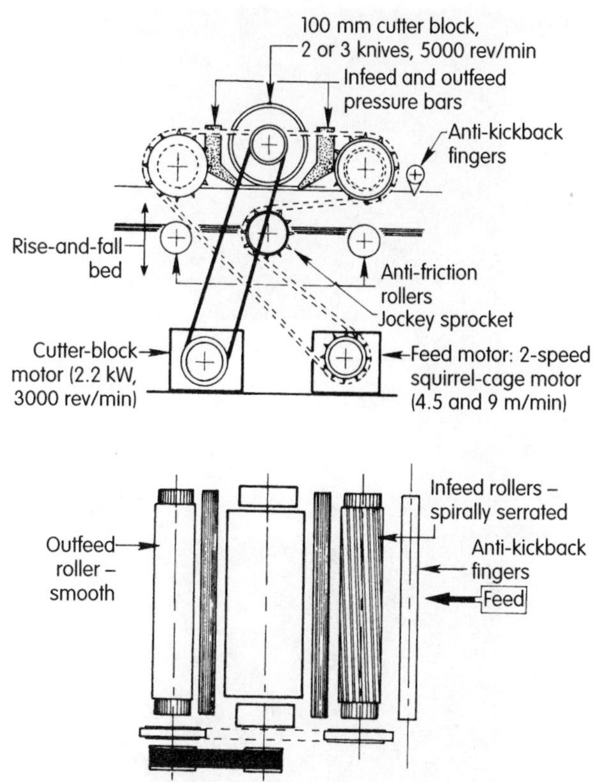

Figure 9.34 General arrangement of a 'Wadkin' BAO 300 roller-feed planer and thicknesser. Note: for clarity, guards are not shown

9.7 NARROW BANDSAW MACHINES

These machines have blades which do not exceed 50 mm in width. The bandsaw illustrated in Figure 9.35 has a maximum blade width of 38 mm.

These smaller machines are used for all kinds of sawing, from cutting freehand curves – the radius of which will depend on the blade width – to ripping, deepcutting and crosscutting. By tilting the table, angled and bevelled cuts can be made. The machines consist of a main frame on which two large pulleys (wheels) are fixed. The bottom wheel is motor driven, while the top wheel is driven by the belt action of the saw blade. The pulley (wheel) rims are covered with a rubber tyre to prevent the blade slipping and to protect its teeth.

To facilitate blade tensioning and alignment, the top wheel can be adjusted vertically and tilted sideways. The amount of tension will depend on the blade width; incorrect tension could result in the blade breaking.

Saw guides give side support to the blade above and below the table while cutting takes place. Back movement of the blade is resisted by a thrust wheel or disc which revolves whenever the blade makes contact. The position of the guides and thrust wheel relative to the blade is critical if efficient support to the blade is to be

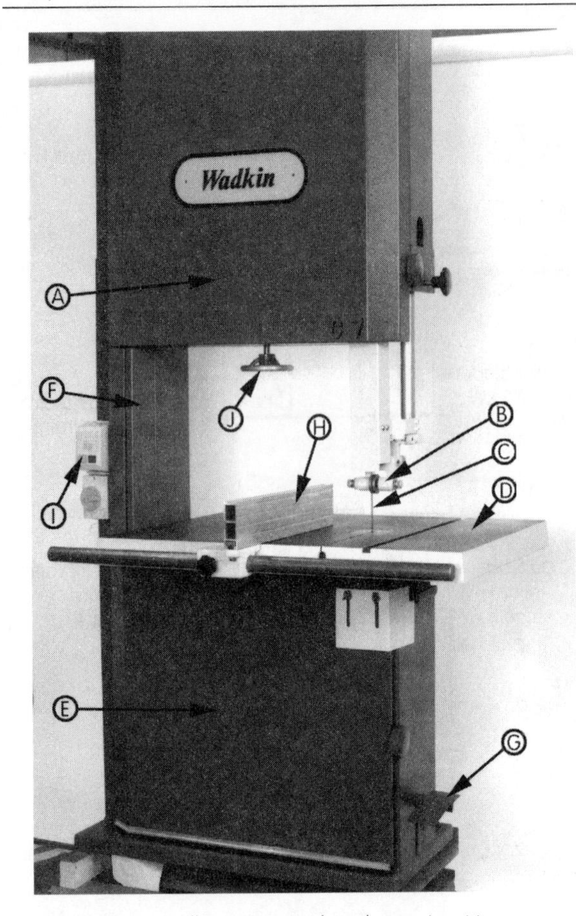

Figure 9.35 'Wadkin' C series bandsaw. **A** – Upper swing-away guard door; **B** – Saw guides and thrust wheel (adjustable guard); **C** – Band saw blade; **D** – Table; **E** – Lower swing-away guard/door; **F** – Main frame; **G** – Foot brake; **H** – Fence; **I** – Control, start and stop buttons; **J** – Handwheel; for regulating saw tension

maintained. To this end, they are provided with a range of adjustments to allow for varying blade widths. The assembly of guides, thrust wheel and blade guard adjusts vertically so that it can be positioned as close as practicable to the workpiece (see Figure 9.37).

For the bandsaw to cut safely and effectively, it should be fitted with a blade which is correctly tracked and tensioned and of a size suitable for the work (Table 9.1).

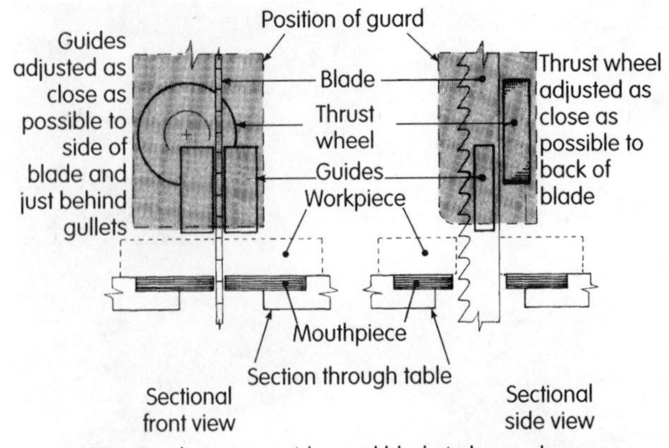

Figure 9.37 Bandsaw guard, guide, thrust wheel and table

These relatively narrow bandsaw blades are supplied and stored in a folded coil consisting of three loops. Figure 9.38 and the following description which follows briefly explain how this is done.

- The blade is held firmly, using gloves, with arms outstretched and palms uppermost.
- When the hands are turned over simultaneously, the blade will twist.
- When the blade is lowered to the floor it will fold into three looped coils.

Warning: Never allow the saw blade to slip or turn in or on to your hand, or severe cuts to the skin could result.

Table 9.1 Minimum radius capable of being cut with a given blade width (see Figure 9.38)

Width of blade (mm)	Radius (mm)
3	3
5	8
6	16
10	37
13	64
16	95
19	138
25	184

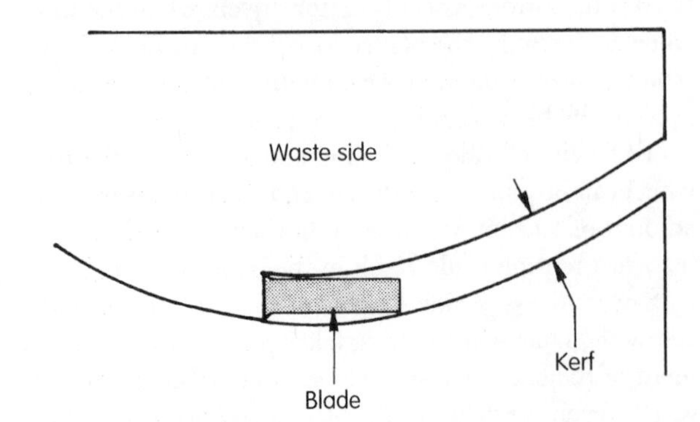

Figure 9.36 Bandsaw blade and kerf (sides of blade must not rub on sides of kerf)

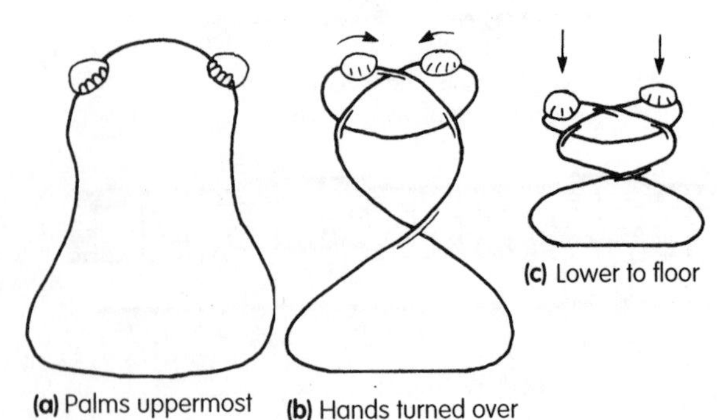

Figure 9.38 Folding a bandsaw blade with gloved hands

9.7.1 Fitting a Bandsaw Blade

Fitting or replacing a saw blade will mean opening the upper and lower guard doors for the purpose of access; therefore, before the start of any such work, the machine must be isolated from its electricity supply. Then, and only then, the work may begin.

1 Open and move aside the top and bottom guard doors.
2 Remove or move aside other obstructions; for example the blade top guards, guides and the table mouthpiece.
3 Using the appropriate handwheel, lower the top wheel enough to enable the blade to fit on to both wheels.
4 Fit the blade on to the wheels, making sure the teeth at the table cutting point are facing downwards, then raise the top wheel sufficiently to hold the blade on to the wheels.
5 Track the saw blade (Section 9.7.3).
6 Tension the saw blade (Section 9.7.4).
7 Reset thrust wheels and guides (Figure 9.37).
8 Reposition and secure guards.

9.7.2 Blade Length

When measuring the length of a band saw blade, the amount of tensioning adjustment should be taken into account (Figure 9.39).

The following formula can be used:

[Length of blade required] = πD + 2 × [maximum distance between wheel centres] − [tension allowance]

For most industrial saws, a tension allowance of 50 mm is adequate.

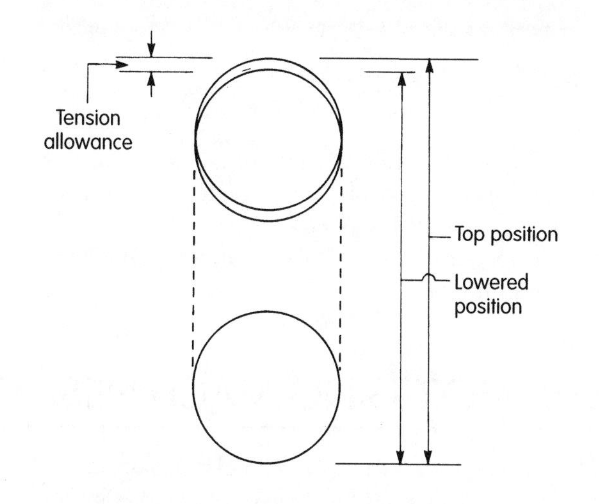

Figure 9.39 Top-wheel allowance for tensioning purposes

For example, to find the length of a band saw blade when the wheels are 500 mm diameter and the maximum distance between wheel centres (top wheel fully raised) is 1250 mm:

Length of blade = (3.142 × 500 mm) + (2 × 1250 mm) − 50 mm
= 1571 mm + 2500 mm − 50 mm
= 4021 mm

9.7.3 Tracking

This is a means by which the blade is made to run in a straight line between the top and bottom wheels, while passing centrally over their tyres without 'snaking'. The effect of snaking would mean that the back of the blade would intermittently hit the thrust wheel (or thrust roller).

Tracking is checked by turning the top wheel slowly, by hand, in a clockwise direction, adjusting the wheel-tilting device until true alignment is achieved. The blade is then tensioned.

Warning: A blade must never be tracked with the motor running.

9.7.4 Tensioning

Bandsaw blades must always be tensioned correctly, otherwise serious damage could result and a dangerous situation could ensue.

The amount of tension applied to the blade via the raising of the top wheel should be such that the blade can be pulled 6 mm away from its running line centrally between the wheels. Alternatively, in the case of machines which have a built-in tensioning scale, tension is applied until the pointer reaches the point on the scale which corresponds with the width of the saw blade being fitted.

9.7.5 Use

During all sawing operations the top guard must be set as close to the workpiece as practicable. This not only protects the operative from the saw blade but also provides the blade with maximum support via the guides and the thrust wheel or thrust roller assembly (Figure 9.37).

Figure 9.40 shows a straight cut being made with the aid of a ripping fence. Towards the end of the cut, the pushstick will be used to push the side nearest the fence.

When cutting into a corner (Figure 9.41), short straight cuts are made first (this also applies when making curved freehand cuts; Figure 9.42). An exception to this rule (unless a jig was used) would be when removing waste wood from a haunch (Figure 9.43), in which case the cuts at A with the grain are made first, to reduce the risk of cutting into the tenon, then the cuts

Figure 9.40 Ripping on a bandsaw

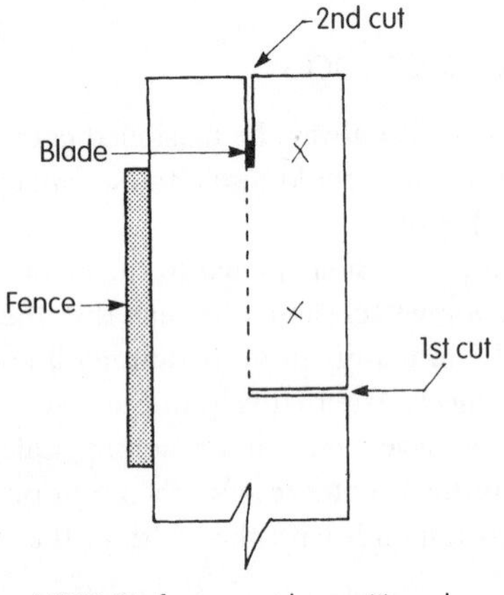

Figure 9.41 Ripping into a corner

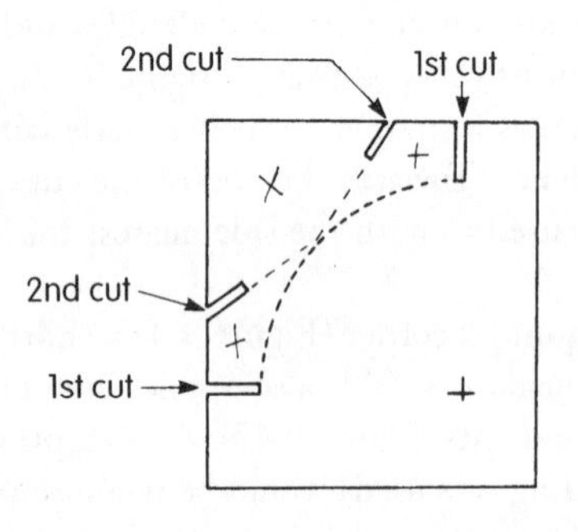

Figure 9.42 Treatment at the corners of a curve about to be cut freehand

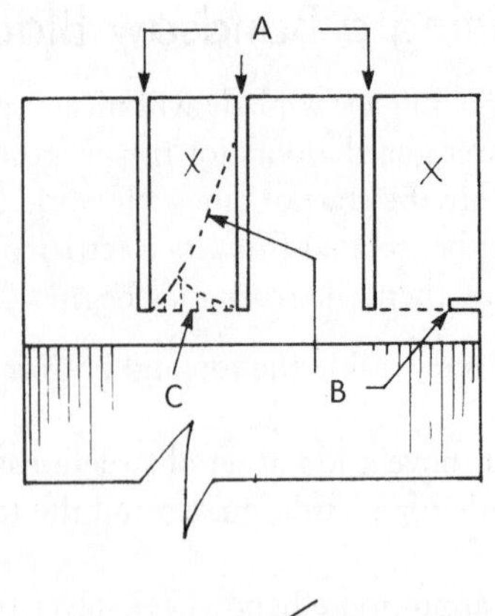

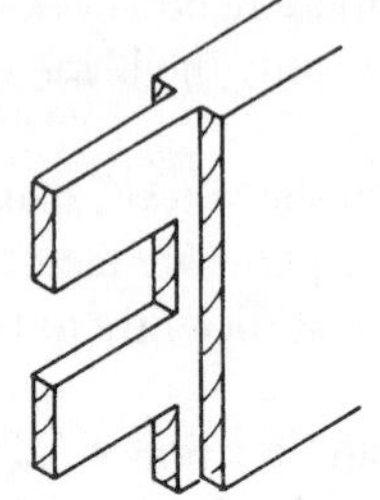

Figure 9.43 Removing waste wood from a haunched tenon

Figure 9.44 Cutting curves freehand

at B. The small portion left at C would, depending on the blade width, probably have to be nibbled away by making a series of short straight cuts.

The operative in Figure 9.44 is using a bandsaw to cut curves freehand.

9.8 MORTISING MACHINES

These machines cut square-sided holes or slots to accommodate a tenon. The hole is made either by a revolving auger bit inside a square tabular chisel or by an endless chain with cutters on the outer edge of each link. Machines are made to accommodate either method or a combination of both.

9.8.1 Hollow Chisel Mortiser

Figure 9.45 shows the various components of this machine (see also Figure 8.12). As the mortising head is lowered, the auger bores a hole while the chisel pares it square. The chippings are ejected from slots in the chisel side. After reaching the required depth, which for through mortise holes is about two-thirds the depth of the material, the procedure is repeated along the desired length of the mortise. At all times during the operation the workpiece must be held secure with the clamp against the table and rear face (fence). It is then turned over and reversed end for end, to keep its face side against the fence, and the process is repeated to produce a through mortise (see Figure 9.46).

In order to cut the square hole cleanly and safely without damaging the cutters or 'blueing' them (as a result of overheating due to friction), a gap of 1 mm must be left between the spur edges of the auger bit and the inside cutting edges of the square chisel, as shown in Figure 9.47. This gap can be obtained by positioning both the chisel and the auger bit in the machine together and tightening the auger bit in the chuck when the chisel is 2 mm below its shoulder seating; then lift the chisel to close the gap and tighten the chisel as shown in Figure 9.48.

Figure 9.45 'Wadkin' hollow chisel mortiser. **A** – Operating levers; **B** – Mortising head; **C** – Hollow chisel and auger; **D** – Clamp (should be faced with a wooden plate); **E** – Handwheel; operates table longitudinal movement; **F** – Handwheel; operates table cross-traverse; **G** – Main frame; **H** – Table stop bar; **I** – Work table (timber packing) and rear face; **J** – Mortising head slideway; **K** – Control; start and stop buttons; **L** – Depth stop bar (mortise depth adjustment)

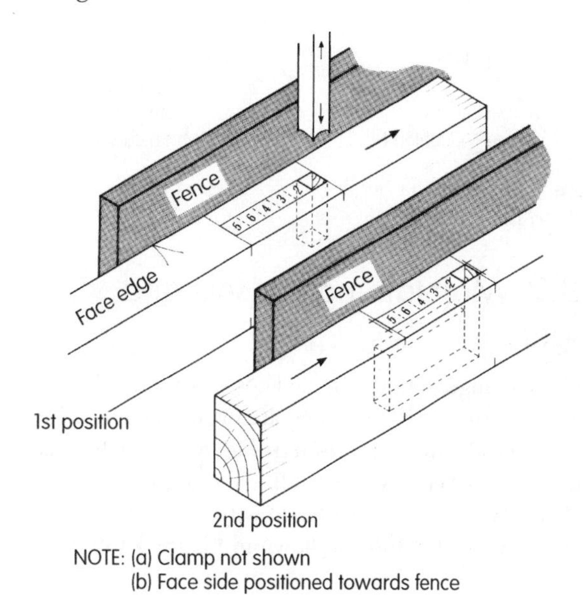

Figure 9.46 Hollow chisel mortiser, cutting a mortise hole

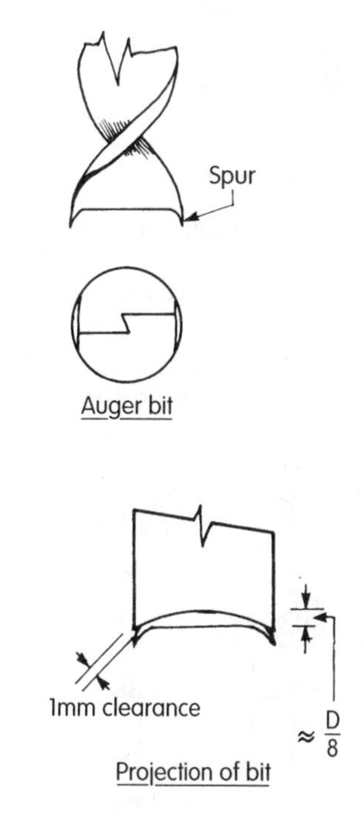

Figure 9.47 Auger bit and square chisel into a mortising machine

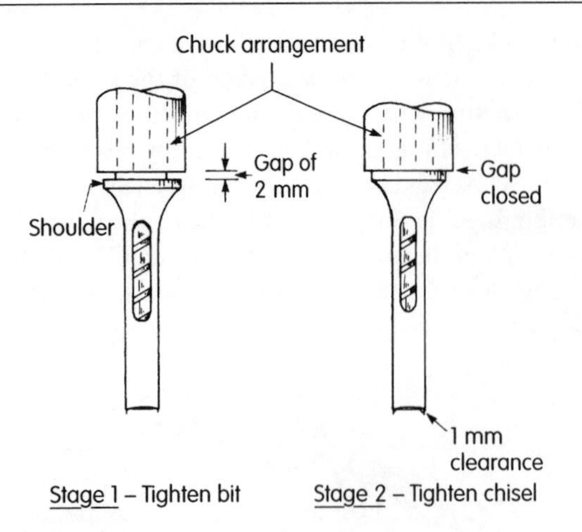

Figure 9.48 Fitting an auger bit and square chisel into a mortising machine

9.8.2 Maintaining Chisels and Augers

Chisel and auger bits should be kept sharp. Chisel sharpening angles (Figure 9.49) are maintained by using a sharpener like the one shown in Figure 9.50, which is used with a carpenter's brace; a fine file is used on the inside corners. Bits are sharpened with a fine file in a similar way to a twistbit, as shown in Figure 5.140.

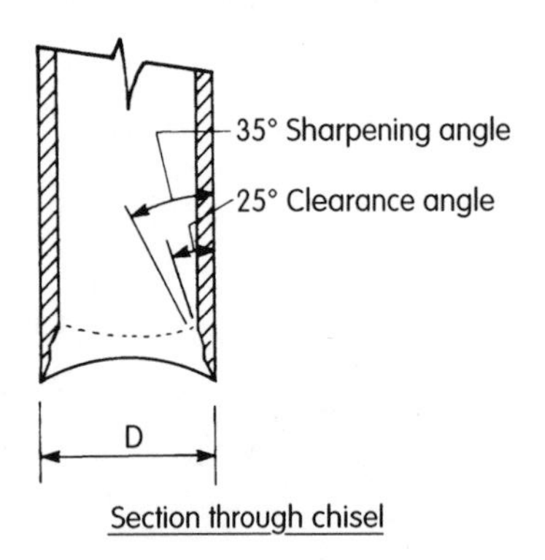

Figure 9.49 Section through a hollow mortiser chisel

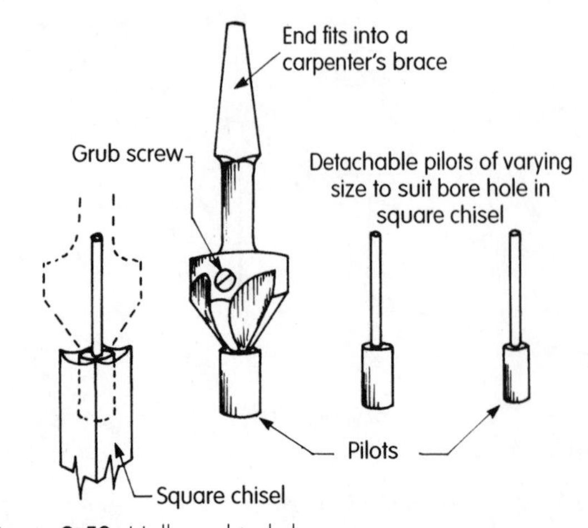

Figure 9.50 Hollow chisel sharpener

9.8.3 Chain and Chisel Mortiser

Figure 9.51 shows a machine which is capable of mortising by chain as well as by hollow chisel.

Figure 9.51 'Thomas Robinson's' chain and chisel mortiser. **A** – Chain/chisel operating levels; **B** – Chisel headstock; **C** – Chisel mortising, depth stop arrangement; **D** – Hollow chisel and auger; **E** – Work table (timber packing) and rear face; **F** – Control reset button; G – Isolator; **H** – Handwheel; table rise and fall; **I** – Handwheel; operates table cross-traverse; **J** – Handwheel; operates table longitudinal movement; **K** – Clamp; **L** – Chipbreaker; **M** – Chain guard and window; **N** – Chain mortising, depth stop arrangement; **O** – Chain mortising headstock

The chain mechanism shown in Figure 9.52 consists of the chain, a guide bar and wheel and a sprocket which turns the chain at high speed. The guide bar section (fully guarded at all times) is lowered into the securely held workpiece to cut a large slot (depending on the chain size) in one operation of the lever arm. To prevent the chain splintering away the surface of the wood on its upward motion, a chipbreaker is used.

Using a chain can be much quicker than a chisel but, because of the semicircular-bottomed hole that is left, it is not suitable for short or stub tenons.

The chain mortiser is regarded as being much more dangerous to use than the hollow chisel machine.

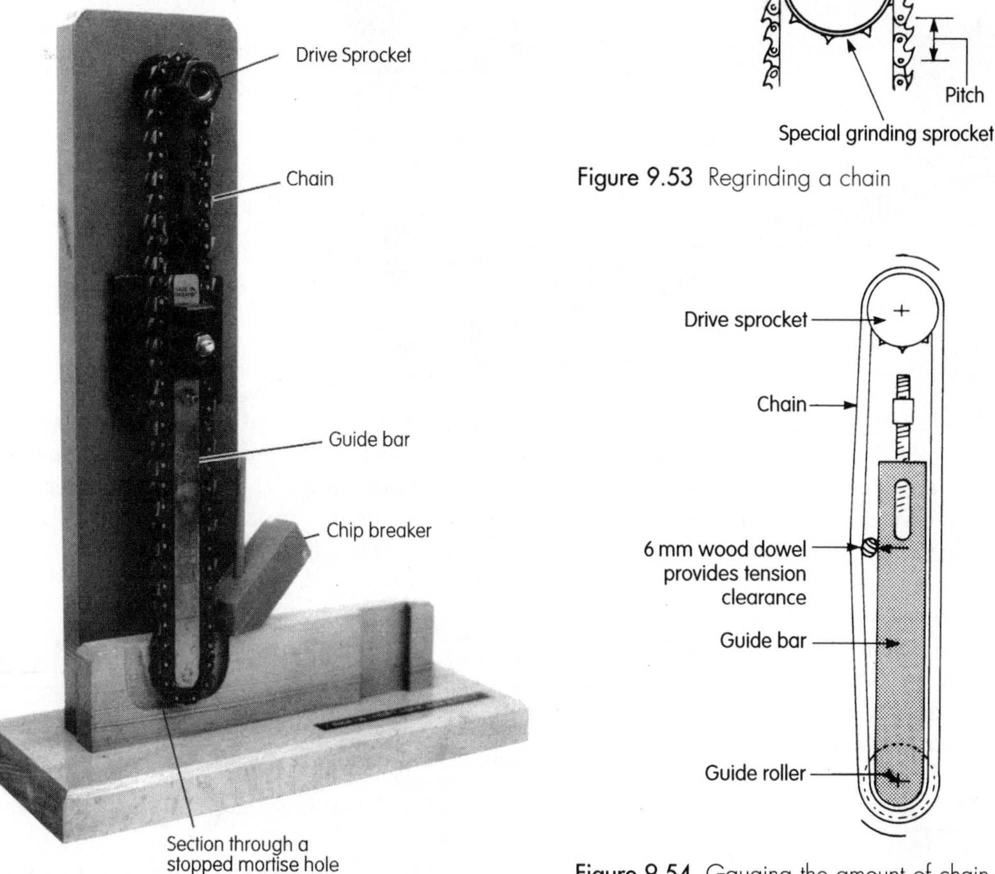

Figure 9.52 Model showing the arrangement of a mortise chain, its mechanism and application

Figure 9.53 Regrinding a chain

Figure 9.54 Gauging the amount of chain slack

9.8.4 Chain Mortising

Chains should be kept sharp, correctly adjusted for tension and effectively guarded at all times.

Chains are sharpened on the cutting face of each link with an oil-coated slipstone. Regrinding (Figure 9.53) is done on the grinding assembly at the rear of the machine. When grinding, care must be taken not to cut too deep into the gullet, as this would weaken the chain. Chain tension between the drive sprocket and the guide bar wheel should be such as to permit 6 mm of chain slack, as shown in Figure 9.54.

The chain must be fully guarded at all times during the mortising operation; while the chain guard rests on the workpiece, shrouding the mortise hole, the chaining mechanism is allowed to rise or fall as necessary. The hardwood chipbreaker should be sited on the workpiece and close to the chain, thus preventing the upward cutting action of the chain from splitting away the end of the mortise hole. Similarly, where through-mortise holes are to be cut, it will be necessary to plant and securely fix a length of 32 mm thick packing into or onto the bottom of the table to prevent the chain from splitting wood away as it passes through the workpiece. Fixing the packing is of vital importance, otherwise it would be driven out from under the workpiece at dangerously high speed as soon as the chain made contact with it.

Note: At no time should the table be traversed while the chain is cutting a mortise, otherwise the guide bar will be strained.

9.9 SANDING MACHINES

The larger of these machines, like the one illustrated in Figure 9.55, are used mainly to remove marks left by the rotary action of the planers and any ragged grain

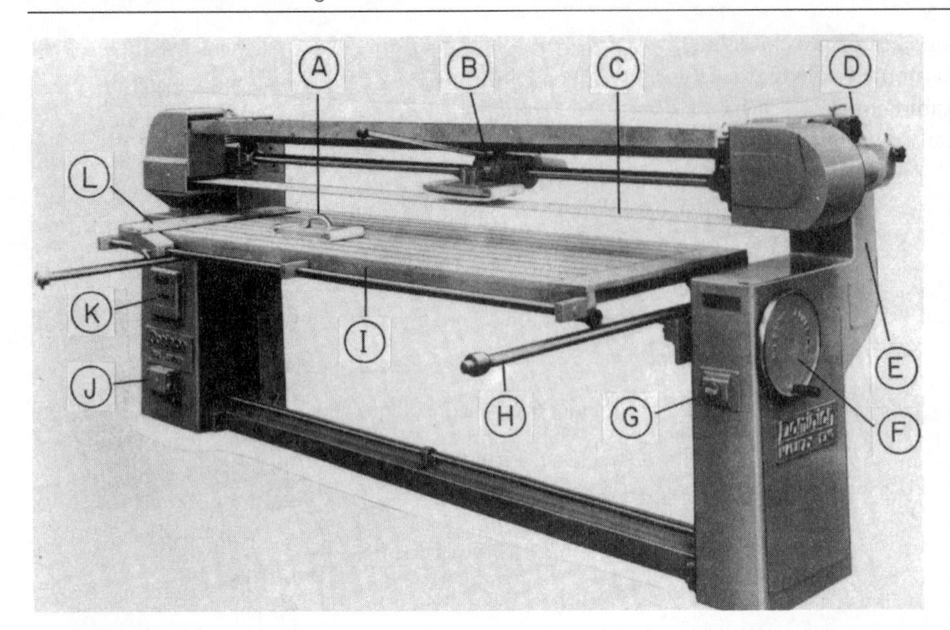

Figure 9.55 'Dominion' under and over pad sanding machine. A – Hand pad; B – Travelling pressure pad; C – Sanding belt; D – Belt tracking and tensioning device; E – Swan–neck; accommodates long work; F – Handwheel; table rise and fall; G – Stop button; H – Table rails; I – Laminated wood table; J – Isolator; K – Control start and stop buttons; L – Table fence

incurred during the planing process and also to dress (flatten) any uneven joints left after the assembly of such joinery items as doors and windows. Smaller machines, like the combined belt and disc sander in Figure 9.56, can be used with great accuracy for dressing small fitments, trueing and trimming endgrain and sanding concave or convex surfaces.

A separate dust extraction system is essential with all sanding machines. Wood dust can not only be offensive but can also, depending on the wood species, cause dermatitis and become a contributory factor towards respiratory diseases. The Control of Substances Hazardous to Health Regulations (COSHH) deems hardwood dust to be hazardous to health. These stipulate that measures must be taken to control the amount of dust allowed in the atmosphere; this can be achieved by efficient extraction at the machine dust outlet.

Figure 9.56 'Wadkin Bursgreen' disc and belt sander. A – Sanding disc (405 mm dia.); B – Sliding adjustable swivel fence; C – Sanding disc table (will tilt −10° to + 45°); D – Dust exhaust; E – Sanding belt table (horizontal or vertical positions); F – Small diameter idler pulley (for sanding internal curves); G – Sanding belt; H – Workpiece stop

9.10 GRINDING MACHINES

Some kind of grinding machine is essential in the workshop, to carry out such functions as:

- reforming the grinding angle on chisel and plane blades
- regrinding and shaping hand and machine cutters
- resharpening screwdriver blades
- resharpening cold and plugging chisels.

All the above operations can be done on a dry grinding machine like the one in Figure 9.57. Using this type of grinding machine requires considerable skill in preventing the part of the blade or tools which comes in contact with the high-speed abrasive wheel from becoming overheated (indicated by a blue colour) and losing its temper (hardness). The end of the blade or tool should be kept cool by periodically dipping it into a dish of water, which should be close at hand.
Because of the small diameter of the grinding wheel, chisel and plane blades will have a grinding angle which is hollow (hollow ground), as seen in Figure 9.58. This profile is preferred by many joiners as it tends to last longer than a flat ground angle.

The problem of overheating can be overcome by using a Viceroy sharp-edged horizontal grinding machine, shown in Figure 9.59. This machine has a built-in coolant system supplied with a special honing oil which flows continually over and through the surface of the grinding wheel while it is in motion. This

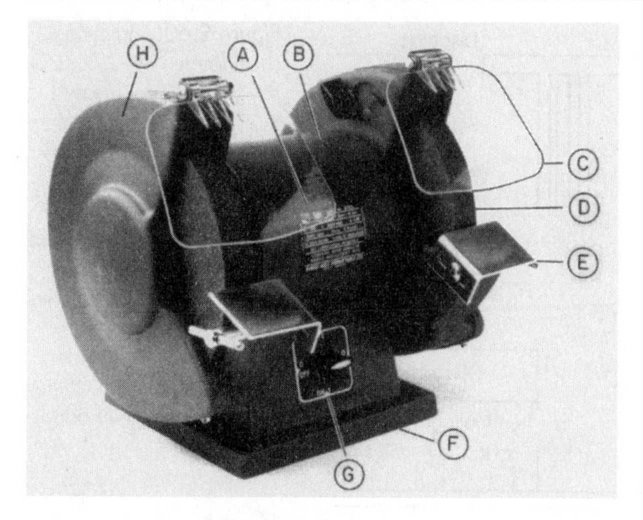

Figure 9.57 Bench grinder. A – Specification plate; B – Motor; C - Eye shield; D – Grinding wheel; E – Adjustable tool rest; F – Base plate (fix to bench or stand); G – Machine control; H – Grinding wheel guard

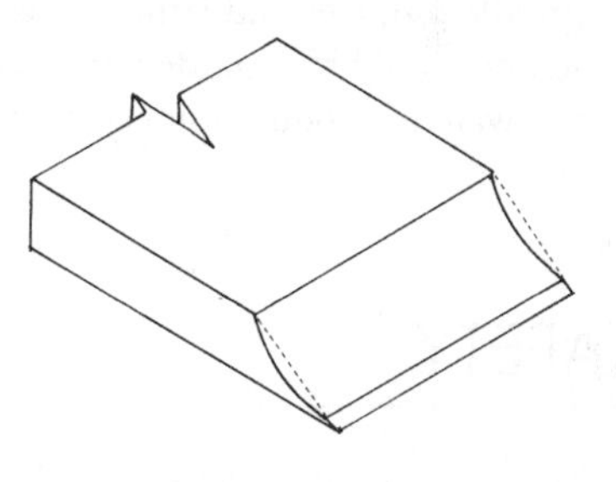

Figure 9.58 Hollow ground blades

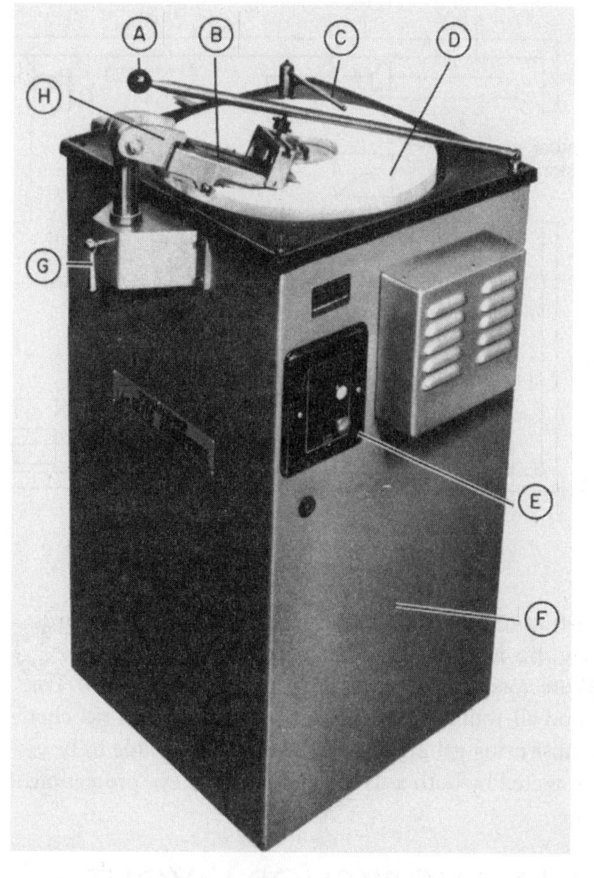

Figure 9.59 'Viceroy' sharp edge machine. A – Operating lever; B – Plane lever holder; reverse side accommodates chisels; C – Honing oil distribution bar; D – Grinding wheel; E – Control; start and stop buttons; F – Cabinet; houses honing oil container and motor; G – Column height adjusts to suit angle of master arm; H – Master arm

machine is primarily used for grinding flat chisel and plane blades; but there are attachments available which allow gouges and small machine blades to be ground.

9.10.1 Legislation and Use of Grinding Machines

As with woodworking machinery, legislation concerning the use of grinding machines is contained within the Provision and Use of Work Equipment Regulations 1992. Prior to this, the Abrasive Wheel Regulations 1970 were specifically for grinding machines. Guarding and safe practices mentioned in the 1970 specific regulations are considered a minimum requirement and to cover the following aspects:

- the maximum permissible speed of wheel to be specified; overspeeding could cause the wheel to burst
- the provision and maintenance of guards and protection flanges
- an effective means of starting and cutting off the motive power
- workrests to be adjusted as close as practicable to the wheel whenever the machine is in use (common practice being no more than 3 mm away), otherwise the workpiece could become jammed between the wheel and rest, causing serious injury
- the condition of the floor where the machine is to be used.

Mounting of abrasive wheels must be undertaken only by qualified and appointed persons.

The Personal Protective Equipment at Work Regulations 1992 stipulate the use of eye protection when carrying out grinding operations where there is a possibility of chippings or abrasive materials being propelled. This would include dry grinding, wheel dressing (an operation for removing metal particles which have become embedded in the wheel) or trueing (to keep the wheel concentric with the spindle). Suitable eye protection, which may be included in the machine design (transparent screen or shield) and/or worn by the

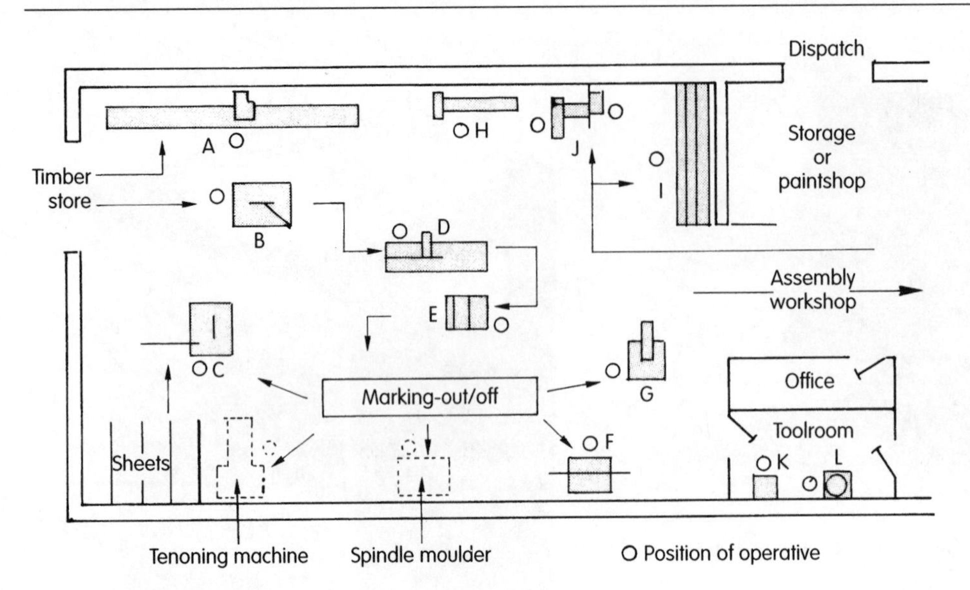

Figure 9.60 Machine shop layout. **A** – Crosscut saw; **B** – Circular saw bench (rip and general purpose); **C** – Dimension saw bench; **D** – Hand-feed planer and surfacer; **E** – Roller-feed planer and thicknesser; **F** – Mortiser; **G** – Bandsaw; **H** – Under and over belt sander; **I** – Belt and disc sander; **J** – Dry grinding machine; **K** – Horizontal wet grinding machine

operator, should conform to existing British Standards (e.g. BS EN 166, 167 and 168 Specification for Eye Protectors for Industrial and Non-industrial Uses). For good all-round protection against unsuspected ricochet whilst using grinding machinery, it is advisable to be protected by both a screen and personal eye protection.

9.11 WORKSHOP LAYOUT

Before any decision is made with regard to workshop layout some, if not all, of the following factors should be considered:

- the size of firm
- the type of work
- the available space (a clear space of 900 mm plus the maximum length of material to be handled should be allowed around three sides of every machine)
- the workforce – whether full-time wood machinists are to be employed
- the number and type of machines likely to be cost effective
- the methods of providing chip and dust extraction systems
- the provision for a tool room (for tool and machine maintenance)
- the storage and racking facilities
- suitable and adequate lighting
- a suitable and sufficient power supply.

Ideally, for a machine shop to work with maximum efficiency, machines should not occupy valuable space unless they are used regularly or contribute towards a steady flow of jobs through the workshop.

Figure 9.60 shows how the machines mentioned in this chapter could be positioned to produce a work flow to suit a small to medium-sized joiner's shop. The overall layout does, however, allow for the inclusion of a tenoner and a spindle moulder and lathe at a later date. These extra machines could be regarded as essential if the machine shop were to produce joinery items on a production basis.

9.12 SAFETY

No machine must be used unless permission has been granted and the prospective operative is qualified in its use and is fully conversant with, and able to act upon, all matters of safety (PUWER Regulation 8).

It is important, therefore, that all trainees make an indepth study of all aspects of woodworking machinery safety. Listed below are areas of study together with some of their relevant safety factors.

9.12.1 Provision and Use of Work Equipment Regulations 1992

These regulations apply to all machines, not just those utilised in the timber/construction industries. They are therefore of general legislative application. However, they are legal requirements and should be understood by all who purchase and/or use woodworking machinery. Guidelines are available which help to illustrate and interpret these regulations.

9.12.2 Noise at Work Regulations 1989

The more traditional woodworking machines illustrated in this chapter are typically found in the joinery workshop. Inherently, if not modified, they are noisy in use and in many cases exceed the noise levels considered 'acceptable' according to these regulations (i.e. below 85 dB(A)). The regulations state that noise measurements should be taken and recorded. Suitable safeguards to the health, safety and welfare of employees must be taken by employers if measurements exceed certain criteria.

9.12.3 Physical Condition of the Machine and Its Attachments

Know the location and operation of:

- the electricity isolating switch
- the machine controls
- guards and their adjustment.

9.12.4 Setting Up and Guarding the Machine

- Know how to isolate the machine.
- Ensure that the blade/cutters are of the correct type, size and shape and are sharp.
- Ensure that all adjustment levers are locked securely.
- Check that all guards and safety devices are in place and secure.
- Ensure that all adjusting tools have been returned to their 'keep'.
- Make sure that pushsticks and blocks are at hand.
- Ensure that work and floor areas are free from obstruction.
- Have the machine settings, etc. checked by an authorised person.

9.12.5 Suitability of the Material to be Cut

- Understand the cutting characteristics of the material.
- Check length and section limitations.

9.12.6 Machine Use

- Always allow the blade/cutters to reach maximum speed before making a cut.
- Know the correct stance and posture for the operative.
- Use an assistant as necessary.
- Never make guard adjustments until all moving parts are stationary and the machine has been isolated.
- Never make fence adjustments when a blade/cutter is in motion.
- Never make fence adjustments within the area around the blade/cutters or other moving parts until these parts are stationary and the machine has been isolated.
- Concentrate on the job – never become distracted while the machine is in motion.
- Never allow hands to travel near or over a blade/cutter while it is in motion.
- Never leave a machine until the blade/cutters are stationary.
- Always isolate the machine after use.

9.12.7 Personal Safety

- Ensure that dress and hair cannot become caught in moving parts or obstruct vision.
- Finger rings should never be worn in a machine shop, for fear of directing splinters of wood into the hand or crushing the finger if the hand becomes trapped.
- Footwear should be sound with non-slip soles of adequate thickness and firm uppers to afford good toe protection.
- Wear eye protection.
- Wear ear protection.

All the above must be in accordance with the Personal Protective Equipment at Work Regulations 1992.

10

Wood Adhesives

An adhesive is a medium that allows the surfaces of two or more items to be attached or bonded together. Adhesives are made from either natural or synthetic (man-made) materials. They come in a liquid (one or two part) or powder form, or a combination of both powder and liquid, or as a semisolid which requires melting.

Table 10.1 lists nine adhesive types together with their classification, main characteristics, moisture resistance, any gap-filling properties and their general usage. The following notes briefly describe these adhesives and some of the technology associated with them.

10.1 ADHESIVE TYPES

10.1.1 Casein

This is derived from dairy byproducts that are dried, treated and mixed with chemical additives to produce a powder which, when mixed with water, is ready for use. It is used in general joinery assembly work and in the manufacture of plywood. It tends to stain some hardwoods.

10.1.2 Urea Formaldehyde (UF)

This is a very widely used synthetic resin adhesive, for such things as general assembly work and binding within some manufactured boards (including MDF). Strong mixes can achieve moisture-resistant (MR) requirements. Most are designed for close contact jointing, but formulations are available to satisfy gap-filling requirements (see Section 10.2.3).

When set (chemically cured) these adhesives will either be clear or lightly coloured so glue lines can be concealed.

10.1.3 Melamine Urea Formaldehyde (MUF)

MUF is an adhesive with more melamine than urea which can enhance moisture resistance to bring it above boil-resistant (BR) rating. Its usage is similar to that of UF.

10.1.4 Phenol Formaldehyde (PF)

The main purpose of these adhesive types is to provide the best moisture resistance (weather and boil proof – WBP) to structural plywood. They are also used as a binder in particleboard and wafer boards. Glue lines may show as slight red/brown staining to the wood.

10.1.5 Resorcinol Formaldehyde (RF)

Because of their WBP properties these adhesives are highly suited to the assembly of external structures and marine applications. They are, however, expensive. Glue lines can result in the staining of wood. Again, the glue lines may show a red/brown staining to the wood.

10.1.6 Polyvinyl Acetate (PVAc)

These are thermoplastic adhesives consisting of a simple to use, one-part water-based emulsion with additives to produce either a standard interior type, cured mainly by evaporation and used extensively for glueing internal joinery components and veneering, or an improved PVAc which gives higher moisture resistance by inducing a chemical reaction. This so-called 'crosslinking' will put this type of PVAc in the class of a thermosetting adhesive.

10.1.7 Contact Adhesives

These are made of natural or synthetic rubber and a solvent which evaporates when exposed to the air, giving off a heavy flammable vapour. They are used in the bonding of wood and plastics veneers (laminated plastics) to wood based materials. Bonding is achieved by coating both

Table 10.1 Adhesives characteristics (general guide only as properties can differ)

Adhesive classification			Moisture resistance			Gap filling	General usage
					Class		
N	Casein			Poor	Int	Yes	Not in general use today – has been used for assembly work and interior plywood
	Urea formaldehyde (UF)	Amino – plastic	Thermo setting*	Fair	MR	Some types	Probably the most common adhesive – general purpose assembly work, manufactured boards (MDF), and veneering
	Melamine urea formaldehyde (MUF)			Good	BR		Moisture resistant properties of UF can be improved by this additive – MUF is used in some particleboards and plywood
S	Phenol formaldehyde (PF)	Phenolic		Very good	WBP	Yes	Structural plywood and binder in particleboards and wafer boards – not often for joints
	Resorcinol formaldehyde (RF)			Very good	WBP	Yes	Outdoor timber structures – but not often used on its own because of its high cost RF may be linked with PF to produce PRF
	Polyvinyl acetate (PVAc)		Thermo plastic	Poor to good	Int or MR†	May be	Internal assembly of joinery items and veneering (a general purpose adhesive group)
S&N	Contact		Solvent or emulsion based	Fair	Int	GAP	Veneers of wood and plastic laminate
	Hot melt		Thermo-plastic		Int		Edge veneering and spot or strip glueing (glue gun) – bond strength is relatively low
S	Epoxy	Thermo setting			Int	Yes	Very limited to specialist use – usually formulated to enable wood to be bonded to metal and glass reinforced plastics (GRP)

N – adhesives in the main are of natural origin; S – synthetic resins; WBP – weather and boil proof; BR – boil resistant; MR – moisture resistant; INT – interior; * Note: All are two part (component) adhesives; † Special formulation for exterior use, depending on formulation may be classed as a thermosetting adhesive

Thermoplastics – Thermoplastic adhesives may be in the form of fusible solids which soften by heat, or in a soluble form. Generally after heating thermoplastics will regain their original form and degree of strength

Thermosets – Thermosetting resins become solid either by a chemical reaction, or via a heat source; however, once set they, unlike thermoplastics, can not be reconstituted by heat

surfaces to be joined, leaving them to become tacky (for a time specified by the manufacturer) and then laying one onto the other while excluding any air. Bonding is instantaneous on contact (hence the name '*contact adhesive*'), with the exception of *thixotropic* types which allow a certain amount of movement for minor adjustments.

Some generally available contact adhesives are emulsion based, non-flammable and less hazardous.

10.1.8 Hot Melt Synthetic (Thermoplastic) Adhesives

These are either semisolid rods or pellets or come in the form of tapes or films, which are melted for application by heat. As they cool, they reharden to their original strength. The most common type of application is with an edge-banding machine, used for applying veneers of wood, plastics, and plastics laminates. Handheld glue guns are quite popular for spot glueing or running narrow joints; glue guns use cylindrical adhesive 'slugs', which are available in various formulations.

10.1.9 Epoxy Synthetic (Thermosetting) Adhesives

The use of this type of adhesive is limited to specialised applications, for example glueing wood to metal or glass reinforced plastics (GRP) and small repair work.

10.2 ADHESIVE CHARACTERISTICS

10.2.1 Form

Adhesives may be of the one- or two-component type – liquid, powder or both. Two component types become usable either by applying them direct from the container, mixing the components together, or by applying each part separately to the surfaces being joined. Some types, however, have to be mixed with water.

10.2.2 Moisture Resistance

This refers to the adhesive's inherent ability to resist decomposition by moisture. Resistance is classified as follows:

- INT (interior) – joints made with these adhesives will be resistant to breakdown by cold water
- MR (moisture (and moderately weather)-resistant) – joints will resist full exposure to weather for a few years when made with these adhesive types.
- BR (boil resistant) – joints made with these adhesives will resist weather but fail when subjected to prolonged exposure. They will, however, resist breakdown when subjected to boiling water.
- WBP (weather and boil proof) – joints made from these adhesives will have high resistance to weather, micro-organisms, cold and boiling water, steam and dry heat.

Reference to the above categories is also made in Table 2.1 with regard to the bonding of plywood veneers; these are British Standard Terms.

10.2.3 Gap Filling

Gap-filling adhesives should be capable of spanning a 1–1.3 mm gap without crazing. They are used in situations where a tight fit cannot be assured and where structural components are to be bonded.

10.2.4 Bond Pressure

This refers to the pressure necessary to ensure a suitable bond between the two or more surfaces joined together. Pressure may be applied simply by hand, as with contact adhesives. With items of joinery, mechanical or manual presses or clamps (see Section 2.1.1) of various shapes and sizes are usually employed. Wood wedges can be used not only to apply pressure but also to retain a joint permanently. The length of time needed to secure a bond will vary with each type of adhesive, its condition and the surrounding temperature.

10.2.5 Assembly (Closed) Time

Time will be required to accurately set cramping devices and correctly position frame components; each type and make of adhesive will state the time allowed for this.

10.2.6 Storage or Shelf Life

This is the length of time that the containerised adhesive will remain suitable for use. Beyond this period, marked deterioration may occur, affecting the strength and setting qualities.

Once the adhesive's components have been exposed to the atmosphere, the storage life may well be shortened. 'Shelf life' therefore may or may not refer to the usable period after the initial opening; always note the manufacturer's recommendations.

10.2.7 Pot Life

This is the length of time allowed for use after either mixing or preparing the adhesive.

10.2.8 Application of Adhesives

Methods and equipment used to apply adhesives will depend on the following factors:

- type of adhesive
- width of surface to be covered
- total surface area
- work situation
- clamping facilities available
- quality of work.

The spreading equipment could be any one of the devices listed below:

- mechanical spreader
- roller
- brush
- spatula
- toothed scraper.

10.2.9 Safety Precautions

All forms of adhesives should be regarded as potentially hazardous if they are not used in accordance with the manufacturer's instructions, either displayed on the container or issued as a separate information sheet. Depending on the type of adhesive being used, failure to carry out the precautions thought necessary by the manufacturer could result in:

- an explosion – due to the adhesive's flammable nature or the flammable vapour given off by it
- poisoning – due to inhaling toxic fumes or powdered components
- skin disorders – due to contact while mixing or handling uncured adhesives. Always cover skin abrasions before starting work.

Where there is a risk of dermatitis, use a barrier cream or disposable protective gloves. Always wash hands thoroughly with soap and water at the end of a working period.

Finally, the manufacturer's instructions on handling precautions should cover the following:

- good 'housekeeping'
- skin contact
- ingestion
- eye protection
- fire risk
- toxicity.

11

Fixing Devices

In addition to forming joints in solid wood to hold the various pieces together, it is frequently necessary to secure timber members in place or to fix them to other materials using metal fixings. The method and devices used as fixings are often left to the discretion of the carpenter and joiner, but it should be remembered that this is not always the case and note should be taken of drawings and specifications (see section 12.9.2) if fixing details are given.

A structural engineer will design timber structures where the methods of jointing and the fixings used are critical to the stability of the structure and must be formed to the exact specification given. Timber structures have failed because the importance of following the designer's requirements has not been fully appreciated.

When using fixing devices the following factors must be kept in mind:

- designed fixings must be used and secured as directed
- fixings must be of the right size and correctly located to provide maximum efficiency
- the fixing device must be of the right type to resist the forces that it will be subject to
- will the presence of moisture or other agents corrode the fixing devices?
- are the fixings hidden or do they improve the appearance of the finished structure?
- cost and availability have some bearing on the choice of the fixing devices used.

Fixing devices can be divided into the following eight broad classifications:

- nails
- screws
- bolts
- anchor bolts
- metal fixings or reinforcing plates
- plugs
- direct frame fixings
- cavity fixings.

11.1 NAILS

Nails are used to hold timbers together at their point of contact, to resist withdrawal (pull out) and lateral movement of the members in the joint (the tendency of one part of the joint to slide over the other) (Figure 11.1).

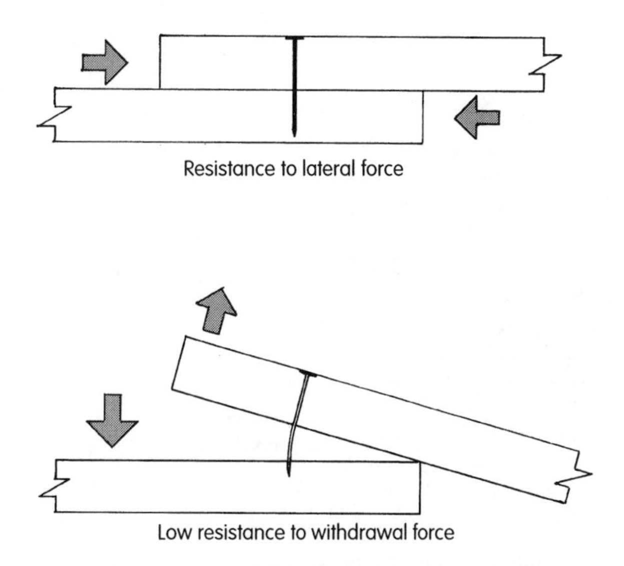

Figure 11.1 Forces acting on nails

Nails can be obtained in a variety of lengths and types (Table 11.1). Wire round and oval nails are made in incremental lengths from 12 to 150 mm with the cross-section of the nail shank being in proportion to its length, although some round wire nails can be ordered to a specific wire gauge to meet structural design requirements, nails required for metal fixing plates is just one example.

Nails offer the quickest, simplest and least expensive method of securing timbers together, provided that you have some knowledge of how nails hold in timber, recognise their limitations and apply good practice to their use.

Table 11.1 Types of nail

Nail type	Material	Finish or treatment	Shape or style	Application
Wire nails				
Round plain-head	S	SC	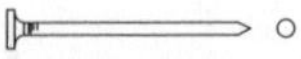	Carpentry, carcass construction, wood to wood
Clout (various sized heads)	S, C	SC, G	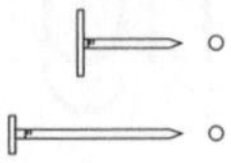	Thin sheet materials, plasterboard, slates, tiles, roofing felt
Round lost-head	S	SC		Joinery, flooring, second fixing (small head can be easily concealed)
Oval brad-head	S	SC	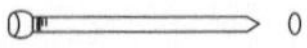	Joinery, general-purpose, less inclined to split grain
Oval lost-head	S	SC		As for oval brad-head
Improved nails				
Twisted shank	S	SC, G		
Annular-ring shank		SH		Roof covering, corrugated and flat materials, metal plates, etc., flooring, sheet materials. Resist popping (lifting). Good holding power, resisting withdrawal
Duplex head	S			Where nails are to be re-drawn – formwork, etc.
Cut nails				
Cut clasp	S	SC	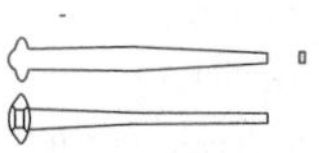	Fixing to masonry (light weight (aerated) concrete and carpentry. Good holding properties
Flooring brad (cut nail)	S	SC		Floorboards to joists (good holding-down qualities)
Panel pins				
Flat head	S	SC, Z		Beads and small-sectioned timber
Deep drive		G		Sheet material, plywood, hardboard
Masonry nails	S/hardened and tempered	Z		Direct driving into brickwork, masonry, lightweight concrete (Caution: not to be driven with hardened-headed hammers. Goggles should always be used.)
Staples (mechanically driven)	S	Z	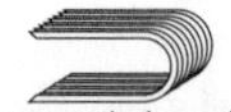 Temporarily bonded	Plywood, fibreboards, plasterboards, insulation board, plastics films to wood battens
Corrugated fasteners (dogs)	S	SC	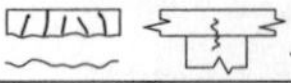 Joint	Rough framing or edge-to-edge joints
Star dowel	A	SC		An alternative to hardwood dowel for pinning mortise and tenons or bridles

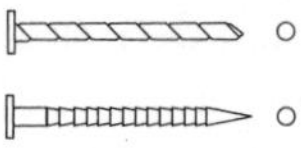

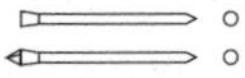

Key to Tables 11.1, 11.2, 11.3 and 11.4

Materials: A – aluminium alloy; B – brass; BR – bronze; C – copper; P – plastics; S – steel; SS – stainless steel

Finish/treatment: B – brass; BR – bronze; CR – chromium; G – galvanised; J – japanned (black); N – nickel; SC – self-coloured; SH – sherardised; Z – zinc; BZ – bright zinc

Head shape: CKS – countersunk; DM – dome; RND – round head; RSD – raised head; SQ – square BH – Bugle head

Drive mechanism: SD – Superdriv (Pozidriver); SL – slotted (screwdriver); SP – square head (spanner)

11.1.1 Selecting Nails of the Right Length

When boards are being fixed to joists the nail should be 2.5 times the thickness of the board (Figure 11.2a). For fixing boards to battens (through nailing) the length of the nails should be the combined thickness of the board and batten plus 4–6 mm to allow for clenching the nail head (bending over the protruding end of the nail and embedding the nail point in the timber) (Figure 11.2b).

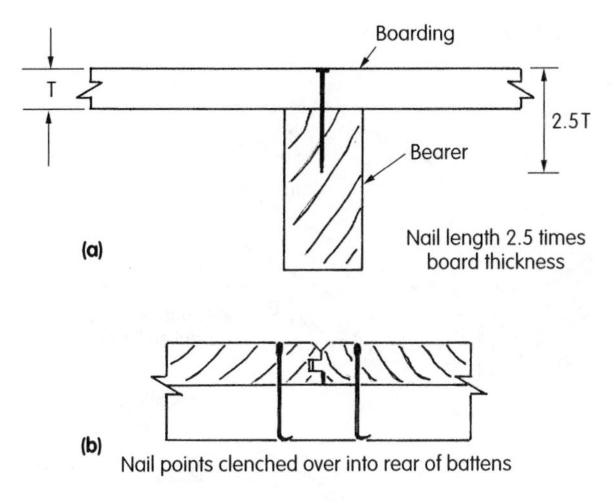

Figure 11.2 Nail lengths

11.1.2 Selecting Nails of the Right Cross-section

Oval nails are mainly used for fixing finishing joinery (architraves, skirtings, linings and cover moulds), as the heads are less obtrusive than those of round nails and can easily be punched below the timber surface. Oval nails should be driven with the major axis of the oval cross-section parallel to the grain of the timber (Figure 11.3a).

The gauge (diameter of round wire nails) should be such that in driving in the nail the timber does not split excessively; this may happen if the ratio of the gauge to the distance between the nails is less than 20 or the nail is too near the end of the timber (Figure 11.7b).

To help reduce the tendency for the nail to split the timber, the nail point is blunted by striking it with a hammer (Figure 11.3b). This has the effect of putting a cutting edge on the point, which cuts rather than cleaves (splits) the wood fibres apart (Figure 11.3c, d).

11.1.3 How Do Nails Hold in Wood?

As the nail is driven into the timber, it cuts and compresses the surrounding wood fibre which, on recovery, tries to expand and reoccupy the space taken up by the nail (Figure 11.4a). In doing so it exerts a frictional grip

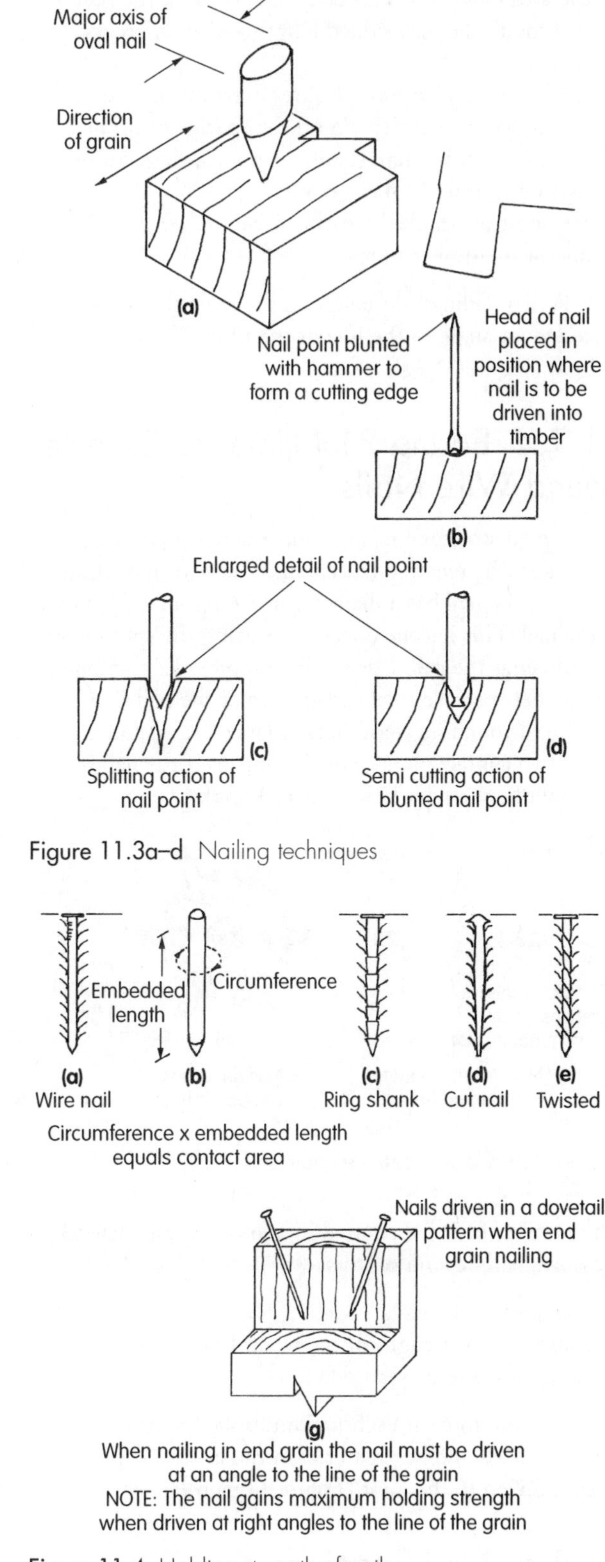

Figure 11.3a–d Nailing techniques

Figure 11.4 Holding strength of nails

on the embedded shank of the nail which offers resistance to withdrawal. The amount of force required to pull the nail out will depend upon:

- the total length of the nail shank embedded in the timber

- the area of contact between the surface of the nail and the timber embedded length × circumference (Figure 11.4b)
- the force the compressed fibres exert on the nail shank, which will depend on the species of timber
- the profile of the nail shank – smooth, ring, cut or twisted (Figure 11.4a, c, d, e)
- the angle at which the nail is driven relative to the line of the timber fibres.

NB: When nailing into endgrain the nails should be driven at an angle to the line of the fibres if the nails are to hold (Figure 11.4g).

11.1.4 Boring Pilot Holes to Receive Round Wire Nails

In designed structural nailed joints the boring of pilot holes is usually required (specified). The pilot hole drilled to receive the nail has a diameter of 80% (four-fifths) that of the nail. This has the effect of reducing the splitting of the timber as the nail is driven in and provides frictional contact between the total surface area of the nail shank and the surrounding wood fibre (Figure 11.5b) and not just at two contact points, which happens if the nail is driven without a pilot hole (Figure 11.5a).

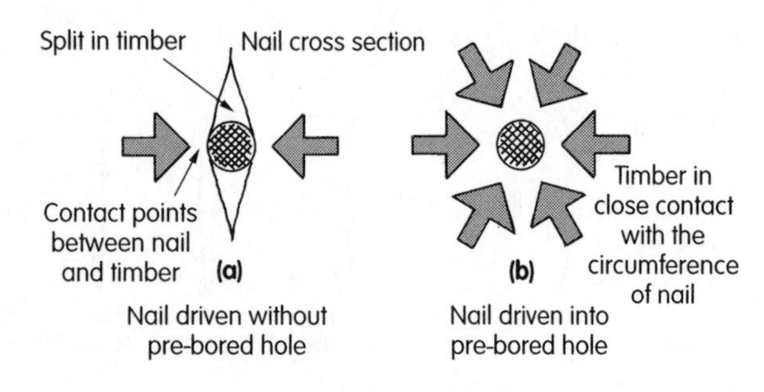

Figure 11.5 Contact between nail and timber

The number of nails required in a joint to resist lateral pressures will depend on the:

- diameter of the nails
- compression strength of the wood fibre
- lateral force to be resisted.

Compression force at each nail multiplied by the number of nails should be greater than the lateral (sliding force) applied to the joint (Figure 11.6a,b).

11.1.5 Nail Spacings

Nails can be placed much closer together if they are driven into pre-bored holes; Figure 11.7a gives recommended spacing. Figure 11.7b shows the spacing for nails without pre-bored holes.

NB: The spacing is between nails and not from nail centre to nail centre.

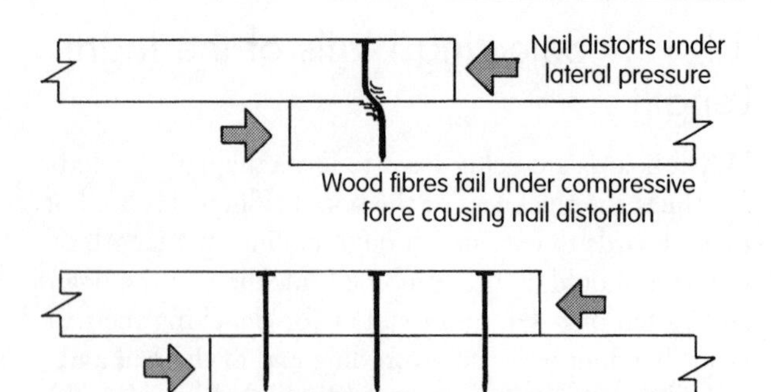

Figure 11.6 Nails required in a joint

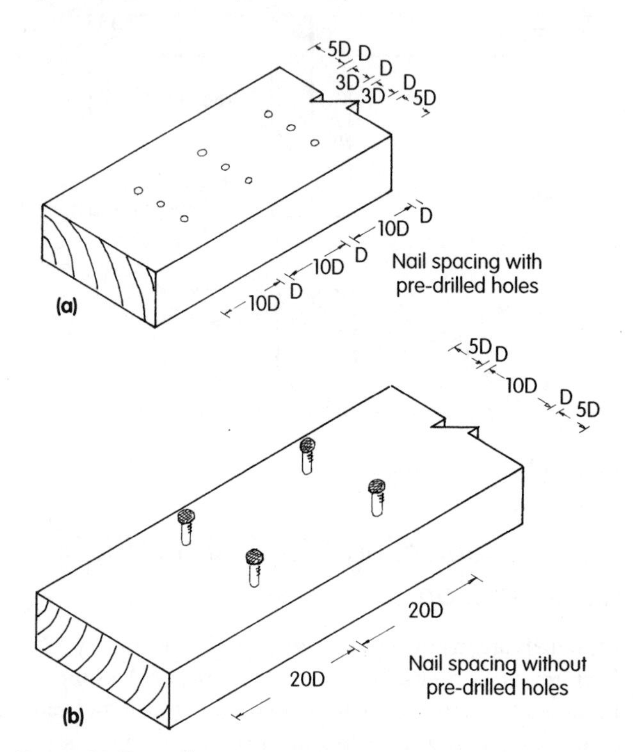

Figure 11.7 Nail spacing

11.2 WOOD SCREWS

Wood screws, like nails, are used to secure timber joints and resist lateral movement. Their advantage over nails is that they have greater holding power against withdrawal and they draw the joint members closer together, acting as a permanent cramp (Figure 11.8). Unlike nails, where the wood fibre round the head is damaged when they are removed, screws can be withdrawn without damage to the wood fibre.

11.2.1 Types and Sizes of Screws

There is a wide range of types and sizes of screws available and although there has been some rationalisation of screw gauges (diameters) because of the introduction of *preferred sizes*, the range of specialist screws available has increased. Table 11.2 illustrates the most common types

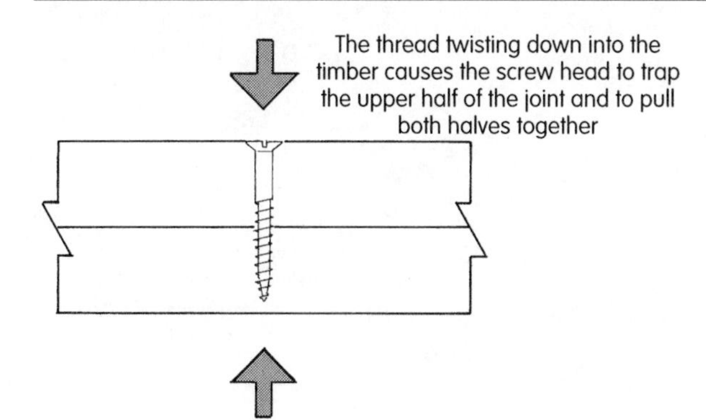

Figure 11.8 Cramping action of screws

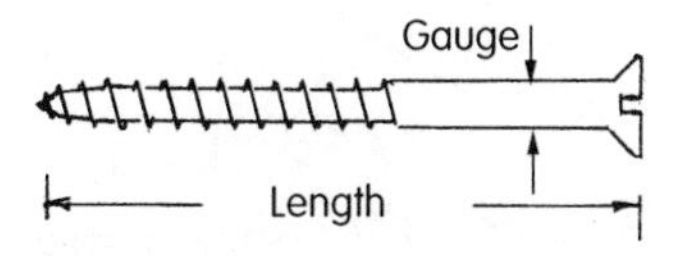

Figure 11.9 Countersunk steel screw

of screw available. Figure 11.9 shows a standard countersunk steel screw.

11.2.2 Purchasing Screws. Identification labels

Wood screws are usually sold in boxes of 100 and 200, although they can be purchased in smaller quantities.

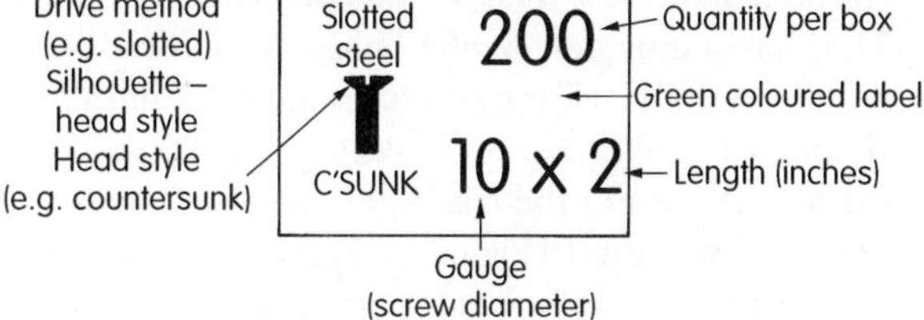

Figure 11.10 Screw identification table

On the end of each box there is a label that, by its colour and screw head silhouette, allows for quick identification of the type, size, gauge, length and metal from which the screw is made. A green label indicates bright steel and a yellow label brass (Figure 11.10).

11.2.3 Modifications in Basic Wood Screw Patterns

There have been four basic changes to the traditional wood screw pattern (Figure 11.11):

a a steeper, double helical, hardened thread cut on the whole length of the screw shank; this increases the speed at which the screw can be driven home by reducing the number of turns required

Table 11.2 Types of screw

Screw type	Material	Finish/treatment	Head shape/style	Drive mechanism	Application
Wood screw	S, B, BR, A, SS	B, BR, CP, J, N SC, SH, Z	CKS RND RSD BH	SL SD	Wood to wood; metal to wood, e.g. ironmongery, hinges, locks, etc.
Twinfast wood screw	S, SS	B, SC, SH, Z		SD only	Low-density material; particleboard, fibreboard, etc. Drive quicker than conventional screws, having an extra thread per pitch for each turn
Supafast (Supascrew and Mastascrew)	S	BZ	CKS BH	SL, SD	All hard and soft woods as well as man-made boards. Spaced thread and sharp point for quicker insertion. Hardened head and body, so virtually abuse-proof
Coach screw	S	SC, Z	SQ	SP	Wood to wood; metal to wood (Extra-strong fixing)
Clutch screw	S	SC			Non-removable – ideal as a security fixing)
Mirror screw	S, B	CP	CKS DM		Thin sheet material to wood – mirrors, glass, plastics
Dowel screw (double-ended)	S	SC			Wood to wood – concealed fastener, cupboard handle, etc.
Hooks and eyes	S	B, CP, SC			Hanging – fixing wire, chain, etc.

Note: See key, Table 11.1

b sharper and harder points (Figure 11.11b), with the shank diameter less than the thread diameter (Figure 11.11), requiring only a pilot hole and countersink, rather than pilot, clearance and countersink hole (Figure 11.12b)

c the introduction of the cruciform indent to take a Posidriv screwdriver (Figure 5.95) and hardening of the metal in the screw head to prevent damage to the slot from the screwdriver

d bugle head screws have a longer taper to the head than countersunk screws (Figure 11.11d), allowing them to be driven in without the need to countersink. (Figure 11.12).

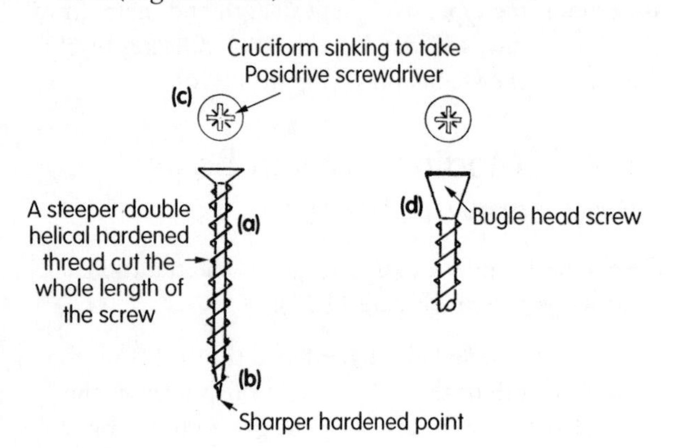

Figure 11.11 Modifications to screw design

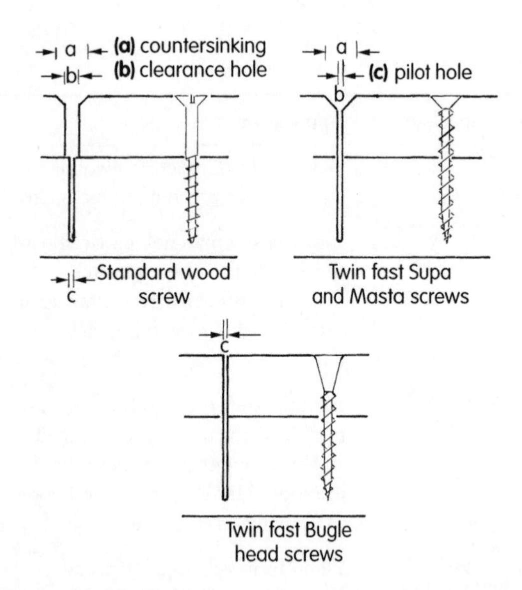

Figure 11.12 Holes to receive screws

11.2.4 Preparation of Timber to Receive Screws

Figure 11.12 shows how the timber is drilled and countersunk to receive a traditional wood screw. In one member of the joint a clearance hole (b) and countersinking (a) are formed. In the other member of the joint the pilot hole (c) is drilled. When using smaller screws the pilot hole can be made using a bradawl (see Chapter 5, Figure 5.73). Also the countersink, clearance and pilot hole can be formed in a single operation by using a screw bit (see Chapter 5, Figure 5.64).

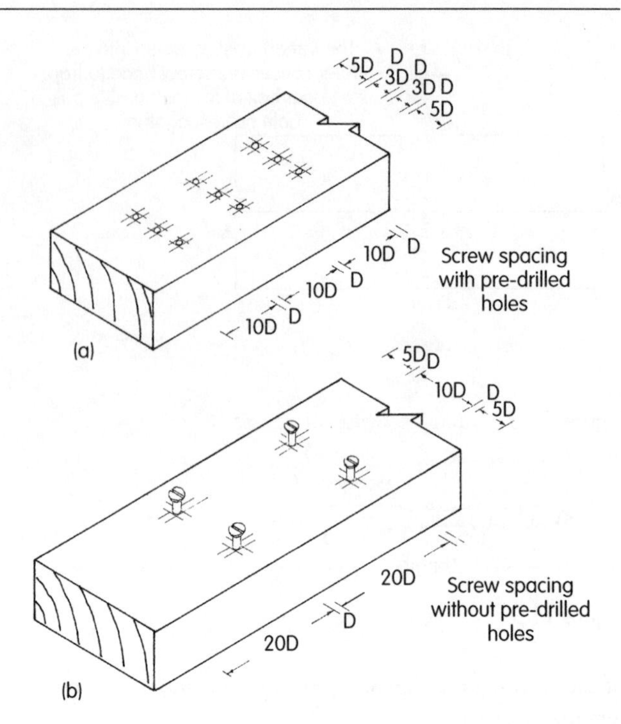

Figure 11.13 Screw spacing in timber joints

11.2.5 Screw Spacing Recommendations

Figure 11.13a, b shows the minimum spacing for screws in a timber lap joint when:

- the screws are placed in pre-drilled holes (the closer spacing of the screws reduces the contact area of the timber in the joint to a minimum)
- the screws are driven directly into the timber without predrilling the holes.

11.2.6 Coach Screws

These have the same profile as the shank of a traditional wood screw but are usually much larger in diameter relative to length and have a square or hexagonal bolt head turned with a spanner (Table 11.2). When fixing, it is necessary to drill both a pilot and clearance hole.

11.2.7 Screw Cups and Covers

Screw cups are used to line the countersinking; they increase the bearing area of the screw head and protect the countersinking from wear when the screws have to be removed from time to time, as in the case of glazing beads. Screw caps and domes hide the screw heads and make for a more acceptable finish (Table 11.3).

Table 11.3 Types of screw cups and covers

	Screw cups		Cover domes		
Material	B, SS		P		
Finish/ colour	SC, N		Black, white, brown		
	CKS	CKS		DM	DM
Shape			DM (a)	(b)	(c)
Application	Countersunk flange increases screw-head bearing area. Used where screw may be re-drawn, e.g. glass beads, access panels, etc.		(a) slots over screw: (b) slots into screw hole; (c) slots into Superdriv screw head. Neat finish yet still indicates location		

Note: See key, Table 11.1

11.3 NUTS AND BOLTS

Table 11.4 shows a range of nuts and bolts and their associated washers used by the carpenter and joiner. Figure 11.14a, b shows the use of a coach bolts (also known as carriage bolts). The square shank directly under the mushroom head is driven directly into the pre-bored bolt hole to give a secure grip so that the bolt does not rotate while the nut is being turned. Figure 11.15 illustrates the use of a handrail bolt.

Note: If the thread on a hexagonal headed bolt is cut along the whole length of the bolt shank it is known as a set screw. Round head roofing and gutter bolts threaded along their whole length are collectively known as machine screws.

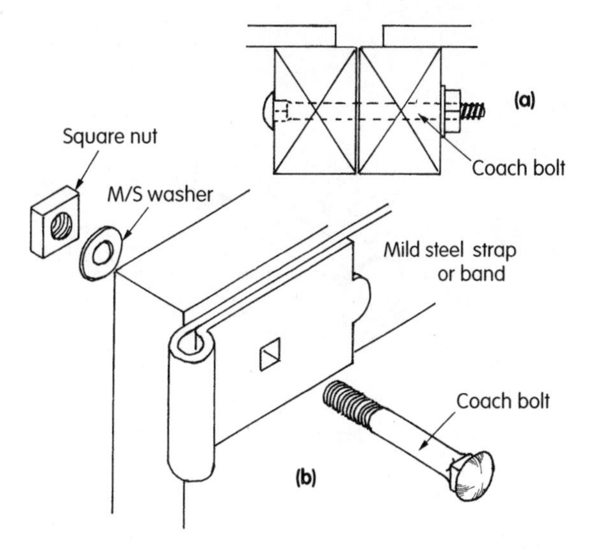

Figure 11.14 Use of coach bolts

11.4 ANCHOR BOLTS

Anchor bolts are used for fixing heavy objects to concrete and are of two basic types.

1. The first is a hexagonal-headed detached bolt which can engage with a captive truncated conical nut and cause the nut to move up to and expand a three- or four-sectioned sleeve, held in place by a retaining collar and wire restraint. This causes the segments of the sleeve to open up and grip against the side of the pre-drilled hole (Figure 11.16).
2. A projecting anchor bolt consists of a bolt with a truncated conical head, which is passed through a three- or four-sectioned sleeve retained by a cylindrical collar and wire restraint. The assembly is placed,

Table 11.4 Types of threaded bolts

Bolt type	Material	Head	Nut and bolt	Application
Hexagonal head	S			Timber to timber, steel to timber (timber connectors, etc.)
Coach bolt (carriage bolt)	S			Timber to timber; steel to timber (sectional timber buildings, gate hinges, etc.)
Roofing bolt	S, A			Metal to metal
Gutter bolt	S, A			Metal to metal
Handrail bolt	S			Timber in its length (staircase handrail, bay-window sill, etc.)

Note: See key, Table 11.1

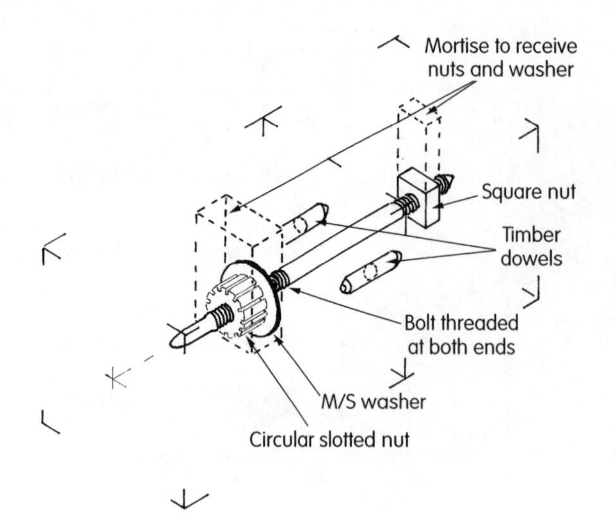

Figure 11.15 Handrail bolt

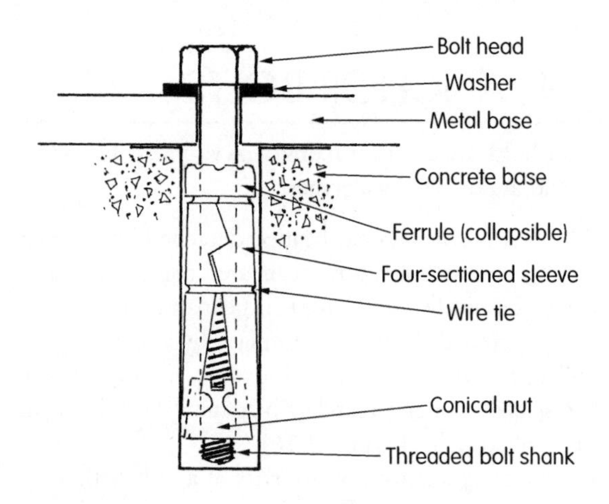

Figure 11.16 Anchor bolt (detached)

conical head first, into the pre-drilled hole and the circular collar is pushed down to engage the sectioned sleeve with the conical bolt head. This causes the sleeve to expand and grip the sides of the hole, leaving the threaded end of the bolt projecting clear beyond the face of the backing. A hole in the fitting to be fixed is passed over the projecting bolt and a washer and nut are then attached and screwed tight, securing the fitting to the concrete background and further tightening the sleeves against the sides of the hole (Figure 11.17).

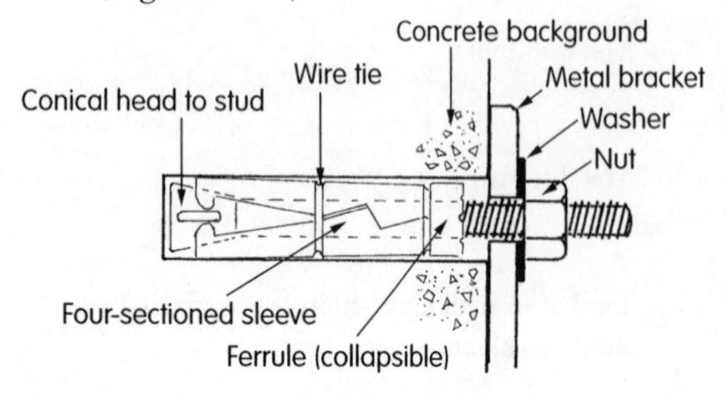

Figure 11.17 Anchor bolt (projecting)

11.5 METAL FIXING PLATES

Metal fixing plates are manufactured from aluminium, brass and mild steel, which may have a surface coating of zinc or other non-ferrous metal. Their primary function is to hold things together and strengthen joints.

11.5.1 Mirror Plates

Sometimes known as glass plates, mirror plates are used for screwing frames or small cabinets to walls and other vertical structures (Figure 11.18).

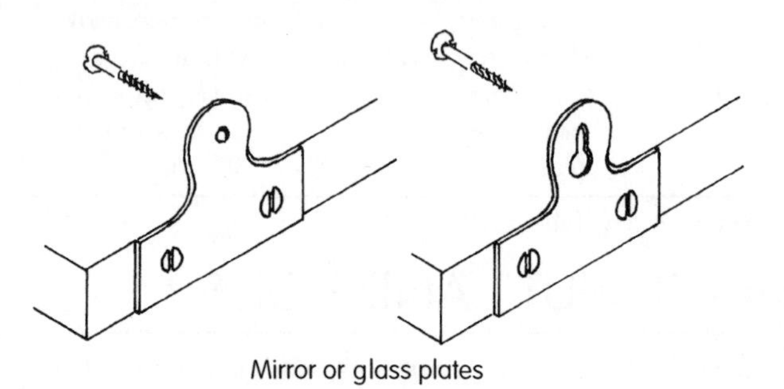

Figure 11.18 Mirror plates

11.5.2 Expansion Plates

Expansion plates are used to hold together members which have differential rates of movement (see Figure 11.19 and Chapter 1, Figure 1.45).

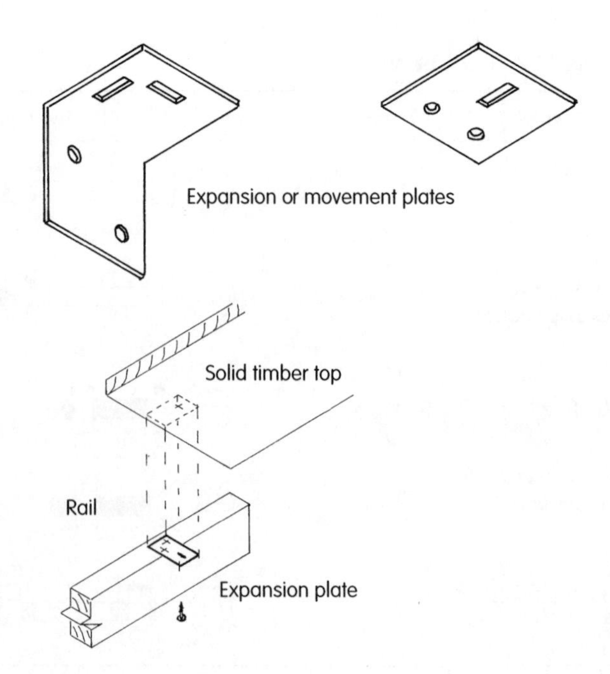

Figure 11.19 Expansion plates

11.5.3 Corner Plates

Corner plates are used to hold framed members together and to reinforce corner joints. In certain types of cabinets and chests, corner plates are made from brass, being a decorative as well as a structural feature (Figure 11.20a, b).

If corner plates are to function efficiently the gauge of the screws used should match the holes in the plates to give a snug fit, and round headed screws used if the holes are not counter sunk.

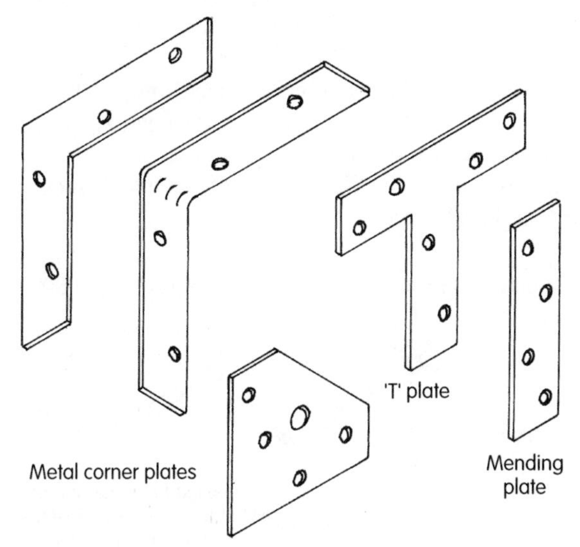

Figure 11.20a Corner plates

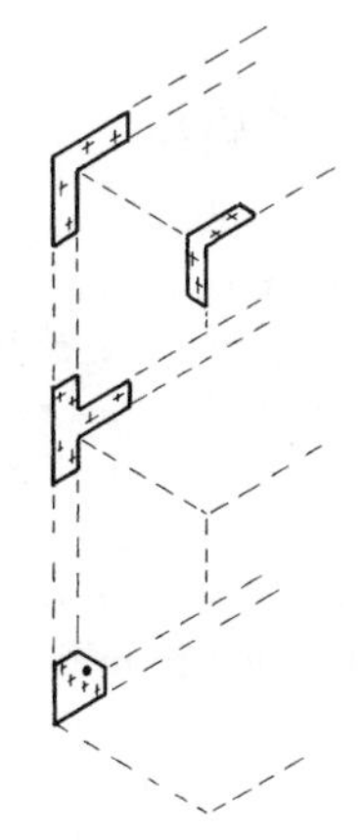

Figure 11.20b Application of corner plates

11.5.4 Anchor Plates, Straps and Hangers

A whole range of these metal fittings is manufactured to suit a variety of fixing and anchoring needs in structural carpentry. They are made from mild steel which is treated with a rustproof coating, usually of zinc (galvanised). The lighter fittings are formed by pressing mild steel sheet (Figure 11.21a, b, c), while the heavier fittings are

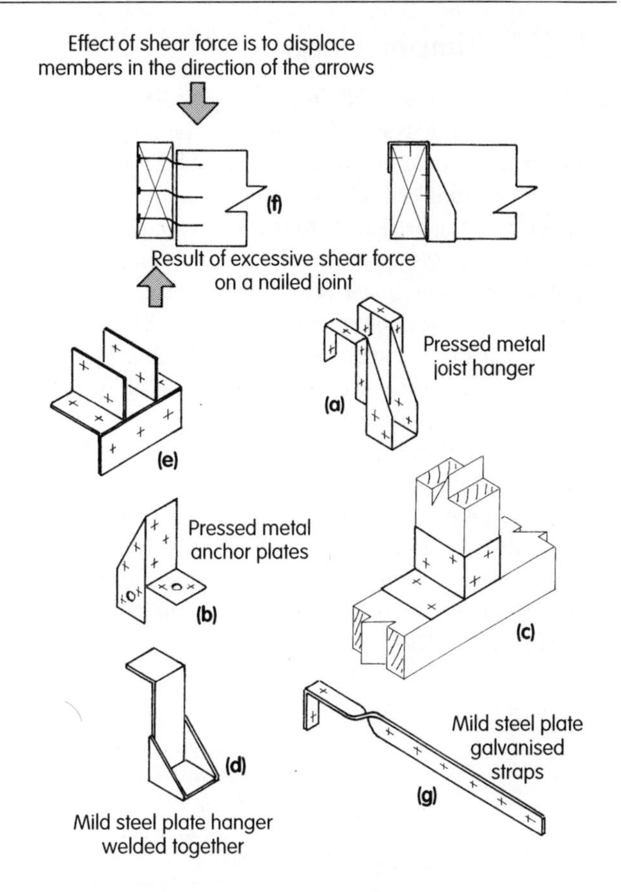

Figure 11.21a–g Anchor plates, hangers and straps

made from mild steel plate which can be bent or welded into the required shapes (Figure 11.21d, e).

These fittings are used to hold and locate structural timbers at their jointing points. Hangers transfer shear forces across the joint (Figure 11.21f), while plates and straps resist lateral and tensile forces (Figure 11.21g).

11.6 FIXING PLUGS

Plugs are used when it is not possible to make a direct fixing to the base or backing material, due to its hardness, brittleness, lack of suitable thickness or for other reasons. The type of plug used will depend on:

- what is being fixed to what, the loads carried and the type of fixing used
- the condition and density of the base material
- whether the fixing is to solid or hollow material.

Materials from which plugs are made include:

- wood
- other organic fibrous materials
- plastics
- soft metals
- chemicals.

11.6.1 Timber Plugs

Timber plugs may be used for plugging drilled holes, but it is now more usual to use them in the horizontal and vertical joints of brickwork (Figure 11.22a). Using a plugging (seaming) chisel, the mortar is cut out of the joint between the bricks where the plug is required (Figure 11.22b). Using an axe, the wooden plugs are then cut from waste material to shape and size, as in Figure 11.22c,d, and driven into the raked-out joint using the back of the axe head (Figure 11.22e). The projecting plug surplus is cut off with a saw just proud of the brickwork (Figure 11.22f).

Note: Brickwork is a gravity structure which depends on its weight to hold it in place (the mortar is a bedding material and not an adhesive). If the plugs are placed too near the top or end of the wall and driven in with too much force, the mortar joints will crack and the bricks will be displaced, allowing the plug to become loose (Figure 11.22a).

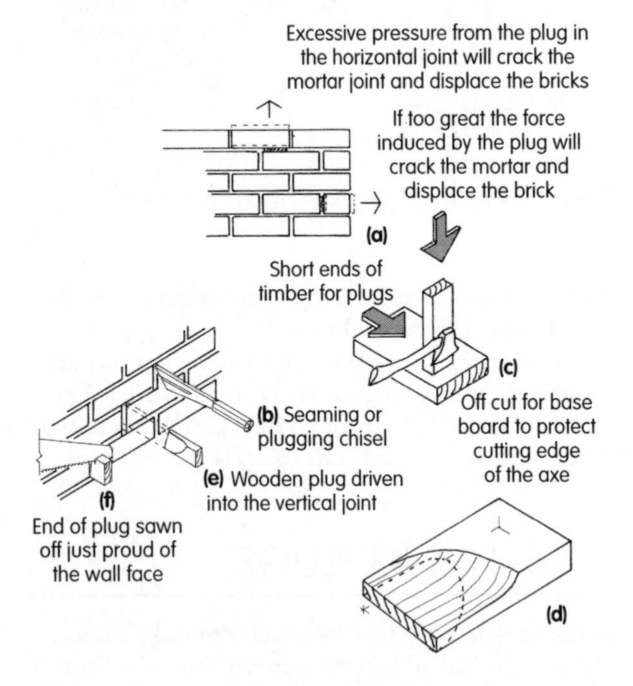

Figure 11.22 Wooden plugs in brickwork

11.6.2 Fibre Plugs

These are circular tubes of fibre of varying diameters and lengths to suit the size of screw being used and the diameter of the hole being drilled in the background material (Figure 11.23). Organic fibre plugs have almost completely been replaced by plastics.

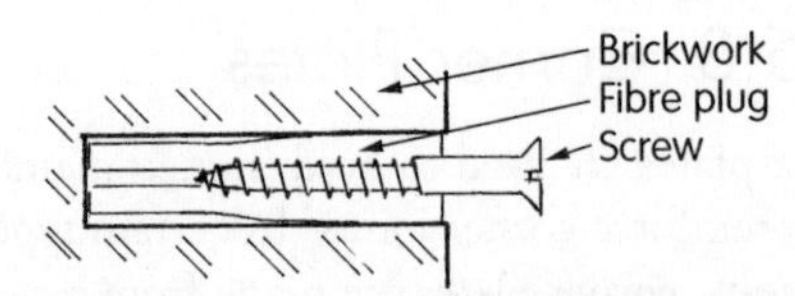

NOTE: As the screw is turned into the plug the threads bite into the fibre and the diameter of the screw being larger than the pilot hole in the plug causes the plug to expand and grip the sides of the hole

Figure 11.23 Fibre plugs

11.6.3 Soft Metal Plugs

Soft metal plugs have also been almost completely replaced by plastics plugs but on external work, lead wool can be used or small pieces of sheet lead can be rolled round a nail to form a tube of suitable diameter (Figure 11.24).

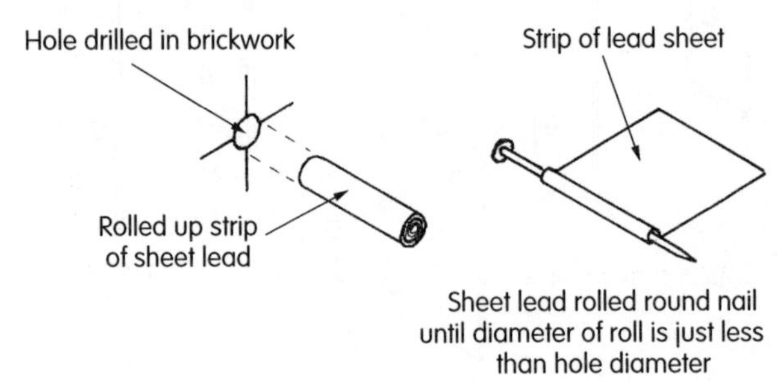

Figure 11.24 Soft metal plug

11.6.4 Plastics Plugs

Plastics plugs are manufactured from polythene or nylon, have good holding properties, but can be affected by temperature and humidity changes but, being hidden in the structure, direct sunlight does not cause deterioration. They are manufactured as individual plugs with a variety of lengths, diameters and profiles or as lengths of tube of varying diameter which can be cut to size. Some plastic plugs are colour coded to indicate their size (Table 11.5).

Table 11.5 Colour code for plastic plugs

Colour	Gauge	Drill diameter
Yellow	6–8	5 mm
Red	8–10	6 mm
Brown	10–14	7 mm
Blue	14–18	10 mm

Plastic plugs are applied as follows (Figure 11.25).

a Drill a hole in the background material of the right diameter for the size of plug required (use eye protection when drilling).

b Remove bore dust from the hole using the head of a round wire nail. *Never blow the dust out using your mouth.*
c Insert the plug which should be a push fit.
d Pass the screw through the material to be fixed to the background and place the point of the screw into the plug.
e Using a screwdriver, turn the screw home into the plug to secure the material being fixed to the back ground.

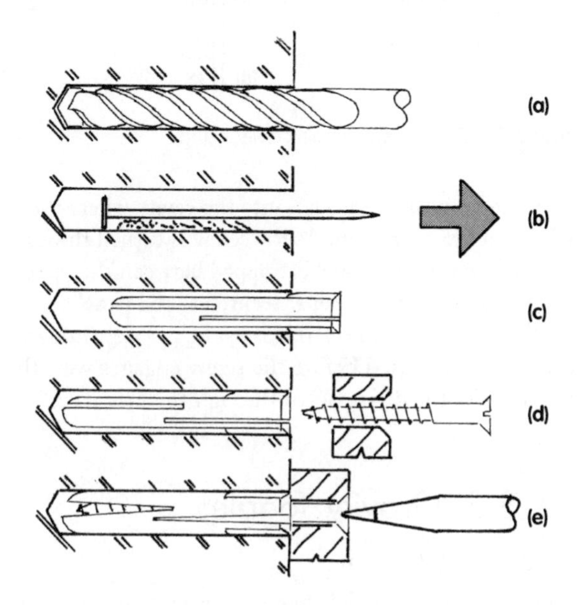

Figure 11.25 Sequence of using plastics plugs

11.6.5 Chemical Plugs (Special Fittings)

The chemicals forming the anchor are liquids encased in a thinwalled glass or plastic phial, which is placed in a hole pre- drilled into the background. On driving the metal fixing into the hole, the phial is broken and a chemical change occurs, bonding the fixing in the hole as the chemicals set (Figure 11.26). No contraction occurs in the process of hardening, making this form of fixing ideal for use in brittle materials and securing fittings close to exposed edges.

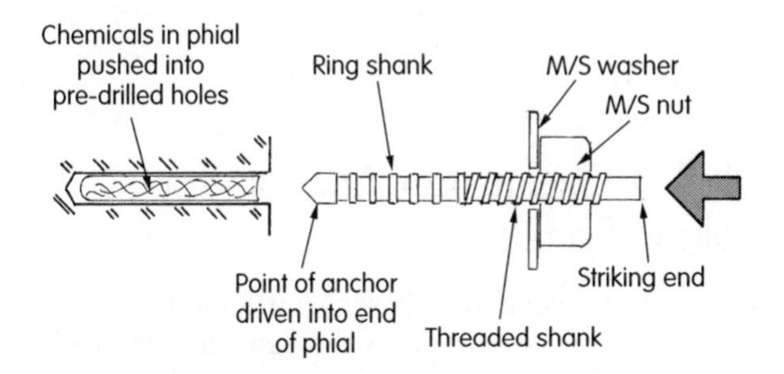

Figure 11.26 Chemical plug

11.7 COMBINED PLUG, SLEEVE AND SCREW FITTINGS

This type of fixing allows the background and material being fixed to be marked out and drilled, the plugs inserted and the fixings screwed or driven home without the item having to be taken down and offered up a number of times. There are two basic types of fixing but both can be used for the fixing of frames or other items to backing materials.

11.7.1 Hammer Fixings

The background and positioned frame are drilled in one operation, using a long shank masonry or multipurpose drill of the required diameter, the drill cutting through the frame and into the backing material (Figure 11.27a). A combined plug and sleeve of suitable length and diameter is pushed home through the frame and into the hole drilled in the background (Figure 11.27b). The fixing is a hardened drive screw, in appearance like a round wire twist nail, with the lead end having a shallow thread formed on its shank for about a third of its length and a head in the form of a countersunk screw slotted for a Posidriv screwdriver (to help with its removal). It is driven home using a hammer and the shallow headed end expands that part of the plug embedded in the background material (Figure 11.27c).

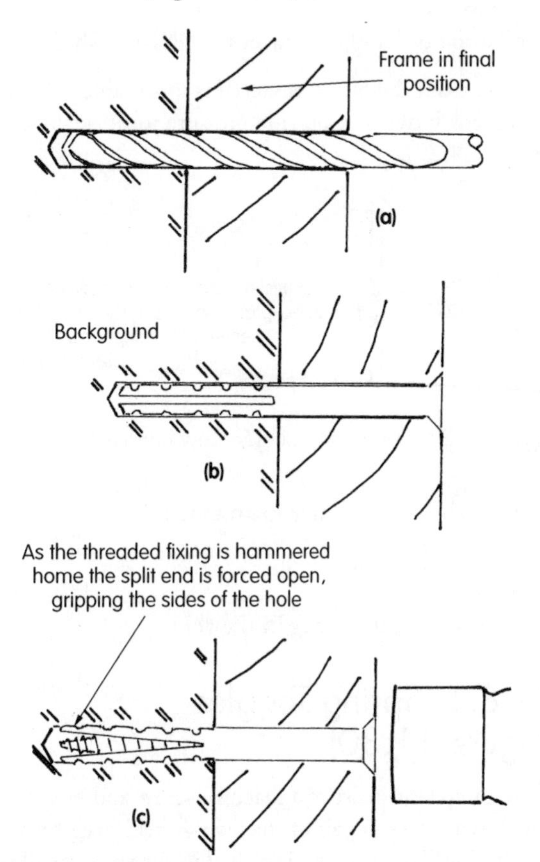

Figure 11.27 Hammer fixing sleeve plugs

11.7.2 Frame Fixings

A frame fixing is similar in appearance to a hammer fixing except that the end is replaced by a true screw threaded end and is driven in using a screwdriver (Figure 11.28).

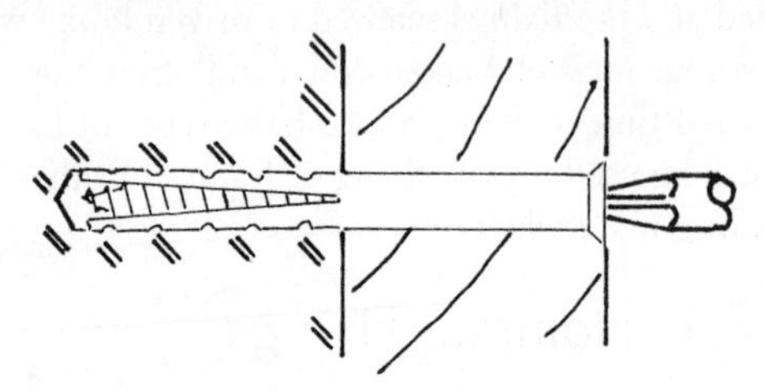

Figure 11.28 Frame fixing sleeve plugs

11.8 CAVITY FIXINGS

Cavity fixings are used for fixing to hollow forms of construction which consist of thin facing materials secured to a cellular or honeycomb core, where access for fixing is only possible from one side. Common examples are:

- hollow core building blocks in walls
- plasterboard and other thin sheet materials lining partitions
- plywood or hardboard faces to cellular flush doors.

In most cases the facing material is unsuitable to take a fixing which depends on side pressure for its stability (Figure 11.29).

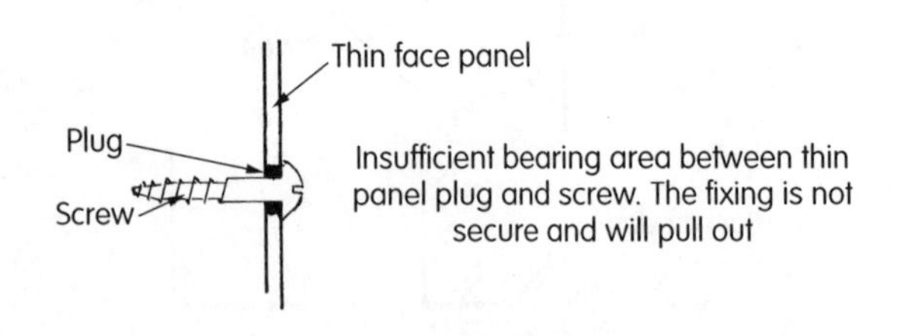

Figure 11.29 Thin panel fixing (hollow construction)

Cavity fixings are of three main types:

1 spring toggle
2 gravity toggle
3 expansion of the fixing in the void.

11.8.1 Spring Toggles (Figure 11.30)

Spring toggles consist of a machine screw and nut to which is attached a pair of steel wings held open by a fine wire spring. To fix, a hole is drilled in the panel to allow the wings, which are folded back along the machine

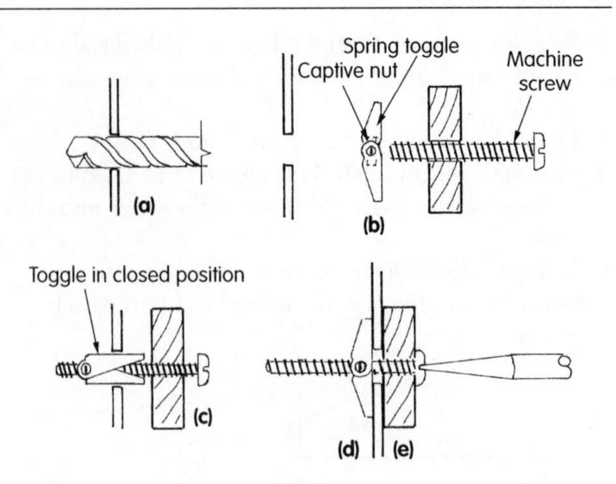

Figure 11.30 Spring toggle cavity fixing

screw, to be pushed through into the cavity where the wings then spring open. As the screw is turned through the captive nut, the panel is trapped between the open wings and whatever is being secured to the panel.

Note: The machine screw must be passed through the item being fixed before the screw engages with the nut and the toggle is passed through the face panel (Figure 11.30).

11.8.2 Gravity Toggles (Figure 11.31)

Gravity toggles function like the spring toggle by gripping the panel, but use gravity rather than a spring to place the toggle correctly in the cavity (Figure 11.31a–e).

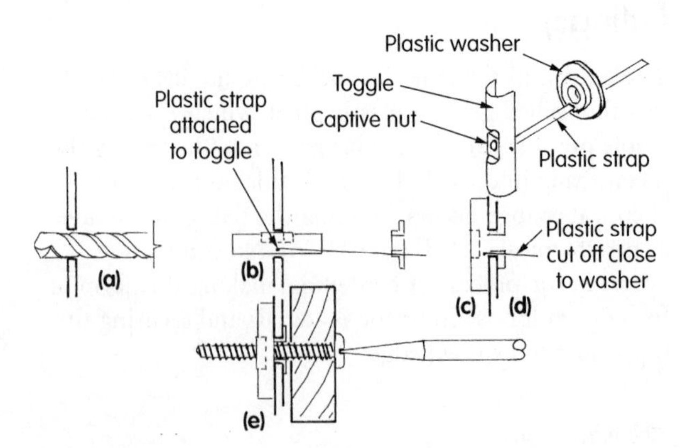

Figure 11.31 Gravity toggle cavity fixing

11.8.3 Expansion Anchors for Hollow Panels

In this type of fixing the tube which will form the anchor is passed into the cavity through a hole drilled in the face panel. By using a wood screw or machine screw the material forming the anchor tube is pulled back on

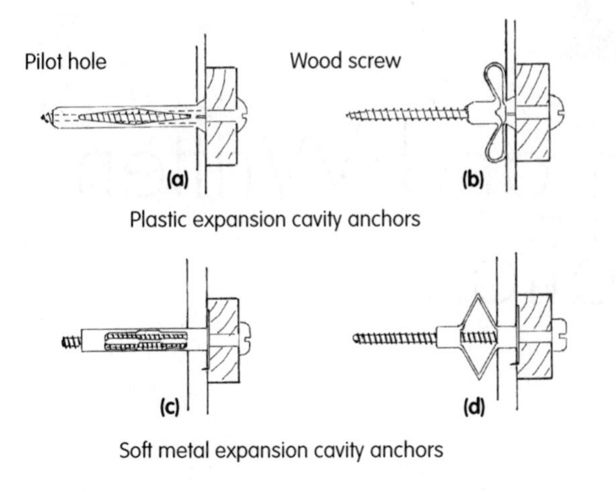

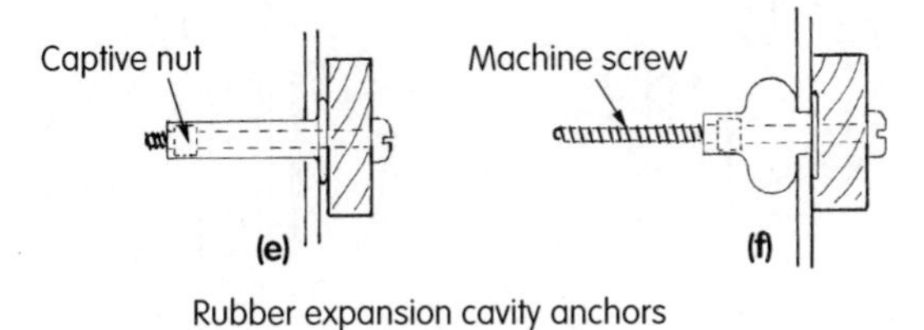

Figure 11.32 Expansion anchor cavity fixing

itself until it grips the face panel. This type of fixing has the advantage that it will remain in place even if the fixing screw is removed. These types of cavity fixing can be made from plastics (Figure 11.32a, b), soft metal (Figure 11.32c, d) and rubber (Figure 11.32e, f).

12

Drawn, Spoken and Written Information

Carpenters and joiners, like everyone else, need to receive and pass on information that is understood by those involved in the process of communication.

12.1 TOOLS OF COMMUNICATION

The tools of communication used by woodworkers are:

- speaking and gesture
- writing (manually or electronically)
- formal drawing and sketching.

Each tool of communication has to be learned, practised and then applied in our working lives. How successful we are in receiving, understanding, applying or passing on information to others will depend on how well we are able to use the tools of communication.

Speaking requires an understanding of language and the meaning of words. Writing requires the ability to form letters, spell, use grammar correctly and read with understanding what is written. Drawing involves understanding how we show a large solid, three-dimensional object on a small piece of paper which has only two dimensions (Figure 12.1).

Each form of communication has its use and the carpenter and joiner selects the form most suitable for a given situation.

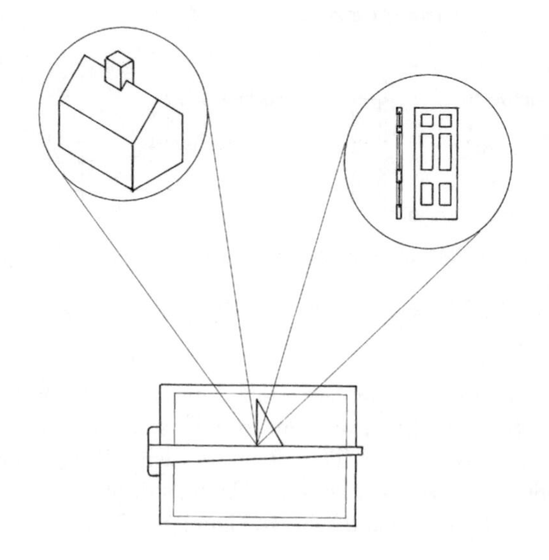

Figure 12.1 Drawing large objects

12.2 SCALED MEASUREMENTS

Scaled measurements allow a proportional reduction in the size of large objects so that they can be drawn on relatively small pieces of paper. The proportional drawing of an object is known as a scaled drawing (Figure12.2).

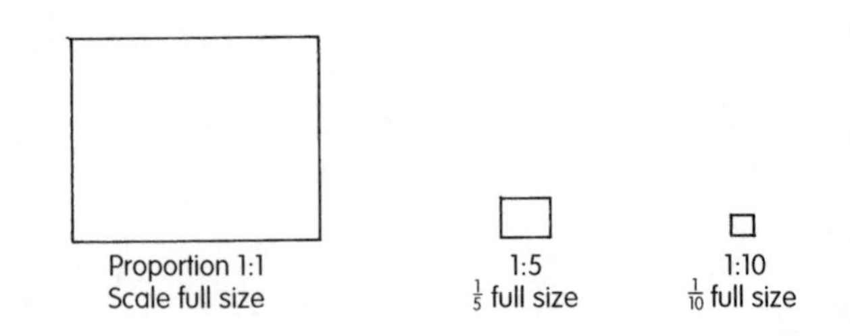

Figure 12.2 Comparison of scales

12.2.1 Scale Rules

To save having to calculate the proportional reduction of every line drawn, as shown in Figure 12.3, scale rules are used (Figure 12.4). These rules are machine engraved so that the dimensions can be taken directly from the scale rule (Figure 12.5). Figure 12.6 represents 1 metre drawn to various scales. Figure 12.7 shows both sides of a builder's scale rule (see also section 5.1.1).

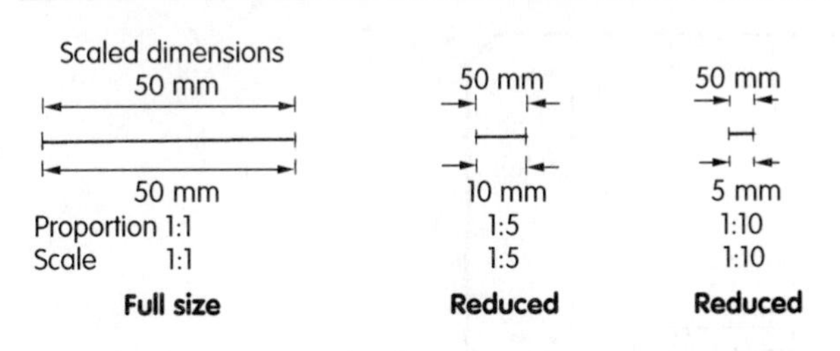

Figure 12.3 Proportional reduction

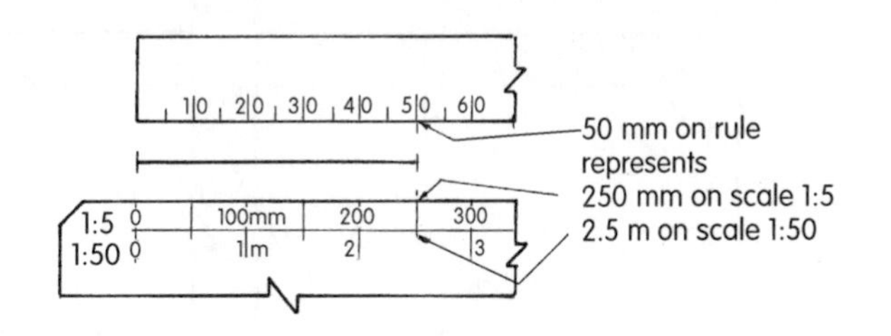

Figure 12.5 Actual and scaled distances

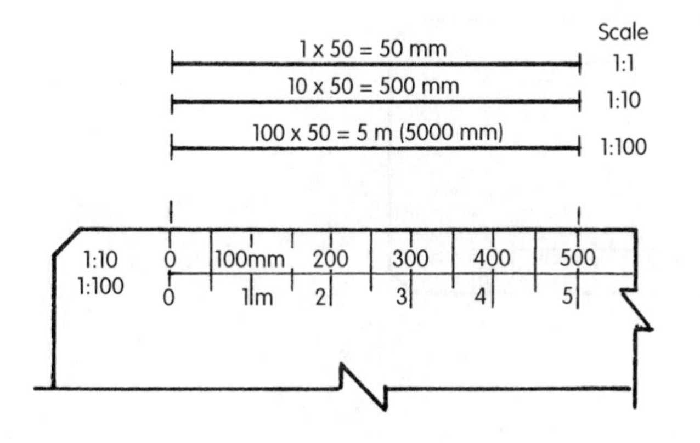

Figure 12.4 Proportional calculations

12.2.2 Scale Proportions

The scale used will depend on the type of drawing being prepared as shown in Table 12.1.

12.3 SCALE DRAWINGS

Drawings used by carpenters and joiners and other building workers use the principles of applied geometry to provide drawn information for the construction or alteration of a building. The drawings will also include written dimensions and printed notes to clarify what is required. Drawings are prepared on paper of various

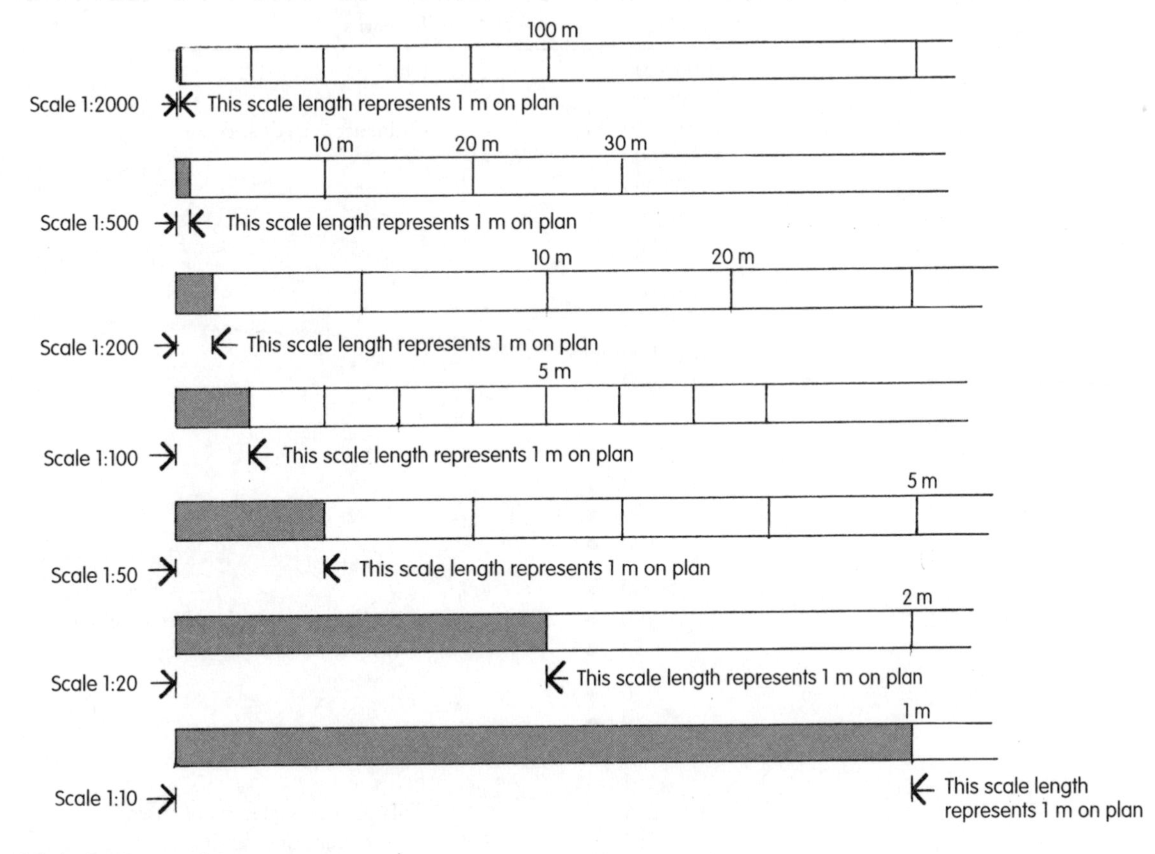

Figure 12.6 One metre drawn to various scales

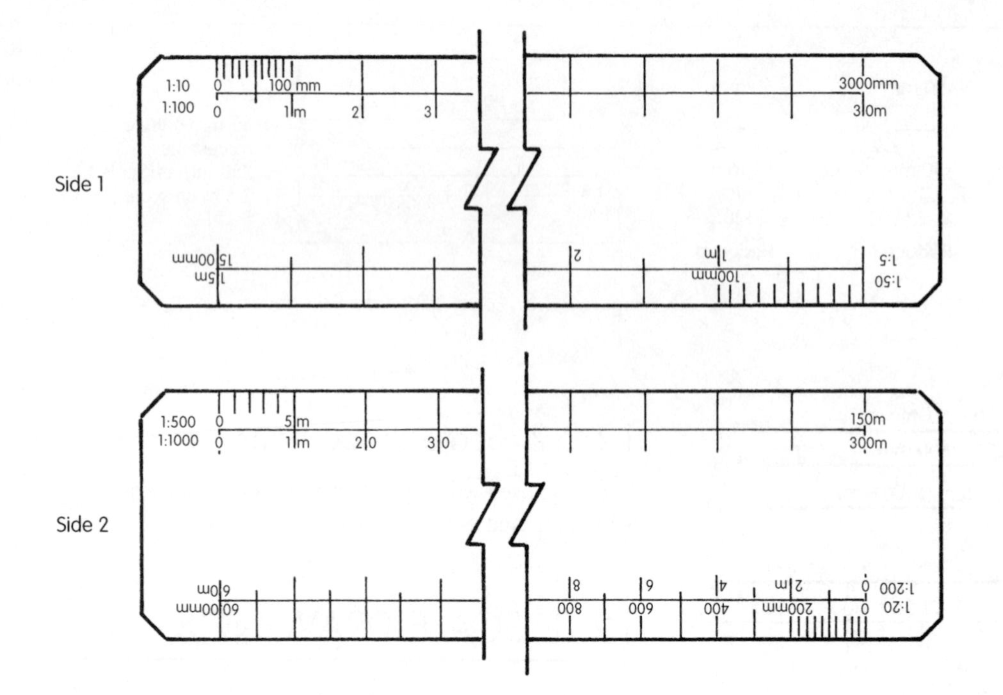

Figure 12.7 Builder's scale rule

Table 12.1 Range of scales

Type of drawing	Scale	Remarks
Maps	1: 1,000,000	
	1: 500,000	
	1: 250,000	Ordnance Survey scale
	1: 200,000	
	1: 100,000	
	1: 50,000	
	1: 25,000	
Urban surveys	1: 50,000	
	1: 20,000	
	1: 10,000	
	1: 5,000	
Survey, layout, site and key plans	1: 2,500	
	1: 2,000	
	1: 1,250	
	1: 1,000 ●	
	1: 500 ●	
Location drawings	1: 200 ●	
	1: 100 ●	● As shown on the builders scale rule
	1:50 ●	(Figure 12.7)
Component and assembly drawings	1:20 ●	
	1:10 ●	
	1: 5 ●	
	1: 1	Full size
	2: 1	Enlargement scales mainly used
	5: 1	to show very fine detail

sizes (Figure 12.8). Each drawing contains a title panel (Figure 12.9) which identifies and provides information about the drawing. Figure 12.10 shows a typical title panel.

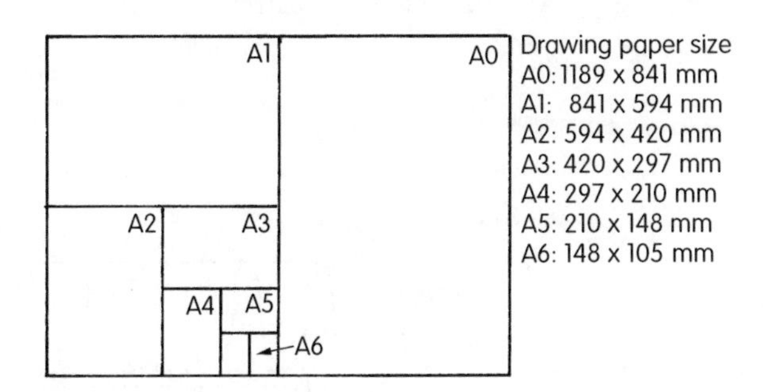

Figure 12.8 Drawing paper sizes

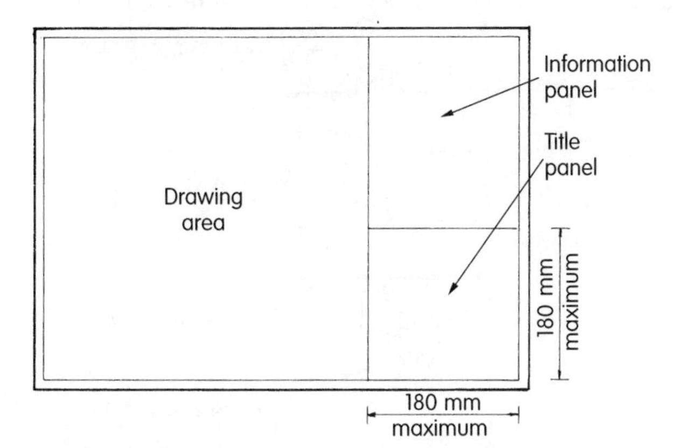

Figure 12.9 Drawing sheet layout

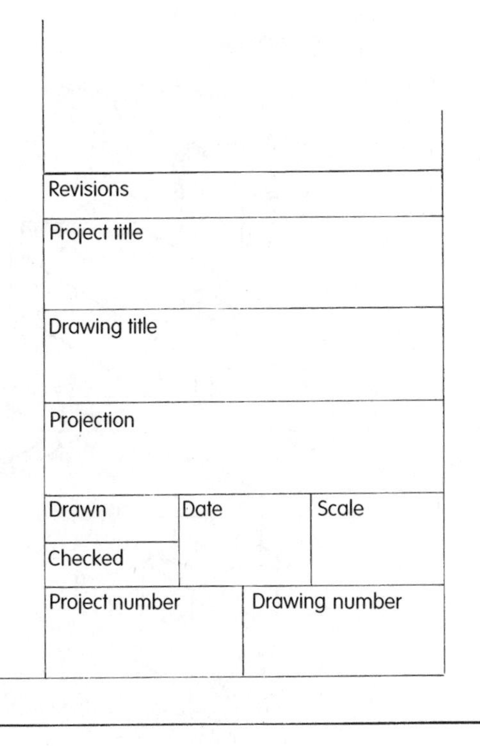

Figure 12.10 Typical title panel

Building drawings, although called plans, also show elevations and sections. Depending on their scale and what they are to be used for, drawings can be divided into the following types:

1 Maps – small-scale drawings showing the plan of large areas of land (Figure 12.11).
2 Layout or block plans – show the relationship of a site and its buildings, or proposed buildings, to the surrounding area (Figure 12.12).
3 Site plans – show the buildings or proposed buildings in relationship to each other and the boundaries of the site (Figure 12.13).
4 Location drawings – show sufficient detail in plans, elevations and sections for the main work to be started (Figure 12.14). Check the accepted scales by referring to Table 12.1.

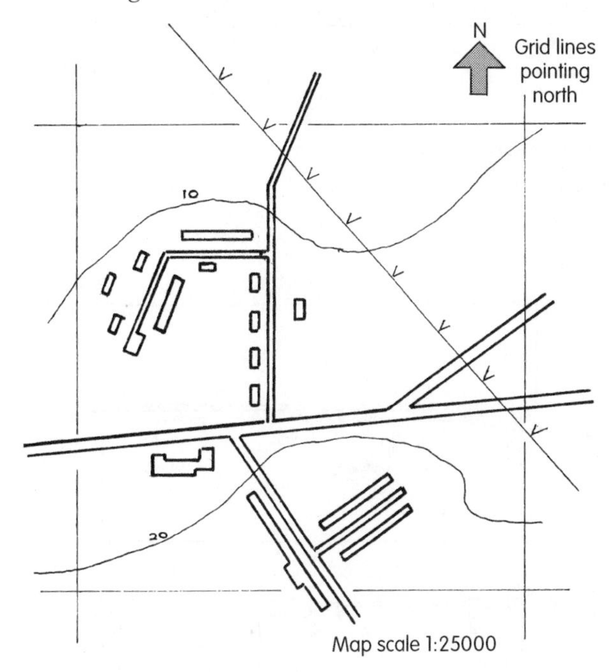

Figure 12.11 Map

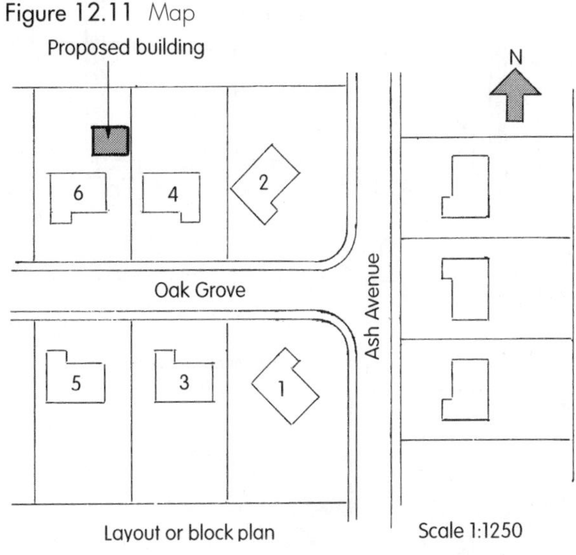

Figure 12.12 Layout or block plan

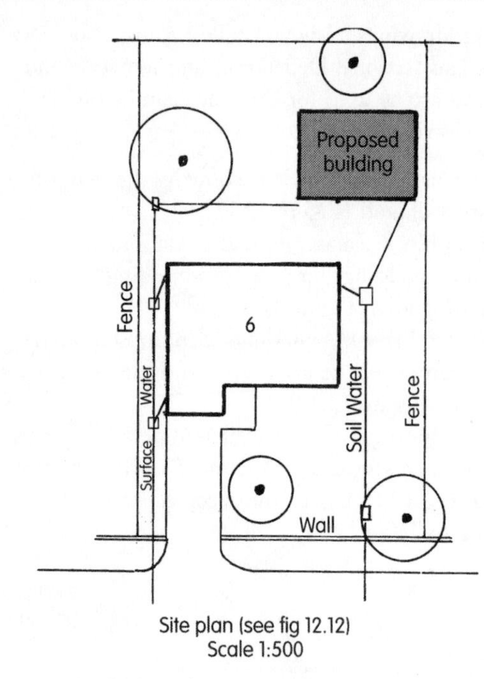

Figure 12.13 Site plan

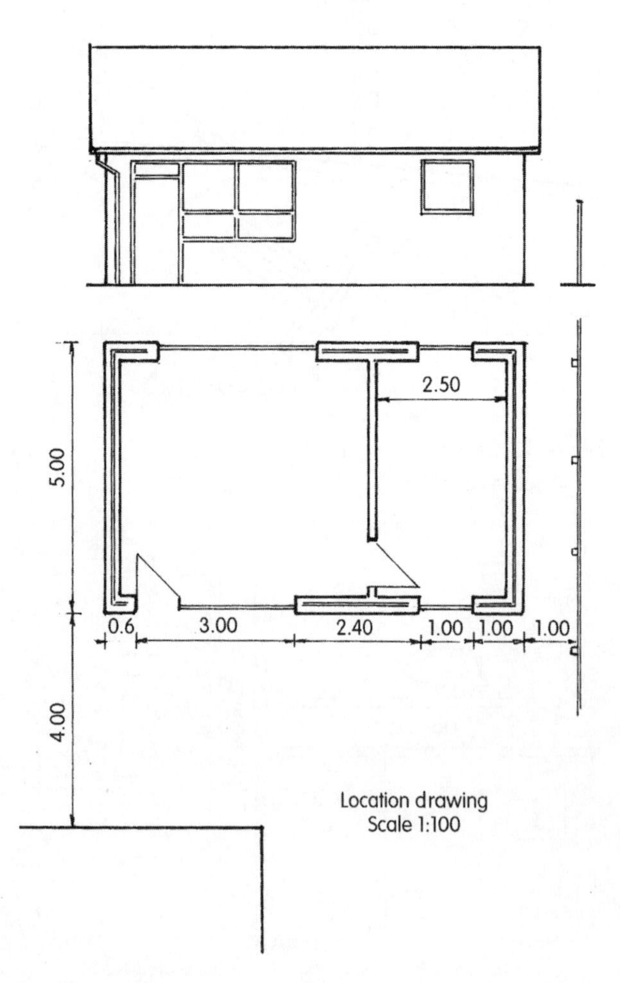

Figure 12.14 Location drawing

5 Component and assembly drawings – show in full detail, to a larger scale, the way in which the individual components in a building fit together (Figure 12.15). Exploded views are used to show how components are assembled (Figure 12.16).

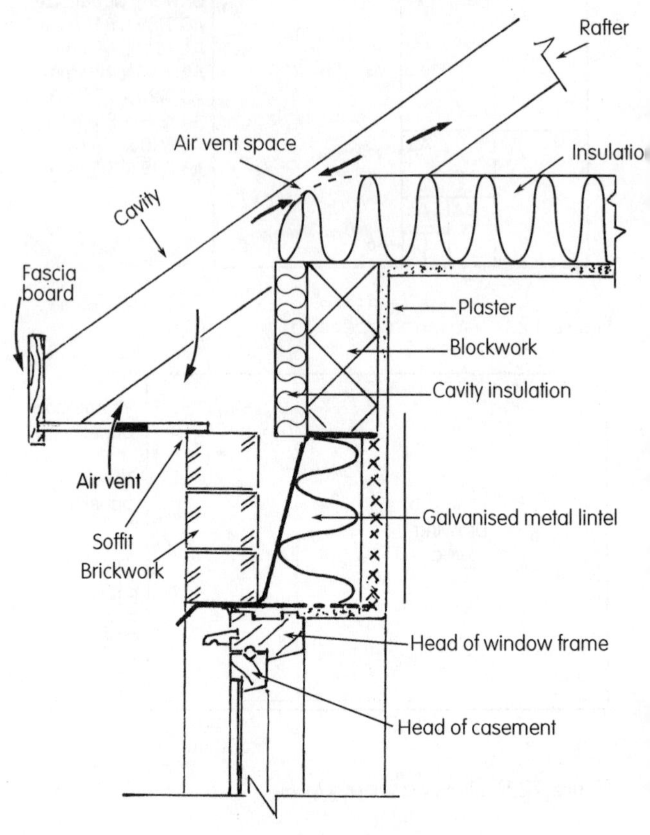

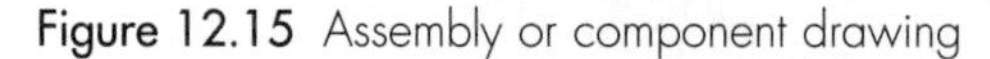

Figure 12.15 Assembly or component drawing

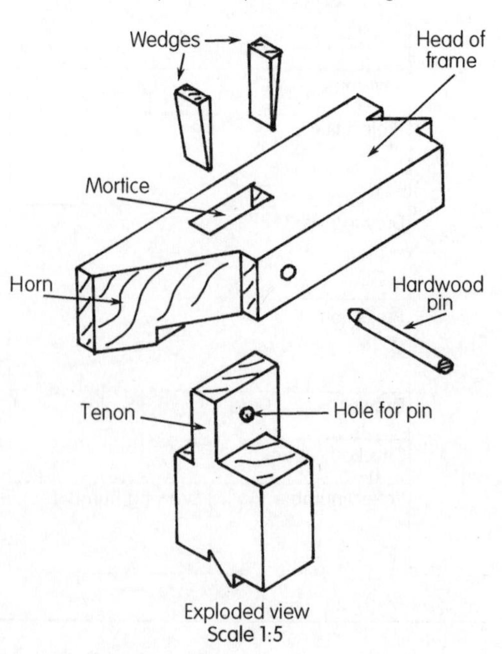

Figure 12.16 Exploded view shows detail of construction

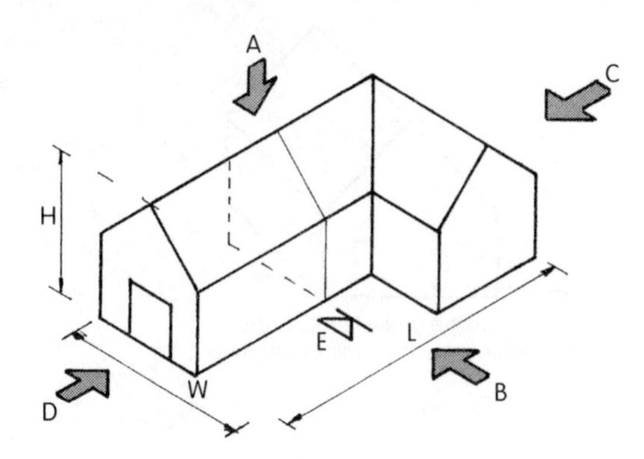

Figure 12.17 Views of a building

12.4 DRAWING PROJECTIONS

The simplest way of showing a three-dimensional object having length, breadth and height, such as a building (Figure 12.17), is a scaled drawing on a two-dimensional sheet of paper to show related views, such as plans having length and width and elevations showing height and width or height and length.

12.4.1 First Angle Orthographic Projection

This is the preferred drawing projection (Figure 12.18).

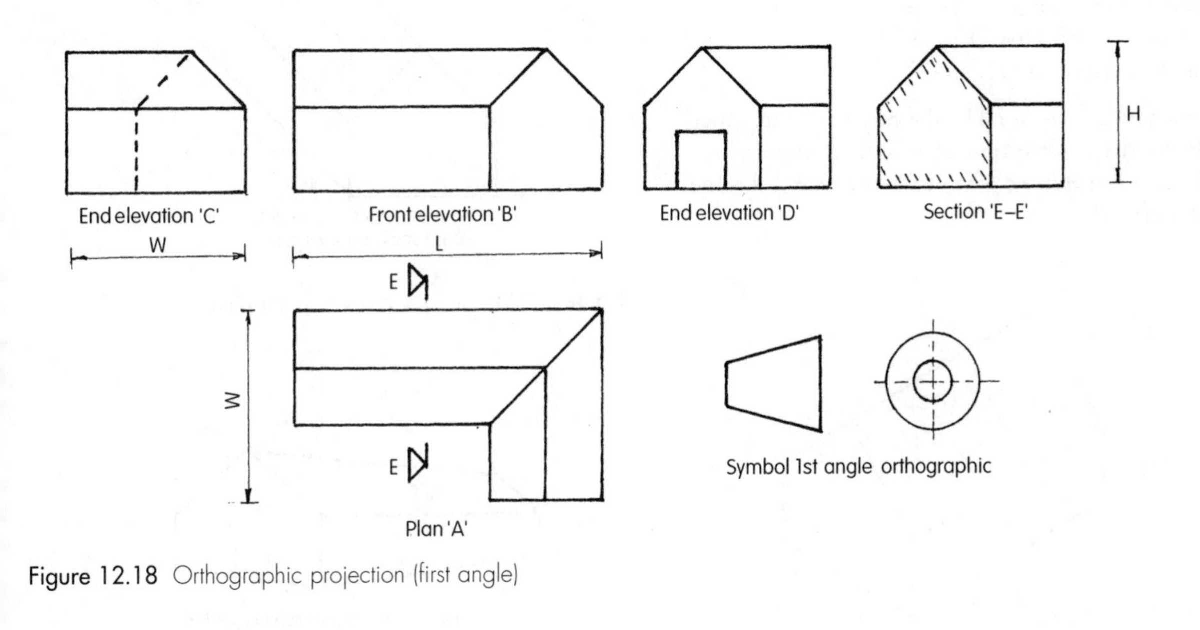

Figure 12.18 Orthographic projection (first angle)

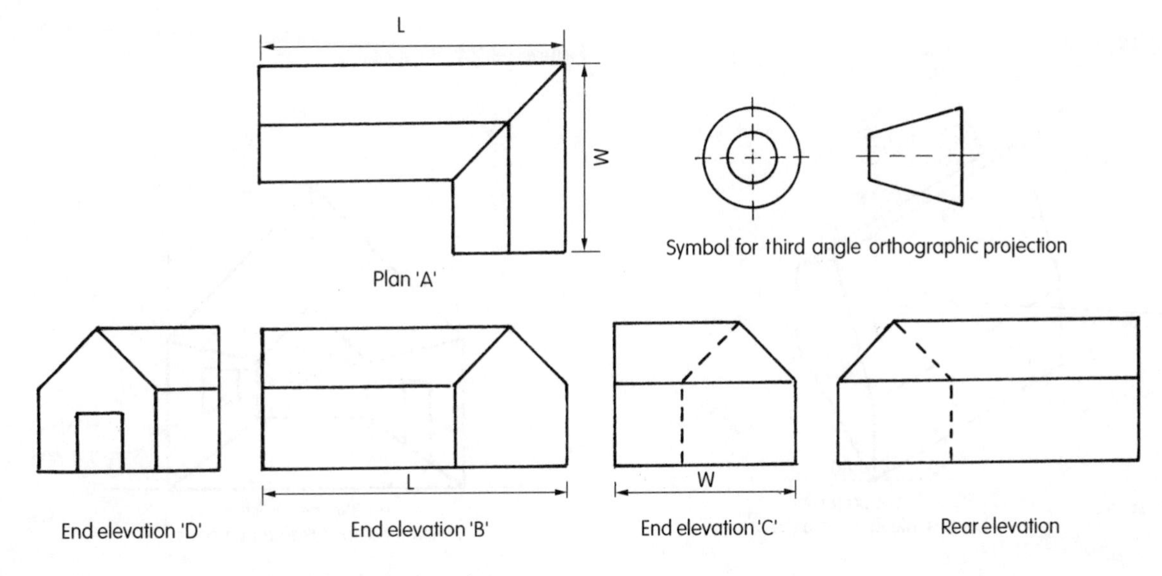

Figure 12.19 Orthographic projection (third angle)

12.4.2 Third Angle Orthographic Projection

The relationship of the plans and elevations is opposite to that used in first angle projection (Figure 12.19). Make sure you know the symbols that represent each of the two projections.

12.4.3 Pictorial Projections

This is a method of showing the three dimensions of a building on drawing paper, as a formal picture, drawn to a set of geometrical rules:

- isometric projection (Figure 12.20)
- planometric projection (Figure 12.21)
- oblique projection (cabinet) (Figure 12.22)
- oblique projection (cavalier) (Figure 12.23)
- two-point perspective (Figure 12.24)
- parallel perspective (Figure 12.25).

The most commonly used by the carpenter and joiner are the isometric projection when scale is important, and its two perspectives when using free hand sketches to communicate.

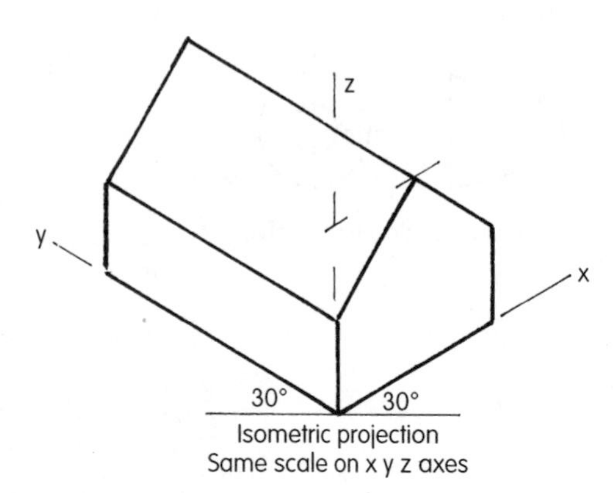

Figure 12.20 Isometric projection

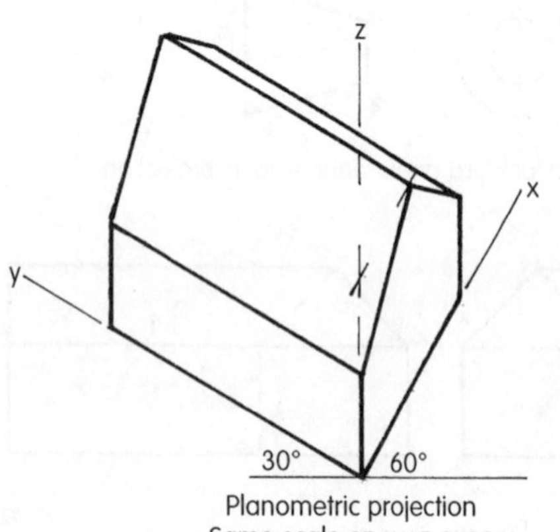

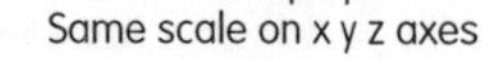

Figure 12.21 Planometric projection

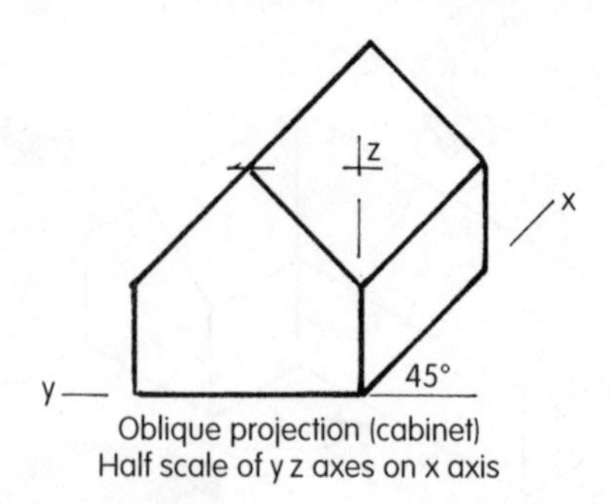

Figure 12.22 Oblique projection (cabinet)

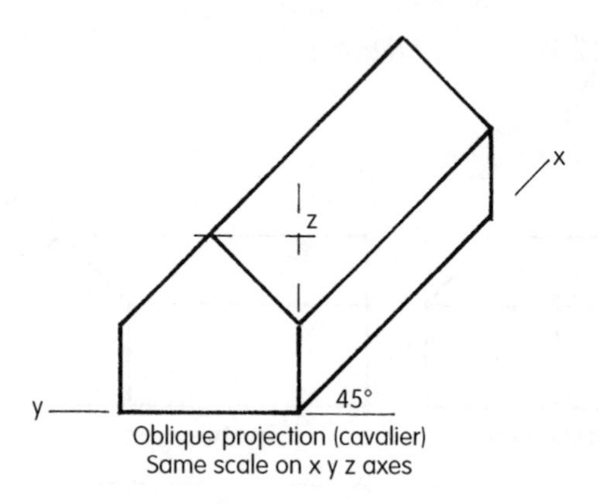

Figure 12.23 Oblique projection (cavalier)

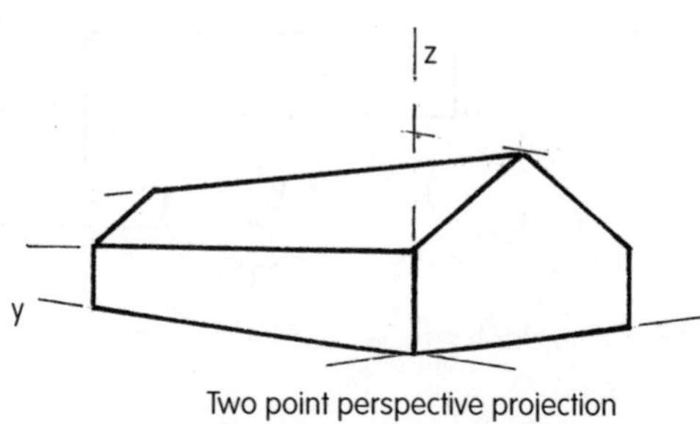

Figure 12.24 Perspective (two-point)

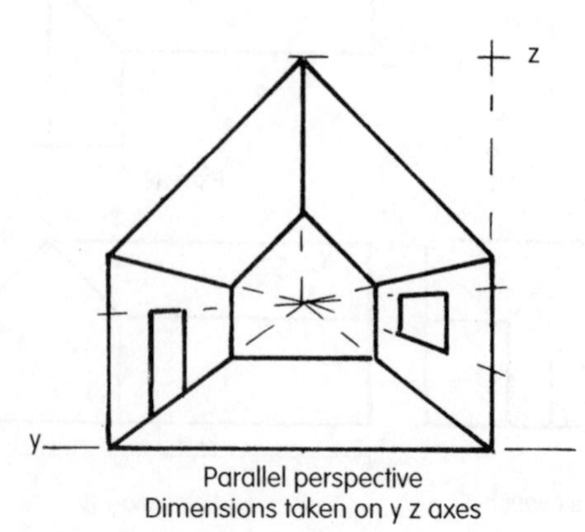

Figure 12.25 Parallel perspective

12.5 TAKING DIMENSIONS FROM A DRAWING

When taking dimensions from a drawing, printed or written dimensions must always take precedence and care should be taken that all given dimensions add up correctly to the overall size. Scale rules should not be used for taking off dimensions, as paper is a hygroscopic material made from wood pulp which will shrink or swell depending on the surrounding humidity. This makes it almost impossible to accurately measure lengths from a drawing using a scale rule.

In Figure 12.26, the actual size of the opening taken from the drawing is 9 – 5 = 4 m using the printed dimensions. The scale rule shows the opening to measure between 3.5 and 4.5 m, depending on the amount of moisture in the paper.

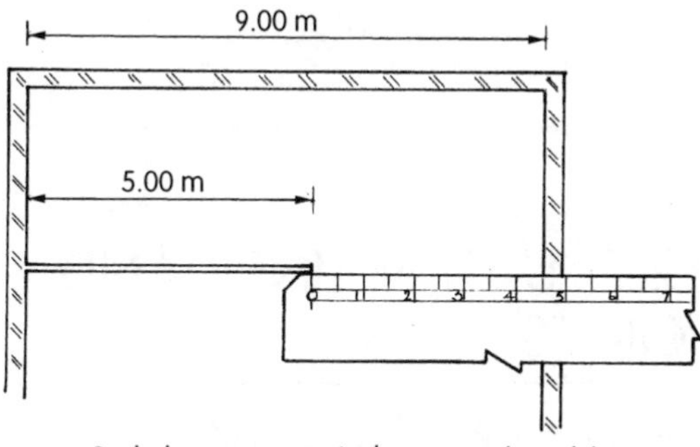

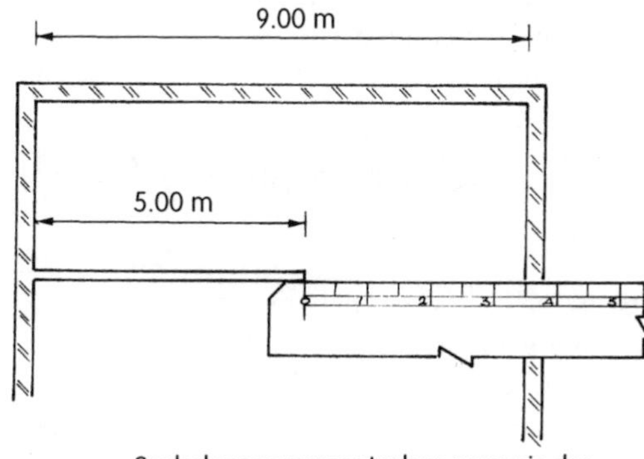

Figure 12.26 Taking off measurements

12.5.1 Reading Dimensions

Care must be taken when reading off dimensions from a drawing not to confuse single with running dimensions. Figure 12.27 shows the difference between the two methods of recording dimensions – also see Figure 6.2.

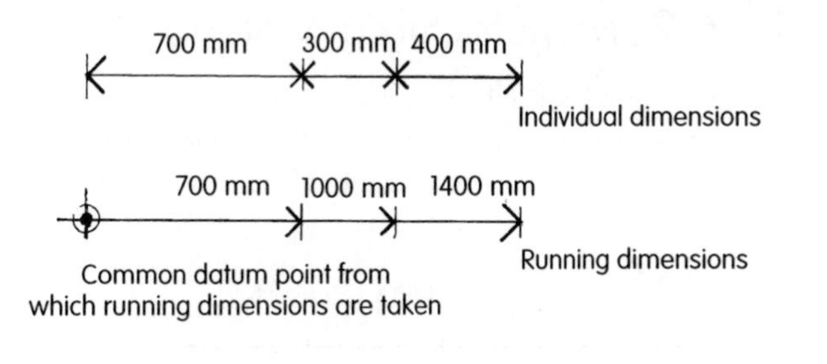

Figure 12.27 Types of dimensions

12.6 SYMBOLS USED IN DRAWING

When preparing drawings, materials and components can be identified by the use of symbols unique to what they represent. There are many symbols in use and reference should be made to BS 1192 part 3 for the complete range. The following are some of the more common symbols that the carpenter and joiner will become familiar with.

12.6.1 Materials

The material is identified by the pattern and shading on the cross-section of the material (Table 12.2).

Table 12.2 Material symbols

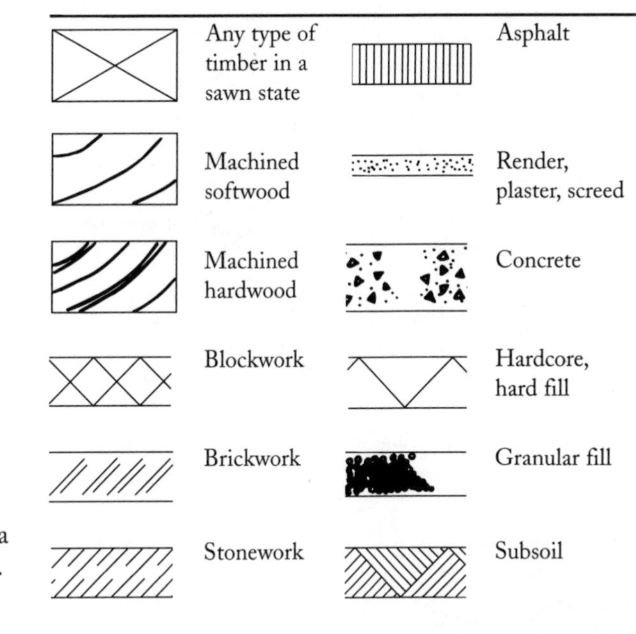

Symbol	Material	Symbol	Material
	Any type of timber in a sawn state		Asphalt
	Machined softwood		Render, plaster, screed
	Machined hardwood		Concrete
	Blockwork		Hardcore, hard fill
	Brickwork		Granular fill
	Stonework		Subsoil

12.6.2 Manufactured Materials

These are raw materials manufactured into sheets, quilts, extruded or rolled sections. They are identified on cross-section as shown in Table 12.3.

Table 12.3 Manufactured material symbols

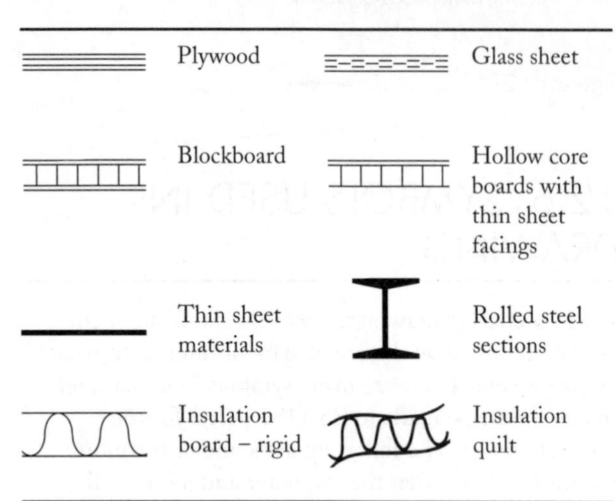

12.6.3 Location Symbols

These are used to give information about levels, falls and rises, orientation of buildings and details for cross-referencing (Table 12.4).

Table 12.4 Location symbols

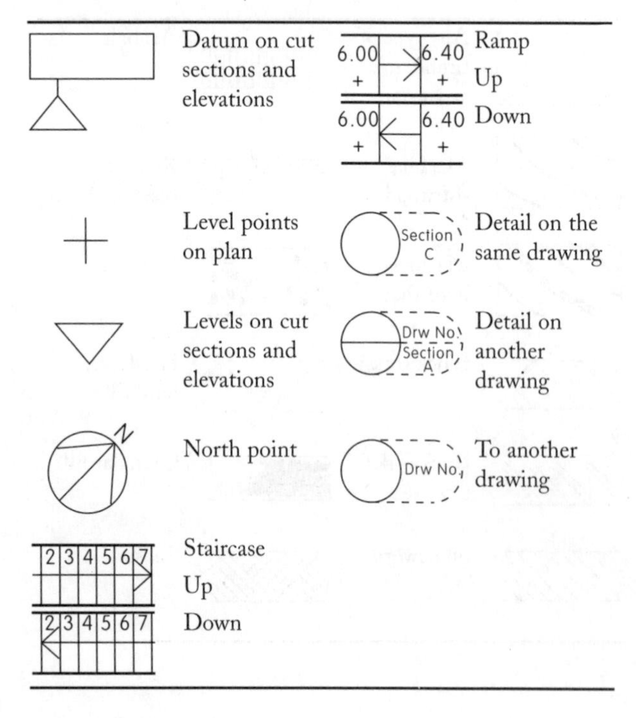

12.6.4 Modification Symbols

These symbols are used to identify alterations and modifications to existing buildings (Table 12.5).

Table 12.5 Modification symbols

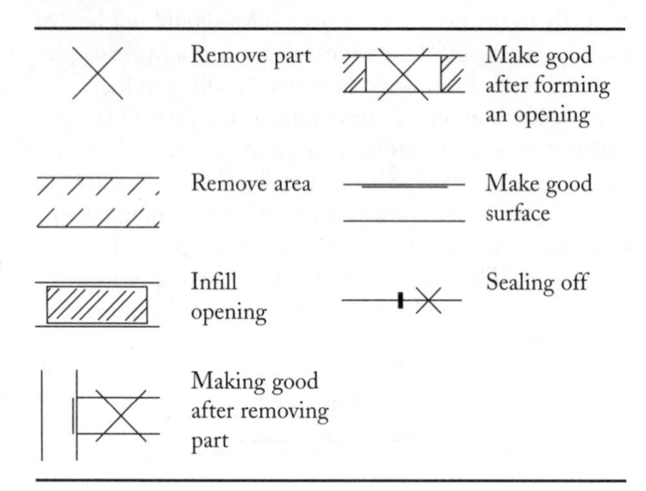

12.7 DATUM LINES AND POINTS

These are identified points from which all other levels are taken. They are of three types:

- Ordnance Survey bench marks
- temporary bench marks
- datum lines.

12.7.1 Ordnance Survey Bench Marks

These are placed on permanent structures (Figure 12.28) by the Ordnance Survey to indicate the height above mean sea level as established at Newlyn in Cornwall. The positions of these bench marks are shown on Ordnance Survey maps (see also Figure 6.22).

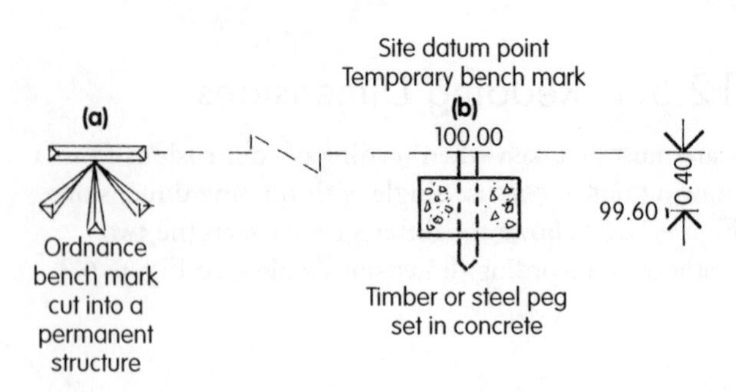

Figure 12.28 Ordnance Survey bench marks

12.7.2 Temporary or Site Bench Marks

From the nearest permanent bench mark, (Figure 12.28a) a level is transferred to the building site and established as the temporary bench mark, (Figure 12.28b) from which all other levels on the site are taken as example Figure 12.29.

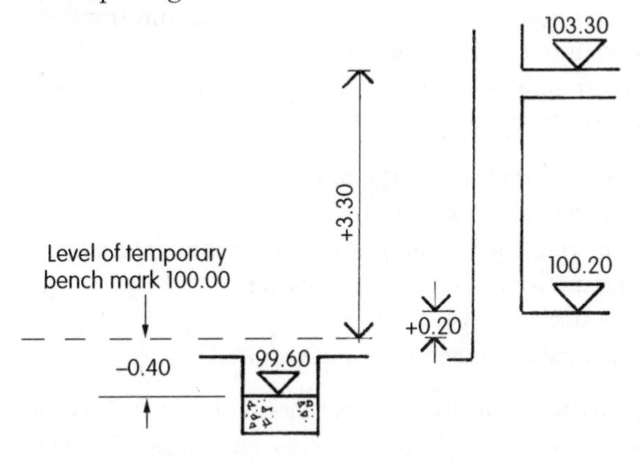

Figure 12.29 Temporary bench marks

12.7.3 Datum Lines

Datum lines can be set up without reference to Ordnance Survey bench marks. These are temporary horizontal or vertical (plumb) lines marked up at some convenient position in a room or building to which points in the structure can be related. In taking measurements without reference to a datum line, wrong assumptions about the shape of a wall can be made even if the measurements taken are correct (Figure 12.30a). Is the floor level and the ceiling sloping or the floor sloping and the ceiling level? Using a temporary datum line allows the true shape of the wall to be determined (Figure 12.30b). Knowing the true shape of a wall is most important if fittings or panelling are to be made accurately. Further examples are shown in Figure 6.23.

12.8 ABBREVIATIONS

All trades and professions write abbreviated (shortened) technical words associated with their work and over time lists of nationally recognised abbreviations have been incorporated into British Standards and others. Table 12.6 gives a list of some of the most common abbreviations.

Table 12.6 Abbreviations

Abbreviation	Meaning
bwk	Brickwork
bdg	Boarding
bldg	Building
dwg	Drawing
DPC	Damp proof course
DPM	Damp proof membrane
Ex	Nominal sawn size (out of)
hdb	Hardboard
hwd	Hardwood
jst	Joist
PAR	Planed all round
Pbd	Plasterboard
SAA	Satin anodised aluminium
SC	Satin chrome
SS	Stainless steel
swd	Softwood
T&G	Tongue and groove

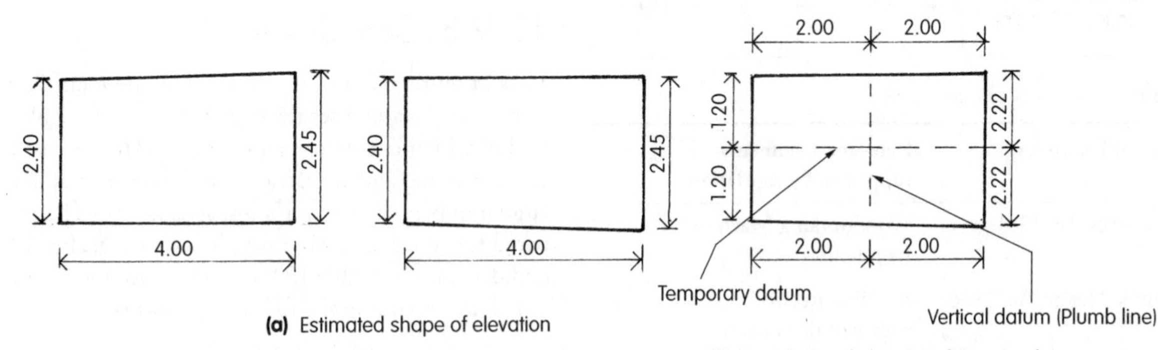

Figure 12.30 Datum lines

12.9 WRITTEN INFORMATION

Although the carpenter and joiner communicates by the use of sketches and drawings, the drawing alone does not convey all the information that is required and so it is supplemented by the use of:

- statutory regulations
- specifications
- schedules
- manufacturer's instructions.

12.9.1 Statutory Regulations

These are sets of rules which have legal status conferred on them by Acts of Parliament (Table 12.7). The regulations can be modified as needs and circumstances change by Orders in Council, without the need for a new Act of Parliament. Table 12.8 shows the Act of Parliament under which the Regulations are given legal status. In addition, European Codes will be increasingly used as the unified legislation of the European Parliament and Council of Ministers is implemented in the countries of the European Community.

Table 12.7 Statutory regulations and preferred documents

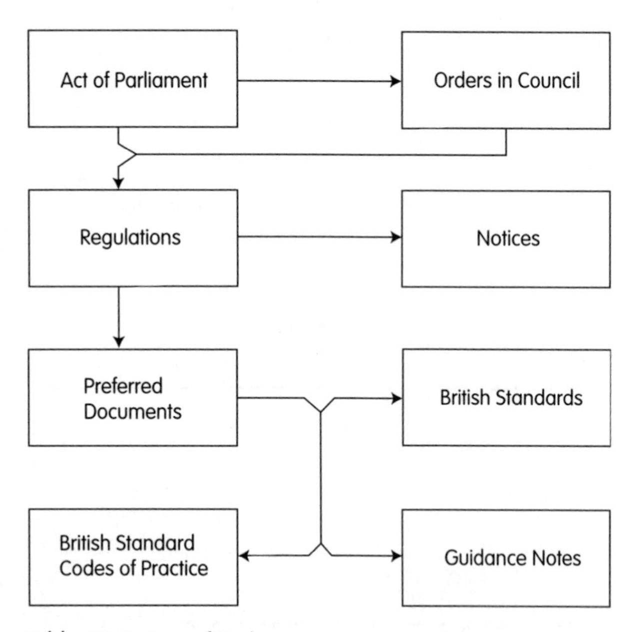

Table 12.8 Acts of Parliament

Act of Parliament	Relevant documents supported by legislation
Factories Act 1961	Woodworking Machinery Regulations
Public Health Act 1936	Building Regulations and approved documents
Health and Safety at Work Act 1974	Appropriate BS Codes of Practice; Health and Safety guidance notes

12.9.2 Specifications

These are detailed descriptions of construction processes, giving precise information about materials and workmanship. The specification is prepared by the designer or architect to provide additional information that cannot be easily shown on a drawing. Figure 12.31 is an example of a specification for internal hardwood door linings and gives the following information:

a the sectional size of the linings
b the shape in cross-section
c the type, quality and moisture content of the timber
d the method of fixing and to what
e the surface finish
f the quality of workmanship required.

This written information, together with the component drawing, will provide all the necessary information for the making and fixing of the linings.

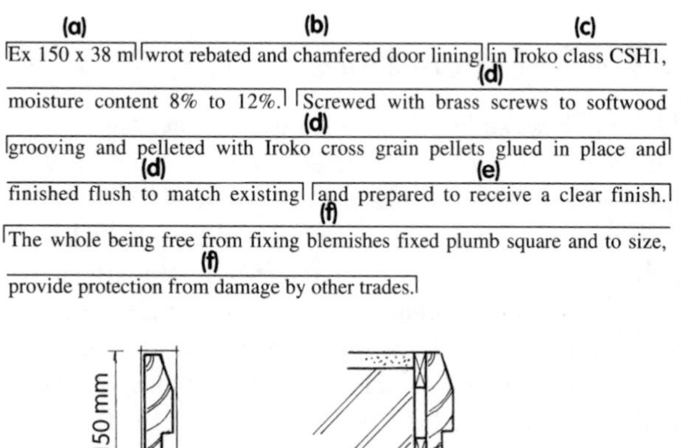

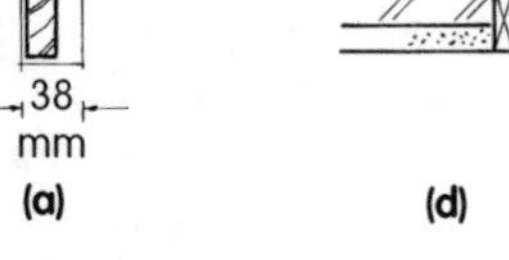

Figure 12.31 Specification

12.9.3 Schedules

Lists of numerous and repetitive items are identified from a drawing and set down in the form of a table or grid in which like items can be grouped for ease of reference, ordering and checking. In buildings which have large numbers of similar rooms, schedules can be prepared for things like windows, doors, fittings, furnishings and decorations. Table 12.9 is a door and ironmongery schedule containing the following information:

- a short description of the items
- the location of these items
- a grouping of like items, making it easier to determine the number and types of components needed.

Table 12.9 Door and ironmongery (hardware) schedule

Description	Door number								
	G1	G2	G3	G4	G5	G6	G7	G8	Total
External flush, two-leaf rebated: 1980 × 760 × 50 mm	X					X			4
Number 1 ½ pairs of 100 mm steel washered brass butts	X					X			6 pair
Flush edge bolts No. 2 per door set Cat. No. 654	X					X			4
Rebated mortise latch and lock including brass lever furniture Cat. No. 432	X					X			2
Internal flush doors 1980 × 760 × 38 mm Cat. No. G 604		X		X		X	X		4
No. 1 pair 75 mm brass butts		X		X					2 pair
No. 1 Mortise latch and brass lever furniture Cat. No. 789		X	X	X	X		X	X	6 sets
Internal flush doors 1980 × 760 × 45 mm with glazed observation panel Cat. No. 2G 704			X	X					2
No. 1 pair 100 mm brass butts			X		X				2 pair
Brass kicking plates 740 mm × 225 mm two per leaf Cat. No. 987			X	X					4
Overhead door closers, brass finish Cat. No. 1098			X	X					2

Note: G1–G8 (location) will be marked on the plan – in this case the ground floor plan (G)

12.9.4 Manufacturer's Technical Literature

This consists of drawings and text which are provided to give the user details of the component. In its simplest form, instructions in the form of drawings and text are printed on the back of the packaging (Figure 12.32) while more complex instructions may be printed on a separate pamphlet. The manufacturer's instructions will contain the following information:

- a brief description of the item
- conditions under which it can be used
- fixing and maintenance instructions
- details about any potential hazards and any safety precautions to be taken in fixing and use.

12.10 SPOKEN (ORAL) INFORMATION

Instructions are frequently given by word of mouth. Keep them brief and to the point, without complicated and confusing detail, making it easier for the recipient to remember the instructions. If you wish to aid your memory take notes of what is being said; it is often useful for correcting misunderstandings and for later reference. In order to record and pass on oral information correctly, keep a small notebook (Figure 12.33) in your pocket and always use it if the oral information is complicated and has to be passed on to others. You may not be able to write down the instructions word for word:

a make lists
b use notes and sketches
c abbreviate notes using key words.

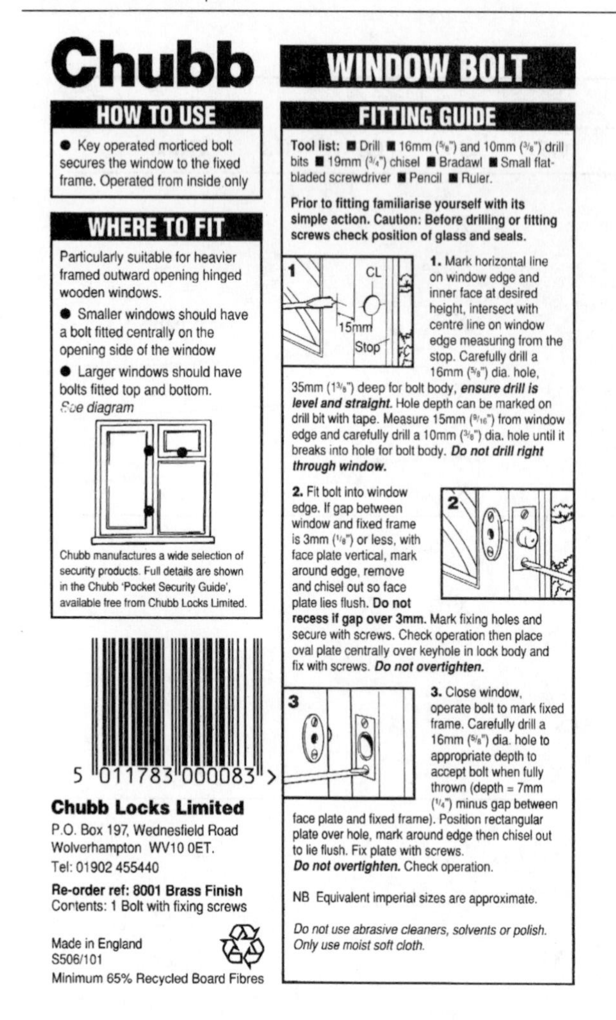

Chubb **WINDOW BOLT**

HOW TO USE

- Key operated morticed bolt secures the window to the fixed frame. Operated from inside only

WHERE TO FIT

Particularly suitable for heavier framed outward opening hinged wooden windows.

- Smaller windows should have a bolt fitted centrally on the opening side of the window
- Larger windows should have bolts fitted top and bottom.

See diagram

Chubb manufactures a wide selection of security products. Full details are shown in the Chubb 'Pocket Security Guide', available free from Chubb Locks Limited.

5 011783 000083 >

Chubb Locks Limited
P.O. Box 197, Wednesfield Road
Wolverhampton WV10 0ET.
Tel: 01902 455440

Re-order ref: 8001 Brass Finish
Contents: 1 Bolt with fixing screws

Made in England
S506/101
Minimum 65% Recycled Board Fibres

FITTING GUIDE

Tool list: ■ Drill ■ 16mm (5/8") and 10mm (3/8") drill bits ■ 19mm (3/4") chisel ■ Bradawl ■ Small flat-bladed screwdriver ■ Pencil ■ Ruler.

Prior to fitting familiarise yourself with its simple action. Caution: Before drilling or fitting screws check position of glass and seals.

1. Mark horizontal line on window edge and inner face at desired height, intersect with centre line on window edge measuring from the stop. Carefully drill a 16mm (5/8") dia. hole, 35mm (1 3/8") deep for bolt body, ***ensure drill is level and straight.*** Hole depth can be marked on drill bit with tape. Measure 15mm (9/16") from window edge and carefully drill a 10mm (3/8") dia. hole until it breaks into hole for bolt body. ***Do not drill right through window.***

2. Fit bolt into window edge. If gap between window and fixed frame is 3mm (1/8") or less, with face plate vertical, mark around edge, remove and chisel out so face plate lies flush. **Do not recess if gap over 3mm.** Mark fixing holes and secure with screws. Check operation then place oval plate centrally over keyhole in lock body and fix with screws. ***Do not overtighten.***

3. Close window, operate bolt to mark fixed frame. Carefully drill a 16mm (5/8") dia. hole to appropriate depth to accept bolt when fully thrown (depth = 7mm (1/4") minus gap between face plate and fixed frame). Position rectangular plate over hole, mark around edge then chisel out to lie flush. Fix plate with screws. ***Do not overtighten.*** Check operation.

NB Equivalent imperial sizes are approximate.

Do not use abrasive cleaners, solvents or polish. Only use moist soft cloth.

Figure 12.32 Manufacturer's instructions

Figure 12.33 Page of a notebook

If you are not sure of what is being said to you, ask for clarification until you do understand

Apply the rules of spelling and grammar as well as you are able; these skills develop with practise. Figure 12.33 shows a typical page in a notebook used by a joiner.

12.11 REPORTING INACCURACIES

Errors can and do occur in drawings and text and in application. It is important that the mistake is identified and immediate action is taken to rectify what is wrong. A delay in taking action can become very costly. If an inaccuracy is noticed, it should be rectified or reported to someone who can decide what to do. The following action should be taken on discovering what you think is an error.

- Check to make sure it is an error and not a misunderstanding of the information.
- If it is within your power to put the matter right, do so.
- Otherwise, report it to your immediate superior, who will correct the error or pass on the information to others for action.
- As correcting mistakes may cost money, prompt action is necessary so that matters may be put right as soon as possible and at least cost.

As an example, a correction to a rebate design is shown in Figure 12.34.

1 The person setting out the hatchway doors sees an error in the design of the rebate at (a) which will bind as the door is opened.
2 The workshop manager is informed and discussion takes place on an alternative shape for the rebate (b).
3 The manager contacts the designer to see if the alternative detail is acceptable.
4 The suggested modification is agreed and incorporated in the door construction.

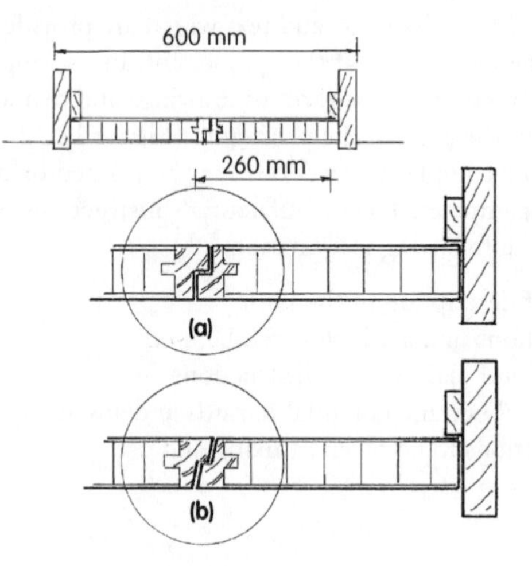

Figure 12.34 Error corrections

13
Calculations

The use of numbers is a further tool of communication which is most effective when it is fully understood. As in all other activities, development of numerical skills is necessary if we are to practise effectively as carpenters. Relevant questions are: How many? What length, area or volume? What time will it take? What will it cost? (Figure 13.1a–f). All these questions are answered by the manipulation of numbers (calculations).

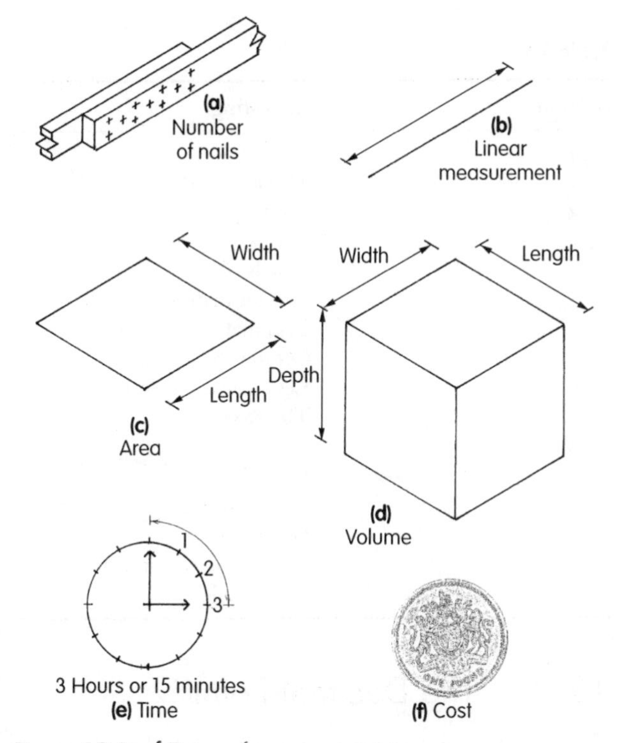

Figure 13.1a–f Types of measurement

13.1 UNITS OF MEASUREMENT

The British units of measurement were based on human proportions, such as 'the king's foot'. The foot was eventually established as a standard length which was no longer dependent on the monarch's shoe size!

At the end of the French Revolution a unit of measurement related to the circumference of the earth was developed. This was the metre, with a length equivalent to one ten-millionth part of the distance from the pole to the equator measured along the meridian (Figure 13.2). Since 1960 the standard length of the metre has been determined by measurement of the wavelengths of orange radiation of the krypton atom.

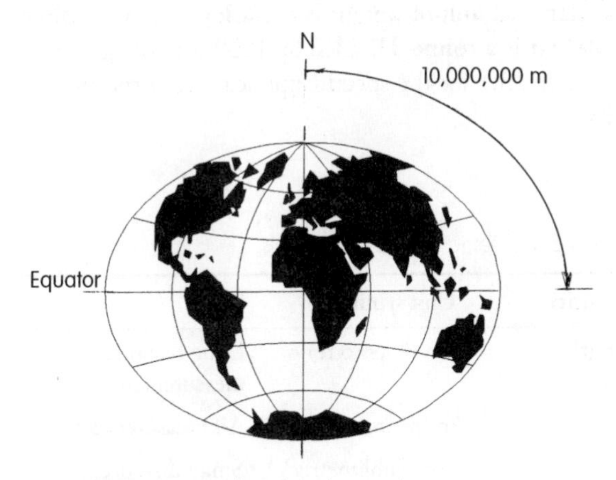

Figure 13.2 Metre related to the earth

Between 1966 and the end of 1972 Britain changed from using anthropomorphic measurements (feet and inches determined by measurements based on the human body) to the metric system by adopting SI units (Système Internationale d'Unités). The only measurements now in everyday use which are not based on a unit of ten are time and angular measurement.

13.1.1 Metric Units for the Measurement of Length

Although the standard unit of measurement is the metre it can be multiplied by 1000 to give a kilometre or divided by 1000 to give a millimetre. A metre divided by ten gives a decimetre(dm) and by 100 a centimetre(cm) (Figure 13.3). The two measurements most used by the carpenter are the **metre** and **millimetre**.

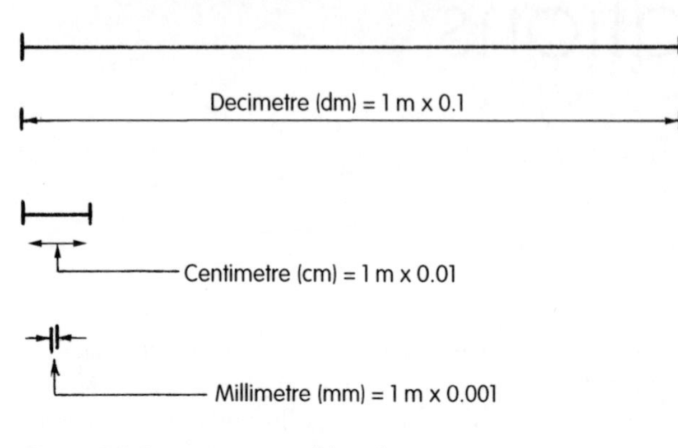

Figure 13.3 Metric units of length

13.1.2 Metric Units for the Measurement of Mass (Weight)

The standard unit of weight is the **kilogram**. Multiplied by 1000 it is a **tonne**. Divided by 1000 it is the **gram**. See Table 13.1 for the specific application of metric units.

Table 13.1 Metric units and symbols

Quantity	Unit symbol	Use
Length	km (kilometre)	Long distance measurement
	m (metre)	All measurement uses
	mm (millimetres)	Small distance measurements, setting out joinery
Area	km^2	Land measure
	ha (hectare)	Land measure
	10,000 m^2	Land measure
	m^2	All uses
	mm^2	Structural calculations
Volume	m^3	Material quantities
	mm^3	Structural calculations
Liquid measure	l (litre) 1 litre = 1 dm^3	Volume of liquids, paint and preservatives
Mass (weight)	kg (kilogram)	All uses
	g (gram)	1000th part of a kilogram

13.1.3 Metric Units for the Measurement of Liquids

The standard unit for the measurement of liquids is the **litre**. It is the volume of liquid that can be contained in a 1-decimetre cube (100 × 100 × 100 mm) or a one thousandth part of a cubic metre. If the liquid is water at its maximum density it will weigh one kilogram.

13.2 BASIC ARITHMETIC OPERATIONS

There are four basic arithmetic operations used in the manipulation of numbers which must be understood and applied correctly if the numerical information is to be accurate. The four functions can be grouped into two pairs, subtraction being the reverse of addition and multiplication the reverse of division. **Addition** and **subtraction** are indicated by the symbols (+) and (−) and **multiplication** and **division** by the symbols (×) and (÷). See Table 13.2 for more detail about mathematical symbols.

Table 13.2 Mathematical symbols

Symbol	Indicating
+	Plus
−	Minus
×	Multiply
÷	Divide
=	Equals
>	Greater than (e.g. 6 > 3)
<	Less than (e.g. 3 < 6)
≥	Greater than or equal to
≤	Less than or equal to
∴	Therefore
%	Percentage
∥	Parallel
⊥	Perpendicular
√	Square root
≃	Approximately equal to
∟	Right angle

13.2.1 The Decimal Point

The decimal point separates the whole numbers (integers) from the decimal fraction (part) of the whole number. When you are writing down metric numbers the figures are set out in units of ten, to the left of the point if integers and to the right if decimal fractions (Figure 13.4). When writing down dimensions, builders show fractions of the metre to three decimal places even if the measurement is in whole metres. This indicates that measurements as small as a millimetre have been taken (Figure 13.5).

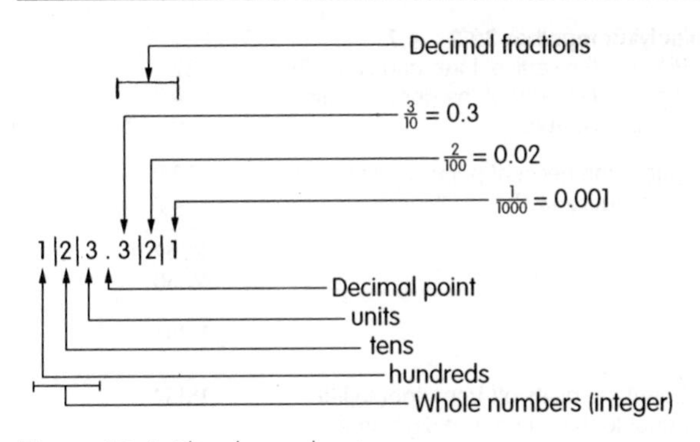

Figure 13.4 The decimal point

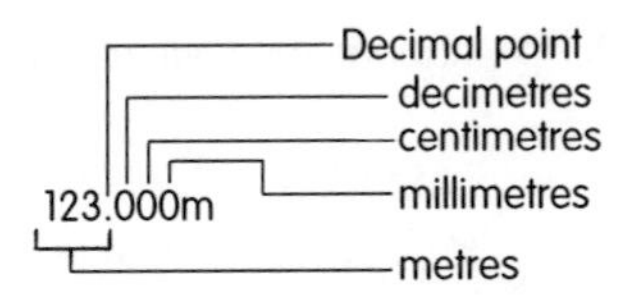

Figure 13.5 Metres and decimal fractions of a metre

13.2.2 Addition (+)

A quicker and more convenient way to find the sum total than by counting individual items is to group the figures together and add (Figure 13.6). As an alternative to setting out the figures and symbols in rows, they can be set up in vertical columns. Many find this much easier when adding together large numbers (Figure 13.7).

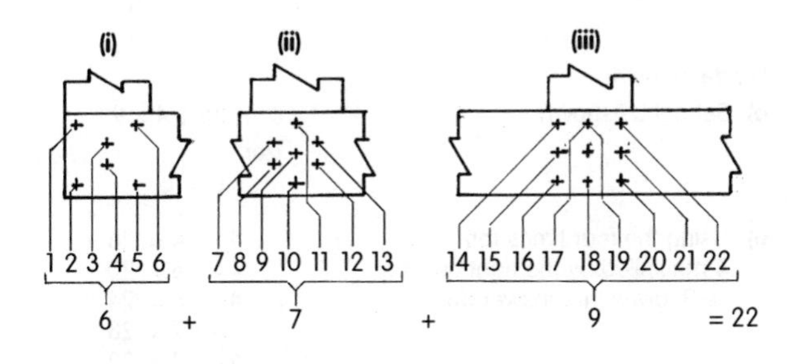

Figure 13.6 Numbers by counting

a) There are three groups in Figure 13.6. Group (i) has 6 symbols, group (ii) contains 7, while group (iii) holds 9 symbols. An alternative to counting the symbols in rows is to add the groups together to get a total of 22.

```
 6
 7
 9
---
22
```

Figure 13.7 Addition

13.2.3 Subtraction (–)

Subtraction is the opposite of addition. How many of the 22 nails are required if joint (iii) is removed in Figure 13.6? Although this is a simple subtraction, for larger numbers the operation is carried out as shown in Figure 13.8.

a) Dealing with each column in turn, starting with the units column.

```
  22
 - 9
```

b) Subtracting 9 from 2 is not possible without leaving a negative number.

c) Borrow 1 from the tens column multiply it by ten and add the two in the units column [(1 × 10) + 2 = 12] From the twelve take away nine [12 – 9 = 3] Place the 3 in the units column.

```
             2 ¹2
           -  ₁9
Answer row   1 3
```

d) This leaves 1 in the tens column and as there is nothing to subtract from it [1 – 0 = 1], 1 is placed in the answer row; [22 – 9 = 13]

e) For small numbers the above subtraction is worked out almost automatically in the head using mental arithmetic.

Figure 13.8 Subtraction

subtraction using larger numbers

a) One minus three gives a negative number.

```
  321
- 123
```

b) Carry out the operation as in 'c' in Figure 13.8: [11 – 3 = 8]

```
  31¹1
- 12 3
     8
```

c) As one minus two gives a negative number borrow one from the hundreds column multiply it by ten and add the tens figure. [(1 × 10) + 1 = 11] then subtract two from eleven. [11 – 2 = 9]

```
  3 ¹1 1
- 1  2 3
     9 8
```

d) As one has been taken from the hundreds column two is left, from which we subtract one [2 – 1 = 1]

```
  3 ¹1 1
- 1  2 3
  1  9 8
```

e) As a check add the answer to the sum subtracted [198 + 123 = 321]

```
  198
+ 123
  321
```

Figure 13.9 Method of subtraction using larger numbers

13.2.4 Multiplication (×)

This is a more efficient method than counting or adding groups of like numbers as shown in Figure 13.10. To multiply numbers you need to know your multiplication tables, at least up to ten, or use a mechanical or electronic method of multiplying. Figure 13.11 shows how larger numbers are multiplied together. When multiplying decimal fractions carry out the operations as illustrated in Figure 13.12.

a) There is a more efficient way to carry out the calculation shown to the right.

(1) (2) (3) (4) (5) (6)
4 + 4 + 4 + 4 + 4 + 4 = 24

b) To multiply the numbers together is a much neater and quicker way than adding four together six times.

6 × 4 = 24

Figure 13.10 Multiplication

Multiply 345 by 123

a) Arrange in vertical columns	345 × 123
b) Multiply the units together [5 × 3 = 15]; place the 5 in the units column and carry the 1 over into the tens column.	345 × 123 —— $_1$5
c) Multiply the unit and tens together [3 × 4 =12]; then add the one carried over [12 + 1 = 13]; place the three in the tens column and carry the one into the hundreds column.	345 × 123 —— $_1$35
d) Multiply the hundreds and unit columns; then add the one carried forward to the hundreds column [9 + 1 = 10]; place the nought in the hundreds column and the one in the thousands column.	345 × 123 —— 1035
e) Place a nought in the units column	345 × 123 —— 1035 0
f) Multiply the units by the tens column [2 × 5 = 10]; place a nought in the tens column and carry the one into the hundreds column.	345 × 123 —— 1035 $_1$00
g) Multiply the tens column together [2 × 4 = 8]; add the one carried over [8 + 1 = 9] place nine in the hundreds column.	345 × 123 —— 1035 900
h) Multiply two by three [2 × 3 = 6] and place the 6 in the thousands column.	345 × 123 —— 1035 6900
i) Place noughts in the units and tens column.	345 × 123 —— 1035 6900 00
j) Multiply the hundreds and units columns [1 × 5 = 5] and place the five in the hundreds column.	345 × 123 —— 1035 6900 500
k) Multiply the hundreds and ten columns together [1 × 4 = 4] and place four in the thousands column.	345 × 123 —— 1035 6900 4500
l) Multiply the hundreds and hundreds column [3 × 1 = 3] and place in the ten thousands column	345 × 123 —— 1035 6900 34500
m) Add the three rows of numbers together giving an answer of 42,435	345 × 123 —— 1035 6900 34500 —— 42435

Figure 13.11 Method of multiplication

Multiplying together 32.2 × 4.7

a) Place in the vertical form and count the digits to the right of the decimal point – in this case **two**.	32.2 4.7
b) Ignore the decimal points and multiply as in Figure 13.11.	32.2 4.7 —— 2254 12880 —— 15134
c) Count the digits off from right to left equal to the number of digits in the decimal fraction of the two numbers being multiplied together, i.e. 2.	15134 151.34

Figure 13.12 Multiplying together numbers containing decimal fractions

13.2.5 Division (÷)

This is the opposite of multiplication, but you still need the ability to multiply single figures together. If you wish to divide 36 nails between four carpenters you can create four piles, putting a nail in each pile until all the nails are used up. How many times will four divide into 36 (Figure 13.13a, b)? Figure 13.14 shows the method of dividing 246.33 by 4.6.

Divide 36 by 4

a) Set out as shown	36 ÷ 4 = 9 or $^{36}/_4$ = 9
b) Using the four times table until 4 multiplied by the right number, i.e 9, gives the answer 36.	4 × 4 = 16 4 × 5 = 20 4 × 6 = 24 4 × 7 = 28 4 × 8 = 32 **4 × 9 = 36** 4 × 10 = 40

Figure 13.13a Division

4 piles of 9 nails in each pile

Figure 13.13b Deriving parts of the whole using division

Divide 246.33 by 4.6

a) Set out as shown

```
4.6|246.33
```

b) Multiply 4.6 and 246.33 by ten to give 46 and 2463.3

```
46|2463.3
```

c) 46 will not divide into 2.

```
46|2463.3
```

d) 46 will not divide into 24.

```
46|2463.3
```

e) As 46 will divide into 246, find a number that when multiplied by 46 will be equal to or slightly less than 246.

$46 \times 1 = 46$
$46 \times 2 = 92$
$46 \times 3 = 138$
$46 \times 4 = 184$
$46 \times 5 = 230$ is the nearest below 246
$46 \times 6 = 276$ is the nearest above 246

```
    5l
46|2463.3
   230
    16
```

f) Subtract 230 from 246 giving a remainder of 16; bring down the 3 to make 163 and repeat step (e); $46 \times 3 = 138$ and $163 - 138$ gives a remainder of 25; place the decimal point after the 3 to keep it in line; and bring down the 3 to make 25 into 253.

```
    53.
46|2463.3
   230
    163
    138
     25 3
```

g) Repeat step (e) $46 \times 5 = 230$ and $253 - 230 = 23$; place the 5 to the right of the decimal point and as there are no more digits to the right of the 3 place a '0' after 23 making 230.

```
    53.5
46|2463.3
   230
    163
    138
     25 3
     23 0
      2 30
```

h) repeat step (e) $46 \times 5 = 230$ and $230 - 230 = 0$; place the '0' in the second place to the right of the decimal point; 4.6 will divide into 246.33 – 53.55 times.

```
    53.55
46|2463.3
   230
    163
    138
     25 3
     23 0
      2 30
      2 30
      0 00
```

Figure 13.14 Method of division

13.2.6 Application of Calculating Skills

These four arithmetic functions are used to calculate the amounts and costs of materials used. The most frequently occurring calculations are those used to determine the amount of timber required for a particular piece of work.

13.3 CALCULATING QUANTITIES OF WOOD-BASED MATERIALS

Sheet materials are calculated in square metres. Small quantities of timber are ordered by length in metres (m), while large quantities are ordered by the cubic metre (m^3).

13.3.1 Measurement of Length, Area and Volume

In calculations involving timber and sheet materials the following should be remembered and reference made to section 1.9.3:

- a number multiplied by length will give **length**, e.g. 1×1 m = 1 m
- two linear measurements multiplied together will give **area**, e.g. 1 m $\times$ 1 m = 1 m^2
- three linear measurements multiplied together will give **volume**, e.g. 1 m $\times$ 1 m $\times$ 1 m = 1 m^3
- area multiplied by length will give **volume**, e.g., 1 m $\times$ 1 m^2 = 1 m^3
- volume divided by area will give **length**, e.g., 1 m^3 $\div$ 1 m^2 = 1^2 m
- volume divided by length will give **area**, e.g. 1 m^3 $\div$ 1 m = 1 m
- area divided by length will give **length**, e.g. 1 m^2 $\div$ 1 m = 1 m.

13.3.2 Linear and Volumetric Measurement of Timber

A piece of timber 1 m long, wide and deep would have a volume of one cubic metre (Figure 13.15a, b). If the volume remained the same and the area of cross-section was halved the length would be doubled (Figure 13.15a, b). The smaller the cross-sectional area, the greater the length that would be required to make 1 m^3 of timber. For example, if we take a piece of timber 100 mm $\times$ 100 mm in cross-section it will require a length of 100 m to make 1 m^3 of timber. Figure 13.16b shows how this figure of 100 m is found. Counting the squares on the end

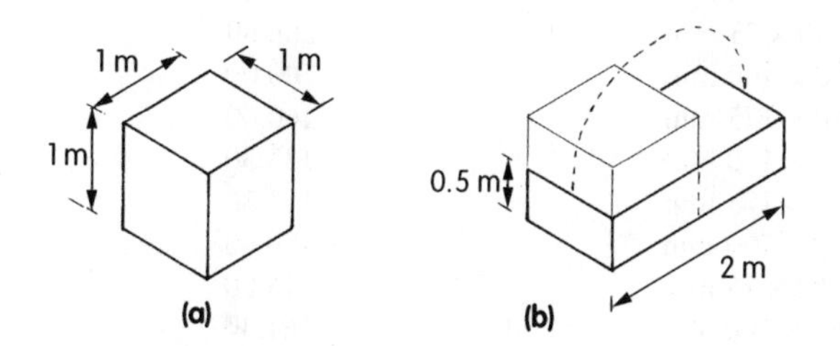

Figure 13.15 The cubic metre

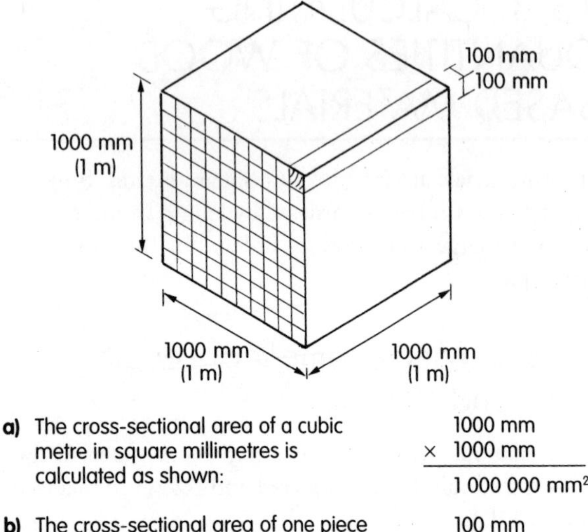

a) The cross-sectional area of a cubic metre in square millimetres is calculated as shown:

  1000 mm
× 1000 mm
1 000 000 mm²

b) The cross-sectional area of one piece of the timber required is calculated as shown:

  100 mm
× 100 mm
10 000 mm²

c) The cross-sectional area of the timber is one hundredth of the cross-section of a metre cube.

$\frac{1\,000\,000}{10\,000} = \frac{100}{1}$

d) There are 100 m of 100 × 100 mm in a cubic metre.

100 × 1 m = 100 m

Figure 13.16 Division of the cubic metre

section of the cube will show that there are 100 and if this number is multiplied by 1 m, it will show, as does the calculation in Figure 13.16b, that there are 100 m of 100 mm × 100 mm in 1 m^3 of timber. This is a theoretical answer as it is assumed the saw cuts have no width. Table 13.3 gives the length in metres for timbers of differing cross-sections that can be theoretically obtained from one cubic metre.

Table 13.3 Number of metres of timber of given cross-section in a cubic metre

Cross-section size	Linear metres in a cubic metre
25 × 50 mm	800.00
25 × 75 mm	533.30
25 × 100 mm	400.00
32 × 75 mm	416.60
32 × 100 mm	312.50
32 × 150 mm	208.30
38 × 75 mm	350.90
38 × 100 mm	263.20
38 × 150 mm	175.40
50 × 75 mm	266.60
50 × 100 mm	200.00
50 × 125 mm	160.00
50 × 150 mm	133.30
50 × 175 mm	114.30
50 × 200 mm	100.00
62 × 75 mm	215.00
62 × 100 mm	161.30
62 × 200 mm	80.60

13.4 PERCENTAGES

When making calculations for cutting and waste, the waste is expressed as a fraction of the total material required and the denominator of the fraction is always 100 (centi). $^{1}/_{100}$ or 0.01 is 1% or one hundredth part. $^{20}/_{100}$ or 0.20 is 20%. A decimal fraction taken to two places of decimals is a percentage. 0.05 is 5%, 0.50 is 50%, 0.85 is 85% and 1.43 is 143%.

13.4.1 Worked Examples of Percentages

Example 1 shows how much useful material is available if 5% has to be rejected.

Out of 100 m of timber, 5% is of no use but 95% (100% – 5%) is. Therefore:

amount of useful timber = 100 m × 0.95 = 95 m

timber of no use = 100 m × 0.05 = 5 m

total amount of timber = 100 m

Example 2 shows the percentage amount of material that is cut to waste if 2.1 m lengths of timber have to be purchased to cut 1.911 m lengths.

Length	[minus]	useful length	[equals]	waste
2.100 m	–	1.911 m	=	0.189 m

Waste expressed as a percentage

$^{0.189}/_{2.100}$ = 0.09 × 100 = 9% waste

Example 3 shows the amount of timber that can be cut from a log of known volume if 48% is lost in cutting and waste.

A log contains 3 m^3 of timber. On conversion into boards, 48% is lost in cutting. How much useful timber is obtained?

Total volume	[minus]	waste	[equals]	useful timber
100%	–	48%	=	52%

Useful timber:

3 m^3 × 0.52 = 1.56 m^3

13.5 AREAS OF REGULAR FIGURES

Regular figures (shapes) can be divided into three groups:

1 squares, rectangles, trapeziums and parallelograms (4 sided)
2 triangles and regular polygons (other than 4 sided)
3 circles, segments and sectors.

13.5.1 Squares, Rectangles, Trapeziums and Parallelograms

These are four-sided figures (Figure 13.17).

- The area of a square is found by multiplying the length by the breadth and as both are the same:

  $a \times a = a^2$ (Figure 13.17a).

- The area of a rectangle is the length times the breadth:

  $a \times b = ab$ (Figure 13.17b).

- The area of trapezium is the average length of side b plus c multiplied by the perpendicular distance a between the parallel sides:

  $\frac{b+c}{2} \times a$ = area of trapezium (Figure 13.17c).

- The area of a parallelogram is the length of one of the parallel sides b multiplied by the perpendicular distance a between the parallel sides

  $b \times a = ba$ (Figure 13.17d).

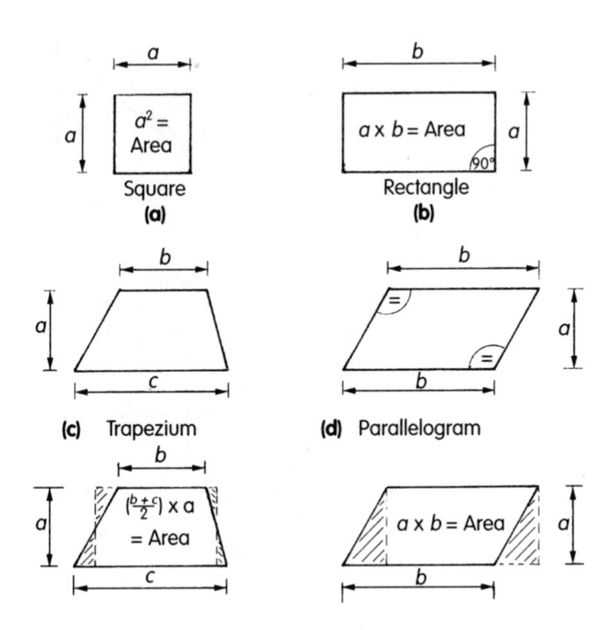

Figure 13.17 Regular four-sided figures

13.5.2 Triangles and Regular Polygons

Triangles are plain figures with three sides (Figure 13.18) and regular polygons have five or more sides of equal length (Figure 13.19). The area of a triangle is

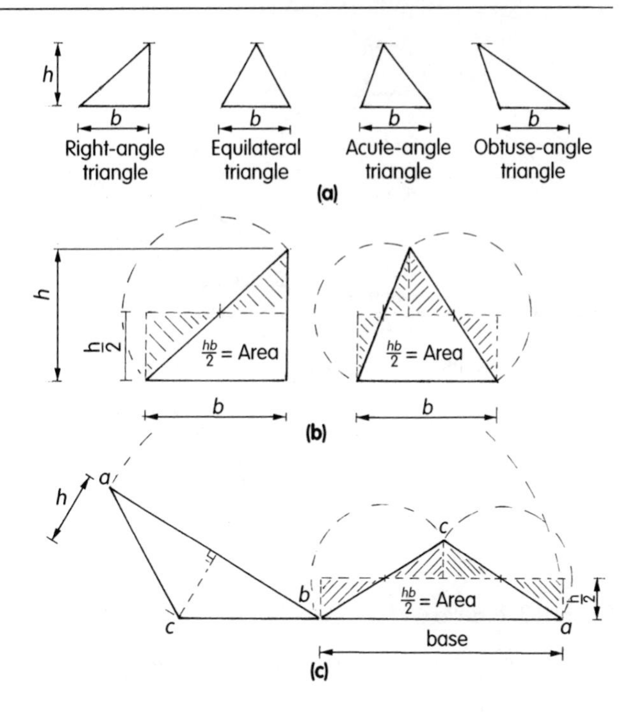

Figure 13.18 Areas of triangles

found by multiplying the base length b of the triangle by half the perpendicular height (h):

$\frac{b \times h}{2}$ = area of triangle (Figure 13.18a, b, c).

A regular polygon is made up of a series of similar triangles equal to the number of sides of the polygon. The area is found by finding the area of one of the triangles and multiplying it by the number of triangles in the polygon, where b = the side length of the polygon (base of the triangle), h = the perpendicular height of the triangle and n = the number of sides of the polygon (Figure 13.19).

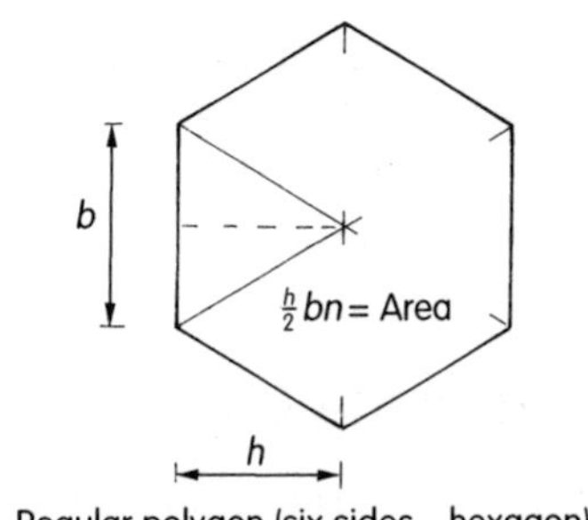

Figure 13.19 Regular polygons

13.5.3 Circles, Segments and Sectors

Figure 13.20a illustrates a circle and shows the name of each part.

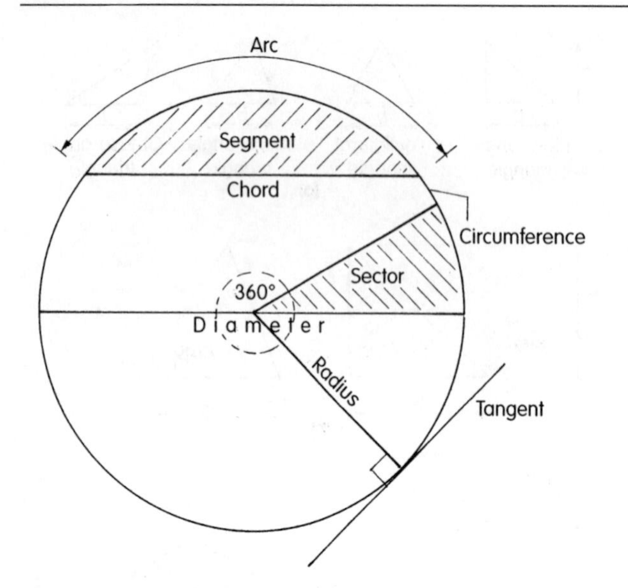

Figure 13.20a Parts of the circle

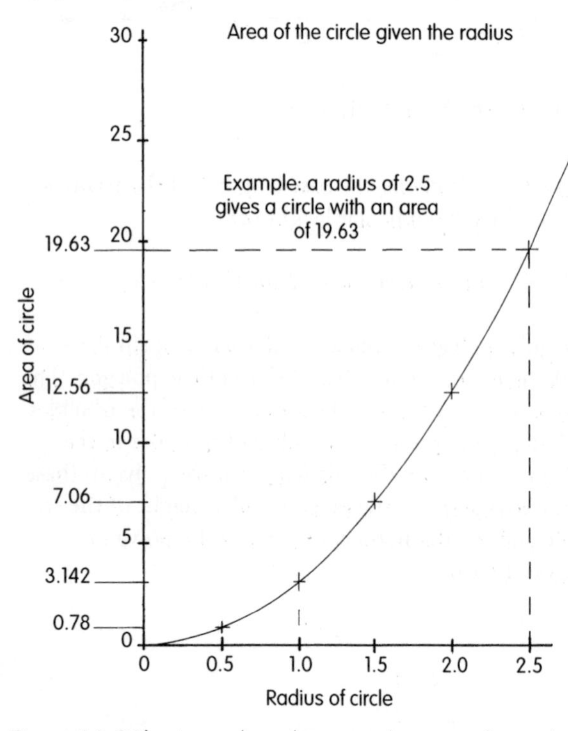

Figure 13.20b A graph to determine the area of a circle knowing its radius

13.5.4 Area of the Circle

The area of a circle is $\pi \times r^2$ (Pi = π). Pi is the 16th letter of the Greek alphabet and stands for the value of the ratio between the diameter and circumference of a circle. Circle circumference÷circle diameter approximately equals 3.142.

The diameter (d) can also be used to calculate the area of a circle. Radius r = half diameter

$\pi \times r \times r$ = area
$\pi \times d/2 \times d/2$ = area
$\pi \times d^2/4$ = area

13.5.5 Area of the Sector of a Circle

A circle contains 360°, a semicircle (half a circle) 180° and a quadrant (quarter of a circle) 90°.

A sector of a circle has a fractional area of the circle that contains it. That fraction is found by dividing the degrees in the circle (360°) by the degrees at the centre point of the sector.

Area of the segment shown in Figure 13.21a:

$\pi r^2 \times 45°/360°$ = area of sector
$\pi r^2 \times 0.125$ = area of sector
$\pi 5^2 \times 0.125$ = area of sector
$\pi \times 25 \times 0.125 = 9.82\ m^2$

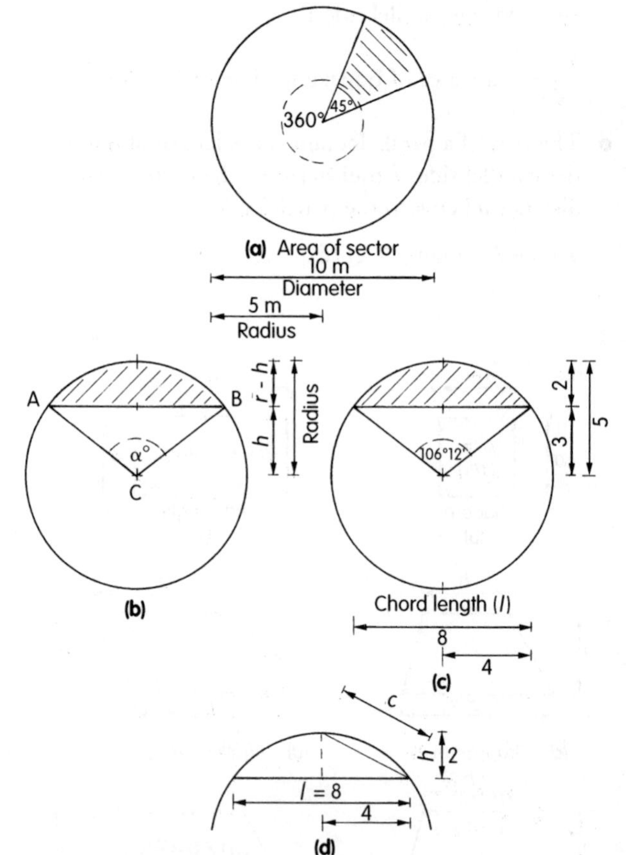

Figure 13.21 Areas of sectors and segments

13.5.6 Area of the Segment of a Circle

The area of a segment is the area of the sector less the area of the triangle ABC (Figure 13.21b).

Calculate the area of the segment shown in Figure 13.21c.

$(\pi r^2 \times \alpha°/360°) - (AB \times h/2)$ = area of segment.
$(3.142 \times 5^2 \times (106°12'/360°)) - (ab \times h/2)$
$(3.142 \times 25 \times 0.295) - (8 \times 3/2)$
$23.17 - 12 = 11.17\ m^2$.

13.5.7 Standard Formula for Area of the Segment of a Circle

To find the area of the segment in Figure 13.21d use the standard formula h/15 (6L + 8C) and refer to Section 13.10 to find the length of c. (Theorem of Pythagoras)

e.g. $c^2 = a^2 + b^2$

using $\quad h = 2$

$\quad\quad\quad l/2 = 4$

$c = 4^2 + 2^2$; $c^2 = 20$; $c = \sqrt{20} = 4.47$

$(2/15)((6 \times 8) + (8 \times 4.47))$ = area of segment
$(2/15)(48 + 35.76)$ = area of segment
$(2 \times 83.76)/15 = 11.17\ \text{m}^2$

13.6 PERIMETERS OF REGULAR FIGURES

13.6.1 The Circumference (Perimeter) of a Circle (Figure 13.22)

This is found by multiplying the diameter by the constant π, which approximately equals 3.142.

πD = the circumference of the circle.

As the diameter D is twice the radius r, $D = 2r$.

$2\pi r$ = the circumference of the circle.

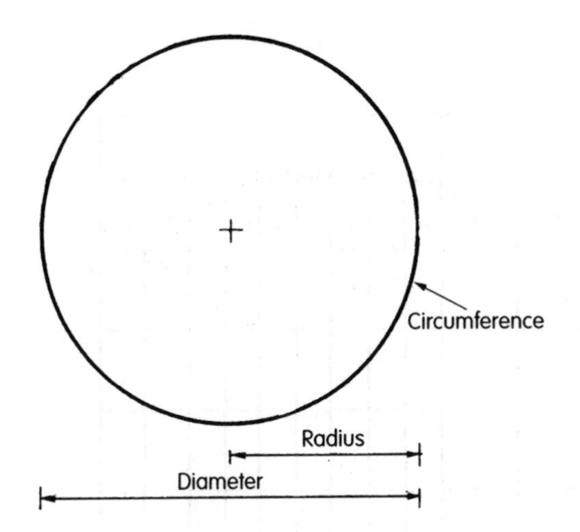

Figure 13.22 Circumference of a circle

13.6.2 The Perimeter of a Square

Side $a \times 4$ = the perimeter (Figure 13.23).

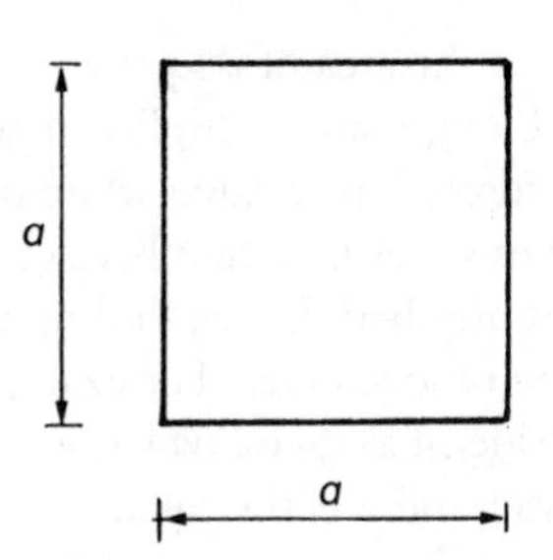

Figure 13.23 Perimeter of a square

13.6.3 The Perimeter of a Rectangle

Long side b plus short side a multiplied by 2 equals the perimeter (Figure 13.24).

$2(a + b)$ = the perimeter

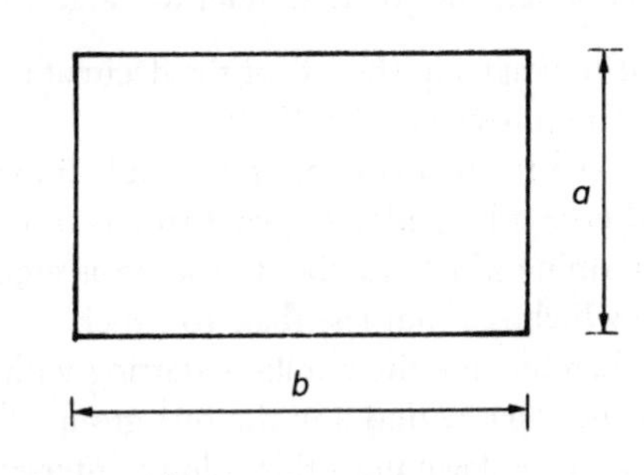

Figure 13.24 Perimeter of a rectangle

13.6.4 The Perimeter of a Regular Polygon

The length of one side a multiplied by the number n of sides equals the perimeter (Figure 13.25).

$a \times n$ = perimeter

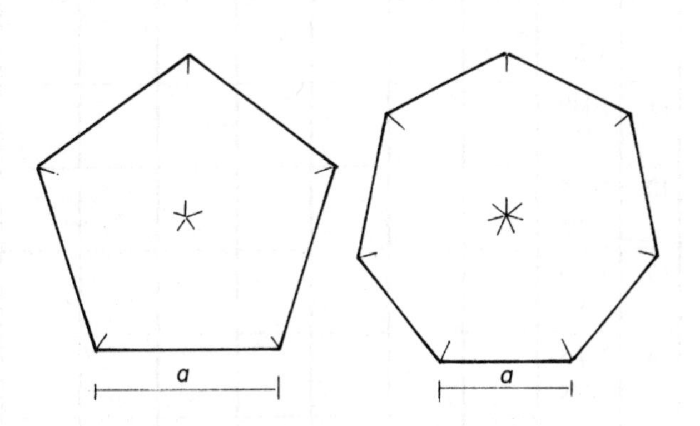

Figure 13.25 Perimeter of regular polygons

13.7 SQUARES AND SQUARE ROOTS

In the calculation of the area of a square, the two sides which are of equal length are multiplied together – $a \times a$ or $2 \times 2 = 4$. The figure 2 is squared when multiplied by itself. The reverse of squaring a number is finding the square root of that number: that is, finding the length of the side of a square of given area. For example, what is the length of the side of a square which is 4 units of area? 2 is the answer and 2 is the square root of 4. To show that we require the square root of a number we use the symbol $^{2}\sqrt{}$ or simply $\sqrt{}$.

What is the square root of 9, or $\sqrt{9}$. The $\sqrt{9}$ is 3 – as $3 \times 3 = 9$. Finding the square root of a number by calculation is a long and tedious process and tables of square roots are available (Table 13.4)

13.7.1 Use of Square Root Tables

To find the square root of 1545 refer to Table 13.4.

1 Pair off the digits to the left of the decimal point in the given number, i.e. |15|45|.0
2 Look down the first column in the table till you find 15 and note it is divided to give two sets of numbers.
3 To determine which number to use, 15 is greater than 9 which is 3^2 but less than 16 which is 4^2. Therefore use the numbers starting with 3.
4 Along the top row find 4 in the first group of numbers and trace down until this column intersects with row 15 containing the number starting with 3 and you will see 3924.
5 Along the top row find 5 in the second block of numbers and trace down until this column intersects with row 15 and in line with the numbers starting with 3 you will find 6.
6 Add the numbers from steps 4 and 5 together – $3924 + 6 = 3930$.
7 In step (1) the number was divided into groups and the number of groups will determine the number of digits to the left of the decimal point, that is, two. Therefore the square root of 1545 is 39.30.

To find the square root of 154.5:

1 Pair off the digits to the left of the decimal point 1|54|.5.
2 Look down the first column till you find 15 and the division into two sets of numbers.
3 To determine the number to use, 1 is greater than or equal to 1 which is 1^2 but less than 4 which is 2^2, therefore the value will be 1 plus some other digits.
4 Along the top row find 4 in the first group of numbers and trace down until this column intersects with row 15 containing the numbers starting with 1 and you will see 1241.
5 Along the top row find 5 in the second block of numbers and trace down until this column intersects with row 15 and in line with the numbers starting with 1 you will find 2.
6 Add the numbers found in steps 4 and 5 together – $1241 + 2 = 1243$.
7 In step (1) the number was divided into two groups and the number of groups will determine the number of digits to the left of the decimal point, that is, two. Therefore the square root of 154.5 is 12.43.

Table 13.4 Square root table

1st column of numbers		1st block of figures									2nd block of figures								
	0	1	2	3	4	5	6	7	8	9	1	2	3	4	5	6	7	8	9
10	1000 3162																		
11	1049 3317																		
12	1095 3464																		
13	1140 3606																		
14	1183 3742																		
15	1225 3873	1229 3886	1233 3899	1237 3912	1241 3924	1245 3937	1249 3950	1253 3962	1257 3975	1261 3987	0 1	1 3	1 4	2 5	2 6	2 8	3 9	3 10	4 11
16																			

13.8 ELECTRONIC CALCULATORS

Throughout recorded history aids have been used in calculating and probably the oldest of these still in use is the Chinese abacus. Other aids have been Napier's bones (logarithms), printed tables, slide rules and graphs (see Figure 13.20b). Today the battery-operated electronic pocket calculator is in general use and even inexpensive models will carry out addition, subtraction, multiplication and division, calculate percentages and give the squares and square roots of numbers. Figure 13.26 shows the layout of a calculator.

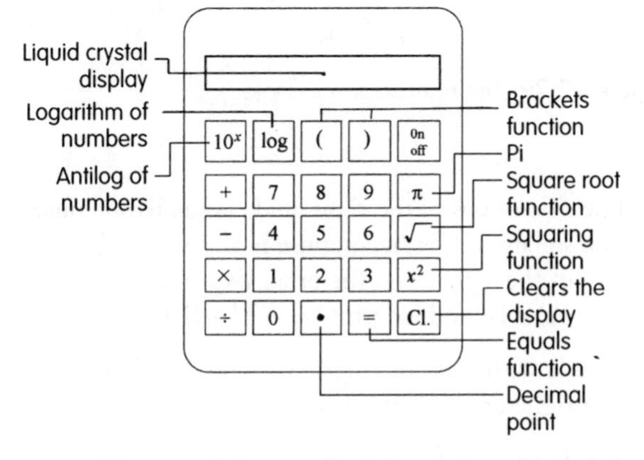

Figure 13.26 Simple layout of an electronic pocket calculator

13.8.1 Using a Calculator

When using the calculator, expressions placed in brackets are worked out first. It is always good practice to follow the 'keying in' instructions supplied by the manufacturer of the calculator.

To calculate the following:

	Key strokes	Ans
a) 8×4	[8] [×] [4] [=]	32
b) $32 - 4$	[3] [2] [−] [4] [=]	28
c) 3.2×4	[3] [.] [2] [×] [4] [=]	12.8
d) $4 \times (8 + 12)$	[4] [×] [(] [8] [+] [1] [2] [)] [=]	80
e) $\sqrt{106}$	[√] [1] [0] [6] [=]	10.3
f) $\pi 2^2$	[2] [×] [2] [×] [π] [=]	12.57
g) 20% of 30	[3] [0] [×] [2] [0] [%] [=]	6
h) 120% of 40:	[4] [0] [×] [1] [2] [0] [%] [=]	48

13.8.2 Important

Although the calculator performs the correct operations when you press the appropriate keys, it is easy to make keying errors and you must have some idea of what the approximate answer is likely to be. Is 10 divided by 4 equal to 2.5 or 25? Always question the answer, as discovering an error at the calculating stage may save you money, time and grief later on.

13.9 USE OF DATUM LINES

When fixing fittings into a building it is good practice to set up datum lines to which you can refer for accurate positioning. A level or datum line is carefully marked on the wall at a convenient height and from this line all related vertical measurements are taken. Figure 13.27 shows a datum line and how measurements are taken from it to establish the positioned heights of a set of base units and wall cupboards.

1 Set up the datum line, in this case 1.500 m from the floor. This is approximately equidistant between the top of the base units and the underside of the wall cupboards.

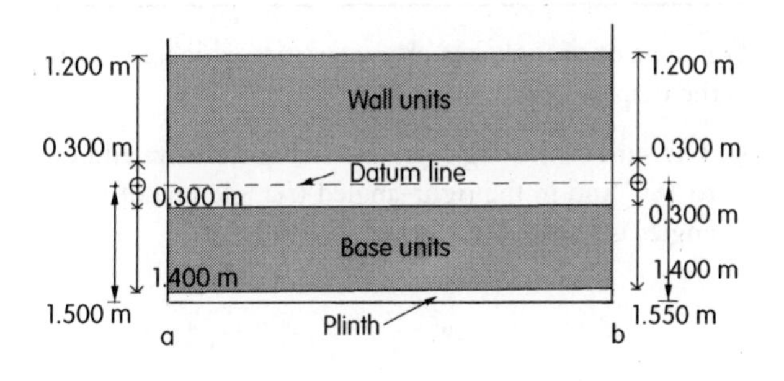

Figure 13.27 Measuring from datum lines

2 From the datum, mark-up running measurements (see Figure 12.27) to show the top and bottom of the wall cupboards.

3 From the datum, mark down running measurements to show the top and bottom lines of the base units. It will be noted that the plinth for the base units will have to be tapered to make up for the floor being slightly out of level.

4 From the datum, the depth of the base unit plinth can be found at a and b:

- Depth of plinth at a = 1.500 − 1.400 = 0.100 m or 100 mm
- Depth of plinth at b = 1.550 − 1.400 = 0.150 m or 150 mm.

5 All measurements above and below the datum must be set off vertically (Figure 13.28).

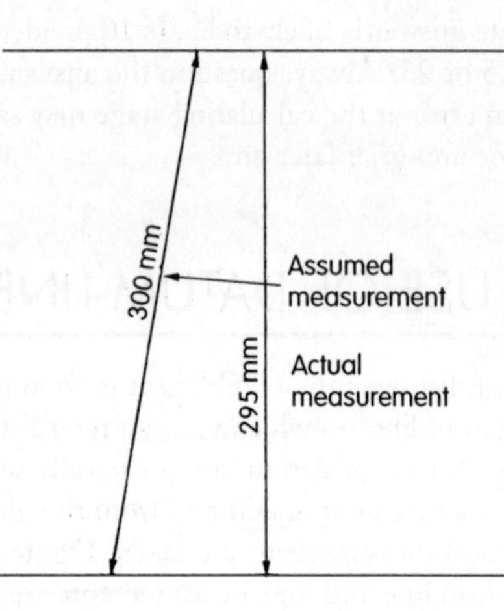

Figure 13.28 Need for vertical measurements

13.10 THE RIGHT-ANGLED TRIANGLE

This type of triangle has characteristics which are useful to the carpenter and joiner.

a The sum of the angles in any triangle always add up to 180° and in the right-angled triangle one of the angles is always 90° (Figure 13.29).

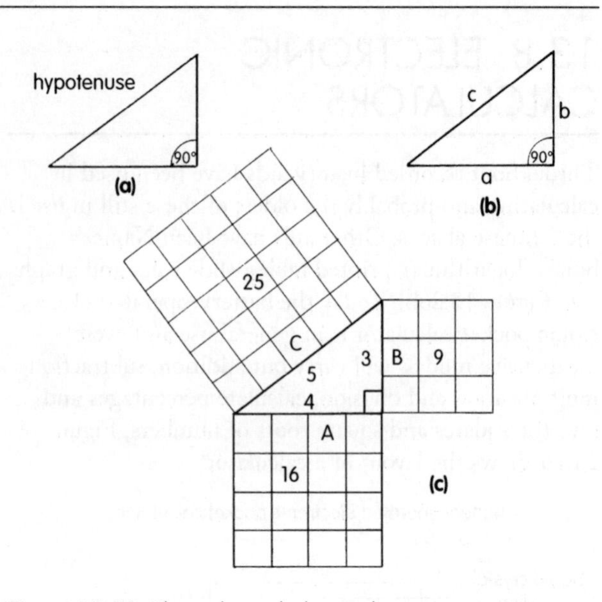

Figure 13.29 The right-angled triangle

b The side opposite the right angle is the hypotenuse.

c The square on the hypotenuse is always equal to the sum of the squares on the other two sides
$C^2 = A^2 + B^2$ and length $C = \sqrt{(A^2 + B^2)}$

A triangle with sides of length 3, 4 and 5 will always make a right-angled triangle as

$5^2 = 3^2 + 4^2$ or $25 = 9 + 16$.

14

Planning and Organising Work

14.1 REASONS FOR PLANNING

How are we to do that? Planning and organising (Figure 14.1) is a process that can be complex and carried out by those with specialist knowledge of the techniques involved. However, it can be a simpler, almost automatic, activity. Planning takes place at all levels, on international, national, local, company and individually. It is a tool or system used to organise efficiently the resources of skill, material and time available to complete some given task.

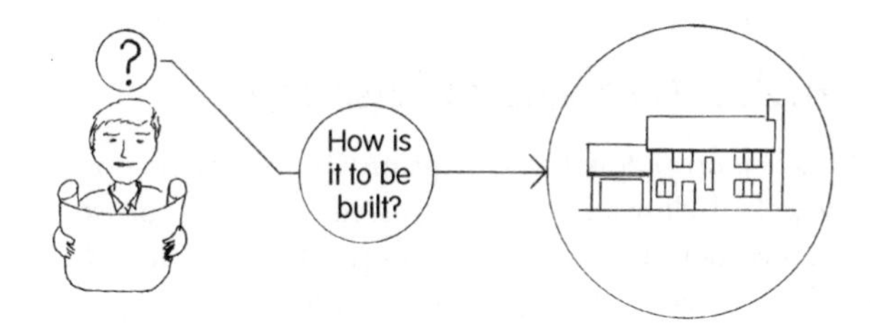

Figure 14.1 Planning

Planning is a process that everyone becomes involved in at some level. We may not analyse the planning process, but we apply its principles to the everyday activities in which we are involved. The ability to organise increases our effectiveness as carpenters and joiners.

Without warning, someone in authority asks you to put a few tools in a bag and go with them. What questions might you wish to ask?

- What is the task?
- When is it to be done?
- Where is it to be carried out?
- How is it to be done?

Finding answers to these questions is part of the planning process. Not knowing the answers will make it almost impossible to organise the work, leading to frustration, lost time and temper if the correct tools, materials and equipment have not been brought. A contingency plan would be to take along everything you think you might need, although this is not always practical.

It is necessary to be as well informed as possible about the work to be undertaken. Ask questions, look at drawings, read specifications, find out about what has to be done (Figure 14.2). When the work is completed, remember for another time any snags that occurred or anything that made the work easier (Figure 14.3). Simple organisation will help us become more efficient and co-ordinate our own work with those of other trades involved in the construction process.

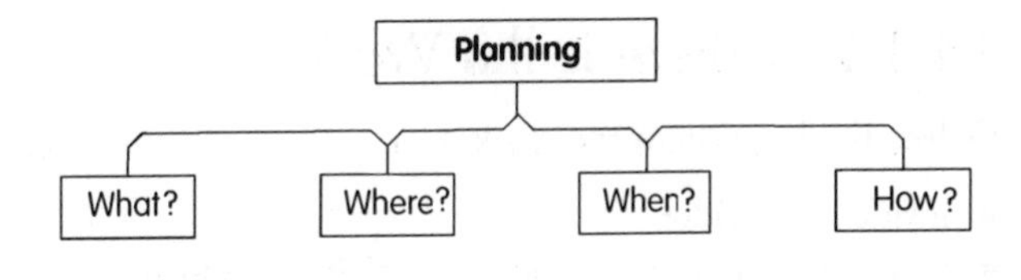

Figure 14.2 Questions to ask when planning

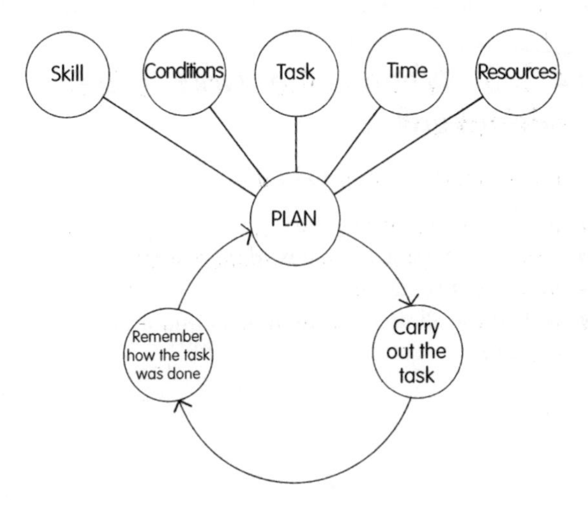

Figure 14.3 The planning process

14.1.1 What is the Work?

Define the work by asking the following questions:

- Is it large and complicated?
- Is it simple and/or repetitive?
- Is it new work, repairs or alterations?
- Are there any restrictions (protection of the surrounding structure, sealing off the work area from other areas) (Figure 14.4)?
- Do I know what the client expects of me?

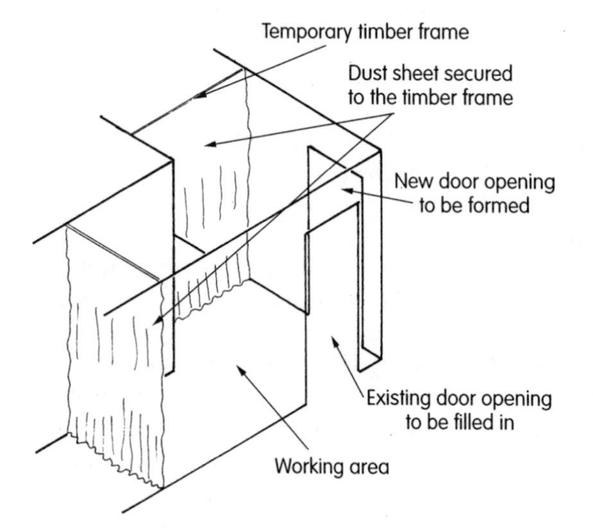

Figure 14.4 Plans include protection of what exists

14.1.2 Where is the Work?

Define the location of the work. Is it:

- inside or outside?
- on or near property boundaries? (When repairing fences take care with the neighbour's climbing rose.)
- in open or confined spaces?
- below, at or above ground level?
- away from or near to public areas?

14.1.3 When is the Work to be Undertaken?

Define when the work is to be done:

- in what season of the year?
- during or outside normal working hours?
- in daylight or darkness?
- is the time allotted for the task an important consideration?

14.1.4 How is the Work to be Done?

Define the procedure for doing the work:

- In what order?
- Are there sufficient resources available (people, equipment and materials)?
- Have I the necessary knowledge, skill or experience?
- Do I understand my part in carrying out the work?

14.2 AIDS TO PLANNING

With practice and experience we know almost automatically what tools and equipment are required to complete a given task and the planning for that task occurs almost as a reflex action.

If the work is a new task, it is useful to spend some time thinking about what has to be done and how (method study) and what tools, equipment and materials are required. This is achieved by using analysis sheets and checklists. Writing things down in lists aids memory and helps us to understand the task and what has to be done. Lists are of three types:

- analysis of task
- tool and equipment lists
- material lists.

14.2.1 Analysis of Task List

This is a breakdown into steps of what has to be done. Table 14.1 shows the steps involved in fixing mortise window locks to casement windows. This may look long and complicated, but after some practice the process is retained to memory.

14.2.2 Tool and Equipment Lists

From the identified steps involved in fitting the window locks we can now identify the tools and equipment required to do the work (Table 14.2).

14.2.3 Material Lists

These are more commonly known as cutting lists (see Figure 5.7) and are written up by reference to the materials identified in the task list (Table 14.3).

14.2.4 Time Required to Perform a Task (Time Study)

The time taken to carry out the task will depend on how familiar we are with it and the level of skill

Table 14.1 Fixing Number 2 Mortise window locks to each of 20 side hung casements in ground floor windows

Elements of the task	Tools and equipment	Materials	Illustrative sketches
1 Mark out the vertical position for the locks above the sill on the stile of the casement and jamb of the frame while in the closed position	Four-fold rule or tape, pencil, saw stool, hop up or pair of steps	No. 40 Chubb mortise window locks complete with striking plates, screws and No. 2 keys	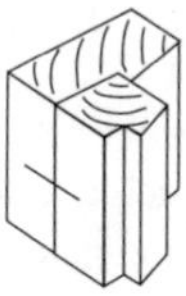
2 Open the casement and from the marks made in (1) square a pencil line round the opening edge and inside of the casement	Adjustable square, pencil		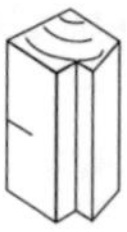
3 Square a line across the inside face of the rebate in the frame jamb from the positions marked out in (1)	Adjustable square, pencil		
4 Mark the centre of the casement edge and the position of the key hole from the closing edge of the casement style	Marking gauge or adjustable square		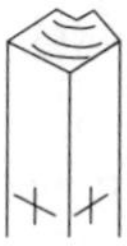
5 Measuring from the inside edge of the jamb rebate mark off the distance from the inside face of the casement to the centre mark on the casement and add 1 mm for clearance	Four-fold rule or tape, Adjustable square or marking gauge		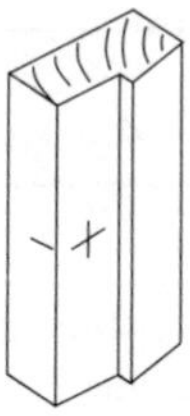
6 Drill a hole for the body of the lock in the edge of the casement style	Carpenter's brace, 15 mm bit, and depth stop		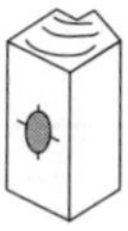
7 Drill a hole for the key on the inside face of the casement style	Carpenter's brace, 9 mm bit		
8 Fit the body of the lock into the hole and mark out the position of the face plate	Sharp pencil or metal scriber or chisel		
9 Cut out the housing for the face plate. Fit and screw into place	Mallet, 18 mm bevelled edged chisel. Screw driver, bradawl		
10 Drill a hole in the rebate of the frame jamb as marked out in (3) and (5)	Carpenter's brace, 12 mm bit		
11 Mark out the position of the keep in the jamb rebate. House and fit as (9).	Sharp pencil or metal scriber, 18 mm bevelled edge chisel, mallet, screw driver, bradawl		
12 Close the casements and check that the bolt of the lock correctly engages in the keep		Key for the lock	

Table 14.2 Tool and equipment list

Tool list
Pencil
Four-fold rule or tape
Adjustable square
Marking gauge
Scriber
Carpenter's brace
Bits
• 15 mm
• 12 mm
• 9 mm
Mallet
Bevelled edge chisel 18 mm
Bradawl
Screwdriver
Equipment
Saw stool
Hop up (step stool)
Pair of steps

Table 14.3 Goods and materials list

Materials list
• No. 40 Chubb mortise window locks complete with striking plates
• No. 160 ½-inch × No. 4 countersunk brass screws
• No. 2 standard keys for the locks

required. In estimating the time it will take to do a piece of work remember to include:

- travelling to the work
- gathering materials, tools and equipment required and other preparatory work
- time spent in natural and formal breaks, checking, adjustment and inspection
- clearing away and tidying up after the work is completed.

A reasonable time for fitting a pair of mortise window locks to a side-hung casement under good working conditions would be from 30 to 40 minutes.

14.3 PROCEDURES FOR TAKING DELIVERY OF MATERIALS

Organising is not limited to the actual process of cutting and assembly but includes inspection, taking delivery of, storing, handling and fixing materials that often have pre-finished surfaces.

14.3.1 Inspection of Materials (Figure 14.5)

When materials are drawn from a store or delivered from a supplier, an inspection should be made to ensure they are what were ordered and that they have not been damaged. Look for the following:

- the number of items are correct and are as described in the order and delivery note
- bowing or twisting in sheet materials or framed items
- the protective coverings, if any, are not damaged or torn
- bruising or compression damage caused on the face edges or corners of items, especially sheet materials
- scratch or score marks across finished surfaces (torn protective coverings may indicate that this type of damage has taken place)
- contamination by other materials such as water, paint, alkaline and powdered materials.

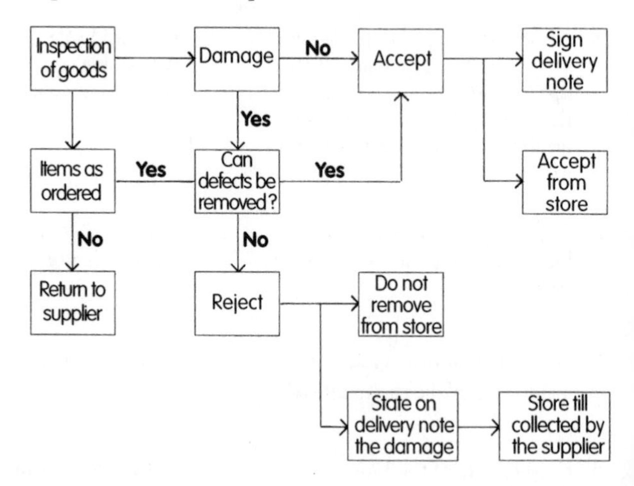

Figure 14.5 Procedure for the inspection of goods and materials

14.3.2 Action on Delivery of Damaged Materials

- The person delivering the materials should have the damage drawn to their attention.
- The type and amount of damage should be recorded on the delivery note, one copy of which will be given to the person making the delivery and the remainder distributed within the company (Figure 14.6).
- If the damaged goods are not returned straight away they should be carefully stored until they are collected by the supplier.

14.3.3 Storage of Delivered Goods

All materials should be safely and correctly stored to the manufacturer's recommendations until required (see Section 1.11 and 2.11).

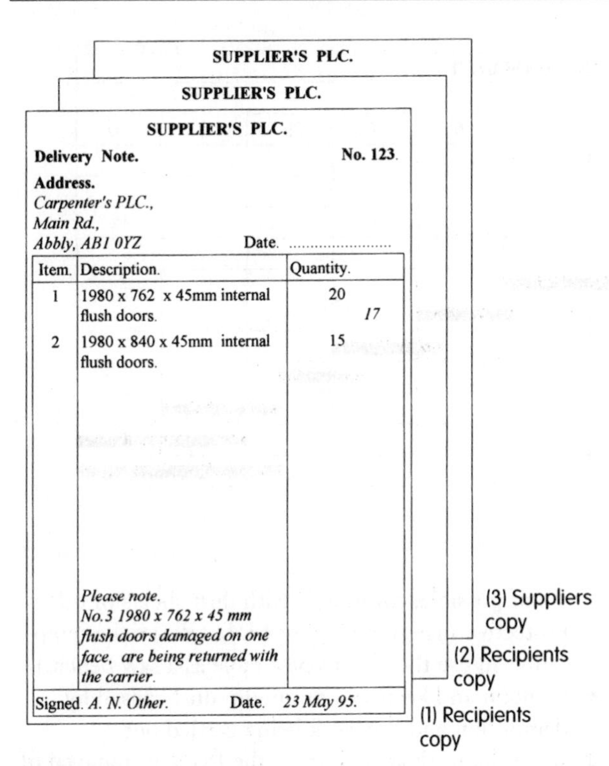

SUPPLIER'S PLC.

SUPPLIER'S PLC.

SUPPLIER'S PLC.

Delivery Note. **No. 123.**

Address.
Carpenter's PLC.,
Main Rd.,
Abbly, AB1 0YZ Date. ....

Item.	Description.	Quantity.
1	1980 x 762 x 45mm internal flush doors.	20 *17*
2	1980 x 840 x 45mm internal flush doors.	15
	Please note. No.3 1980 x 762 x 45 mm flush doors damaged on one face, are being returned with the carrier.	

Signed. *A. N. Other.* Date. *23 May 95.*

Figure 14.6 Delivery notes

14.3.4 Drawing Materials from Store

Check the materials being drawn from the store for any damage. If the damaged area can be cut away in the waste, accept it. If not, reject it and ask for undamaged material. Under certain circumstances repairs to damage is acceptable. Only draw materials from store when you are ready to use them.

14.3.5 Handling Materials

Proper handling of materials includes:

- the proper organising of temporary storage of the materials as near to the work area as possible
- lifting off rather than dragging sheet materials; when moving sheet materials be careful not to catch and damage the corners
- using correct cutting techniques to avoid splintering finished surfaces
- not using finished and fitted horizontal surfaces as temporary resting places for tools or other materials
- protecting finished surfaces from damage; this may require the use of plastic sheeting or more substantial protective material
- never sliding or dragging other materials over finished work (e.g. scaffold boards over the unprotected bottom sash rails of unglazed casements). See also Sections 1.11 and 2.11.

14.3.6 Organise the Work Stages to Meet the Required Finish

Inferior work, hammer marks, twisted linings, out of square frames or poor fitting are often excused, as the next stage or process will put things right. This never happens and errors or poor work should be put right as soon as they occur. Leaving errors only creates larger problems later that can become expensive to rectify. The standard of the finished work is reflected in the quality of the preliminary work.

Fittings and fixtures should be put into a building at the optimum time to save them from unnecessary damage. This optimum (best) time will be determined by the planners in their overall plan for the construction of the building.

14.4 CO-ORDINATION OF THE CRAFT OPERATIONS IN BUILDING

Building is the co-operation between individuals practising different trade skills – bricklayers, carpenters, plumbers, electricians, plasterers and painters. Each trade has its own tasks to perform, but they must be done in a particular order and in co-operation. The carpenter must do what is required in the time given for the task.

14.4.1 Programming of Building Work

Figure 14.7 shows a programme of work in the form of a bar chart showing:

- the tasks that need to be done
- the order in which they are to be carried out
- the time allowed for each task.

You will see that the carpenter's work is divided into three distinct areas:

- **carcassing work** (floor joists and roof construction) – activities (2) (4) and (6)
- **first fixing** (floor boards, partitions, door and window linings and other preparatory work required to take the finishing joinery at a later stage – activity (7)
- **second fixing** (finishing work), hanging doors, fixing timber trim and other fittings – activity (9).

BAR CHART OF THE BUILDING PROGRAMME											
	Activity	Time in weeks									
		1	2	3	4	5	6	7	8	9	10
1	Substructure										
2	Ground floor joists										
3	Structure to 1st floor										
4	1st floor joists										
5	Structure to eaves										
6	Roof structure and trim										
7	First fixing trades										
8	Plastering										
9	Second fixing trades										
10	Decorating										
11	Landscaping										

Figure 14.7 Co-ordinating and programming the work of various trades

14.4.2 Co-operation Between the Different Building Trades

Good pre-planning in terms of detailed drawings and programmes of work will help co-operation between trades.

Study the drawings and specifications carefully and discuss any problems that may arise, so that everyone concerned will know what is required in making provision for other trades to carry out their work.

14.4.3 Quality of the Work

Good workmanship makes for good relationships with others. Good finishes on joinery, nails punched in, no hammer marks on the surfaces, well-fitting joints and rough surfaces smoothed off will encourage good working relationships between the carpenter and the painter.

14.5 CLIENT AND CARPENTER CO-OPERATION

When asking for work to be done, the client will have a mental picture of what the finished project will look like and how the work will be carried out, causing the least inconvenience. This ideal image tends to ignore technical problems. The carpenter will also have an image of how the finished work will look and how he will go about achieving it, based on his technical experience. It is important that both parties come to a common understanding of what is being done and how, before the work begins. Understanding the client's wishes and carrying them out as far as it is possible are paramount. In sensitive areas like the client's home emphasis must be given to the following:

- Protect the client's portable property (furniture and fittings) by removing it from the work area or if this is not possible, covering it with dust sheets or other protective materials (Figure 14.4). (Resist the temptation to use the client's best table as a saw bench.)
- Contain and keep to a minimum dust caused by demolition and the work being carried out.
- Keep the work area clean by the frequent removal of debris.
- Keep noise to a minimum.
- Clean and tidy up at the end of each working day.
- Exhibit good personal manners – you are in another person's home and are an ambassador for the business you represent.

No matter how good the quality of the work, there will be some dissatisfaction if the execution and finish of the work are not as perceived in the mind of the client (Figure 14.8).

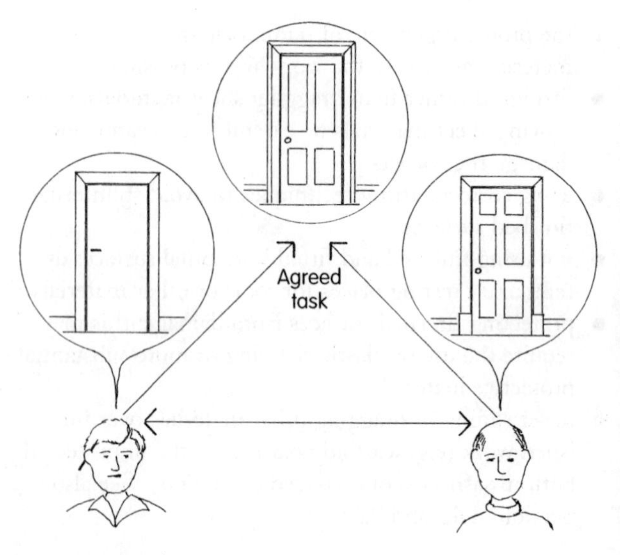

Figure 14.8 Client–carpenter co-operation

14.6 ORDERS AND REQUISITIONS

When requesting components and materials, ensure that the person supplying what you require understands exactly what you need. If the request is made orally make sure there is no misunderstanding of what is said. If the order is written care must be taken to write clearly and provide all the necessary information about what you are ordering. Books of printed order forms are usually supplied in duplicate or triplicate.

14.6.1 Ordering Components (Figure 14.9)

The following information must be given when ordering components:

- a the actual number of items
- b their size, if applicable
- c the hand of the items, if they are handed
- d the name and/or a description of the items
- e a reference code if ordering from a catalogue or a schedule.

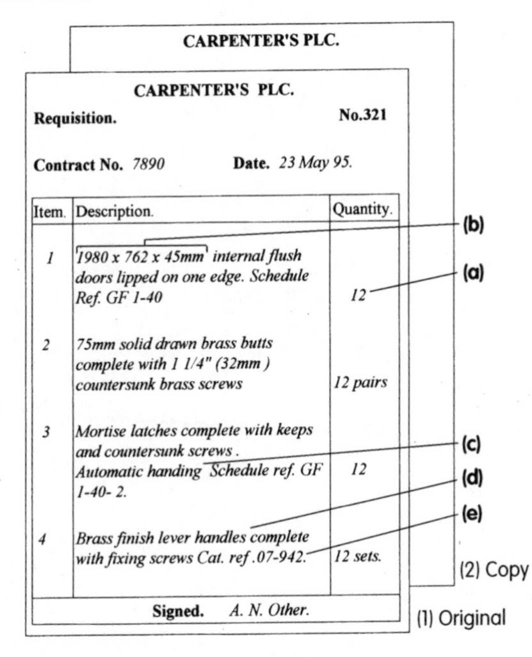

CARPENTER'S PLC.

CARPENTER'S PLC.

Requisition. **No.321**

Contract No. *7890* **Date.** *23 May 95.*

Item.	Description.	Quantity.
1	*1980 x 762 x 45mm internal flush doors lipped on one edge. Schedule Ref. GF 1-40*	*12*
2	*75mm solid drawn brass butts complete with 1 1/4" (32mm) countersunk brass screws*	*12 pairs*
3	*Mortise latches complete with keeps and countersunk screws. Automatic handing Schedule ref. GF 1-40- 2.*	*12*
4	*Brass finish lever handles complete with fixing screws Cat. ref.07-942.*	*12 sets.*

Signed. *A. N. Other.*

Figure 14.9 Requisition for components

14.6.2 Ordering Materials (Figure 14.10)

It is important to remember several things when ordering materials:

- a when ordering materials the amounts ordered must include allowances for cutting and waste
- b if special rather than random lengths of material are required, this must be made clear
- c the cross-sectional size must be stated and the profile shown if other than rectangular.
- d any special finish must be given – rough, sawn or primed.

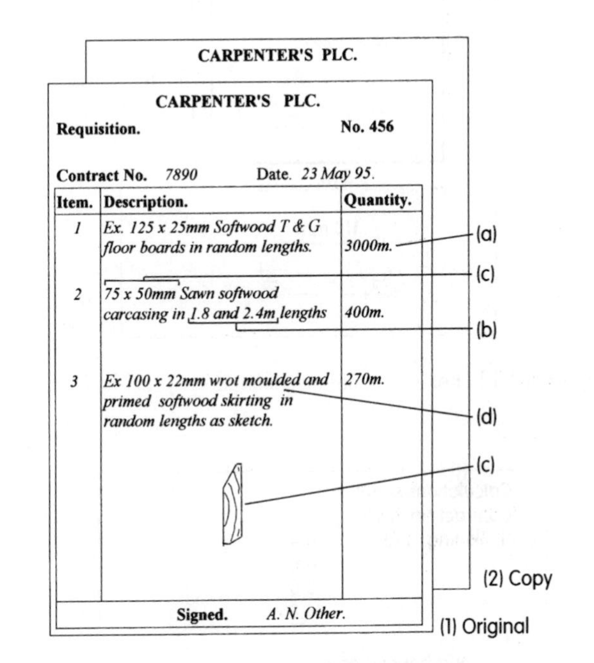

CARPENTER'S PLC.

CARPENTER'S PLC.

Requisition. **No. 456**

Contract No. *7890* **Date.** *23 May 95.*

Item.	Description.	Quantity.
1	*Ex. 125 x 25mm Softwood T & G floor boards in random lengths.*	*3000m.*
2	*75 x 50mm Sawn softwood carcasing in 1.8 and 2.4m lengths*	*400m.*
3	*Ex 100 x 22mm wrot moulded and primed softwood skirting in random lengths as sketch.*	*270m.*

Signed. *A. N. Other.*

Figure 14.10 Requisition for materials

14.6.3 Information for Ordering

Information for ordering will be obtained from the drawings and schedules or actual measurements taken on the building site.

To prepare an order from site measurements for skirting board and floor board:

- measure and record the length and breadth of the room (if the rooms are of the same size you only need measure one room and then multiply by the number of rooms (Figure 14.11)
- determine the net lengths you require (Figure 14.12)
- make an addition for waste as indicated
- check and double check your measurements and calculations to make sure they are correct
- write out the order or requisition using the figures you have calculated (Figure 14.13).

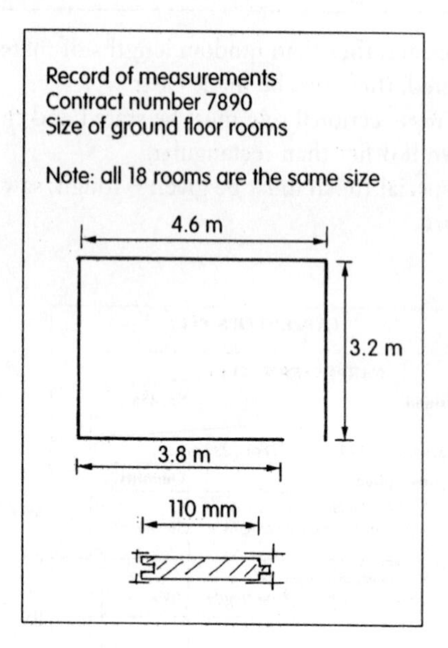

Figure 14.11 Record of site measurements

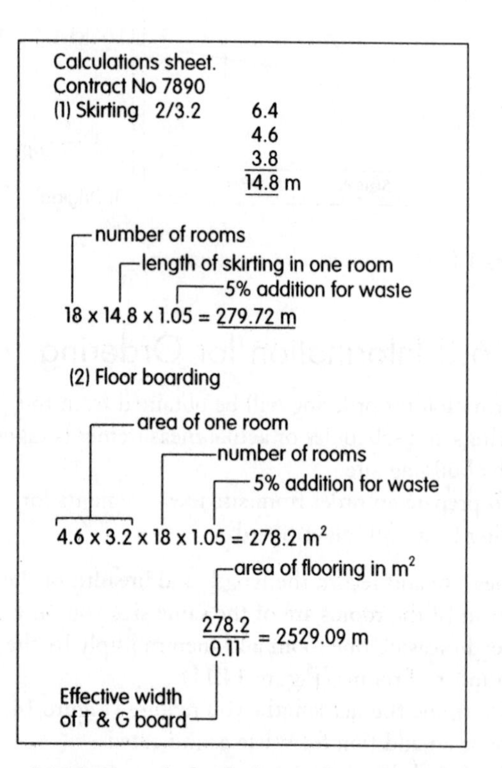

Figure 14.12 Record of calculations for materials

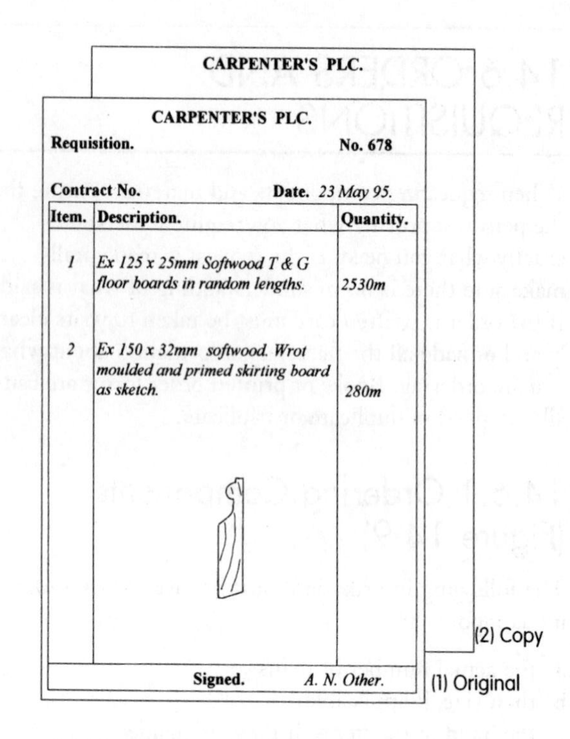

CARPENTER'S PLC.

CARPENTER'S PLC.

Requisition. No. 678

Contract No. Date. *23 May 95.*

Item.	Description.	Quantity.
1	*Ex 125 x 25mm Softwood T & G floor boards in random lengths.*	*2530m*
2	*Ex 150 x 32mm softwood. Wrot moulded and primed skirting board as sketch.*	*280m*

Signed. *A. N. Other.*

(2) Copy

(1) Original

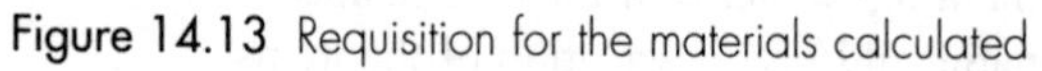

Figure 14.13 Requisition for the materials calculated

Index